Buhmann / Ervin / Pietsch (Eds.)

Peer Reviewed Proceedings of
Digital Landscape Architecture 2012
at Anhalt University of Applied Sciences

Buhmann / Ervin / Pietsch (Eds.)

Peer Reviewed Proceedings of
Digital Landscape Architecture 2012
at Anhalt University of Applied Sciences

All explanations, data, results etc. contained in this book have been made by the authors to the best of their knowledge and have been approved with care. However, some errors could not be excluded. For this reason the explanations etc. are given without any obligations or guarantee by the authors, editors, and publisher. They cannot take over any responsibility for eventual erroneous contents.

Bibliographic information published by the Deutsche Nationalbibliothek
The Deutsche Nationalbibliothek lists this publication in the Deutsche Nationalbibliografie. Detailed bibliographic data are available in the Internet at *http://dnb.d-nb.de*.

ISBN 978-3-87907-519-5

All rights reserved. No part of this book may be reproduced in any form, or any means, electronic or mechanical, including photocopying, recording, or by any information storage and retrieval system, without permission in writing from the publisher.

Cover: Image by Sebastian Kluth 2012, Hochschule Anhalt, including one image by H.-G. Schwarz-v.Raumer, University of Stuttgart

© 2012 Herbert Wichmann Verlag, VDE VERLAG GMBH, Berlin and Offenbach
    Bismarckstr. 33, 10625 Berlin
    www.vde-verlag.de
    www.wichmann-verlag.de

Printed by H. Heenemann GmbH & Co. KG, Berlin, Germany

Preface by the President
of Anhalt University of Applied Sciences

On behalf of the editors, I am happy to present you this publication of Anhalt University. This year's peer-reviewed proceedings combine the contributions to the 12th conference on Digital Landscape Architecture 2011, and the 13th conference DLA 2012. This issue also includes the papers given in workshops on "Teaching Landscape Architecture" held parallel during the LE:NOTRE Summer School at our Bernburg and Dessau campuses in May 2011. The Bernburg campus hosts our English language Master of Landscape Architecture Program and the German language Bachelor Program in Landscape Architecture, as well as the Bachelor and Master Programs in Nature Conservancy. The Dessau campus hosts the English language Master of Architecture Program DIA; the Master of Integrated Design, MAID; our German language Bachelor Programs in Architecture and Design; and the Bachelor and Master Programs in GeoInformation.

The series of conferences on Information Technology in Landscape Architecture is organized by Prof. Erich Buhmann. As well as being the Scientific Director of the Digital Landscape Architecture Conferences, Prof. Buhmann is also the Chair of the European Council of Landscape Architecture Schools (ECLAS) Committee on Digital Technology. The conference series is also supported by the Vice President for Information Technology of Anhalt University of Applied Sciences, Prof. Einar Kretzler, who heads the IT working group of the German Association of Landscape Architects (BDLA).

I would like to thank the editors of these proceedings Prof. Erich Buhmann, Dr. Stephan Ervin, and Matthias Pietsch for processing all the submitted papers and compiling them into sixty one peer-reviewed scientific publications totaling 608 pages. The editorial work shouldered by these colleagues and the international review board for this peer review process is equal to a total of ten months of editorial work over the period of two years. We also thank the academic partners within our university, as well as the ECLAS Digital Technology Committee and the thirty-two external reviewers who make this international annual scientific conference happen.

Thanks go to the Deutsche Forschungsgemeinschaft DFG for the financial support within the program for „Internationale Wissenschaftliche Tagungen" and special thanks to the sponsors from the software industry who have generously supported these conferences over the years.

We are very happy to have the honor of welcoming a number of new keynote speakers in both years. In 2011, Prof. Ralf Bill acted as opening speaker of the Digital Landscape Architecture Conference for the first time. Prof. Dr. Ralf Bill from the University of Rostock is the most published scientist in Geoinformation in Germany. We were very happy to welcome him together with another first time keynote, Dana Tomlin, the "Father of Raster GIS" from the University of Pennsylvania.

We rely on the continuing support of many outstanding scientists in the areas of geospatial planning and design. The long term and continuous support given by Prof. Carl Steinitz, Harvard University was honored when we awarded him the Honorary Senator Award of

Anhalt University in 2010. In 2012 we are happy to welcome "the Father of City GML" Prof. Dr. Thomas Kolbe, TU Berlin to the DLA conference series for the first time.

We see our annual conference on digital landscape architecture as part of the trend towards integration of geospatial disciplines, such as summarized by the industry with the catchphrase GeoDesign. We are very pleased to have some of the world leading GeoDesign developers presenting during the 2012 conference, and to present William R. Miller's technical paper on GeoDesign in the closing chapter of these proceedings.

Anhalt University currently has over 7,900 students. With more than 24 percent international students, it continues to be one of the universities in Germany with the highest level of international enrollment. Holding this conference in the State of Saxony-Anhalt in Germany is a gesture to all the international collaborators working with Anhalt on the concept of worldwide training. The great number of international speakers indicates the level of international awareness the conference has developed. The conference series also serves as a conference of the ECLAS Committee Digital Technology and the IT working group of the European Network of Landscape Architecture Schools. This LE:NOTRE thematic network in landscape architecture already includes over one hundred universities in Europe and is cooperating with a number of selected schools outside of Europe.

In 2011 Anhalt University established the interdisciplinary KAT competence center of "Digital Planning and Designing" with the aim of fostering CAD and GIS research cooperation in the region. Numerous colleagues of several departments are cooperating in the area of applied information systems in regard to design and building. The annual conference of Digital Landscape Architecture complements this research cooperative.

International conferences open the window of the world for our students and researchers. Having positioned the international conference in Digital Landscape Architecture at our university fills us with pride. We have always given new technologies education a special emphasis. Therefore we are very happy about the numerous research collaborations with the software industry. This international conference will strengthen the existing contacts and will be the platform for new partnerships worldwide.

Finally, I do not want to miss this opportunity to thank Professor Buhmann for his strong personal commitment and for the work that he has done over the last 17 years in helping establish Landscape Architecture in Saxony-Anhalt, for which he has received national and international acknowledgment.

I wish the international group of landscape architects and environmental researchers an exciting and successful conference in Bernburg and Dessau. You are always welcome to return to our Bernburg and Dessau campuses. We extend a special invitation to you for next year's DLA conference from 30 May to 1 June 2013.

Bernburg, April 2012

Prof. Dr. Dr. habil. Dieter Orzessek
President Anhalt University of Applied Sciences

Foreword

These proceedings document most of the peer-reviewed papers presented at the twelfth Digital Landscape Architecture (DLA) Conference. It was held jointly with the pan-European LE:NOTRE and ECLAS Summer School in Dessau and Bernburg, Germany, at the end of May, 2011. It also documents the papers given at the thirteenth DLA 2012 held both in Bernburg and Dessau 31 May – 2 June 2012. The summer school in 2011 focused on the theme "Landscape Architecture Education", while the DLA conferences focused on teaching, practice and research issues in the world of "Digital Landscape Architecture" by way of keynote addresses, lectures, workshops, and meetings engaging an international audience. The two chapters on Teaching in these proceedings are also an outcome of the LE:NOTRE III emphasis on Education. As all authors of papers submitted in the year 2011 have updated their papers, we are able to label these proceedings "DLA 2012".

When Anhalt University organized the first international conference on information technologies in landscape architecture thirteen years ago, computers had just started to replace drawing boards in landscape architecture practice. Today, landscape architects work digitally in all project phases, often in international and interdisciplinary settings. Responding to today's international planning and design practice, as well as to the international cooperation in software development for landscape architecture, the tenth annual Digital Landscape Architecture DLA Conference in 2009 was held for the first time outside the Anhalt campus. Local responsibility for that conference was assumed by the Faculty of the Built Environment at the University of Malta. The success of the first DLA held outside our home university may inspire other faculties to apply for hosting the conference for one year at their campus. In fact, we are pleased to announce that the DLA 2014 will be hosted by the teams of PLUS, Planning of Landscape and Urban Systems, Prof. Dr. Adrienne Grêt-Regamey and ILA Institute of Landscape Architecture, Prof. Christophe Girot of the ETH at the Science City Campus, Zurich-Hoenggerberg in Switzerland.

Relevant research and development involving information and communications technology for landscape architecture is only possible when research and teaching units at universities of different disciplines cooperate with each other and with the software industry as well. The DLA conferences provide an international platform for this exchange of research and teaching. We are honored that so great a spectrum of our profession has spent so many hours of effort to document today's state of the art in digital work-flow practices, to share their own experiences, and to point out further needs for research and improvement.

A growing number of professionals are close to achieving a truly paperless digital work-flow; able to analyze, design, review, construct and submit documents in digital form. Individual landscape architects and firms are routinely working around the world as web-connected design teams; CAD, GIS, and imaging software on a laptop computer with ample memory and a fast internet connection are the only requirements. Beyond simply replacing analogue utilities, digital tools enable us to confront the challenges of exploding complexity in 'geo-design' projects by taking advantage of powerful new tools for both creativity and collaboration. Those of us still accustomed to sketching ideas with pencil and paper or making corrections on printed plans today watch in astonishment as even these operations are accomplished by digital natives on hand-held tablet-computing devices using 'gestural-interfaces', software 'mash-ups', and increasingly 'crowd-sourced' data ...

One of the objectives of this conference series is to present these opportunities and challenges and to anticipate future developments. As these proceedings document, the digital landscape architecture work we can do today is outstanding. Exploring virtual worlds with geospatial tools and techniques reveals forms of creativity that are at once both new and familiar. Our challenge is to engage the best of each.

Please see our call for papers for the DLA 2013 on the last two pages. The 2013 focus will be on **"GeoDesign – Teaching and Case Studies of Integrated Geospatial Design and Planning"**. The conference Digital Landscape Architecture DLA 2014 will be focusing on: "Landscape Architecture and Planning: Developing Digital Methods in GeoDesign".

Every one of you can help define the focus of the next year's conferences by contributing their own papers and posters or by offering workshops. While giving a platform to the evolving concept of Geodesign in Europe, we look forward to contributions in all areas of digital landscape architecture research and applications.

Acknowledgments

Contributors to the DLA Conference
This conference is the thirteenth in a series of conferences headed by Scientific Director Prof. Erich Buhmann at Anhalt University of Applied Sciences. These conferences were initially organized in cooperation with the keynote speakers and the LE:NOTRE Information Technology working group. As documented in the closing chapters, a formal DLA Program Committee has been established for the further scientific development of this international conference series. Members of the DLA Program Committee will also serve as local chairs for further DLA Conferences. More than fifty experts are working at the moment on the DLA Review Board. At the General Assembly Meeting of the 2009 ECLAS Conference, the LE:NOTRE Information Technology working group was additionally established as the ECLAS Committee on Digital Technology. Future DLA conferences will continue to focus on topics identified by ECLAS needs for improvements in teaching digital methods.

We would like to thank all of the authors, reviewers, moderators, and keynote speakers for their great efforts and contributions. We also thank the 2012 sponsor ESRI and support from the *Deutsche Forschungsgemeinschaft DFG* in the years 2009, 2010 and 2012.

We especially thank all of our contributors. Renate Geue made the printing of the 2011 interim proceedings for 2011 possible just in time. Their assistance is greatly appreciated. Then there are those who are not named in this book but without whose help it would not have been possible. First we need to thank Prof. Joachim Kieferle of the Rhein-Main University in Wiesbaden, Germany for running the conference review system and answering email requests by authors and committee members 24 hours a day for many months during the last years. Second, we need to thank Sebastian Kluth for the internet processing of the conference contents and the design of the posters and handouts in 2011 and 2012, and our English language consultant Jeanne Colgan, and the layout team headed

by Judith Mahnert and the MLA students Lu Yangwa and Wenjing Wang. The quality of all their work will be appreciated by many readers.
And finally we'd like to thank Andre Schlecht-Pese for organizing the logistics in Dessau in 2011.

To all of them thank you for work, once again, above and beyond the call of duty!
And a big thank you to all our families for their patience and support.

Bernburg, April 2012   *Erich Buhmann, Stephen Ervin and Matthias Pietsch*

Table of Contents

Preface by the President of Anhalt University of Applied Sciences............ V

Foreword VII

Introduction 1

Steinitz, C. (Keynote):
"Getting Started" – Teaching in a Collaborative Multidisciplinary Framework 2

Ervin, S. M. (Keynote):
Geodesign Futures – Nearly 50 Predictions 22

Teaching Methods in Digital Landscape Architecture 31

Paar, P. and Rekittke, J.:
Wheeling a Trojan Horse to Teach MLA Students Geoinformation Methods 32

Leiner, C. and Stemmer, B.:
Teaching Landscape Planning – Landscape Perception and Analysis 41

Mertens, E.:
Action – Structure and Teaching Methods for Lectures and Seminars 50

Thurmayr, A. M. and Straub, D.:
On the Preparation of Images 59

Teaching Planning and Design in Landscape Architecture 65

Rekittke, J. and Paar, P. (Keynote):
There Is no App for that – Ardous Fieldwork under Mega-urban Conditions 66

Lenz, R., Rolf, W. and Tilk, C.:
Teaching Digital Methods in Landscape Planning – Design, Content and
Experiences with a Course for Postgraduate Education 76

Schmidt, R. (Keynote):
Nature by culture – From „Romantic" to „Organic" 86

Petschek, P. (Keynote):
LandscapingSMART 99

Fetzer, E. and Kaiser, H.:
Computer-Supported Collaborative Learning with Wikis and Virtual Classrooms across Institutional Boundaries – Potentials for Landscape Architecture Education 105

Kircher, W., Messer, U., Fenzl, J., Heins, M. and Dunnett, N.:
Development of Randomly Mixed Perennial Plantings and Application Approaches for Planting Design 113

Mclean, R.:
The Diagrammatic Landscape 126

Fricker, P., Girot, C., Melsom, J. and Wemer, P.:
"From Reality to Virtuality and Back Again" Teaching Experience within a Postgraduate Study Program in Landscape Architecture 130

Polyzou, E., Hasanagas, N. and Tamoutseli, K.:
A Gender-based Typology of Determinants of Video Games Use by Primary School Children 141

Thurmayr, A. M.:
Materials and Digital Representation 147

Buhmann, E., Palmer, J. and Pietsch, M.:
Managing the Visual Resource of the Mediterranean Island of Gozo, Malta for Tourists – A Studio Approach for International Conversion Students Bridging Different Levels of English 155

Wissen-Hayek, U., Melsom, J., Neuenschwander, N., Girot, C. and Grêt-Regamey, A.:
Interdisciplinary Studio for Teaching 3D Landscape Visualization – Lessons from the LVML 156

GeoDesign Concepts and Applications 157

Ervin, S. M. (Keynote):
A System for GeoDesign 158

Warren-Kretzschmar, B., v. Haaren, C., Hachmann, R. and Albert, C. (Keynote):
The Potential of GeoDesign for Linking Landscape Planning and Design 168

Tomlin, C. D. (Keynote):
Speaking of GeoDesign 180

Schwarz-v.Raumer, H. G. and Stokman, A.:
GeoDesign – Approximations of a Catchphrase 189

Rekittke, J., Paar, P. and Ninsalam, Y.:
Foot Soldiers of GeoDesign 199

Jombach, S., Kollányi, L., Molnár, J. L., Szabó, Á. and Tóth, T. D.:
GeoDesign Approach in Vital Landscapes Project 211

Albert, C. and Vargas-Moreno, J. C.:
Testing GeoDesign in Landscape Planning – First Results 219

Sztejn, J., Łabędź, P. and Ozimek, P.:
Visual Landscape Character in the Approach of GeoDesign 227

Schaller, J. (Keynote):
Applying 3D Landscape Modeling in Geodesign 235

Glander, T., Trapp, M. and Döllner, J.:
Concepts for Automatic Generalization of Virtual 3D Landscape Models 240

GeoDesign and Participation 241

Steinitz, C. (Keynote):
Public Participation in Geodesign – A Prognosis for the Future 242

Mülder, J. and Strickmann, M.:
Ems3D – Communicating Landscape Change 249

Harwood, A., Lovett, A. and Turner, J.:
Extending Virtual Globes to Help Enhance Public Landscape Awareness 256

Kramer, H., Houtkamp, J. and Danes, M.:
Things Have Changed – A Visual Assessment of a Virtual Landscape from 1900
and 2006 263

*Hehl-Lange, S., Gill, L., Henneberry, J., Keskin, B., Lange, E., Mell, I. C. and
Morgan, E.:*
Using 3D Virtual GeoDesigns for Exploring the Economic Value of Alternative
Green Infrastructure Options 273

*Wissen Hayek, U., Neuenschwander, N., Kunze, A., Halatsch, J., v. Wirth, T.,
Grêt-Regamey, A. and Schmitt, G.:*
Transdisciplinary Collaboration Platform Based on GeoDesign for Securing
Urban Quality 281

Styliadis, A. and Hasanagas, N.:
Adaptive e-Learning DLA Course – A Framework 289

*Manyoky, M., Wissen Hayek, U., Klein, T. M., Pieren, R., Heutschi, K. and
Grêt-Regamey, A.:*
Concept for Collaborative Design of Wind Farms Facilitated by an Interactive
GIS-based Visual-acoustic 3D Simulation 297

Stemmer, B.:
Collaborative Landscape Assessment and GeoDesign 307

Bujaidar, F., Santa Cruz, D. and Seah, G.:
"Beep-Scape" – Using Applications for Mobile Devices to Communicate the
Landscape 315

Landscape Modeling 321

Bill, R. (Keynote):
Interdisciplinary Research and Education in a Virtual Cultural Landscape
Laboratory 322

Hahn, H.:
Visualizing Wetland and Meadow Landscapes 333

Fricker, P., Girot, C., Kapellos, A. and Melsom, J.:
Landscape Architecture Design Simulation – Using CNC Tools as Hands-On Tools 343

Łabędź, P. and Ozimek, A.:
Detecting Greenery in Near Infrared Images of Ground-level Scenes 354

Wei, S., Fleurant, C. and Burley, J. B.:
Replicating Fractal Structures with the Reverse Box Counting Method – An Urban
South-east Asian Example 364

Garcia Padilla, M. A.:
Landscape Analysis Using GIS for Ecologically Oriented Planning in Costa Rica 371

Formosa, S.:
Soaring Spaces – The Development of an Integrated Terrestrial and Bathymetric
Information System for the Maltese Islands 378

Pé, R.:
Hyper-localism and Parametric Mapping for Collaborative Urbanism 389

Visualization in Landscape Architecture 397

Bishop, I., Handmer, J., Winarto, A. and McCowan, E.:
Survival in Dangerous Landscapes – A Game Environment for Increasing Public
Preparedness 398

Schroth, O., Pond, E. and Sheppard, S. R. J.:
Integration of Spatial Outputs from Mathematical Models in Climate Change
Visioning Tools for Community-Decision Making on the Landscape Scale 406

Gulev, E.:
Aesthetic Evaluation of Forest Landscapes within the Training and Experimental
Forest Range (TEFR) Yundola 415

Joye, R., Verbeken, J., Heyde, S. and Librecht, H.:
Location-aware Mobile Devices and Landscape Reading 425

Czinkóczky, A. and Bede-Fazekas, A.:
Visualization of the Climate Change with the Shift of the So-called Moesz-line 437

Egginton, Z.:
Hollywood Landscapes – An Exploration of Hollywood Styled Visual Effects
Techniques for Landscape Visualisations 445

Kim, M.:
Modeling Nightscapes of Designed Spaces – Case Studies of the University of
Arizona and Virginia Tech Campuses 455

Ozimek, A., Łabędź, P. and Michoń, K.:
Discrimination of Distance-dependent Zones in Landscape Views 464

Pokladek, M., Thürkow, D., Dette, C. and Gläßler, C.:
Geovisualization of the Garden Kingdom of Dessau-Wörlitz 472

Partin, S., Burley, J. B., Schutzki, R. and Crawford, P.:
Concordance between Photographs and Computer Generated 3D Models in a
Michigan Highway Transportation Setting... 482

Virtual Reality in Landscape Architecture 491

Morgan, E., Gill, L., Lange, E. and Dallimer, M.:
Integrating Bird Survey Data into Real Time 3D Visual and Aural Simulations 492

Barbarash, D.:
The Communication Value of Graphic Visualizations Prepared with Differing
Levels of Detail 499

Houtkamp, J. M. and Toet, A.:
Who's Afraid of Virtual Darkness – Affective Appraisal of Night-time Virtual
Environments.. 508

Chandwania, D. and Verma, A.:
Digitalized Re-Rendering of a City's Landscape 516

Standardization and BIM 523

Goldberg, D. E., Holland, R. J. and Wing, S. W.:
GIS + BIM = Integrated Project Delivery @ Penn State 524

Ahmad, A. M. and Aliyu, A. A.:
The Need for Landscape Information Modelling (LIM) in Landscape Architecture 531

Invited Contributions 541

Miller, William R.:
Introducing Geodesign – The Concept 542

Kader, A., Pinkau, S., Kretzler, E., Dießenbacher, K. and Heins M.:
Introducing the KAT Competence Center "Digital Planning and Design" at Anhalt
University 568
Vorstellung des KAT-Kompetenzzentrums " DIGITALES PLANEN und
GESTALTEN" an der Hochschule Anhalt 572

Dießenbacher, C. and Walter, M.:
NAEXUS – Virtual Space Scope – A Space of Illusion 577
NAEXUS – Virtual Space Scope – Ein Illusionsraum 583

DLA Awards 589
DLA 2011 Award 589
DLA 2012 Award 590

Committees Contributing to the DLA Conference 593
Scientific Program Committee of DLA 593
Review Committee 596
ECLAS and LE:NOTRE Committee Digital Technology 601

Addresses of Authors 603

Early Conference Announcement & Call for Papers
**for International Conference "Digital Landscape Architecture DLA 2013":
GeoDesign – Teaching and Case Studies of Integrated Geospatial Design and Planning** to be held 30 May – 1 June 2013, in Bernburg and Dessau Campus, Germany 607

Introduction

"Getting Started" – Teaching in a Collaborative Multidisciplinary Framework

Carl STEINITZ

Keynote: 26 May, 2011

1 Introduction

I have led and taught collaborative, multidisciplinary, semester-long studios on large and complex landscape planning and design problems for more than 40 years at Harvard, and sometimes in collaboration with other universities. I also have organized and taught many one-to-five day workshops. I have written about the framework within which I organize most of my work (1990, 2003) and about teaching strategies. In this paper I want to focus on the most difficult stage, "getting started" on the change-designs which will be proposed as the main "product". I consider this stage to be the most important of any academic or professional project because if the beginning is unsatisfactory, then the ending must also be.

The reasons for my teaching in a manner which requires students to work in teams, and frequently in large multidisciplinary teams, are many but normally center upon the scope and complexity of the problem around which the workshop or studio is focused and the need for many individual tasks to be coordinated. Sometimes teams have been as small as three persons, and sometimes they have involved a studio class of 12 to 18 persons acting as "a team of the whole". I would generally characterize these experiences as being successful and positive ones, both from my perspective and those of my students (though I concede that these experiences are not without pain). I would like to share with you some of the issues which I raise and some of the techniques which I use to ensure a higher probability of success than failure.

An initial field trip is indispensible. It is always an intensively scheduled working period with both group and individual responsibilities. The tasks associated with becoming familiar with issues, geography, and people are of prime importance. There are presentations by knowledgeable persons, and these are frequently in conflict with each other. The entire group meets every evening and there is a high level of debriefing and other communications. Of critical importance during the field trip is the absence of any collective attempt to define the study. I make a major point of telling the students that we are on the field trip to observe and ask questions, not to decide anything. I do not want the students to informally negotiate the scope and responsibilities of the project.

2 But what then? How Does one Start?

I teach my students that there is no such thing as "THE Design Method" or "THE Planning Method" (and I consider a plan to be a design). Rather, there are many methods and they must be chosen and adapted to issues and questions raised by the problem at hand in the second iteration of the framework. Every landscape design regardless of size or scale has three groups of influences which should be considered: the history of the place and past proposals, the "facts" of the area which are not likely to be changed, and the "constants" which should be incorporated into any proposed alternative.

There are two fundamentally different ways of getting started on a design, and choosing wisely is especially important when making a large and complex landscape plan that has serious spatial and temporal consequences. These two paths are "anticipatory" and "exploratory" and they can be seen in Figure 3. Both approaches are vulnerable to the uncertainties about the fundamental assumptions and contingent choices that we recognize at the beginning of our design processes.

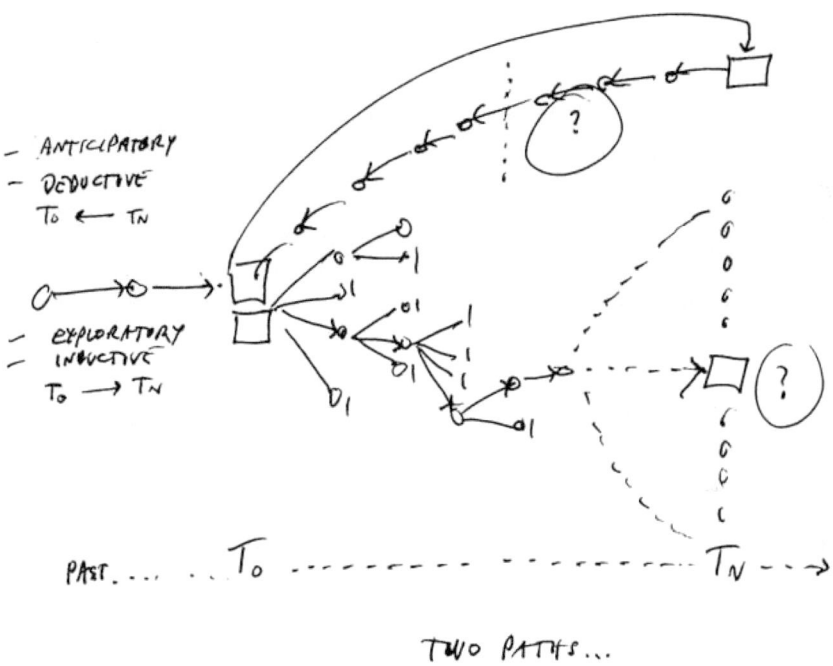

Fig. 1: Two paths into a design (and then several different ways through them)

The anticipatory approach embodies the idea that the designer is expected to make a heroic leap forward in time and implicitly make correct choices among the many assumptions and contingencies inherent in the situation. The designer then will represent and present a proposed future change – the plan. This anticipatory method then requires the use of deductive logic in order to figure out how to get from the desired future state back to the present. This is relatively easy to think about if the present circumstance is based upon a "clean slate" but if the present circumstance is large and complicated this approach frequently faces the problem of implementation. It is too often difficult or impossible to connect the future back to the present. And an early wrong decision can be fatal to the plan.

The alternative exploratory method requires the explicit development of a scenario, a sequence of the assumptions which will shape the plan. It requires the use of the inductive logic. Again, this approach is relatively easy if the problem is a simple one but if it is large and complex, and if each single assumption has several options to be considered, this combinatorial approach can also fail in that it cannot achieve a sufficient level of precision and detail. There are too many combinations to consider. There are too many risks of taking the wrong path. Therefore, the essential initial steps must be to "sensitivity- test" the combinatorial sets of the most important assumptions … and not worry too much about the details.

Most frequently, one skips back and forth between these extremes. But which way should we begin? Here the issues of "size and scale" and "risk" must be considered. Scale matters. In my view, the smaller project types such as a residential site plan present fewer real risks of being wrong than do the larger design projects and regional landscape plan studies which involve enormous cost, large numbers of people, many unknowns and a longer time horizon. The smaller projects are easier to change while the larger frequently require fundamental institutional change. The smaller projects end in working drawings and constructed physical change, while the larger regional projects rarely are directly built. Rather their aim is to influence the way society values and changes its landscapes, including aspects of water and land use. At the extremes these varied scales require different initial strategies. For the larger size and complexity of the problems given in my studios, exploratory methods are the more appropriate starting strategies. These exploratory methods using diagramming methods can result in one dominant design strategy or they can produce several made by smaller teams or they can produce individually differentiated designs.

Technically, the methods which I most frequently have the students use rely on the making of simple and clear diagrams to represent ideas, be they physical changes or policies. All ideas, whether invented by the students, proposed by local persons, derived from historic examples, etc. and included, without pre-judgment as to their value. These diagrams are used in several ways, often via variations on the anonymous Delphi assessment technique. A core concept is making a distinction between the generation and "ownership" of ideas, and their use. The diagrams are presented anonymously; the process of selecting, combining and interpreting diagrammed ideas into a design is available for any group or individual.

3 Bermuda

The diagramming methods and their organization derive from a studio which I taught in 1982. Bermuda had recently achieved independence from Great Britain. The first Prime Minister, John Swan, requested a study of the future of the garbage dump of that small island nation. There was a plan to build a new waste incinerator but it would take three years (and, in reality, many more) for that project to become operational. The garbage dump was surrounded by civic institutions, a large wetland, the well fields which supplied drinking water to most of Bermuda, and important play fields. It was in the midst of the residential area of the poorest people in the country. A promise had been made as part Mr. Swan's election campaign to transform the dump area into a central park for Bermuda. I offered to teach a studio which would illustrate different assumptions regarding what kind of park and ancillary facilities might be developed for the site, and this offer was accepted and the studio was financed. Students volunteered for the studio knowing that it would be organized with some aspects of a design competition and that not all of their individual designs would be carried forward to the end.

Fig. 2: The Bermuda dump study area

The studio traveled to Bermuda and visited the study area (Figure 3). There were several presentations and several open meetings for interested persons during which records were kept of the issues which were raised, and ideas for program elements, physical designs and policies which were presented to the students. Each evening I met with the students and had them list and categorize the issues which had been raised, and also to prepare simple diagrams of every idea and proposal which they had been offered or which they themselves had. These diagrams were all simple line drawings to a standard scale. They were anonymous and were intended to be shared, and all students knew this.

Upon returning to the University, and in the first working session of the studio, the students agreed on a final list of about 20 issues which had to be resolved in any design. These were of two kinds: the constants which had to be incorporated into every design, and the variables, for which there might be alternative diagrammatic solutions. Pairs of students were assigned by their choice to the variable issues and were asked to produce between two and five alternative strategies regarding each issue. There were approximately 80 diagrams each drawn with permanent black marker on thin clear plastic so that they could easily be selected, overlain, and looked at together as a set or, as the students called them, "a sandwich".

Fig. 3: The array of the constants and the variables and their options

The next exercise was to rank order the issues and alternatives, and this was done using a modified Delphi technique. Figure 4 represents what was actually the laying out of the small diagrams on a very large table. The constants all are in the left most columns. The variables are listed along the top row, but they are in the rank order of their perceived importance, with the most important being to the left. The alternatives diagrams are below their heading and are also in rank order of likely success as the judged by the class using Delphi methods. Thus one can interpret the positions of the diagrams on the table in the following way. Every constant diagram must be included and in addition, the most likely successful design strategy would be to select the top row of issue alternatives starting from the left. If compromises were to be made, they could be made the two ways: first, by choosing to ignore the less important issues, or, by dropping to second or third best alternatives for any issue but preferably not the more important ones. This process was completed at the end of the third studio class.

In the next phase of the studio, each individual student was required to prepare an initial design by selecting an appropriate set of the diagrams. A lottery was held and the number one winning student had first choice among the variables diagrams. Each subsequent student in the lottery ranking was required to choose a different set from all previous students. Thus, there were 14 substantially different initial diagrammatic designs and these were available after the fourth studio class, at the end of the second week.

A special issue of pedagogic ethics had to be discussed with the studio students in this phase. Even in the most organized or faculty led studio, there is an absolute right for a student to explore his or her own ideas, and in his or her own way. This is an issue which must be discussed openly and the students must understand that an objective of the faculty member is the teaching and testing of a method which is expected to be of interest and use to the student. The priority in this studio is not the encouragement of the students' idiosyncratic creativity. I am well aware that some colleagues and some students do not agree with this position but it is the one which I hold. The ethics of being a teacher require this to be openly stated, openly discussed, and somehow managed within the social contract between student and teacher.

Each student then prepared a physical model of his or her initial design. At the end of the sixth week, these were presented in a standard scale using standard and mass produced materials, and in a representational style organized by a student subcommittee. Each model could be segmented and placed in a shipping carton.

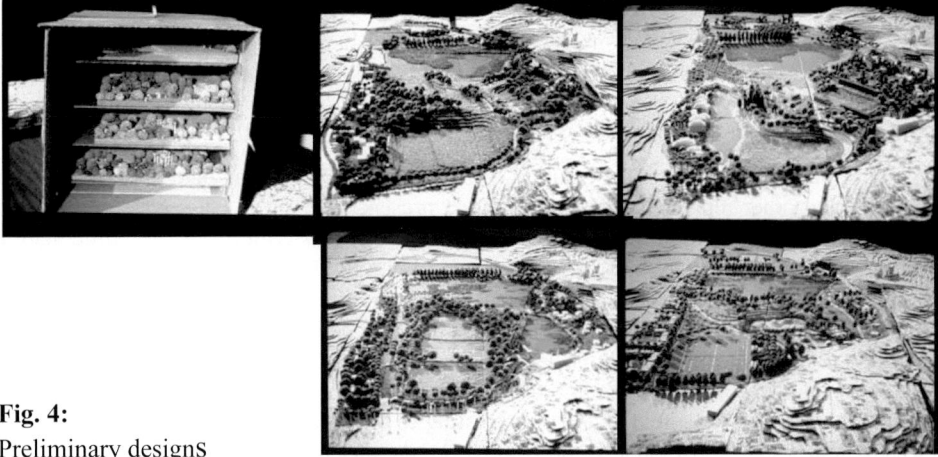

Fig. 4:
Preliminary designs

Fourteen designs were flown to Bermuda along with three students who presented these designs to the group of persons actually responsible for the redevelopment of the site. After careful consideration, the Bermudian committee decided that three of the designs should be moved forward to the next stage.

Fig. 5: Final designs A, B and C

This was reported to all in a presentation and discussion by the students who went to Bermuda. The students whose designs were not chosen to go forward then had to join the team of one of the three designs which would be presented at the end of the semester. The teams were of approximately equal size and were organized on a volunteer basis, and the studio continued with three very different design strategies.

At the end of the semester, a presentation of these three designs was held at Harvard with Mr. Swan and other representatives of Bermuda present. The entire class was then invited to return to Bermuda to present the three final designs. Bermuda at that time had a population of approximately 90,000 persons and about 10,000 persons saw at least one of the several presentations made by the students. The committee and Mr. Swan then decided to place the choice of one of the three park concepts before the electorate in a special election. The intent was not to build one of the student designs but rather to identify the preferences of the general public for the strategies which were embedded in the design options. This election was held and it is interesting to note that the winning design C, figure 7, was the one which most closely conformed to the upper row and left hand section of the diagram layout with which the studio got started. A version of that student design was eventually built.

Fig. 61 Design C

4   Tepotzatlan, Mexico

The following example, Tepotzotlan Mexico, is an application of the diagramming method to a problem at different scales and with different products, but using a digital adaptation of the basic approach. This was a graduate-level studio which I taught in 2004-5 at Harvard with Juan Carlos Vargas-Moreno, and in collaboration with a faculty/student team from the Universidad Autonoma Metropolitana (UAM) in Mexico city led by Professor Anibal Figueroa, and with the full cooperation of the municipal government.

Tepotzotlan is a municipality at the northern edge of what may be the largest city in the world. It is facing enormous development pressures, as it is on the main highway to the north of Mexico. It has, and is surrounded by, considerable amounts of 'social housing and distribution warehouses. There are increasing amounts of 'informal housing'. The untreated sewage of the entire Mexico City area flows via Tepotzotlan, some in canals and some in a broken pipe system under Tepotzotlan. Yet the municipality retains the character of a group of relatively small settlements, with some agriculture and large adjacent National Forest lands.

The main attraction of the town of Tepotzotlan is the church and monastery of St. Francis Xavier, founded in 1584. This extraordinary complex is now the national museum of colonial art, and a major Mexican tourist attraction. The sponsor, FUNDEA, is the leading Mexican environmentally-oriented NGO. It has a large landholding in the municipality, and this area is partly developed and operated as an environmental-education park. Tepotzotlan has been identified by the Mexican Ministry of Tourism as one of 10 national priority areas for tourism development.

Thus the several potential conflicts needing resolution: housing, transport, water, sewage treatment, tourism, conservation, and recreation, and all in a rapidly changing, environmentally degraded, politically complex and relatively poor economy. There were issues and "projects" ranging from metropolitan-regional to very detailed scales, and always the need to play a constructive part in helping the municipality to shape its future.

During the five day visit to Tepotzotlan, Mexico; the students participating in the studio created a list of projects and policies as reactions to the daily meetings, discussion, visits and information that had been gathered. Each project was proposed by one or more students and presented in brainstorming sessions that were held at the end of each day. But they were not edited or rejected. By the end of the eight-day field visit, the students had identified around 200 projects.

The project proposals had a specific protocol in order to be considered and entered in the system. The technical process had been designed by Juan Carlos Vargas-Moreno. The projects were first entered in a "project list" composed in EXCEL spreadsheet and then diagrammed by hand on a large regional map of 3 by 6 meters size that was placed in the studio work space. The table-map was a large print of the most recent high-resolution orthophotography and several layers of transparent plastic sheet. While the orthophoto allowed the students to locate and describe the geography of each proposed project, the plastic sheets allowed the sketching of projects over the orthophoto in independent sheets. In the EXCEL spreadsheet, each project had a number, the name of student who proposed it, a classification that determined if the project was a spatially specific physical change or a policy. Furthermore each project had to be classified in one or more of eight color-coded categories: national or municipal government related, neighborhood related, transportation, industrial, ecological (including hydrology), heritage, utilities or wildlife restoration.

During the last day of the site visit, students were divided in groups corresponding to each category, and were asked to act as experts by selecting up to 20 of the most significant projects in each category. This limit certainly focused the students on the issues of strategy and priority. A new short-list of around eighty projects was selected for further development. These projects were then digitized as diagrams in a GIS employing ESRI ArcMap 9.0. Each project diagram was digitized as a separate layer in the color code of its assigned category and the full spectrum of attributes entered in the EXCEL spreadsheet. With this electronic data base of individual projects, and by simply selecting the number of wanted layers via the spreadsheet, the students created different clusters of projects as overlays in a 3-D visualization generated by ESRI's ArcScene. The visualization featured the orthophotography draped over the digital elevation model and covered by the individual project layers in inverse order of presumed importance. Different clusters such as tourism or ecological-related projects were created as initial explorations. This allowed the students to visualize the cumulative effect of different projects and categories in the region of study. Later, the through class discussion, three scenarios were developed by combining different projects. The three scenarios were: tourism, ecological and economically-driven alternatives. Each scenario was presented in a 3-D visualization and coded as a group of project numbers (e. g. projects: 2, 6, 26, 54, 55, 43). These visualizations were presented to the local collaborators and government representatives, and discussed for future refinement. This had been accomplished within the site visit.

Later, Several more scenarios of more complex objectives were prepared and compared before the studio team decided to focus on one. This was developed further into an alternative municipal plan, and several projects developed at much more detailed scales.
Note the sequence of project 55, the conservation of riparian "green corridors" located between the village-scale settlements of the municipality.

Fig. 7: Making diagrams

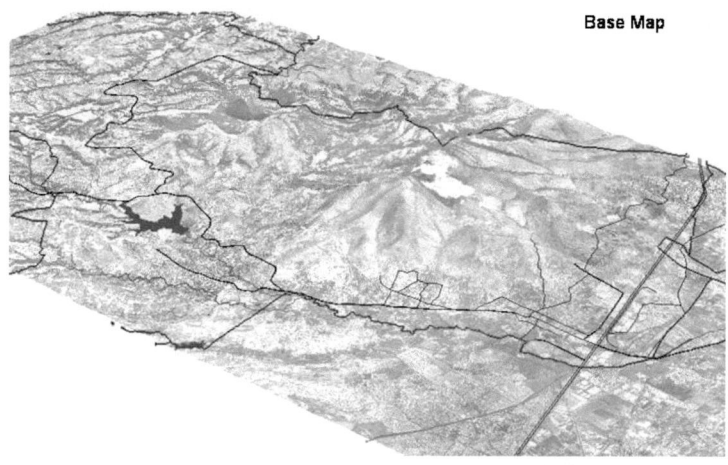

Fig. 8:
The base map

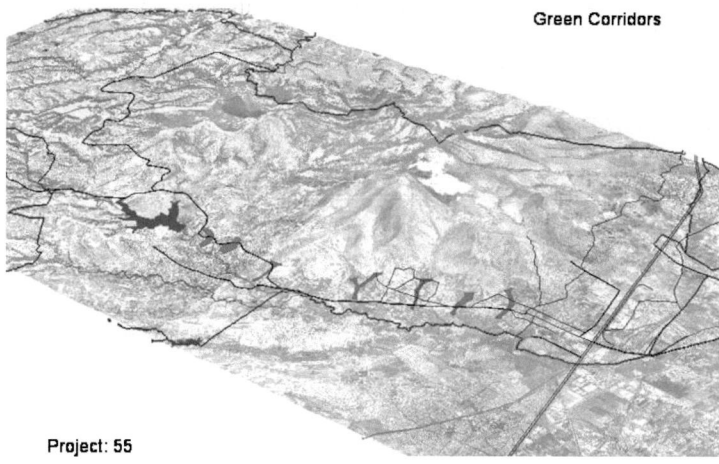

Fig. 9:
Project 55,
green corridors

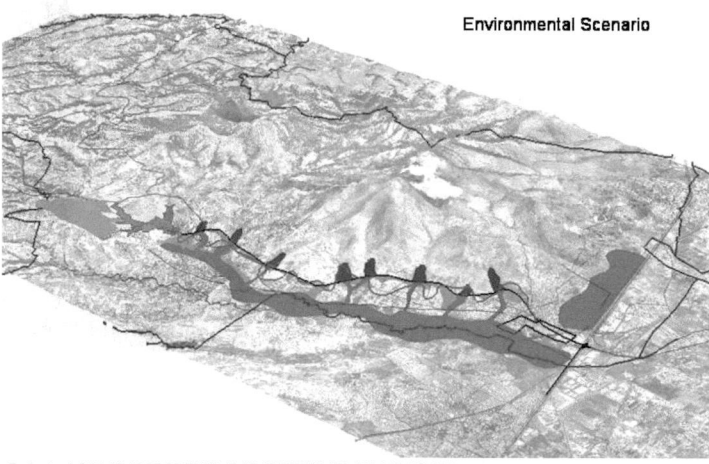

Fig. 10:
Environmental
scenario, note
project 55

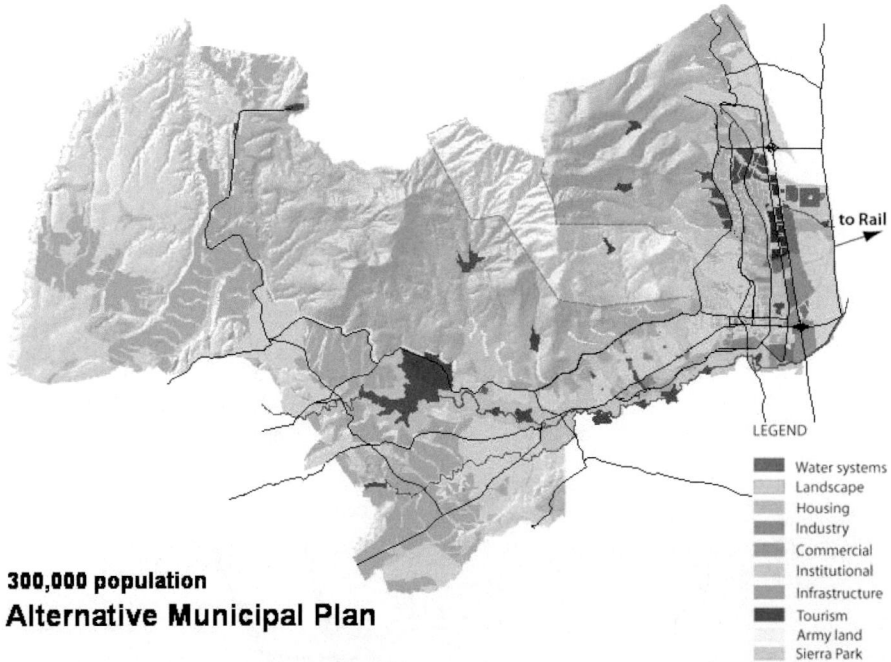

Fig. 11: Alternative municipal plan, note project 55

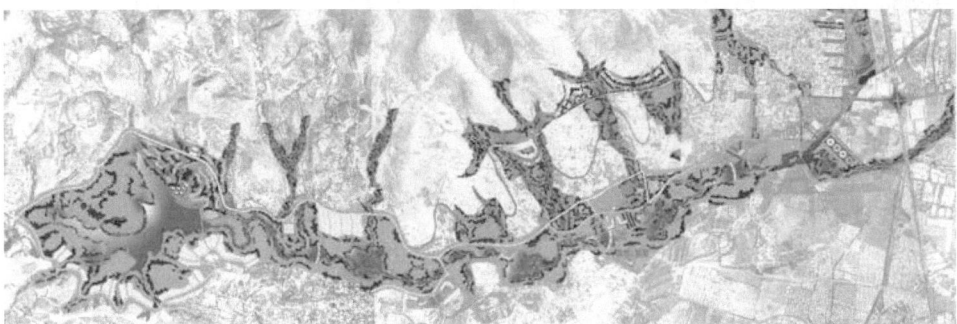

Fig. 12: River and riparian zone restoration, note project 55

5 Workshop at University of Cagliari, Sardinia, Italy, Alternative Futures for Cagliari, Sardinia

Cagliari is the capital city of Sardinia, a region of Italy. It has about 160,000 residents, and about 500,000 people live in the metropolitan region. Sardinia has a substantial tourism industry, with millions of visitors annually.

The international workshop "Alternative Futures for the Metropolitan Area of Cagliari, Sardinia", was organized by Prof. Emanuela Abis , Claudia Palmas and Stefano Pili, University of Cagliari, and Professor Christina von Haaren, University of Hannover,

Germany, and Christian Albert and Daniela Kempa, Leibniz University, Germany. It was held in Cagliari during five days in March 2009. The participants were 20 architecture and engineering students from Cagliari and 12 landscape architecture students from Hannover. The teaching team for the part of the workshop in which students had to develop alternative proposals for the study area was led by me (Carl STEINITZ), with Juan Carlos Vargas Moreno and Christian Albert. All activities in the workshop were to be undertaken by mixed teams of Italian and German students (some of whom come from different countries) and that all publicly accessible work would be conducted in English, except for the final presentation which would be conducted in Italian because of the persons who were invited to review the work.

The workshop began with two intensive days devoted to a general orientation to the study area, its history, its current characteristics and future projections. The format for this orientation was a series of lectures, each taking approximately 20 minutes and followed by questions, and a well organized, half day, guided bus and walking trip throughout the study area.

Fig. 13: The Cagliari study area

Despite the fact that Cagliari is a well studied city with a modern system of planning, very few data which were likely to be needed by the workshop were actually available in digital form. This is not an uncommon circumstance and one must always adjust expectations to the available resources. We had access to good data on current land-use which we generalized into a simpler classification, a terrain model, several sector plans and in particular the applicable section of the Sardinian Regional Landscape Plan, a large number of photographs taken by the students on their site visit, and Google Earth.

There is a necessary distinction which must be made between a short workshop and a semester long studio and it relates to both the time available for the activity and the freedom of students to define what they are exploring. In a workshop many more decisions must be made by the organizing faculty whereas in the studio an important part of the education of the students relates to the definition of the problem itself and methods which should be applied. In the Cagliari workshop, and because of the constrained time schedule, the faculty decided on the scope of the problem, its methods and its expected products.

It was decided that there would be 10 teams devoted to evaluating processes which were deemed central to the future of the region; habitat, the visual landscape, the cultural and recreational landscape, residential development, tourism, transport, hydrology, and because of a special interest on the part of the German students, geothermal energy, solar and wind energy, and biomass energy. Each team was to produce two things. The first was a map in two colors where green represented highly valuable elements which sustain that process and which should be protected and wear red showed areas of problem or threat to that process and which should be improved. The second product was a set of diagrams which represented ideas for projects to change that process, either via the protection of valuable areas or the improvement of problem areas. These were to be drawn on thin plastic sheets, color-coded with a different color applicable to each of the 10 evaluation teams. They were to be rank ordered in terms of their importance and efficacy edge as judged by the team.

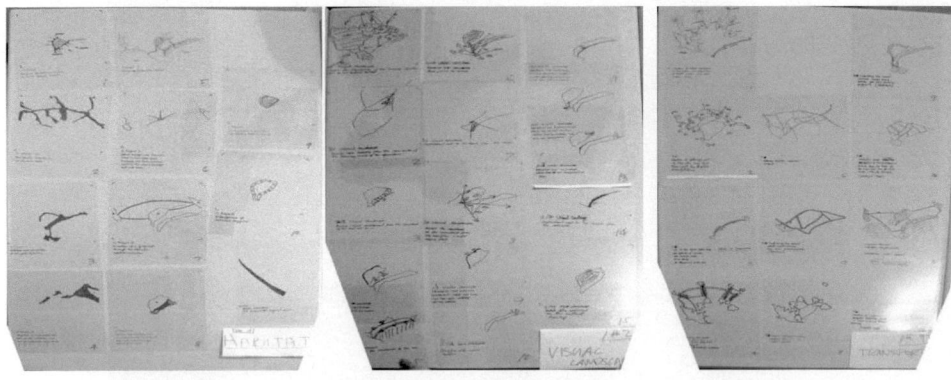

Fig. 14: Potential project and policy diagrams for habitat, visual and transport

I cannot emphasize enough the importance of standardizing color codes graphic scales and styles of representation. In a workshop but also in a semester long studio it is very important that students and visitors rapidly comprehend each other's work and a shared graphic language is essential to assure this task. The reader will also note the crudeness of the initial diagrams. This is a normal outgrowth of speed in production. It has little to do with the thinking behind the diagram and everything to do with time and the technologies of data management and representation.

Each of the teams made a concise presentation of how it understood its assigned process how it defined the priority areas of conservation and change, and the initial rank ordered set of potential projects. The audience was the entire workshop group because each participant knew that the teams would be reorganized for other stages of the workshop and that they would be required to integrate all of the process assessments and evaluations. In addition all of the local experts who had made preliminary lectures to the students were invited, and they met with the student teams to present their critiques, proposed revisions, and additional potential projects. A second and brief presentation was made of all changes and additional projects and I cannot overemphasize the importance of having instituted a standardized representation process. This enabled students to understand what was happening without extensive discussion.

At the end of this stage in the workshop, all of the proposed projects were uniquely numbered by the team and the rank ordering number and systematically placed on a long table organized. This long table was set up to enable many additional potential projects to be added in the next stage of the workshop.

Fig. 15: Developing the designs

The faculty team decided to organize the students into six larger teams each representing a different "stakeholder group" with interests in the future of the Cagliari metropolitan area. These "change teams" were 1 – conservationists, 2 – residential commercial and industrial developers, 3 – regional planners emphasizing the Sardinian Regional Landscape Plan,

4 – a foundation for the support of renewable energy, 5 – the tourism development board, and 6 – the several local governments in the area each seeking reelection.

Each change team had to come to an internal agreement as to the policy objectives of its stakeholder client, it had to make a proposal for changes which would support those objectives over the next 20 years it had to accommodate a 4% growth in population and its concomitant land-use changes and it had to be as self-sufficient in energy as possible. The resulting design would be based on a selection of the projects presented by the process evaluation teams and any variations or additional projects proposed by the change team. The new or varied projects were to be drawn in the same graphic format as the prior sets: they were made known to all other students via an announcement and then numbered and placed on the long table so that they were available for use by anyone else.

Each change team then had the difficult task of selecting no more than about 10 of the available projects which numbered approximately 150. In some cases change teams grouped projects related to a particular process, read through them and thus limited the number of plastic drawn sheets which they had to overlay and combine. The process of overlaying and combining the drawings was rapid and easy, using an overhead projector as a light table and a digital camera for recording. The projects selected by each change team had to be rank ordered so that the most important were most clearly and graphically represented.

Fig. 16: Comparing the impacts of the six designs

Each team then made a concise presentation to the rest of the class. The students knew that they would be reorganized into their original process evaluation teams for the purpose of comparatively assessing the impacts of each of the proposed designs.

The impact assessments consisted of each process evaluation team evaluating each of the six stakeholder change proposals in a simple six level scale but one which required thoughtful internal discussion and judgment: +3 represented a much better circumstance for that process, +1 meant a better situation, 0 meant no change, -1 meant a worse situation, -3 meant that it was very much worse, and -5 meant that the process was "lost". An example of -5 would be if the resident flamingo population of Cagliari was no longer present. The evaluations were then placed on a chart where a green circle meant the team was doing relatively best and a red circle meant the team was doing relatively worst among the six alternatives.

This impact assessment was not subject to public discussion but rather private consultation between the change teams and the impact assessment teams with the intent that the designs would be improved in a second stage.

An important technical byproduct of this first round of design and comparative assessment was that some projects were clearly more significant either because they were central to the change proposals of one stakeholder team or they were recovering in the designs of several teams. A small group of students, one representing each process evaluation team, then digitally redrew these projects in a manner similar to that of developed for Tepotzotlan studio.

A second design cycle was then begun with the teams again reformed into their stakeholder change teams. Each team could rapidly drop or add projects. Many of them made variations and new projects which again were brought to the attention of the entire group, numbered, and placed on the long table for common use.

A second presentation was held of the designs, but this time it was entirely silent because of everyone's having understood the graphic conventions.

A second round of comparative impact assessments was then made.

At this time all work was assumed to be finished except for a very few last-minute changes that could be made within the briefest of time. Each change team then delegated one person to make a digital representation of each project that would be combined into their alternative future for Cagliari and to make a set of digital graphics for a final presentation.

The graphic product for the final public was specified by the faculty. It would consist of

1. 3 to 5 images presenting the team's principal policy objectives in order of importance,
2. an image of the accumulated projects for each process with the processes presented in rank order,
3. the existing conditions and
4. the proposed changes,
5. the future alternative for Cagliari (the existing and changed conditions) and
6. a summary graphic for comparative purposes that showed the process-projects and the proposed alternative future.

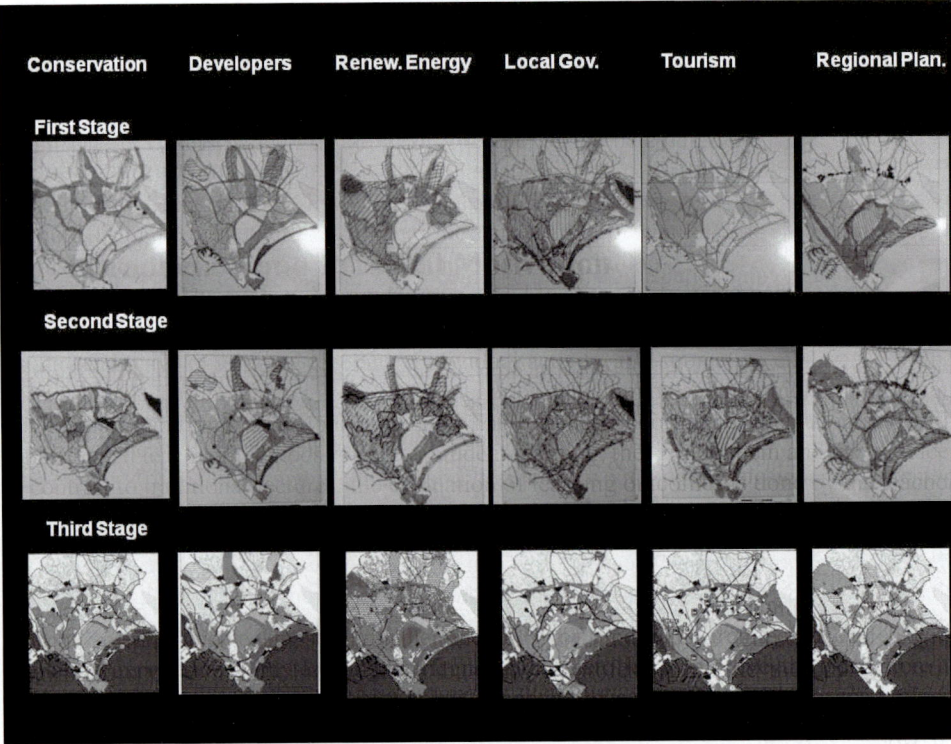

Fig. 17: The stages of the six designs

Each team was then asked to organize a presentation in no more than 10 minutes which would be given to the entire workshop group, all of the local experts, who had participated, and many additional faculty and students from the University of Cagliari. This entire presentation would be conducted in Italian. The following are the graphics from two of these presentations.

Each presentation was followed by a period of questions and answers and sometimes argument, and the entire set of presentations was also followed by discussion. Each of the local experts was invited to judge which of the six alternatives best met his or her expectations for the future of the Cagliari metropolitan area. In some cases people chose to reflect their institutional stakeholder interests and some chose a broader personal perspective.

Juan Carlos Vargas Moreno then led a discussion related to the question of whether one should choose among the six alternatives, thus seeing the exercise as a zero sum game, or whether it was possible to be selective from among the alternatives and generate a new plan which compromised the interests of the stakeholder groups. Christian Albert had made a frequency assessment of the number of times each of the projects had been selected for integration into the change proposals and we did a very rapid exercise in real time. We selected the most frequently used projects from those proposals which the local experts had praised the most and overlaid their hand-drawn plastic diagrams the overhead projector.

Juan Carlos then created and displayed a composite digital image from the selected project graphics. In a following discussion it was generally agreed that this was a very good proposal.

This was the end of the working sessions of the workshop. It was followed by a reception and a very fine party/dinner for all participants, and especially for what I believe to be a very satisfied and tired group of students.

6 Discussion and Summary

I have illustrated several variations of the application of the framework to a workshop or studio's beginning phase which relies on a mix of diagramming and Delphi methods and of judgment. I have several times led similar workshops as part of the beginning of a semester long studio, and also as a beginning phase of a large research program. In all cases, the participants were made fully aware that workshop was intended as an exploration with the objective of identifying the issues and needs for a more thorough use of the framework over a longer period of time, with better data and with application to projects at different scales of design.

I cannot claim that this approach and its several teaching methods will always work efficiently and well. I fully acknowledge that they are potentially open to diagrammatic exaggeration and errors of judgment, and that they are dependent upon the participants having sufficient comfort in working together and in making rapid diagrams and judgments. At worst, they can raise questions for further research, data development and alternative design strategies. However, in my experience and at best, they provide clear, rapid and robust ways of "getting started".

In making assignments people should recognize that there are normally two reasons for undertaking a task (other than project needs). The first is that you can do something, enjoy it, and want to do more. The second reason is that you can't do something and want to learn how to do it. Even though project efficiency and effectiveness may favor the former reason, the latter is a better rationale for an educational institution and it is usually highly respected by student self-managers. Learning from other students in small group tasks is clearly an advantage which students find in this organizational style.

A crucial social paradigm is that your word is your bond. If you say you are going to do something you'd better do it, and if you realize that you can't you have to let others know as soon as possible so that they can help. And people should help. The project belongs to the students as a group. Credit is shared by the team in alphabetical order. There may be internal tasks in which certain pieces are broken down under individual authorship. Indeed it is expected that every project have some component in which each student can say, "I did this." But these situations are known in advance because of the study design. Everything else is "we".

The space which the studio occupies is organized for group activity, with one very large central table for meetings, and smaller individual work spaces. All classes start at the table, and all important discussions and presentations, including reviews of individual work, take place there and are public.

It is obvious that one of the individual costs of this highly organized team-based studio structure is that each student cannot and does not do all tasks even though each student participates in each phase of the study. Because of this students are encouraged to make presentations to the class meetings of things which they are doing and which may be of interest to others. Final presentations have always included the need for a relatively nervous group of students to clearly explain their work to an interested and critical audience of peers, local experts and citizens. This is not an easy task when expectations are high (and the expectations are very high).

A considerable amount of innovative and high quality work gets performed during this type of workshop or studio, despite the fact that students are placed in an unfamiliar situation for which their previous education has not fully prepared them. The major reason for this is the self motivation and the peer pressure associated with the fact that it is "their problem", with both the credit and the blame to be part of their own sense of accomplishment and self confidence.

And what about the faculty? The faculty roles are varied and challenging. Clearly there is the role of producer – of organizing the situation in which the studio project can occur. It takes a long time. It carries a certain personal interest and commitment and it is not always fully successful. There is a considerable consultant-in-chief role in which advice is sought both teams and individual students. It must be either given, or students directed to other expert consultants, frequently other faculty members. Without doubt, there is a substantial "hidden hand" role, in which constant observation is necessary to ensure against disaster. Students are often overly ambitious. They frequently underestimate the impacts of problems which are unforeseen but which experience knows may occur. There are also important mediation roles, frequently around social questions and issues of organization. There is the real responsibility of ensuring that the individual educational needs of individual students are met within the scope of the team organization. There is the faculty role as "critic", but only after the students have reviewed and discussed the work in progress. Finally there is the legal responsibility of oversight, ensuring that the project is not a disaster and that it is completed within the constraints of time and money available. Yet the most difficult faculty role is that of purposefully abstaining from controlling the many difficult managerial and design decisions and letting the group learn by experience. After all, it is these experiences which are among the critical educational lessons to be gained by the students.

Finally, a caveat-A framework is not a theory. It can be a useful aid to the organization of a complex design problem in a workshop, studio or in an applied- research program. It is only as useful as it is seen as useful by the user(s). This framework has been adapted and used many times and in many contexts, and it seems to be useful and robust.

References

Steinitz, C. et al. (1986), *Alternative Futures for The Bermuda Dump, Graduate School of Design, Harvard University, and Government of Bermuda, and Department of Planning.* Government of Bermuda, The Pembroke Marsh Plan.

Steinitz, C. & Figueroa, A. (Eds. with Vargas-Moreno, J. C. & others) (2005, 2010), *Futuros Alternativos para Tepotzotlan/Alternative Futures for Tepotzotlan*. Universidad Autonoma Metropolitana-Azcapotzalco, Mexico, 2005 (in Spanish and English).

Steinitz, C. (2010), *Teaching in a Multidisciplinary Collaborative Workshop Format: The Cagliari Workshop*. In: Steinitz, C., Abis, E., Haaren, C. v., Albert, C., Kempa, D., Palmas, C., Pili, S., Vargas-Moreno, J. C., FutureMAC09: Alternative Futures for the Metropolitan Area of Cagliari, The Cagliari Workshop: An Experiment in Interdisciplinary Education/FutureMAC09: Scenari Alternativi per l'area Metropolitana di Cagliari, Workshop di Sperimentazione Didattica Interdisciplinare, Gangemi, Roma.

Additional References

Steinitz, C. (1990), *A Framework for Theory Applicable to the Education of Landscape Architects and Other Environmental Design Professionals*. In: Landscape Journal, 9, 136-143.

Steinitz, C. (1995), *Design is a Verb; Design is a Noun*. In: Landscape Journal, 4 (2), 188-200.

Steinitz, C. (2000), *On Teaching Ecological Principles to Designers, Chapter 10*. In: Johnson, B & Hill, K. (Eds.), Ecology and Design: Frameworks for Learning. Island Press, Washington, D.C, 231-244.

Steinitz, C et al (2003), *Chapter 3, The Framework for Alternative Futures Studies*. In: Steinitz, C. Arias, H., Bassett, S., Flaxman, M., Goode, T., Maddock, T., Mouat, D., Peiser, R. & Shearer, A. (Eds.), Alternative Futures for Changing landscapes: The Upper San Pedro River Basin in Arizona and Sonora. Island Press, Washington, D.C.

Geodesign Futures – Nearly 50 Predictions

Stephen M. ERVIN

Keynote: 02 June, 2012

1 Abstract

Geodesign is a rapidly evolving set of technologies and approaches, covering a spectrum of disciplines that interact in the built and natural environment, and planning, design, construction, and monitoring processes [ESRI] . This paper speculates, sometimes wildly, on possible future developments in geodesign and in some other disciplines that impinge upon it – accompanied by an estimation of probability, ranging from 'unlikely' to 'certain' – with time frame not specified, and some 'short-term reality constraints' ignored

2 Introduction

I'm well aware of the inherent dangers in making predictions about the future. I know of several famously bad predictions over time, including DEC CEO Ken Olson's in 1977 "There is no reason anyone would want a computer in their home"; and a perhaps less well-known 1955 assertion that "nuclear powered vacuum cleaners will probably be a reality within 10 years". I will try to avoid any such spectacularly bad calls, at either end of the spectrum; but will try slip in a few provocative predictions into this survey of alternative geodesign futures.

I refer to 'futures', not simply 'future', intentionally. One night in 2011 while watching TV forecasts about the approach of Hurricane Irene, I realized that one of the things I've always been fascinated by is how weather predictors combine risky predictions with real impact, geospatial content, and great graphics – and in particular rely upon *probability maps* with zones of widening uncertainty, as in their predicted tracks of tropical storms – and this reminded me of Carl Steinitz's memorable diagram showing, from a moment in time and space, a branching tree of alternative futures stretching out into the future. (Figure 1.) From which of course the geodesigner's challenge is to pick one (or several) to make the most likely. It's with that diagram in mind that I entitled this 'Geodesign Futures' (plural).

In exploring the 'future studies' enterprise, I came across the idea of "3 P's and some W's" for the "possible, probable, and preferable futures, plus wildcards" [WIKIPEDIA]. This seemed like a suitably safe but challenging rubric for constructing my thoughts about this topic, and so I'll use this idea of ranging from 'the dramatically unlikely' to 'the absolutely certain' in the rest of this speculation about some alternative geodesign futures, in five broad categories: 'Big Picture', 'Technologies', 'Software', 'Connections', and 'Warnings'.

3 Predictions

Predictions 1 – 12: Big Picture

Prediction #1. I consider it *possible* that geodesign will be remembered merely as a marketing buzzword – a low probability, to be sure, but possible. #2. More likely in my view is that geodesign will in fact evolve and mature as a useful fusing of G.I. science with design; and that of course is the future I prefer to work towards.

#3. I do consider it *probable* that geodesign will always have the same cloudy, fuzzy-boundaried, "in the eye of the beholder" definitional vagaries that, for example, landscape architecture has enjoyed all these years; and that the question "what is geodesign?" will forever be open to interpretation and discussion. I think that's a good thing.

#4. I do consider it *unlikely* that any agreed-upon suitably crisp definition will emerge. My colleague Mike Flaxman's definition now somewhat abroad in the literature – "Geodesign is a design and planning method which tightly couples the creation of design proposals with impact simulations informed by geographic contexts." [WIKIPEDIA] – is, I think, good and suitably crisp; but "simulations informed by geographic contexts" doesn't quite nail it for me. I propose to replace the last two words with "systems thinking and digital technologies".

So: "Geodesign is a design and planning method which tightly couples the creation of design proposals with impact simulations informed by systems thinking and digital technologies ". Those two essential aspects are what make geodesign projects unique, in my view, worthy of a distinct name, and not just any old design or planning project, or "what some of us have been doing for many years".

So I'll start prediction #5 with a certainty and a very large number: *10 billion*; about the number of people predicted to be on the planet in the next century of the 'anthropocene era', that we are in today. It is for these people that we do geodesign; not for the forests or dams or transportation systems, but for the people who live, work, and play in the environments we build, modify, or conserve, and for whose well-being we are at least partially responsible in our geodesign activities.

#6. Also certain: a lot of geodesign will be on the Asian continent. Today the average human is a 28-year-old Han Chinese male; but according to the National Geographic moving towards an even younger and somewhat more female individual located closer to the Indian subcontinent; and this indeed is a hotspot for geodesign. Of course we need geodesign everywhere but we especially need it in Asia. Some of the rapid urbanization happening in that region needs desperately to be informed by the aesthetics and the science that geodesign can bring. They have the digital technologies, but systems thinking seems often to be lacking. At the same time, China is a fertile ground for imaginative playful architectural and urban design.

#7. It's probable therefore a lot of geodesign projects will be *urban*, along with two thirds of the Earth's population. Just recently several things caught my eye in this regard. I noted that the Ted prize is being awarded this year not to an individual but rather to "the city 2.0" – this idea that "over the next 90 years we will build more urban living space than in all prior centuries combined and we had better get it right". [cite Ted Talk] I notice also that the Rotterdam architectural Biennale will feature an exhibition themed "smart cities". And

I've heard of the "Ubiquitous city" or "U-city" phenomenon – in which urban regions with ubiquitous information technology transform much of what we have understood about urbanism in previous centuries. This also is an important part of geodesign.

#8. Another thing for certain: *water* will be important. Whether its availability (or shortage) and security and cleanliness, or its overabundance in the wrong place at the wrong time, managing the distribution of water in 4D space time is a major project for geodesign in the future. How we integrate land, water, vegetation and structures is the age-old challenge of all environmental design; and no less so for the future.

#9. And so I consider it probable that *public and environmental health* will increasingly be a concern for geodesign projects, as witnessed by a range of literature coming out including the book "Designing healthy communities" [JACKSON]. The idea that public health is a determinant at the top of the list – harkening back to Frederick Law Olmsted's concerns, and the origins of GIS at the 'cholera well' – will be an important part of any geodesign future.

#10. I consider probable that *biology* will be increasingly important, especially in the form of bioengineering, as we are beginning to use biological structures and changes in the biological environment to support geodesign projects; and also 'biomimicry' as a source of design inspiration [BENYUS]. For those of you haven't seen some of the incredible work coming out of this line of thinking, from architects and engineers looking at how natural systems provide for such things as waterproofing, lubrication, insulation, and structural efficiency, this is an important thread of science contributing to the future of geodesign.

#11. I consider it certain that *agriculture* will also be increasingly important, as a component in sustainable and futuristic geodesign efforts. There is much to be invented, yet, in this regard; and much to lose if we don't pay attention to it!

#12. I consider it probable that we will evolve in geodesign efforts from the *pattern-making* which has characterized some landscape architecture, architecture, and urban design in the past, to what I call *pattern finding* or *pattern -reinforcing*. Using remote sensing, and tools as simple and ubiquitous as Google Earth, we can see large and complex patterns at work in the landscape, to which intelligent geodesign can and must respond.

Predictions 13 – 21: Technologies
#13. Just as medicine in the clinical sense requires stethoscopes and thermometers, I predict that geodesign will increasingly involve *sensors* – remote sensors like satellites, and embedded sensors in roads, people, and trees, providing not just static base data but real-time interactive data, and giving rise not just to the familiar remote sensing that has so shaped planning and design in the last several decades, but also giving rise to a newer form of "intimate sensing".

There is an example of a new system for flood- control infrastructure proposed by RPI, that I think demonstrates the real power of geodesign [RPI]. In the project "Development of a Multi-Scale Monitoring and Health Assessment Framework for Effective Management of Levees and Flood-Control Infrastructure Systems", they propose physical earth dikes as well as concrete retaining walls and steel structures all connected with motors and actuators, tied together into a network of sensors, computers, and satellites; a very clear example of a geodesign future.

#14. I consider it unlikely that our current *visualization tools* will be sufficient. No matter how many rectangular computer display screens we may have, these are incapable of capturing the complexity and the depth of environmental knowledge and perception that we will need. I consider it probable therefore that geodesign representation will require a move beyond maps, graphs, pictures, and movies, and evolve into more 3-D and 4D visualization systems.

#15. For example I consider it probable that *augmented reality* interfaces will become increasingly vivid and useful, especially coupled with mobile devices and new interfaces. Many of us today are carrying cell phones capable of overlaying a video image of the real world with additional information whether about coffee shops or land prices or soil types or vegetation diseases ... These devices are transforming the way we can understand and therefore design the world.

#16. I consider it probable that *gestural interfaces* are here to stay. I wish I could say that a long-held vision of mine for a digital clay grading system, for manipulating terrain by thumb-swipe, was actually in its prototype stage – but it's still only in the dream (or fake-demo) stage. I look forward to the developments that will come out of touch screens, tablet computers, and gestural interfaces, and their impacts upon geodesign practices.

#17. I certainly consider it probable that *virtual realities and immersive displays* will continue to become more and more available, useful, and compelling in the process of visualizing and understanding landscapes and environments. Holographic projection systems are already on the market.
Who for example didn't want the 3-D immersive display system as seen in James Cameron's 'Avatar' movie – just think about the possibilities of this kind of display, many technological components of which are already available.

#18. And I should mention that I consider it likely that geodesign will continue to inherit technology from the military and from Hollywood; and unlikely, sadly, that the military or Hollywood will learn much from geodesign. This is an example where I hope I'm wrong.

#19. I have no doubt that the technology for *simulations* will be improve rapidly and be central as the geodesign enterprise moves forward. Hydrological and fluid dynamics simulations in particular have come a long way towards being useful in everyday design situations, with desktop computers.

#20. I don't exclude the physical or tangible from this fascination with the virtual. Indeed I predict that *physical and tangible interfaces* and simulations will be an important part of geodesign moving forward, whether it's in the familiar style of a 3-D model of the city with an analog camera physically moving through it; or in the more complex possibility that each of these model buildings is endowed with sensors, actuators, or other 'intelligence', capable of interaction and adding complexity to simulations.

#21. I consider it possible that *ontologies* for geodesign will become more comprehensive and embedded in design practice and therefore will enable complex 'geo-reasoning – not just 'geographic analysis' as we have today, but if we are to engage in collaborative reasoning between people, and indeed between people and machines, we'll need agreed-upon vocabularies and shared semantics. The SDSS geodesign knowledge portal available on-line thanks to University of Redlands [SDSS] is an example of this emerging possibility.

Predictions 22 – 29: Software

#22. I consider it unlikely that any dominant software will emerge in this space. It might be great if we could all buy and use "Geodesign Suite 2012", but I actually don't think that's going to be the future. Rather I consider it more probable that there will be "apps for that", and web services – whether in the cloud, open-source, closed-source, or other – that will provide the new paradigm for how software is designed, developed, distributed, and used.

#23. Here's a low-probability, high-payoff wild card: some open-source consortium will begin to the make and develop industrial-strength geodesign apps which will become available to users all over the globe with access to high-quality local data, and in so doing enable a better planetary future!

#24. And here's a possibility that I've talked about at this conference before: the necessary complements of a useful geodesign software toolbox. Let me describe 15 essential components briefly:
- all geodesign in my view is fundamentally the process of arranging *objects* – such as trees, forests, houses, cities, transportation systems, et al. ;
- over a *base* – a geographic context that has some coordinates, and some existing unchangeable features such as topographical, hydrological and others;
- in some particular *configuration* or layout, both geometric in 3-D and also over time in 4D;
- subject to some *constraints* – houses must be parallel to the road, roads can't be in the water, and so on;
- and in so doing we make access to a *library* of objects and configurations, of constraints and base data and other precedents from which we can learn;
- often in *collaboration* with others over space-time, asynchronous and distributed;
- producing along the way multiple *versions*, some of which need to be kept, combined, and branched, some of which are dead-ends and die, and some of which become viable for the future;
- and in so doing we engage across a range of *levels of abstraction,* from high-level abstract concepts such as radial or symmetric, to much more specific and concrete dimensions in particular materials;
- and in this regard I consider a particular class of graphics, to wit *diagrams,* to be essential to the design project – that's where abstract ideas meet specific sites and thereby become plans and proposals;
- and we've learned that our plans and proposals need not be limited to static printed drawings; that they can be like the Web, *hyperlinked* with connected information, whether it's videos, or specifications, or more detail in a range of ways
- also we need to have *modeling* and *scripting* capabilities – all the power of modern computer science – for describing relationships between objects and configurations and constraints, so that we can see both predictions for the future and also modeling configurations according to algorithmic design;
- and our designs must be cognizant of time, whether long-term seasonal changes or just the schedule of a construction project;
- and, especially with the added dimension of time, simulation is an essential part of what makes geodesign (as in Flaxman's definition); and by this I include not just scientific (spatial and a-spatial) simulations, but also visual simulations – renderings and animations are in my model included in this box;

- and for managing complex projects we have talked in previous geodesign conferences about the idea of dashboards – summary displays which capture key performance metrics, and allow for rapid evaluation of complex designs according to the outputs often of the simulations we have made;
- and, finally, in a nod to my colleague and teacher Carl Steinitz – although I've never been a big fan of the animated paperclip avatar injecting its self into one's thought process at the computer – the idea that we might have knowledge of multiple methods of geodesign which could be invoked, and indeed perhaps a non-intrusive kind of *'methods coach'* who might observe: "gosh, I notice that for the last couple of days there's been no real improvements on your important key metrics ; either you should consider this project done, or possibly consider another method altogether ..."

Now from all those 15 items, I want to pick out 5 points that are particularly worthy of attention in this talk.

#25. I consider it likely that we will have *object-oriented design* and I talked about this last year at the 2011 Geodesign Summit conference. [ERVIN (2011)]

#26. With respect to the objects of design, I consider it probable that we will have *full 3-D integration* in our geodesign software and that we will no longer need to distinguish between building and landscape, or CAD or GIS or BIM or LIM.

#27. I consider it possible that *constraints* will be more embedded in software. Many geometric modeling tools already enable simple constraints such as perpendicular or collinear, and also gravity and other dynamics are making their way also into these tools. I do also recognize that having constraints raises serious issues such as what to do when conflicts arise – more silly paperclips making suggestions? – and this is why this is not in the 'certain' or even 'probable' category in my view.

#28. I consider probable that *version management* will make its way into geodesign. Software developers have now at their disposal simple tools for managing and merging and diverging multiple versions with shared code repositories and expanding this idea from just linear textual code to more complex hypermedia documents with drawings, etc., is not that daunting a technological problem, and surely will come to pass.

#29. Sadly I consider it unlikely – although I also consider it highly desirable – that we will get computational support for *diagrams*. I consider this to be such a key part of all design thinking – as I've also talked about at the ESRI Geodesign Summit 2010 [ERVIN (2011)] – unfortunately this is real hard, and may remain the domain of 'wetware' (solely the domain of human designers) into the foreseeable future.

So these are some of the elements that I consider will or should be part of the future of geodesign software, one way or another.

Predictions 30 – 45: Connections

#30. Another thing is for certain: *systems thinking* will have to prevail if we are to be successful. System thinking is simple deceptively – practices such as "seeking to understand the big picture", "considering short and long-term consequences", "changing perspective to increase understanding" – and taught in elementary school nowadays, thanks, for example to the Waters foundation – but also sadly lacking in many aspects of the public environment that geodesign must take on.

And it's not just for grade-schoolers; but also informs complex scientific thinking. I recommend to you Donella Meadows book "Thinking in systems", [MEADOWS] in which she describes her experience and insights starting back with the "Limits to Growth" exercise she was a part of. You can also Google "Dancing with Systems", for a very nice short, poetic essay about systems thinking.

#31. Thinking about the essential roles of biology and agriculture in geodesign, I consider it possible (and desirable) that "resilient design", "permaculture", and "cradle to cradle" thinking, along with other related ideas, and as I mentioned above, agriculture, will become essential complements of geodesign as we move forward into the 21st century.

#32. I consider it certain that geodesign and geodesigners will need to be *connected* and to build *connections* – whether they are social Facebook connections or hydrological underground connections, landscape ecological stepping stones or coherent global transit systems … "Connection" is a necessary by-word for geodesign.

#33. It's certain that the so-called "Web of things" will affect geodesign. This is the world of sensors and controllers and objects and systems that I've alluded to earlier.

#34. I consider it probable that "Robo-dozing" (robotically controlled earth-moving machines) will become more widespread – and more generally: there will be more *robots* in all aspects of geodesign. Caterpillar is already committed to making a range of robotic earth moving equipment and you can imagine the conversations between artificial intelligences, as bulldozers and cellphones communicate with the 'mother computer' at the engineering office.

#35. I consider possible that "terraforming" will become more of a reality. As a science-fiction reader I've long been intrigued with the idea of transforming other planets and moons into more earthlike habitable ecological systems – and the sad truth is we will very likely have to do it to our very own planet!

#36. I certainly consider it likely that flying robots will become a part of our systems – notably obviously for surveillance but also in the future for construction, as these devices such as these available from micro-drones.com become more and more powerful, available, and controllable.

#37. I do consider it unlikely that we'll ever have the walking city envisioned by Ron Herron and Archigram in 1964, in which large robotic mega structures prowled the surface of the earth. That's too bad – I've always liked the drawing.

#38. More probable is that the new term *geo-fencing* will become a part of the geodesign vocabulary. This is the idea that rather than having to build a physical fence, we can simply draw a polygon on a screen, which when coupled with GPS and various other kinds of tracking or restraint systems, such as you might put on a dog or a prisoner, can become an effective virtual fence. The market today is for businesses tracking "employee accountability and productivity", but there's also a dark side to this technology known as "geo-slavery"; the idea that individual rights may be threatened or even lost in the future subject to location-based systems and technologies. This is something that geodesigners will have to take into account and hopefully not be complicit in …

#39. It's for certain that geodesign is already *multi-scale* – ranging from global concerns that really impact thinking locally, to the emergence of microscale nano-technological devices which can make a real impact on the world.

At the broadest scale, I recommend to you Bruce Mau's book and project called 'Massive Change' [MAU] (http://massive change.com) which talks about "not just the world of design" but as the book cover would have it, " the design of the world".

#40. Another intriguing new possibility is 'planet-hacking'. See for example Eli Kintisch's book 'Hack the Planet' [KINTISCH] in which he talks about interventions at the global scale, such as carbon sequestration technologies and artificial volcanoes among other things … Geodesigners may have learn to hack the planet (without breaking it!)

#41. And it's probable that *nanotechnology* will also be a player; witness the growing landscape of "synthetic biology" which includes the development of microbial viruses that may well impact ecological systems in ways that we can't either understand or predict right now …

#42. And so it's probable that some geodesign *ethics* will emerge … and be controversial, with no single set of geodesign ethics. As we engage in transforming our planet and our societies, we will inevitably and repeatedly come up against the complications of cultural diversity and multiple points of view …

#43. It's possible that we will one day do some geodesign on the moon or some other planet (would we call it luno-design?) and in so doing we will have access to Hargrove's book "Beyond spaceship Earth; environmental ethics and the Solar System." [HARGROVE] Now that's thinking big!

#44. It's possible then that our education system will be able to keep up with the demands of budding designers who will need access to information about everything from ethics of geo-slavery to the technology of nano-bots – but also possible that our education system will NOT be able to keep up … This is a very big challenge all by itself.

#45. Here's an unlikely wildcard but something that is happening already: direct brain computer interface technology, in which no mouse or keyboard is required. Right now today it's used simply for managing games like Pong, but it's from these tiny acorns, mighty and complicated oaks (direct brain-interface geodesign systems) may yet grow.

Predictions 46 – 48: 2 Warnings
Finally I started with a big number; and I will close with two small ones. "Two plus Two" – which makes "Four" of course; but may also make for a very complicated world, as we have: (#46) nearly certain global warming of 2 or more degrees C.; and (#47) very possible ocean level rise of 2 meters more or less. In that nearly certain future we will have a very complicated 'FGDC' to respond to (that's a "Future GeoDesign Challenge" – for those who may know some other meaning for that 4 – letter acronym!)

#48. And in so doing we will certainly need to take advantage of the incredible array of satellites now orbiting the earth – and increasingly, exploring the whole solar system!

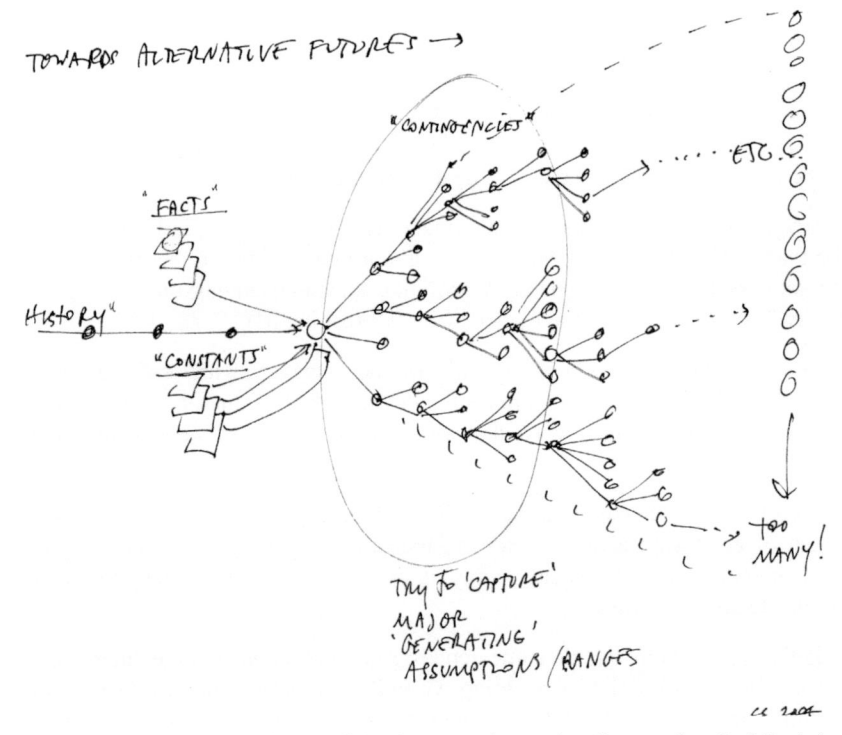

Fig. 1: The branched tree of (too) many alternative futures, by Carl Steinitz, 2004

References

Benyus, J. M. (2002), *Biomimicry – innovation inspired by nature.* New York, Perennial.
Ervin, S. M. (2010), *On the Necessity of Diagrams,* http://video.esri.com/watch/55/on-the-necessity-of-diagrams.
Ervin, S. M. (2011), *Object Oriented GeoDesign,* .http://video.esri.com/watch/195/object-oriented-geodesign
ESRI, http://www.geodesignsummit.com/index.html
Hargrove, E. C. (Ed.) (1986), *Beyond spaceship earth : environmental ethics and the solar system.* San Francisco, Sierra Club Books.
Jackson, R. J. (2012), *Designing healthy communities.* San Francisco, Jossey-Bass.
Kintisch, E. (2010), *Hack the planet: science's best hope-or worst nightmare-for averting climate catastrophe.* Hoboken, N.J., Wiley.
Mau, B. (2004), *Massive Change.* London/New York, Phaidon.
Meadows, D. H. (2008), *Thinking in systems: a primer.* White River Junction, Vt., Chelsea Green Pub.
RPI, *Development of a Multi-Scale Monitoring and Health Assessment Framework for Effective Management of Levees and Flood-Control Infrastructure Systems,* http://www.nist-rpi.org/index.html.
SDSS, *Geodesign Knowledge Portal,* http://www.spatial.redlands.edu/geodesign/ontology/?n=Software:SDSS.

Teaching Methods in Digital Landscape Architecture

Wheeling a Trojan Horse to Teach MLA Students Geoinformation Methods

Philip PAAR and Jörg REKITTKE

1 GIS newbies

The early days of Geographic Information Systems (GIS), the 1960s and 70s, were strongly tied to landscape architecture in both research and education. GIS was largely inspired by MCHARG's (1969) spatial concept in design and manual overlay techniques (SUE 2008). In the academic field, Harvard University Graduate School of Design (GSD) had been a very early innovator in terms of integration of landscape computing, particularly the use of GIS. Their "Laboratory for Computer Graphics and Spatial Analysis" attracted ambitious students, faculty, and other researchers (COPPOCK & RHIND 1991, CHRISMAN 2006). The Lab attracted students, faculty and other researchers including such luminaries as Jack Dangermond, landscape architect, supposable multi billionaire and founder of ESRI.

Meanwhile, GIS are established software tools, well established in the daily work of planners, geographers, engineers and scientists to display and analyse geo-referenced and geo-referenceable data from multiple sources about the health, status, history and future of our environment. Beyond these traditional application areas of geo information science, GIS is ubiquitously used in business, for example in energy sector, financial and telecommunication industry as well as in marketing. Nevertheless, in today's practice of landscape architecture computer-aided design (CAD) systems, graphics design-based and site-oriented approaches are still prevalent. The use of GIS has not yet gained wide acceptance compared to CAD software. Findings of BUHMANN & WIESEL (2003) show the discrepancy between the adoption of CAD (98%) and GIS (35%) in German landscape architecture firms, even though GIS is catching up (Fig. 1).

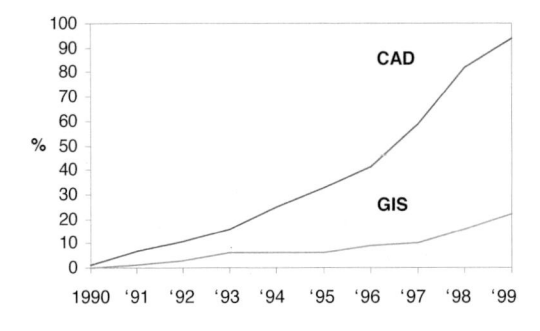

Fig. 1: Slow adoption of GIS in German landscape architecture offices (BUHMANN & JÜHNEMANN 2000)

Early workstation GIS was too costly and complicated for most of the predominantly small offices. The advent of desktop GIS in the early 1990s made the software more affordable. Spatial concepts in GIS are relevant and compatible to certain design tasks; especially at the landscape level because of the cartographic heritage of GIS (e.g. overlay function) rooted in landscape design SUE (2008). Nevertheless, GIS is still deemed of being very complex and difficult to learn – at least for 'visual animals' such as designers. Therefore, e.g. HANNA (1999) and ESRI try to "make GIS more comfortable for reluctant landscape architects".

Most of our master students of landscape architecture (MLA) in Europe or Asia have already become masters of the Adobe Design Collection, frequently flavoured with remarkable skills in Google SketchUp and professional CAD systems. Designers have recognised the fundamental importance of spatial thinking in their curriculum (DILLION et al. 2003). In some MLA programmes such as in Harvard GSD, GIS is at least used in studio projects for an initial site analyses or as a map-making tool. Often GIS specialists assist the studio coordinator to develop basic datasets for the studios. In other faculties or countries, students may not yet have been 'GIS-enabled'. These white spots on the GIS map almost remain a niche bastion, even if the license server runs ESRI ArcGIS, unnoticed by the potential users.

Anecdotal evidence suggests that MLA students rather apply 2D and 3D graphics based workflows than GIS enabled ones and prefer photo and video collages for design and design communication. In this way, students might never get to know the advantage of a GIS assisted iterative design-analysis process. What is the issue with this? While CAD, Illustrator, and Photoshop are suitable for design, they are weak on landscape analysis and assessment. In contrast, GIS is strong on handling large spatial datasets and on analysis but traditionally weak on design tools, and photorealistic 3D visualisation (CLASEN & PAAR 2007; FLAXMAN 2010). Robust landscape models and tools need to take into explicit account both geographic location and the passage of time (ERVIN 2006). Non-spatial and non-temporal approaches are of increasingly limited value. "This is true both of scientists' models and designers' models" (ibid.).

The authors are teaching master students in landscape architecture at an internationally top-ranked and engineering-oriented university. However, many of these students have, according to their own statements, never touched, sometimes never heard of GIS. This paper deals with teaching strategies directed on the procurement of geospatial thinking to design-oriented students. How can we prepare them to become more 'GIS savvy' and overcome perceived flaws? We intend to present findings concerning teaching basic GIS literacy based on a 'hands-on' approach at site level (Oct 2009 and 2010) at National University of Singapore (NUS).

2 GIS in Disguise

These days, first semester students enter the university with often remarkable though time-killing digital and social media skills. Cloud computing 'breeds' web services such as Facebook or Gmail, where technical details are abstracted from end-users, who no longer have need for expertise with or control over the technology infrastructure that supports them (DANIELSON 2008). However, most of professional planning and design software still runs on 'old-fashioned' desktop PCs.

In Architecture/Engineering/Construction (AEC) semantically rich modelling of buildings (BIM) using tools such as Autodesk Revit or Bentley Microstation are fundamentally changing the process of building design and construction. BIM also has changed the way data are organised, described, and stored. However, BIM software ignores the broader (urban landscape) context. ERVIN (2006) looks for 'Landscape Information Models' (LIMs) serving the needs of multiple constituencies, analyses and evaluations, based on a single robust model.

Around 2008, the term 'Geodesign'[1] was invented to describe the concept of a progressive framework, which "[...] brings geographic analysis into the design process, where initial design sketches are instantly vetted for suitability against a myriad of database layers describing a variety of physical and social factors for the spatial extent of the project. [...] GeoDesign lets us design and test various alternatives, helping us make the most educated and informed decisions about the best possible future" (DANGERMOND 2009a). In other words: the creation of planning and design proposals should be coupled with impact simulations informed by geographic context. Geodesign is not a software application; it is rather the notion that computer-aided planning and design processes should be interfaced with impact simulations informed by geographic context. Recently, some research projects in land use and landscape planning have introduced components of a Geodesign framework (FLAXMAN, 2010). ESRI ArcSketch (DANGERMOND 2009b) is worth mentioning, a kind of ad hoc "geo drafting" tool that refreshingly stands out from the complicated and traditionally clumsy user interfaces and user experience of established desktop GIS. In addition, Autodesk's project Galileo[2] and their Cloud-based tool Green Building Studio fit to the Geodesign framework supporting the early-stage design process.

Since 2010, a module of the MLA programme at NUS is named Geodesign. In the lectures of the module, attention is given to some of the building blocks of Geodesign such as geographic information science, CAD, visualisation, landscape architecture and environmental planning. How can students learn to use spatial data, information sources, and advanced digital tools both effectively and responsibly?

3 Trojan Horse

In the computer domain, Trojan horse is often associated with 'malware' computer programmes presented as useful or harmless to induce the user to install and run them.[3] The core of the Trojan horse approach implies that GIS comes packed in a catchy label – Geodesign – and that the question is not CAD versus GIS but learning integrated workflows. Our digital toolbox comprises collaborative fieldwork, web mapping, students' smartphone devices and mapping Apps, desktop GIS – 'in small doses' – as well as 3D software, including popular software applications such as Google Earth and SketchUp. MLA Students at NUS had to collect and map data on site, organise spatial information, do some 'geo drafting', analysing, visualising, exploration of phenomena and information sharing with others.

3.1 Walk on the Map Site

The private IT equipment pool of young students is impressive. It just has to be applied for the right purpose. In 2010, students were thrown in at the deep end, equipped with some smartphones inclusive free mapping Apps and a few additional GPS devices (Fig. 2).

---

[1] http://en.wikipedia.org/wiki/Geodesign
[2] http://labs.autodesk.com/utilities/galileo
[3] http://en.wikipedia.org/wiki/Trojan_Horse

Fig. 2: Gathering for very first Geodesign encounters after a sunrise breakfast inside West Coast Park (by ALICE FISEROVA Oct 2010)

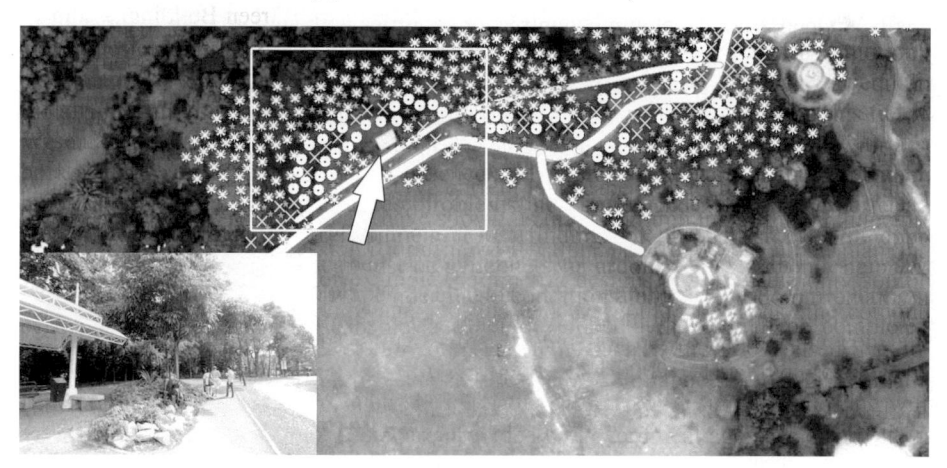

Fig. 3: Example of a student's fieldwork and mapping result (by YVONNE TSUI SUEN YUNG Oct 2010)

Instead of a malware, the Trojan horse sneaks in as beneficial and cool freeware: GPS-enabled Apps such as Google Maps, Camera or MotionX GPS lite. The aim was to take georeferenced photos and map footpaths, and solitary trees (incl. plant identification) at West Coast Park, a public park near the NUS campus. Suddenly, the students turned into "volunteer geographers" (GOODCHILD 2007).

3.2 Cool and Analytic Maps

Back at the campus, students got themselves an Open Street Map (OSM)[4] account and uploaded their GPS data or edited existing OSM data. The OSM server takes care on the cartographic representation.

---

[4] http://www.openstreetmap.org

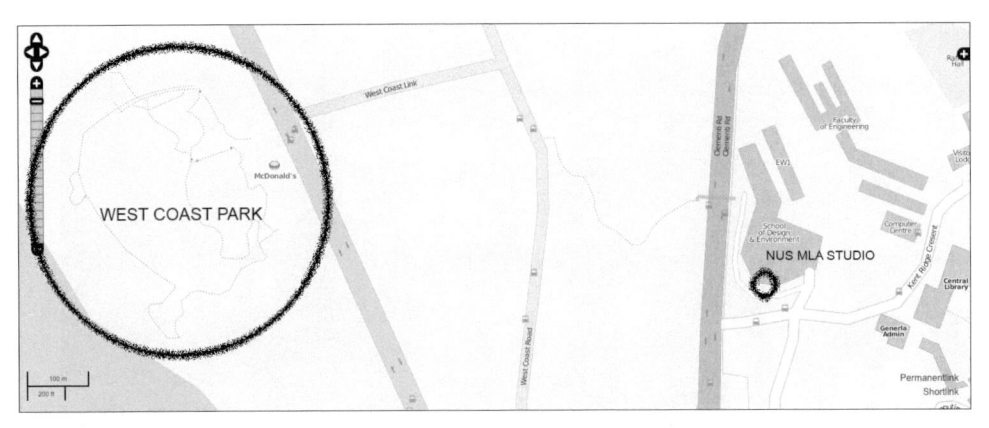

Fig. 4: Upload of a newly mapped footpath connection between NUS Kent Ridge campus and West Coast Park in Open Street (osm.org)

Especially students from developing countries or countries with authoritarian regimes liked to learn about such free geo data sources and mapping tools. The geodata was also loaded to ArcMap to create a geodatabase, cartographic maps, and to learn some basic GIS functions (Fig. 3), e.g. making queries. Some students got more in depth into analytical and 3D GIS (Fig. 5) creating, visualising and exploring models of phenomena and relationships.

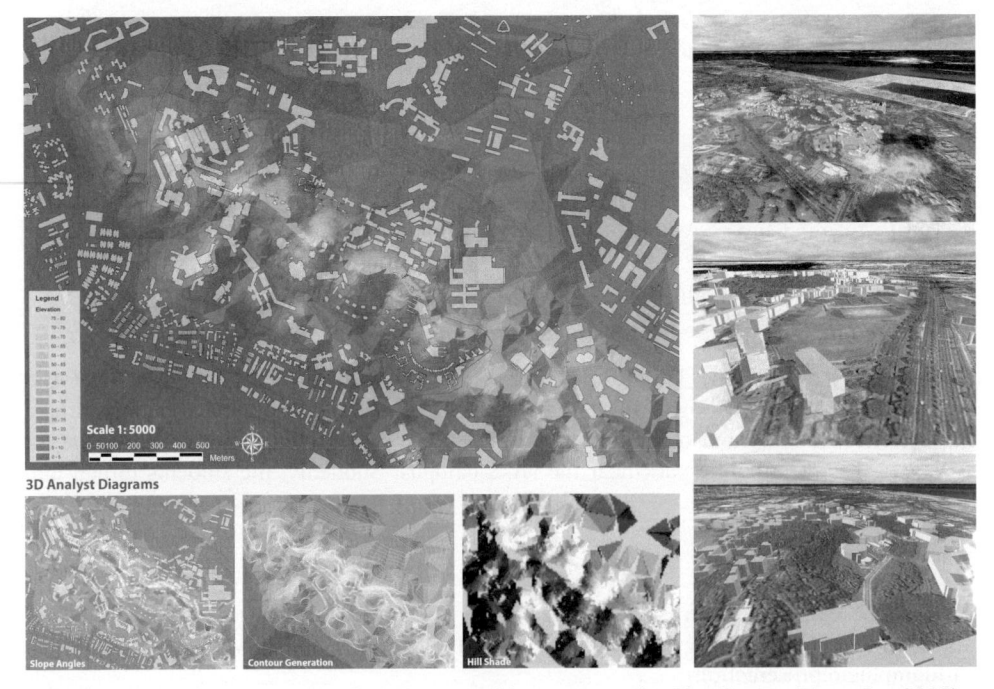

Fig. 5: Topographic study of NUS Kent Ridge campus (by YIT CHUAN TAN, Oct 2009)

3.3 Go 3D but Geo-referenceable

In 2009, the Geodesign module topic was synchronised with the MLA studio theme 'Urban Jungle Singapore'. Students had to screen grab a Google Earth areal picture, georefence and georectify it using ArcGIS, create a rough mass model from footprints and design step-by-step an interactive 3D park model (Fig. 6 and 7).

Fig. 6: Step-by-step interactive 3D park design at Orchard Road (by YIT CHUAN TAN, Oct 2009)

Fig. 7: Geo-drafted Urban Jungle park at Orchard Road (by SHIJIE TAI, Oct 2009)

These models were based on a height map, texture maps, Shapefiles and 3D plant models. The 3D model provided real-time navigation and enabled interactive editing of parameters such as sun position, depth of field, water level and adjustment of colours by layer.

Interactive 3D visualisation supports greater spatial comprehension by allowing, e.g. students to position themselves within their designs, to view from multiple angles, to zoom into and out of various spatial connections and details, and to see if it is making sense (DANAHY 2001). This approach supports better understanding sites and site designs in broader context.

3.4 Real-time Collage

In 2010, the students applied their fieldwork geodata of West Coast Park to create an interactive 3D model of the park. Additionally, the free NASA SRTM terrain dataset, SketchUp models, 3D trees, Shapefiles, and texture maps were added or generated to the landscape model. The first result was an interactive landscape model based on virtual globe (PAAR & CLASEN 2007). After that the students began to render stills and decided to edit the images in Photoshop, adding people, more plants or different skies. The resulting images apparently do not much differ from outcomes of traditional digital methods in landscape architecture. What is different is the workflow. The additional application of billboards and photomontage might be both a reflex of accustomed 2D Photoshop approach and designers' fashion (Fig. 8).

Fig. 8: Rather conventional Photoshop collage except that being created on top of an interactive GIS databased 3D landscape model (by SUSANTO SOENJOYO Oct 2010)

4 Facebookers into Geodesigners

Students should learn how to use geospatial data, data sources, and mainstream and advanced digital tools both effectively and responsibly. SUI (2008) argues that common thread for a new curriculum, which stimulates spatial thinking in both GIS and design will be a renewed emphasis on creativity.

Our students' feedback and results such as OSM inputs and edits gave evidence that the hands-on and workflow-oriented approach helped to clear the hurdle of understanding geoinformation methods. The creation of 3D visualisations – bringing together GIS techniques with advanced real time visualisation – is capable to sweeten students' commitment to look into the subject – as long as the PC is good enough to handle the 3D landscape models fluently.

Advanced digital modelling, GIS analysis, and simulation tools as well as multi-scale and more ecological landscape planning approaches promise to become essential methods in landscape architecture, extending the digital toolbox (STEINITZ 2010). The first attempts entering this new territory were successful but didn't yet make it into students' day-to-day business in the design studios. In fact, the introduction to digital tools doesn't automatically lead to a new pictorial language. However, it became apparent that developing and applying concepts of embedding GIS based spatial analysis and interactive 3D visualisation methods into landscape design processes and curricula would still be innovative and consequent. In that case, the Trojan horse approach should already induce understanding of cultural and physical contexts of architecture and landscapes sites at Bachelor level.

The common recording of the private lives of our students – 'Generation Facebook' – is latently more location-based and geo-enabled than their conventional digital design methods. Therefore we advocate meeting them at their (geo) level and try to get beyond. They won't easily mutate into 'Geodesigners' and not every graduate student has to become a GIS expert. However, FEE (2011) makes the point, that "Place is critical to any planning and thus whether you are a GeoDesign believer or GeoDesign agnostic […] the concepts of GeoDesign matter even if calling it that makes your skin crawl."

References

Buhmann, E. & Jünemann, P. (2000), *Die digitale Bearbeitung landschaftspflegerischer Begleitpläne.* In: Straße-Landschaft-Umwelt. Schriftenreihe des Landschaftsverband Rheinland, 9 ,127-140.

Buhmann, E. & Wiesel, J. (2003), *GIS-Report 2003: Software, Daten, Firmen.* Karlsruhe, Harzer.

Chrisman, N. R. (2006), *Charting the Unknown: How computer mapping at Harvard became GIS.* Redlands, CA., ESRI Press.

Coppock, J. T. & Rhind, D. W. (1991), *The History of GIS.* In: Maguire, D. J., Goodchild, M. F. & Rhind, D. W. (Eds.), Geographical Information Systems: Principles and Applications (Vol. 1). Harlow, U. K., Longman Group, 21-43.

Danahy, J. W. (2001), *Technology for dynamic viewing and peripheral vision in landscape visualization.* In: Landscape and Urban Planning, 54 (1-4), 125-137.

Dangermond, J. (2009a), *GIS: Designing our Future.* In: ArcNews, ESRI, Summer 2009.

Dangermond, J. (2009b), *GIS: Design and Evolving Technology.* In: ArcNews, ESRI, Fall 2009.

Danielson, K. (2008), *Distinguishing Cloud Computing from Utility Computing*. http://www.ebizq.net/blogs/saasweek/2008/03/distinguishing_cloud_computing (2011-02-18).

Dillion, D., Kapur, A. & Carter, H. (2003), *Teaching spatial concepts in the architectural design process*. Tech Directions, 63: 18-19.

Ervin, S. (2006), *Landscape Meta-Modeling*. In: Buhmann, E., Ervin, S., Jørgensen, I. & Strobl, J. (Eds.), Proceedings at Anhalt University of Applied Sciences. Heidelberg, Wichmann, 3-15.

Fee, J. (2011), *Where I lose my mind thinking about geodesign*. Blog. Posted on January 20, 2011. http://www.spatiallyadjusted.com/2011/01/20/where-i-lose-my-mind-thinking-about-geodesign (2010-02-18).

Flaxman, M. (2010), *Fundamentals in GeoDesign*. In: Buhman, E., Pietsch, M. & Kretzler, E. (Eds.), Peer Reviewed Proceedings Digital Landscape Architecture 2010, Anhalt University of Applied Sciences. Berlin/Offenbach, Wichmann, 28-41.

Goodchild, M. F. (2007), *Citizens as Sensors: The World of Volunteered Geography*. In: GeoJournal, 69, 211-221.

Hanna, K. C. (1998), *GIS for Landscape Architects*. Redlands, CA, ESRI Press.

McHarg, I. (1969), *Design with Nature*. Garden City, NY, Natural History Press.

Paar, P. & Claßen, M. (2007), *Earth, Landscape, Biotope, Plant. Interactive visualisation with Biosphere3D*. In: Proc. at CORP, Vienna, 207-214.

Steinitz, C. (2010), *Landscape Architecture into the 21st Century – Methods for Digital Techniques*. In: Buhman, E., Pietsch, M. & Kretzler, E. (Eds.), Peer Reviewed Proceedings Digital Landscape Architecture 2010, Anhalt University of Applied Sciences. Berlin/Offenbach, Wichmann, 2-26.

Sui, D. (2008), *From "GIS for Design" to "Design for GIS": Preliminary thoughts on designing a curriculum for spatial thinking*. Santa Barbara, California: Specialist Meeting on Spatial Concepts in GIS and Design, December 15-16, 2008. http://ncgia*.ucsb.edu/projects/scdg/docs/position/Sui-Design-Position-Paper.pdf.

Teaching Landscape Planning – Landscape Perception and Analysis

Class LEINER and Boris STEMMER

1 Introduction

Landscape architecture education programmes at Kassel University where reformed in 2006 to comply with the Bologna Accord. In this process new educational courses were developed, and these were combined to form clusters called 'Modules'. The first cohort of Bachelor students has graduated in 2010 and it is time to analyse and evaluate the new system. As an example this paper focuses on a module that, at first glance, appears to be working well. The example is a set of first year (second semester) courses that are integrated in a module called 'Landscape Perception and Analysis'. The aims of theses course are for students to learn how to analyse landscapes by perceiving them from three different perspectives, and also to apply, as their analytical tool, a geographic information system (GIS).

2 Objectives of the Module 'Landscape Perception and Analysis'

2.1 An interdisciplinary Learning-Module for Landscape and Urban Planners

At Kassel University the interdisciplinary education in the fields of architecture, urban planning and landscape architecture and planning has a long tradition. It started when the School of Architecture, Urban and Landscape Planning was established in 1971, and earlier initiatives for integrated didactic approaches may be traced back as far as 1948/49 when the first 'Master Class of Landscape Culture' was established at the Kassel Art Academy. Today there still is a strong interdisciplinary philosophy at Kassel University that integrated education is based on. This thinking is reflected in the 2006 Reform as even some of the basic first year 'integrated modules' are obligatory for all students, including students of architecture, planning and landscape architecture. The module described in this text is offered to students of landscape architecture and urban planning. The module integrates three courses and is taught by three departments: Ecological Site & Vegetation Studies, Landscape & Land Use Planning, and Urban & Regional Sociology. Therefore it can be regarded as a "core-competences" module (BRUNS et al. 2010).

2.2 Objectives of the faculties teaching 'Landscape Perception and Analysis'

Every one of the three departments involved in this module is, in some way, specialized in the collection and analysis of landscape related data. Each represents a different view on landscape.

Ecological Site & Vegetation Studies traditionally focus on ecological aspects of the landscape. Objective of their course contribution are to gain an overview on biotope-types and apply traditional ecological ways to describe and analyse the landscape: this is the landscape perceived from the ecologists' point of view.

Urban & Regional Sociology, on the other hand, holds a sociologist's position. Their contribution to the module is focused on people and human behaviour. Observing people in the landscape and describing their activity also offers an opportunity to describe the kind of space people are using. Most of all this course contribution shows what space people prefer and to try and explain why they might prefer specific locations over others.

Landscape & Land Use Planning assists students in learning to understand and analyse landscapes according to their visual appearance and atmosphere. Differences between landscape type and character are emphasized, and techniques of understanding landscape history are presented.

All together, some very different methods and techniques of landscape analysis are taught:
- Biotope and land use classification, using a typological approach;
- Landscape characterisation and landscape atmosphere;
- Observation of space and people behaviour.

On the one hand, all of these approaches are merged, partly for didactic reasons of integrated learning, partly by the needs of the GIS-System. GIS is a fourth component of this module. A GIS specialist takes the responsibility for teaching this component (As suggested by Watson 2010). Didactically, the most important aim is to combine results produced from applying all of the different methods and techniques in a final synopsis. Students are asked to answer the question how all methods they have learned may be used in concert, and which specific questions may be answered by which method individually? And, finally, how do these methods complement each other?

2.3 Didactic approach

The learning aims of this module are to teach different approaches of landscape analysis, using a GIS and giving an overview of the purpose and properties of each of the different methods. Usually a high number of students (about 200) enrol into this module every year.. Didactic approaches must be selected that accommodate the need for many students to achieve their learning goals. 'Learning by doing' has been shown to be the most effective way of learning, particularly in spatial and landscape planning. Also it has turned out that working in groups (collaborative learning) assures best learning effects (f. i. YAMARIK 2007, HWANG 2008). Field work techniques were selected assuming they would provide the best learning effects also in these courses. A field based landscape analysis can only be a success if it is well prepared. The students must be prepared, as well as the chronological and spatial distribution of students and teachers in the field (Chapter 3).

Besides the methods used in GIS also need to be considered. GIS requirements were a permanent component of the semester cycle, not only due to the complexity, but also because it affected all other parts of the course.

3 Organisation and Programme

3.1 Semester Programme

The semester consists of 13 weeks of active teaching time, and several weeks of self study. The programme and thematic organisation for the 'Landscape Perception and Analysis' module is shown in Table 1. The programme is separated into three thematic blocks that relate to what the three participating departments are offering. The programme also begins with a comprehensive introduction and it ends with a closing event. Parallel GIS courses are held using 'moodle' (Chap 3.2). GIS was also a major topic of discussion during many of the meetings.

Week	Day	Content	
15	Wed.	Introduction	Introduction to the module and its courses
	Thur.	Landscape planning	Theory I: land use and land use types
16	Wed.		Field visit I: land use and land use types
	Thur.		Theory II: landscape character, atmosphere, cultural landscapes elements
17	Wed.		Field visit II: landscape character, atmosphere, cultural landscapes elements
	Thur.		Theory III: Landscape history, mapping changes in time
18	Wed.	General Excursion and Field trip week	
	Thur.		
19	Wed.		Field visit III: Landscape change mapping
	Thur.	Holiday	
20	Wed.		Presentation No. 1: land use, character and atmosphere
	Thur.	Landscape ecology	Theory I:
21	Wed.		Field visit I:
	Thur.		Theory II:
23	Wed.		Field visit II:
	Thur.	Holiday	
24	Wed.		Presentation No. 2: landscape ecology
	Thur.	GIS	Consolidation of part No. 1 and No. 2
25	Wed.	Sociology	Theory I & II
	Thur.		Field visit preparation
26	Wed.		Field visit: Observation
	Thur.		Presentation No. 3 Part 1
27	Wed.		Presentation No. 3 Part 2
	Thur.		Formalities, requirements on the written report
28	Wed.		Comprehensive perspective, lessons learned
	Thur.		Preparation of written report (Group work)

Table 1: Thematic organisation of the semester programme summer 2010

3.2 Logistics and Contents

Meetings were normally held in the auditorium on Thursdays' evening, while field visits were arranged on Wednesdays during the morning. As students worked in groups of six they had to start the field trips on their own and then meet instructors at a certain time and place during their ongoing work.

The city of Kassel was subdivided into planning areas of 500 × 500 meters. (Fig. 1) Every quadrant was worked on by five groups. So it was possible to compare the outputs.

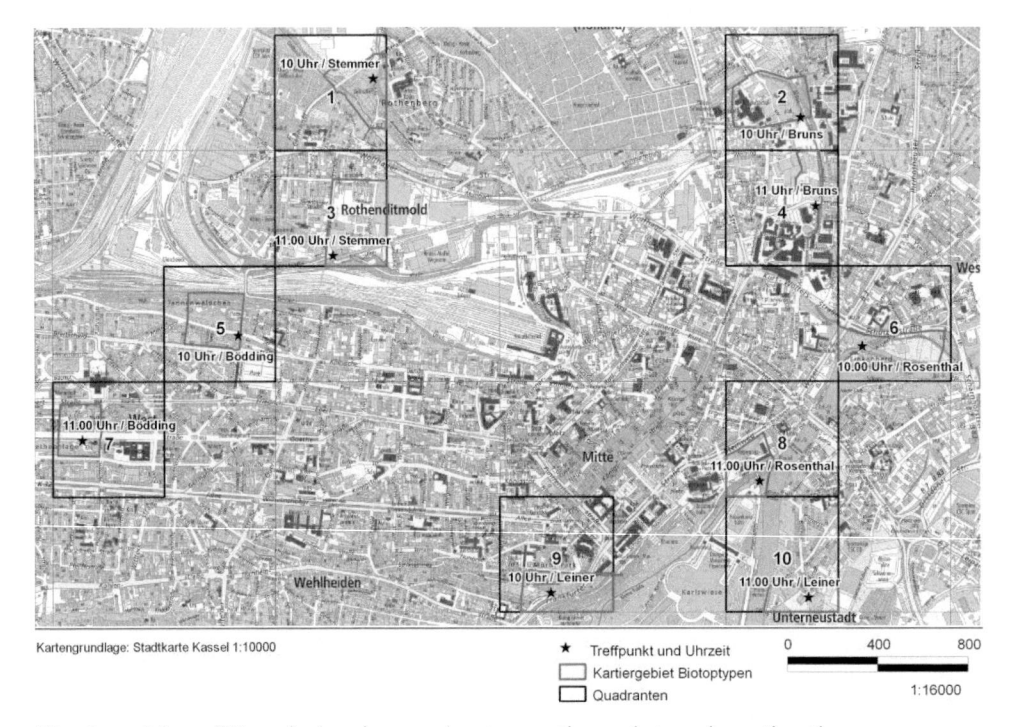

Fig. 1: Map of Kassel, showing quadrants, meeting points and meeting times

4 Teaching Methods

According to the decisions on didactic approaches (Chap. 2.3) a number of different teaching methods are integrated to run the courses of this module. Two major challenges are:
- A high number of students (ca. 200);
- A complex topic and a large number of different views on landscape

4.1 Lectures and Learning Papers

Lectures are part of the teaching programme. Lectures are used to give introductions to the theoretical basis and to outline analytical methods and techniques to prepare for the next field trip. Lectures also include small exercises that are designed to help students in transferring theoretical knowledge into practical applications of methods, and to build and improve practical skills. Homework assignments are given to the students, in most cases the purpose is to prepare for field trips (4.2, Field visits). Lecture sessions are usually held

Thursday evenings. About one and a half hour of lecture appears to be the maximum in an evening according to the students' ability to concentrate on a single topic.

4.2 Field visits

Field trips and field visits are frequently used learning methods. This module included five field visits (Table 1). For their perception and analysis students are required to refer to theoretical knowledge taught during lectures and apply methods and techniques prepared during lecture exercises. Field work is done by groups, and every group must work mostly independently. The need to work independently is one reason why good preparations are important. To support students during their work in the field, on Wednesday morning between 10.00-13:30, instructors meet with every group to help them with problems, to give advice, and to discuss the output produced so far.

4.3 Homework Assignments

Frequently, homework assignments are given to students at the end of lectures, and also at the end of field visits. Two types of homework are:
- Preparation for field visits;
- Preparation of presentations after field visit (output).

Preparations for field visits include the assembling of maps as well as readings of theoretical and practical literature. Maps are arranged with help of geo-data that are provided on 'Moodle'. Texts were also available on this learning management system.

Preparing landscape analysis outputs after field visits mainly includes the digitalization and layout of hand drawn maps, scanning of sketches, arranging photographs and completing questionnaires with external information.

4.4 Presentations

Presenting one's work is considered important in landscape architecture and in planning. As a key competence presenting skills are repeatedly trained. Students are asked to present results at the end of every thematic block, and special presentation events are arranged for this purpose (Table 1). Due to the large number of students it is necessary to select some groups to present their work and to use presentation meetings mainly to discuss standards and quality that is expected from landscape analysis work.

The presentation meetings are prepared, by the instructors, by reviewing the output of the ongoing work. A selection is made of those maps and sketches that would be most suitable to illustrate certain aspects of the assignments. Presentations are given spontaneously by the groups whose work are selected to serve as examples. To minimize frustration effects students were not asked to prepare for presentations ahead of time, but for all students to be aware that they could be asked to comment on their work at any given time.

5 Excursus: GIS and 'Moodle'

Landscape architecture and planning students who graduate with an emphasis on landscape planning or spatial planning are expected to have gained substantiated knowledge and expertise in GIS. However, universities appear to place less value on GIS related education as professionals in practice might expect them to. For students to prepare for future business requirements is one of the main challenges that institutes of higher education are facing (WINKELMANN 2007). At Kassel University the decision was made to include GIS-courses from the start, beginning in the first semester in first year, and to continue in the second semester including GIS as part of the module 'Landscape Perception and Analysis'. During this course students get a first chance to apply knowledge gained in the first semester and place it in a professional context. The mayor aims of the GIS-training in the course are:

- Understanding the overall concept of desktop GIS;
- Understanding properties, analysis functions, visualisation and presentation of geo-data;
- Identifying errors and manipulations in geo-data;
- Ability to design readable maps.

The students GIS-Infrastructure is established during the first semester GIS course. In this course the open source Software 'Quantum GIS' (www.qgis.org) is used and students are made familiar with working in 'moodle'. Therefore, during the second semester, it may be assumed that a basic knowledge of GIS is available and 'moodle' is well known. Students are used to receive distributions of PDF tutorials, scripts and geo-data. They are also used to receive mentoring using the learning management system. 'Moodle' is a well known learning management system that is used by many universities and other institutions of higher education. In this course the system was employed for central organisational and teaching purposes:

- Grouping of students and assigning of planning area (quadrants);
- Providing materials (e.g. maps, instructions, scientific texts);
- Forum for discussions, especially regarding GIS;
- Handing in of assignments and presentations.

Concerning GIS most of the problems occur when students are asked to create presentation maps. Obviously it is not hard to understand how a GIS works and how to use the technique; but choosing an easy to read layout appears to be much harder. Particularly the choice of colours for certain types of land use or biotope often appear to be difficult, and representation and legend layout require lengthy discussions.
Many but not all issues can be solved using 'moodle' forums. And in many lectures GIS issues are also integrated (Chap 4.1)

Nevertheless, the learning management system (LMS) 'Moodle' as well as the individual availability of GIS software are important preconditions to make a success out of teaching this complex module.

6   Output and Outcomes

The semester output is a booklet. Each student group designs and produces a brochure booklet that contains results from all exercises and assignments. The deadline for handing in this booklet is a few weeks after lectures end. This allows for enough time to mainly design the booklet. Individual assignments such as maps had to be ready after every mapping exercise. This procedure assures that students continue working on assignments and exercises throughout the semester. Excerpts from an example booklet are included in Figure 2-4.

As may be expected, outputs qualities are quite diverse. Some products are excellent, while others are hardly able to meet minimum standards. Good working results clearly indicate that learning requirements are not being set at a too high a level, however the wide range of grades earned also shows that all of the exercises also help identifying differences in group capabilities.

Some comments students made on the design of the course programmes are noteworthy. For example, students remarked that, when faced with problems in trying to adjust their group perceptions on landscape to what they remember was said during the lectures, it would have been helpful to receive more intensive tutoring during field visits. Other students comment on problems they encountered while using the GIS software, especially regarding the stability of it running smoothly. Unfortunately, it is difficult to evaluate such issues when using e-learning based methods.

7 Conclusions

The newly created 'module' serves to illustrate how to effectively organize a course designed to learn and understand basics of landscape assessment, and how to combine these not only with methods used in landscape planning, ecological and perception and cognitive methods, but also with modern digital techniques of data processing and communication.

Fig. 2: Atmosphere Sketch of a working class street in Kassel-Nordstadt (by Viola Bornscheuer, Carolin Jakschik, Katja Krüger, Mareike Wingenfeld)

Fig. 3: Sketch of a historic landscape element in Kassel-Unterneustadt (by Chanda Winter)

This 'module' is now one of the best organized ones of the School, and it working quite well. Still there is room for improvement. Students should be encouraged more strongly to work independently and to rely more on the theoretical knowledge and on the methods presented to them in lectures. Additionally, while the 'Moodle' system presents a huge

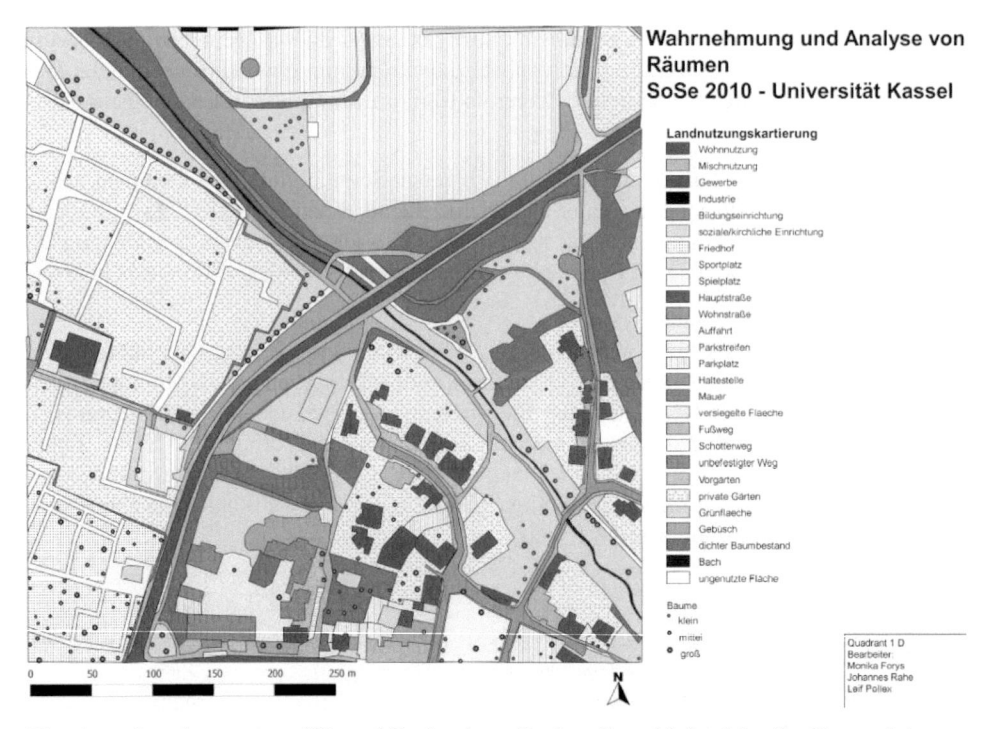

Fig. 4: Land use map of Kassel Rothenberg/Rothenditmold (by Monika Forys, Johannes Rahe, Leif Polex)

organisational benefit, communications and coordination could be improved even more. It can be observed that the strength gained from personal exchanges between teachers and students is important and, due to a lack of face-to-face communication, no any e-learning system is able to be a good substitute.

Nevertheless the approach presented here is capable to ensure that learning goals are met and that participants are enjoying to produce good results.

References

Bruns, D., Ortacesme, V., Stiles, R., de Vries, J., Holden, R., Jorgensen, K. (2010), *Tuning Landscape Architecture Education in Europe*. ECLAS – LE:NOTRE. www.lenotre.org.

Hwang, N.-C. R. & Lui, G. (2008), *Cooperative learning in a passive learning environment. A replication and extension* / Nen-Chen Richard Hwang, Gladie Lui, and Marian Yew Jen Wu Tong. In: Issues in accounting education,

Watson, D. (2010), *Are Landscape Programmes Meeting the Challenge of Educating the Second Generation of Digital Landscape Architects*. In: Buhmann, E. Et al. (Eds.), Peer reviewed proceedings of digital landscape architecture 2010 at Anhalt University of Applied Sciences. [... contributions from the 10th International Conference on Information Technologies in Landscape Architecture, held in May, 2009 at Valletta, Malta and contributions from the 11th conference held from 26 to 30 May 2010 at Aschersleben, near our Bernburg Campus ; 10th and 11th International Conference on Digital Landscape Architecture]. BerlinOffenbach, Wichmann.

Winkelmann, H. P. (2007), *Hochschule und nachhaltige Entwicklung*. In: Michelsen, G. & Godemann, J. (Hrsg.), Handbuch Nachhaltigkeitskommunikation. Grundlagen und Praxis. 2., aktualisierte und überarb. Aufl. München, Oekom.

Yamarik, S. (2007), *Does cooperative learning improve student learning outcomes?* In: The journal of economic education.

Action – Structure and Teaching Methods for Lectures and Seminars

Elke MERTENS

1 Introduction and Personal Motivation

Activating teaching methods can support the learning achievement of students and therefore are attested ways to improve teaching. Individual activity and subjective experience of the purpose of the learning matter are central categories, which may lead to a higher motivation for learning and improved performance. The teachers coordinate their courses with sophisticated teaching methods for better student learning, they guide them and give advice. In contrast to traditional lectures, the evaluation of learning outcomes is done by the teacher and the students. For the use of activating methods the teacher needs enough time for the preparation and a good piece of flexibility.

Since I had the chance to take part in a special teacher's training five years ago, I improved my courses and adapted these and other methods for landscape architecture, I'd like to share the experience with teaching staff and students to help improve the education in landscape architecture in Europe. Not only the content of courses, but also the methods of teaching and learning should be discussed in meetings of landscape architecture schools. I would be glad if this text would be a start for a serious discussion.

2 Teaching Landscape Architecture

There is no discussion that design studios are a very suitable method for teaching complex problem solving techniques in landscape architecture. But typically, also lectures and seminars are very common in teaching basic and advanced knowledge. We know that in general, students capture only 20 to 40 percent of a lecture's main ideas in their notes (KIEWRA 2002, p. 72). Without reviewing the lecture material, they remember less than 10 percent after three weeks (BLIGH 2000, p. 40). It must be called a big mistake in courses typically focused on teaching that one (the teacher) speaks the whole time alone and the group (of students) is expected to listen all the time and get all the information as they were intended to be understood by listening to them for the first time. There are possibilities to improve the learning process and it is crucial to change the focus from teaching to learning. It is necessary to identify relationships, to transmit connections, to solve problems and to train key competences. Is learning the main focus, also the courses need to be planned with the learning skills of students in the focus.

The most important part in teaching is not to give as many facts as possible – but make the group understand and able to learn what is needed to know. Students and everyone who is learning remember best and most permanently the knowledge they have been actively developed and processed. To improve the learning process, activating teaching methods that appeal to multiple senses and promote actions of students should be also used in lectures and seminars.

Lectures usually mean that one teaching person gives most of the input and learning means listening for the given time. Lectures are usually longer than the attention of students can be kept. Also, the groups of listeners are usually relatively large and the students come with different knowledge and interests which make it difficult to respond to everyone's expectations.

Seminars are structured with more active methods but still mostly mean a big part as input from the teacher and from the students only when they are asked (activated). For both types of courses it is important to find the right balance between input (= lecture) and individual learning/processing phases. The structure called Sandwich Method (see below) helps to plan a course with alternation of these types of teaching.

3 Teaching Methods

Methods to support activating the students can be the Advance Organizer as an inspiration in the introductory part of a course unit, the Partner Interview for the motivation, the Group Puzzle as an assignment, Mind Mapping for the discussion or solution, and the Feedback Method during the conclusion. In the following, these methods are described in more detail as examples.

3.1 Advance Organizer (AO)

Advance Organizers are devices that are used to present global summarizations of content to be learned to the student (COFFEY 2005).

Especially for students of landscape architecture as visual discipline which produces and leads to a good understanding of images, a graphic design as Advance Organizer may help to bridge known knowledge with new contents of a course. This can be very useful in courses about construction, history or planning methods.

An Advance Organizer is information that is presented prior to learning and that can be used by the learner to organize and interpret new incoming information (MAYER 2003). It should mark the beginning of a new learning phase, i.e. the start of a new module, the beginning of a workshop or project. An Advance Organizer makes it easier to link and connect the new knowledge with the already existing (pre-) knowledge or the fundamentals to be activated by a fairly general conceptual structure (organizer).

As the capacity of the human reception of information is limited, we help ourselves by perceiving and receiving selectively. The Advance Organizer is a method to focus the "cognitive tuning" and help to select the relevant information. With the help of Advance Organizers a new problem area is first applied in a scientific context, although students at this time are not yet able to understand details. The explanation of the survey is initially limited to the affiliations of the new with the familiar knowledge and to the knowledge to be acquired. The academic depth of knowledge is achieved gradually and by dealing with the problem and its structures.

Beginning of the first teaching unit   beginning of a later teaching unit

Fig. 1: This is an example of a beginning of an Advance Organizer I used to go with the agenda of each teaching unit every week, it got more and more luggage = knowledge visible to the students during the semester. This picture doesn't visualize the content of the module "garden and landscape architecture" but was a nice companion and actually made it easier to focus attention at the beginning of each teaching unit.

Research on the effects of Advance Organizers states that they facilitate meaningful inclusion of information. They integrate new knowledge into the existing mental structure. They create a framework that helps to anchor especially detailed information. This may increase the learning performance in the same study time, depending on the research study by 10% to 18%. The differences in retention performance in comparison to courses without Advance Organizer are measured to be better the longer the course lies back. The reason is the stable, coherent conceptual framework, which is resistant to be forgotten easily. Since the AO emphasizes fundamental insights and relationships, the benefits are especially in the range of transfers, e.g. in the application of newly acquired knowledge to new problems, on average by 10% to 50% higher (depending on the research study) than if no AO was used.

The elements used in an Advance Organizer should be rather simple and clear, it may start with an easy-to-understand and remember graphic that could be added by pictures, graphs, words, short texts, and later combined according to the principles of a presentation to a "learning map".

Objectives of the Advance Organizer
- Overview and networking of new subject areas
- Embedding of new learning content on existing knowledge
- Focused attention
- Better understanding
- clarification of misunderstandings
- Long-term retention
- Improved transfer performance

To arrange an Advance Organizer
Collect the key terms of the thematic unit and write down about 20 to 40 tags.
Arrange those terms in a graphic design or in form of a mind map. Use both verbal and graphic elements.

Create a large bulletin board display (big size/flip chart) of your AO (alternative: set of slides).

Plan 10 to 15 minutes for the explanation of the AO.
Include the Advance Organizer in your teaching and learning setting and use one of the following or another technique to continue with the subject matter.

Advice
- A list of themes and a timetable are not *Advance Organizers*. They only mention topics, but don't sufficiently clarify the conceptual relationships.
- The Graphic Organizer as the visual representation of knowledge is related to an *Advance Organizer*, it is often much more complex than an *Advance Organizer*.

3.2 Sandwich Method

Phases of learning in universities are usually too long to be suitable to teach the subject matter in the form of lectures. The duration of attention is not as long as the lecture time and during long lectures there is no time for reflection and application of the teaching matter. The Sandwich Method organizes learning processes in such a way that periods of collective learning alternate with periods of highly individual learning. In this way, new information is particularly well integrated into the student's own unique mental structure.

For example, in teaching planning methods, the teacher could ask the students to collect methods they know and discuss them in teams, then structure the methods the students already know and give new input. Using the sandwich method in this way, the students also learn more about different methods by talking and discussing with their fellow students.

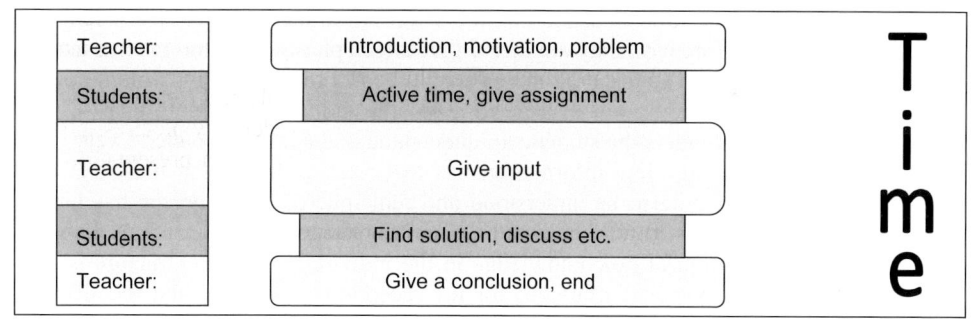

Teacher:	Introduction, motivation, problem	T
Students:	Active time, give assignment	i
Teacher:	Give input	m
Students:	Find solution, discuss etc.	e
Teacher:	Give a conclusion, end	

Fig. 2: Visualization of the basic structure of a teaching unit according to the Sandwich Method, it can be modified e.g. the input can be split and divided by more active phases/ assignments for the students and different techniques for activating the students may be employed

The phases of a teaching and learning unit according to the *Sandwich Method*

It is important that in the phases of collective learning a high density of information is offered. Similarly, the phases of individual learning must be designed demanding, so that the learner is given a significant incentive to deal with the content. In the case of obvious differences in performance between the individual learners, various tasks should be offered. Then, everyone can benefit and use the available learning time best.

How long should the phases of a Sandwich be?

This depends on many factors and cannot be stated for every teaching/learning unit in the same way. In general, the phases of collective learning should not be too long, it is useful to change to a individual learning phase after about 20 to 30 minutes.

The transition between the phases of a *Sandwich*

Significant are the transitions between the phases of collective and individual learning, they have to be well planned also. The beginning of a thematic unit needs an introduction, e.g. the specific structure or the demonstration of thematic contexts, a possible method is the Advance Organizer. The change from the collective learning phase to the individual learning and application phase needs clear assignments, maybe different tasks for different students. If a method is used that needs extra material, e.g. a group puzzle, the material and space for the students must be prepared that the students can start working without a loss of time. At the end of an individual learning phase and at the beginning of the next collective phase, it is important to reflect on the results of the students which they gained during the individual phase. Here it is important to compare solutions of the students, to present their results, to clarify any questions and to share their opinions. In this transitional phase the teacher has to keep in mind the topic and the time in addition to the learning outcomes of students. At the end of the whole teaching unit or the whole course the teacher needs to secure the learning process with suitable methods and/or with a concluding statement.

The methods used to introduce and implement individual learning phases can vary, they depend much on the size and homogeneity of the student group. In many cases, teachers state that they cannot apply activating methods in large groups of students as the noise would get too loud. Also, they are afraid that the transition to the collective learning phase would be impossible. Moreover, they are afraid that theses phases need too much time and the delay seems too large. The teaching matter would be too substantial and voluminous to be minimized and cut at any point to save time for the individual learning phases. This attitude is very sad and hinders the students to understand and apply the matter – very often a requirement for learning. New information has to be integrated in the individual and personal conceptual structure to be understood and remembered. Each person has her/his own conceptual structure with all the nerves, nerve synapses and connections between nerve cells. The integration of new knowledge in the individual memory structure is self-paced learning through personal strategies for the reception, processing and recollection. Consequently, periods should always be included in which the individual has the opportunity to translate the new information into its own internal memory, i.e., to anchor new information in one's own subjective mental structure. If these times are not made available, the learning rate for the same period is much lower, although, paradoxically, more information can been provided!

To apply the sandwich method in large groups of students and to bring the individual learning phase to an end, the help of student tutors may be helpful. They can collect written results quickly so that these can be used in the following phase. Also, students can be asked about their personal opinion by raising hands, the alternatives should then be the subject of the further lecture. It is also possible to print statements on large sheets of paper and ask students either to stick adhesive dots on the paper they agree most/least or to ask them to move to the statement they agree most/least. Both ways are suitable to teach the students different statements which should be well explained in the following phase of the lecture.

Advice for planning a lecture or seminar according to the *Sandwich Method*

- Give compact information at a high level in the presentation/collective learning phases, give demanding tasks for the processing/individual learning phases, maybe in different levels.
- Set a close time limit for individual learning phases but give as much time as possible in the overall setting of the time available.
- Use different methods for the individual learning to meet different learning strategies of the students.
- Plan the transitions between different phases well ahead and don't lose time.

Variations of the Sandwich

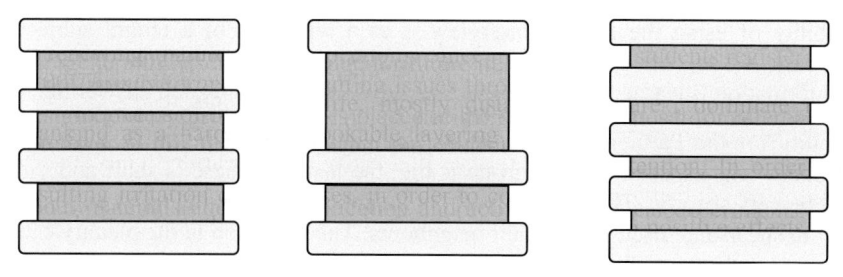

Fig. 3: The basic structure of a Sandwich can be altered according to the length of a teaching unit and according to the needs of the teacher and students. Possible variations to Fig. 2.

According to the length of the whole session, the middle parts of the Sandwich can be divided in more phases of input and different phases for active transfer, reflexion, and problem solving with appropriate methods applied. The beginning and end should never be skipped.

3.3 Partner Interview

In the Partner Interview the students get into contact and gradually acquainted with the large group. They learn to cooperate with different persons and find out about other people's learning habits, thinking and concluding. Also, not less important, the teacher gets information about the motivation of the students for that special course, their level of knowledge and their ability to apply what they have learned before. If the group knows each other person, e.g. the individual learning phases of the Sandwich Method can be a lot more fruitful than if they didn't have the chance to know each other at all. Also, it is helpful to give each student the possibility to talk and express their expectations and get a clear idea about their interest in the specific course. It is my experience that it is advantageous if the students stand up, use the room with all corners and move instead of sitting at the same place during the whole course. They come back more motivated and are curious to listen to the answers of the rest of the group.

Since landscape architects always have to do with different people and with their different objectives, students quickly understand to use the chance to get to know someone else's attitude towards certain facts. Another possibility is a role play if different motives should be shown. So, the partner interview is helpful within the learning process as well as educating landscape architects.

A Partner Interview brings two persons to a talk, the partners should not be friends or people who know each other well already. Use the alphabet or other means to mix the students and bring together people who don't sit together anyway. Give each group a clear and specific assignment and tell everyone the time you plan for them to talk. Announce the time and ask them to switch roles at half time and tell them early enough that the time will run out in one minute, 30 seconds...

The questions to be answered or discussed at the beginning of a course unit, workshop or seminar should be reflected by the partners and then shortly in the plenary. The teacher may give a conclusion if appropriate.

Another possibility of using the Partner Interview is as a repetition of a taught subject matter. In this case, the teacher gives questions to the subject and also answers to secure that the right information is learned. The partners take turns in questioning and answering.

A third possibility for the Partner Interview is the individual leaning phase in a lecture arranged as a Sandwich. Especially if the group is big, the learning phase is short and the questions must be very specific. The students will not be able to move in the room but have the opportunity to speak and discuss with their neighbours. The reflection in the plenary can be short by raising hands to possible answers or attitudes of the auditorium.

The Partner Interview can be used in any size of a group, but the group should not be larger than 16 if the results of each couple should be reflected in the plenary. It may be possible to divide the group into smaller penuries but this has to be well planned. It is already helpful for the learners to be able to talk to one other person in the group.

The Partner Interview takes about 10 to 15 minutes for the partners, depending on the size of the whole group it takes another 30 (to 45) minutes in the plenary. A big room is advantageous, tables are not necessary, movable chairs help to arrange a suitable setting. If the students are asked to answer questions, they should be provided with papers, if they are asked to prepare a flipchart paper, they should be given the necessary material.

3.4 Group Puzzle

The Group Puzzle Method is developed out of the expert group method. Here, the students learn to present results of their work and discuss them with the other students directly. Different learning contents are assigned to different groups. Students learn by themselves first and afterwards in the group. The teacher gives the necessary input, the group becomes an expert group for this aspect of the whole theme. This method can be used within a project work where each student or small groups get tasks to become experts in a special field and to report to the rest of the group. Also, longer but important texts can be divided with this methods into smaller pieces, and in the end, every students has got a deeper understanding of the whole text than if everyone had read the text by her-/himself.

For the Group Puzzle, the expert group is a necessary input. Here, the aim is also to present work results and talk about them in the peer group (learning by teaching). Each student reports about what he just learned, each student learns the results of the other expert groups.

The role of the teacher is important, she/he must complete the results if they lack information and has to make sure that questions are answered correctly.

The way I use Group Puzzles in a group up to 30 students: I divide a text into 4 single and understandable parts and copy each part on different coloured paper, each paper is marked with an individual combination of numbers and letters, see Fig. 4. During the course, I divide the group in small groups of 4 (-6) students. With a plenary of 32 I get 8 groups of 4, a plenary of 16 makes 4 groups of 4, any number in between will have to be planned in that way that each small group gets about the same number of students.

1A	1B	1C	1D
2A	2B	2C	2D
3A	3B	3C	3D
4A	4B	4C	4D
Step 1: Individual reading	Step 2: groups with texts of same colour	Step 3: groups A, B, C, and D to gain a sound understanding of the entire text	
☐	≡≡≡≡	↓↓↓↓	

Fig. 4: Scheme of the numeration of the papers for the group puzzle, here for a plenary group of 16, and the steps to read and discuss the texts in different expert groups

In the first step, every student gets one part of the whole text and reads it individually. The teacher answers individual questions and gives support if there are questions or misunderstanding of the text. After everyone has read her/his text well, the expert group, everyone has the same text (and the same colour to make it easy to distinguish between the groups), gets together. They discuss the content of their text and possibly draw a visualization of the main aspects on a flipchart. In the third phase, all As, Bs, Cs and Ds get together and each person presents her/his part of the full text to the other members of the group. Each group gives a conclusion and/or may draw a visualization. As a result, 4 final presentations can be compared and discussed, each student has to get the full text to read again in context.

Students like the Group Puzzle a lot because they work on the special assignment and understand the content of a longer text without having to read it entirely in the beginning. This method takes time, depending on the length and complexity of the text, but it is a way of making the students understand important issues. It is my experience that the results of learning using the Group Puzzle are much better compared to assignments of reading and preparing a presentation of a text as homework.

3.5 Feedback Method

The Feedback Method gives a response from the students to the teacher of a learning unit. It allows a better and more effective design of future courses. This method calls for and promotes criticism according to fixed rules. These rules can be applied also for presentations, discussions, and role-playing games (RPG) and they help to learn to articulate criticism as well as listening to it, both important factors in communication. The feedback rules must be known by every student in the group and everyone has to follow them.

Feedback rules include respect and empathy, feedback is personal, specific and refers to certain events and behaviours, it concentrates on behaviours that can be changed by the recipient and should in teaching and learning contexts focus on the best results for everyone involved.

Initiated by the teacher, the feedback should be started with precise introductory questions such as "at what point am I progressed?", "is there something that has disappointed me?". The students have the opportunity to express their opinion, it is important that the statements will not be commented or justified. A discussion may follow this Feedback session but doesn't have to.

References

Ausubel, D. P. (1980), *Psychologie des Unterrichts*. Bd. 1und 2. Weinheim/Basel.
Bligh, D. A. (2000), *What's the Use of Lectures?* San Francisco, Jossey-Bass.
Buzan, T. & Buzan, B. (2002), *Das Mind-Map-Buch. Die beste Methode zur Steigerung ihres geistigen Potenzials.*
Center of Teaching and Learning, Stanford University (2005), *Speaking of Teaching.* Newsletter, 14 (1).
Coffey, J. W. (2005), *Knowledge and Information Visualization.* In: LEO: A Concept Map Based Course Visualization Tool for Instructors and Students, 285-301.
Kiewra, K. A. (2002), *How classroom teachers can help students learn and teach them how to learn.* In: Theory into Practice, 41, 71-80.
Kohn, W. (2005), *Statistik, Datenanalysis und Wahrscheinlichkeitsrechnung.* Berlin/Heidelberg/New York, Springer.
Mayer, R. (2003), *Learning and Instruction.* New Jersey, Pearson Education, Inc.
Michel-Schwartze, B. (2007), *Studentenaktivierendes Lehren und Lernen.* Unveröff. Manuskript, Hochschule Neubrandenburg.
Pädagogische Hochschule Heidelberg, Institut für Weiterbildung (Hrsg.) (2005), *Methodenreader*. Heidelberg.

http://lehrerfortbildung-bw.de/unterricht/sol/03_grundlagen/organizer/ (28.1.2011).

http://www.lernen-heute.de/mind_mapping_grundlagen.html (1.3.2011).

On the Preparation of Images

Anna M. THURMAYR and Dietmar STRAUB

Abstract

This essay is a plea to activate the aesthetic potential that computer graphics have so far failed to exploit. While teaching and learning using digital methods and tools is beneficial it is also important to think about media references and media comparison as true modernity can only grow from the oldest techniques.

1 The New Is there for the Taking

The myth that everything that is created must commence from point zero with no prior models, provokes people to deceive themselves or refuse to explore the source of new formulations. The new is there for the taking.

Nothing unsettles a creative person more than having to follow formulas as using prescriptive methods generally seems incompatible with the creation of something genuinely new. Derivative processes seem to be in stark conflict with the ability to create something from nothing.

Anxiety over working with formulas can melt away if the formula in question serves to facilitate the individual, the unmistakable, or if it perhaps even becomes the prerequisite for such a creation. In this interpretation, the term formula is a synonym for concept or program. The essential idea is that it should encourage us to expand boundaries, renew forms, and nurture and maintain curiosity.

Vivienne Westwood, the British fashion designer and a teacher at the Berlin Fashion School starts her course by asking her students to copy styles from costume history. Her clear and simple argument is that if somebody intends to become a famous chef, they do not begin their career by inventing new dishes, but rather by learning the trade and copying well-known recipes.

One must have the courage and the opportunity to question and develop experiences without being inhibited by professional intervention. One should be allowed to become a dilettante. The access to advanced media, especially affordable computers, makes it possible to acquire vast amounts of information, but it is important that students *evaluate* that information. That way they are able to explore what is possible in principle.

Discussing the techniques and media used to produce images will inevitably lead into areas outside the scope of landscape architecture, areas that touch upon the present, but also bring the future and the apparent past into the picture. The preparation of images, meaning the inherent skill and techniques employed, is an important subject matter within the professional study of landscape design. Ultimately, courses in "Landscape Architecture and Communication" are primarily a means of training students to see and to perceive.

2 Painting by Binary Numbers

Today's visual world based on the binary system creates images, which no longer require reference to reality. Both film and Internet use encourage the worship of technical images for which the public is clamouring. There are some operators, especially those with a sharp sense of detail, a trained eye, and capable of sensing microscopic nuances, who are able to satisfy this craving for images in a particularly creative way. Let us learn from them. The fascinating thing about the work of these inventive people, however, is that their products are beautiful precisely because their creators have given them human defects.

2.1 Finding Nemo, produced by PIXAR Animation Studios, 2003

To create their spectacle, animators working on the film "Finding Nemo" were sent, accompanied by underwater scientists, on diving trips and into large aquariums. Their first-hand experiences gave the film an extraordinary density and richness. They learnt from the solitary little creatures they saw and realised that whole movements can be expressed just with the blink of an eye or twitch of a mouth. They were determined to put so many things into the movie that it could not all be taken in on first viewing, things that, in some cases, only scientists could enjoy and be amazed by. They allowed themselves the luxury of sheer extravagance, and it is these tiny elements that make the movie so great. Yet it has plenty of visual value in other respects. Just think of the timid expression on Nemo's face, the greedy eyes of the shark or the furrows in its top lip. These are the tiny, barely perceptible details with which they were wrestling. "Our goal is always to make things believable not realistic. By stylizing the design of things, adding more geometry and pushing the colors, we were able to create a natural and credible world for our characters" (LASSETER 2003).

2.2 Kaya a 3D CG virtual female by the artist Alceu Baptistao, 2001

The Silicon Generation, which emerged in the US at the end of the twentieth century, uses plastic surgery to optimise their bodies as if they were Formula One racing cars, creating better performance, siphoning off the unnecessary and spraying on a new look, all of which is now considered perfectly normal. They make their skin as smooth as a sleek fender, distancing themselves from the fact that real, original skin is covered in hairs, moles and tiny marks of all kinds. And their beautiful, yet all-too normal bodies are surgically optimised to give them the kind of surface tension that is reminiscent of a surf board.

These Silicon beauties and the trigger-happy Tomb Raider star Lara Croft are roughly cut, especially in contrast to the virtual supermodel Kaya. The fascinating thing about Kaya and other Internet nymphs, however, is that they are beautiful precisely because their creators have given them human defects.

Everything about Lara Croft, the first internationally-known avatar of popular culture, is perfect and symmetrical: her eyes, her pneumatic chest, her artificially smooth hair. In contrast, Kaya has irregular freckles, little wrinkles around her eyes and chapped lips. Kaya and her digital friends are top candidates for the title of "Miss Digital World".

The advantage these models have over flesh-and-blood colleagues like Isabella Rossellini and Naomi Campbell is that they are easier to manage than extravagant, moody real-life supermodels. They don't get hungry, don't need expensive hotels and will never need to enter into contract negotiations with their producers. It is interesting to note that Kaya's fashionable beret is actually just a sneaky way of not having to model a full head of hair, which is extraordinarily difficult on a computer.

3 Strokes as Fleeting as the Scent of Perfume

Returning to a different age means returning to a different media. The women whom René Gruau, an Italian-born illustrator created were almost beyond the confines of everyday life. They were dream creatures, fluid goddesses of elegance, who consisted of nothing more than a few lines, patches of red and black and dark, heavy borders: as fleeting as the scent of a perfume, yet capable of leaving the same impression.

Gurau's clients included couturiers such as Christian Dior. The woman with the concealed eyes ceated by Gurau was on every advertising column in the 1950s. Today the sparing use of line and shape in Gaurau's work still impresses. It is the incredible ability to depict so much with so little which makes the difference: the contours of a perfectly moulded face with concealed eyes, a full, slightly pouting mouth, a Doris Day-type snub nose, and a curved line that outlined the neck as a movement of the head (KILROY 2010).

Capturing energy, elegance and audacity by bold lines and fluid style – René Gruau's work indicates what the brush, pen and ink achieved before digital tools and methods took over.

4 The Curiosity of the Matchstick Man

So why draw with only a pencil, drawing ink, brushes and paints today? It is not a question of whether somebody can draw well or whether everyone should be an artist. It's more to do with learning to see. It doesn't really matter whether or not students are capable of drawing something that could be hung in a gallery at the completion of their courses. What matters is the ability to appreciate and experience. Students can learn to draw if they are motivated and thus responsive to the message inherent in images. Drawing makes sense because it teaches students to look carefully and trains their ability to perceive things that otherwise may only have received a fleeting glance (JONAK 2009).

The students' attempts to reproduce with their own hand what they have seen with their eyes, changes their relationship to the object. Instead of just looking at it, students acquire a deeper understanding of the object's components and this enables them to commit it to memory with greater precision. More important than learning techniques is learning to see and foster creative thought processes. Leonardo da Vinci wrote in his notebook that just by pausing from time to time and examining spots on a wall or clouds one can come up with amazing ideas (JUNG 1991).

Drawing has an additional benefit. It helps us to understand *why* we respond to certain things, regardless of whether we are considering landscapes, gardens or buildings. "Ideas conceptualized in your head need to be translated onto paper so that they can be tested." (ZELL 2008, 7). By looking and drawing students find a means of explaining their tastes and can thus develop an "aesthetic sensibility", the capacity to make judgements, which will help them to make decisions within the design process.

5   Provence Has Taken on Different Colours since Van Gogh

It is important to mention Provence in this discussion, a place that has provided a treasure trove of examples illustrating landscape, climate and colour.

As an artist Van Gogh discovered Provence in 1888, although this unique area of France had already been the subject of paintings for more than a hundred years. According to Van Gogh, however, most of his painter colleagues had not done their subject justice; they had failed to give a realistic rendition of the qualities of the place.

"So what did Van Gogh see that other people had overlooked? For one thing, he noticed the way in which the olive trees moved in the wind. He also saw the colours in a different way to other observers. This is due to the area's specific climate. The mistral wind that blows from the Alps through the Rhone valley regularly clears the sky of clouds and moisture and scrubs it to a pure, deep blue without the slightest trace of white. At the same time, a high water table and good irrigation support plant life that is both unique and abundant for a Mediterranean climate. Since the air contains no moisture, there is no mist to obscure or interfere with the colours of trees and flowers. Due to this interplay of a cloudless sky, dry air, water and abundant vegetation, Provence exhibits a lively, contrasting array of primary colours. Van Gogh's predecessors paid little attention to these contrasts and painted exclusively with complementary colours"…and it was this contempt for the natural Provencal colour palette that angered Van Gogh." (DE BOTTON 2002, 212)

By studying Van Gogh students can learn a great deal about shading, colour schemes, proportions, perspective and image composition. And perhaps they might also find something to admire in the Provence sky when it is pointed out, however superficially, that it all comes down to that special shade of blue.

6   Everything Is Design

Tulips, tomatoes, sweet corn: creative manipulation is everywhere, constantly present in our everyday life. It is so matter-of-fact that this type of design has been virtually driven out of our consciousness. This instance also questions what can actually be taught and learnt in creative disciplines.

Genetic engineers decide whether strawberries made in petri culture dishes should look like strawberries and even taste like strawberries but they often end up just looking like strawberries. But apart from genetic manipulation, we should also consider the many extraordinary examples of cross-breeding undertaken by "flower pioneers", whose designs landscape architects utilize so enthusiastically.

The success of any new invention relies on many criteria. But most of all it has to withstand the comparison with familiar traditional applications.

7 Dreaming the Old Avant-garde Dream

It seems to be easier to create monsters than to work on the illusion of reality. Most computer graphics appear so smooth and flat, as if the were made out of plastic. Research into digital methods and tools in landscape design should not neglect this aspect of the topic.

MANOVICH (2001) explains that the digital image technique was originally developed for military purposes in order to train pilots. The entertainment industry adopted the graphic language of military surfaces, which focuses on flight simulation and not on drape illusion. This might be one of the reasons why no Vermeer of computer graphics has yet emerged. Anyone wishing to speak of a new medium must inevitably allude to an old one. According to the fundamental laws of media theory only media comparison creates media reference.

Digital communication expands the possibilities of depiction. With just a few clicks in Photoshop an enormous variety of visual effects can be applied and revised. Never before has it been so easy to work with a multitude of aesthetics, atmospheres or combinations of stylistic devices (ULLRICH 2002). With the immense resource of images and especially through digital access, the Dadaist, Surrealist and Constructivist dream of art as an assembly process has come true.

But the process of creating something new has been left by the wayside. Entire image cultures lie fallow. Images are easier to remember than words. The Danish architect and author Nils Ole Lund is an expert at collage-making and has been creating artificial realities by experimenting with photomontages. "When I start making collages in most cases I have one or two basic pictures, two contrasting images, the rest is added, ideas, details, accessories" (LUND 1990, 5). Through his collages Lund achieved "discussions in a more direct way than by writing articles, giving lectures or even designing houses" (LUND 1990, 8).

Although the old avant-garde dream of art as an assembly process (the universal language of images) appears to have come true designers are not taking full advantage of the enormous creative potential of digital communication.

8 A Language of Image

Working as designer means to stimulate internal images. "Images are means to creating architecture, and may even be my most primary means. Memory is stored in images. It can, of course, be stored in smells, too, but these smells then immediately turn into an image, memory does not remain abstract." (WIDDER CONFURIUS 1998). By stimulating these images during the design of spaces people who look at a building or landscape are led to experience the same feeling. "With time we discover that our personal images, conjured by memory, are actually not so special and that we all share them" (WIDDER CONFURIUS 1998) For prospective landscape architects, learning the language of images is not merely a form of communication for their leisure time, but an indispensable vocabulary that can be used throughout their professional careers. What we teach and research could boldly be redefined as design communication, and even manipulation, through images. The science of images discussed in this paper deserves equivalent status to that of the science of language. Thus, students use the language of images on their courses in order to convince their professors of the quality of their design. Equally, in the everyday world of architectural practice, pictures are used to communicate with a range of participants, from clients to competition judges and local people, and designers obviously hope that their images will fill everyone with enthusiasm for the content and quality of their thoughts.

Just as other specialists seek out effective languages with which to communicate, landscape architects, too, need to create genuine opportunities for communication between their inner senses and the various protagonists in the design process. We expect language to involve a dialogue, and a similar type of dialogue with similar benefits can be conducted between our eyes and our mind.

The media revolutions of the 20th and 21st centuries have established a visual environment in which simple site plans, models and even photographs increasingly appear as hopelessly unspectacular. Even with everything that is known about the psychology of effect, intelligent image production still represents an attack on the recipients' senses. Many people today can draw nice pictures, especially since the introduction of MS paint or Adobe Photoshop, but to work to scale needs professional education. Site plans and site models, digital or not, are still a very important communication method for landscape architects working with three-dimensional illusion and professional assessment.

So instead of offering up a flood of images containing promises that are hard to keep, landscape design representation is most likely to succeed by retaining an original image throughout the whole process in the form of a drawing (the sketch) and the closest correspondence between the desired look (the plans) and the reality (the finished space).

References

Burkardt, L. (1995), *Design = unsichtbar*. Ostfildern, Cantz.
De Botton, A. (2002), *Kunst des Reisens*. Frankfurt a. M., Fischer.
Kilroy, R. (2010), *Rene Gruau*. http://decoymagazine.blogspot.com/2010/09/rene-gruau.html.
Lasseter, J. (2003), *CGNetworks presents The Making of Finding Nemo*. http://features.cgsociety.org/story_custom.php?story_id=1389.
Lund, N. O. (1990), *Collage Architecture*. Berlin, Ernst & Sohn.
Jonak, U. (2009), *Grundlagen der Gestaltung*. Stuttgart, Kohlhammer.
Jung, C. G. et al. (1991), *Der Mensch und seine Symbole*. Olten, Walter.
Manovich, L. (2001), *The Language of New Media*. Cambridge/Mass., MIT Press.
Widder L. & Confurius, G. (1998), *Questioning Images: Interview with Peter Zumthor*. Daidalos 68
Ullrich, W. (2002), *Die Geschichte der Unschärfe*. Berlin, Wagenbach.
Zell, M. (2008), *Architectural Drawing Course*. London, Baron's Educational Series. New York, 7

Teaching Planning and Design in Landscape Architecture

There Is no App for that – Ardous Fieldwork under Mega-urban Conditions

Jörg REKITTKE and Philip PAAR

Keynote: 27 May 2011

1 Big City Struggle

This paper describes the second of three fieldwork studies in the context of the research project "Grassroots GIS. Development of low cost mapping and publishing methods for slums and slum-upgrading projects in Manila", financed by the School of Design and Environment, Department of Architecture, National University of Singapore (NUS). The project follows from preparatory work and the first investigative fieldtrip by the authors in 2010, which had been published the year before (AUTHOR & AUTHOR 2010). The research is conducted in combination with special landscape design studios in the context of the NUS Master of Landscape Architecture programme. These studio projects are related to and organised in cooperation with the Philippine grassroots movement Gawad Kalinga (GK). The mission of Gawad Kalinga – meaning *to give care* – reads as follows: "Building Communities to End Poverty". The method is simple: "Land for the Landless. Homes for the Homeless. Food for the Hungry" (GK 2010). Beneficiaries work hand-in-hand in *bayanihan* (Filipino term for teamwork and cooperation) with GK volunteers in building the infrastructure and structures of the community (Fig. 1). The *kapitbahayan* (association of GK homeowners) composed of the beneficiaries themselves, take on multiple roles and undergo various leadership trainings. The beneficiaries learn to take ownership of their community and are empowered to help themselves and help others (AUTHOR & AUTHOR 2010).

Fig. 1: NUS MLA students and staff during a volunteers' building day in GK Telus Village, Manila 2011 (photos: Nur Syafiqah)

In 2010 the students worked on the topic *Needle in a Haystack Gardens – Manila*, focussing on designs of urban gardens or garden elements that contribute to a healthy environment and to improved living conditions in the selected GK slum-upgrading project areas, denoted as *villages*. The students had been asked to specialise on productive forms of gardens that can provide precious food for the table (AUTHOR et al. 2010). This year, 2011, the 'designs for the real world' of the students (PAPANEK 1985/2000) will focus urban farming and livelihood ideas for two other selected GK villages in Metro Manila – *Concepcion*

Village, Barangay Buayang Bato (administrative district) in Mandaluyong City and *Telus Village* in Quezon City.

Being able to invest research energy and money into the informal parts of the often chaotic context of mega urban Metro Manila means both a blessing and a curse. A blessing because slums and slum-upgrading projects are the most unsurveyed and untraceable parts of megacities, constituting white spots on official maps and planning materials – which makes them excellent and challenging scientific targets. A curse because the mandatory fieldtrip actions can be affected by some loony unpredictabilities. This time we had bad luck. During a city tour – beyond our GK research areas – one of our students had been assaulted by two hooded motorcycle bandits, who finally shot him in his legs two times, injuring him seriously. This happened in a corner of the city Moloch, where no mobile network can be reached, no ambulance is available and no public transport is offered. The dramatic story had a good end thanks to many helping Filipinos. The student underwent a six-hour surgery and could travel home with us after our ten-day fieldwork. We are aware of handling a scientific paper and this kind of detail might be unusual in such a text. The description of the incident is not meant as an emotional anecdote but rather provides a realistic glimpse on the unspeakable conditions that we try to work in – the routine environment for more than 16 million stalwart people in an *endless city* (BURDETT & SUDJIC 2007). It was not only the extreme urban environment that we had to struggle with, also our technical equipment, tools, applications and our methods caused us some serious headache during this fieldwork. We are conducting this research parallel to the guidance of a design studio with a social background, a tightrope walk with bottlenecks. Therefore we regard this article rather as a report and reflection on experiences, problems, necessities and ideas than a perfect document of research success or fulfilment. Most findings of our Grassroots GIS approach (AUTHOR & AUTHOR 2010) can just be generated via ardous fieldwork, only be *perceived through the feet* (INGOLD 2004). Under these conditions learning by doing remains indispensable, disappointment inescapable.

2 Technical Armament

2.1 GK Village Anomaly

Slums are usually places where detailed fieldwork with the aid of costly equipment is virtually impossible. High-capacity cameras inclusive associated equipment, sophisticated surveying tools, expensive phones or the like are out of place. People and communities in slums constantly monitor their personal environment and they usually inhibit any attempt of outsiders to gather revealing data or information. These extreme places are representing hard-fought territory, the daily struggle for survival causes rough customs and those who tune in to criminal activities understandably don't like to put their cards on the table – one of the reasons for the low rate of documentation or mapping of slums (DAVIS 2007). Referring to this the Gawad Kalinga villages feature a positive anomaly. Although they form just tiny islands in the sea of vast urban slum areas, they offer a largely safe and practicable working environment for our purposes. To become dwellers in the GK housing projects, the beneficiaries have to form strong communities and swear off any form of crime, violence, drug abuse and other negative behaviour (GK 2010). Living in and owning their new houses unequivocally changes the lives of these people and they impress everybody by acting friendly, cooperative and social. Seeing the children in these communities creates hope and confidence, they have a realistic chance to get to know the promise of a civilized childhood, parental care and school education. Only this context allows our detailed analysis, design work and research activity in the megacity slum and slum-upgrading environment. Thus we can apply some necessary fieldwork technology, not being afraid to finally leave the place without this equipment.

2.2 Equipment Pool

Slum-upgrading work of non-profit organizations like Gawad Kalinga faces small or diminutive budgets, thence Grassroots GIS tries to build on unexpensive technology, easy or free access to applied tools, geodata and georeferenced design data as well as open source, open standard and cost-free software and data storage possibilities (AUTHOR & AUTHOR 2010). In 2010, a low cost GPS device and an ordinary digital compact camera had represented our only technical equipment. This year we invested some research money in a slight rearmament. In addition to the GPS device we bought iPhones 4, equipping them with some free or low cost applications ("Apps") for our fieldwork. Also a better and primarily faster digital reflex camera found its way into our luggage. This upgrade was related to three estimations: 1) High-capacity smartphones are congesting the markets worldwide and will find their way even into the hands of the poor. In our research we are trying to anticipate this inevitable development; 2) We constantly have to cater for not being outdistanced and compromised by the common technical progress; 3) Smartphone and 'App'-developers lay claim to deliver this legendary *all-in-one*, the *jack of all trades* or what Germans mock as 'Eierlegende Wollmilchsau'. Even the quarter of fulfilment of this promise would accommodate us very much with our intended grassroots approach.

Fig. 2: The complete fieldwork equipment: digital reflex camera (l.), pen and paper (m.), iPhone 4 inclusive selected Apps (r.)

Currently we are not planning to fundamentally expand our equipment pool, any arms race would thwart our basic claim. Many things could be imagined regarding other additional handheld devices for the fieldwork – like a laser telemeter for example, but we don't plan to come off our chosen path. We regard our research as design-oriented basic service, paving the way for subsequent outdoor design inclusive planting design, integrating 2D and 3D geospatial data and tools for the purpose of landscape design activities. We are not trying to poorly copy the work of professional geodesists, who definitely have better methods and tools. We are searching for a way to become independent from non-existent maps of informal urban settlements, testing and applying common technology to use it as fast and effective do-it-yourself geometer tools. Main aim remains the development of a toolbox and user-generated geospatial content process that supports mapping, storing, interactive design, disseminating, and interactive visualizing of landscape architectural interventions in the context of urban informal settlements (AUTHOR & AUTHOR 2010).

2.3 Employed Mobile Apps

From more than 300.000 available iPhone Apps (APPLE.COM 2011) we have chosen the modest number of seven Apps for four topical chapters of the fieldwork:

- **Surveying and Mapping**
  1) 'MotionX GPS' (pro version, 2.99 S$ = Singapore Dollars). According to the developer, 'MotionX GPS' is the leading GPS App for the iPhone and iPad. "Over eight million iPhone or iPad users have chosen MotionX" (MOTIONX.COM 2011).

  2) 'GyroSurveyor' (freeware). According to the developer, 'GyroSurveyor' "(…) can help you estimate distances within visual range with the help of camera and orientation sensors on your device. It uses data from orientation sensors and the camera image to determine 3D coordinates of the object you selected and then give you the distance from the object to you or another object" (IDEAMATS.COM 2011).

- **Photography**
  3) 'ProCamera' (2.99 S$), a powerful, award-winning tool, which might currently be the most sophisticated camera tool for iPhone.

  4) '360 Panorama' (1.99 S$), a tool for realtime panorama creation (APPSTORE 2011).

  5) 'Camera' (Apple iOS), the inbuilt standard camera App of the iPhone.

- **Sound recording**
  6) 'SoundCloud' (freeware), an App which "(…) lets you easily access, browse and listen to the sounds shared to you while you're on the road and away from your computer" (APPSTORE 2011).

- **Colour check**
  7) 'Color Set' (2.99 S$), a digital colour chart, based on the German RAL colour systems 'RAL Classic' and 'RAL Design'. The human eye is able to distinguish about ten million colour shades. "Since 1927, RAL has (…) standardized, numbered and named the abundance of colours. These standards are easily understandable and applicable – worldwide" (RAL.DE 2011).

After some pre-fieldtrip testing of 'MotionX GPS Lite', the restricted but costfree version of MotionX GPS, we decided to rely on MotionX GPS during fieldwork. The App offers an *OpenStreetMap* download feature and caches the map tiles. This way users can avoid expensive data roaming on site. SoundCloud, in contrast, requires Internet access to store the recorded data, which can cause immense costs under international roaming conditions.

2.4 Brothers in Arms

We might be two of the few academics in the field of landscape architecture who dedicate their research and design work to informal settlements in the megacity context, but all in all we are only two scattered foot soldiers in the vast world army of *volunteer geographers* (GOODCHILD 2007, OVER et al. 2009), exploring untrodden paths of urban public realm and (re)mapping the world on own initiative. In Manila we met two of these activists, Emmanuel Sambale – a professional geographer and GIS expert, and Rally de Leon, a businessman, running a messenger service in Manila, desperately missing exact and non-ambiguous delivery addresses of many clients. It was them who established the contact, identifying us via our year 2010 *OpenStreetMap* uploads of GK village maps of *Baseco* in Manila City, and *Espiritu Santo*, part of Sitio Pajo in Quezon City. They are volunteer mappers since more than five years, and they joined us during our mapping activities in *Telus Village*, Quezon City. When they usually do their fieldwork, they celebrate 'mapping parties'. The social part of such parties begins, when all data of a site are collected, processed and uploaded – the same day. They came well equipped – laptop with wireless connection, Garmin GPS devices, laser telemeter and a separate compass. The telemeter was an indoor device, a 'by-product' of the on-going private house building activity of Rally de Leon, and not really suitable for outdoor conditions and distances (Fig. 3).

Fig. 3: Volunteer as well as professional geographer E. Sambale (l.) with indoor laser telemeter. Advanced field equipment for editing and upload of data on site (r.)

Their compass came into operation when using a detailed hand-drawn ground plan of our NUS students to merge it with GPS tracks and a rough ground sketch of their own. The compass was positioned on the paper map, digitally photographed, loaded into *Java OpenStreetMap Editor* (JOSM) and scaled and rotated with the JOSM plug-in *Piclayer*. The GPS data were edited in Garmin's desktop software *MapSource* and imported into JOSM. After a manual calibration of the laser telemeter data, GPS data and the paper maps, they generated an on-screen fair drafting in JOSM, added useful attributes and were ready to upload the new map puzzle piece. They abstained from the immediate upload – presumably because their ambition had been egged by our vivid exchange of knowledge. The principle of gathering, processing and upload of all data on one day and onsite, offers an essential advantage: Mistakes or doubts can immediately be checked *in situ*. After their session in

Telus Village, our Filipino colleagues went straight to the next nearby GK village. GK officials showed vivid interest in the collective mapping and we are pleased that we could initiate a meeting of local actors for a good cause – breaching the isolation and addresslessness of the GK villages by publishing the neighbourhoods via online maps at the public platform *OpenStreetMap* (OSM). The uploaded material of SAMBALE and DE LEON (Fig. 4) forms a small step of progress concerning our work for GK. They included some samples of the tiny GK house footprints – each footprint is about 18 square metres. This detailed scale just could be reached by including the hand drawings of our MLA students, pure GPS mapping doesn't allow this specificity (Fig. 5).

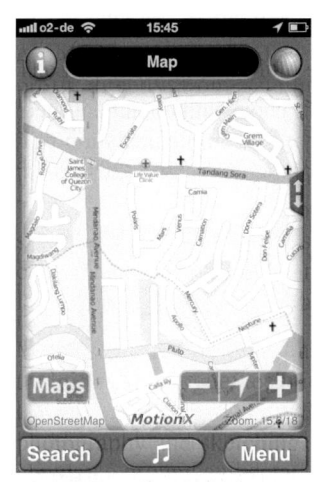

Fig. 4: Results showing the development of night-time lights in China; a: initial (1995/1996) state, b: total increase by factor 5, c: total increase by factor 10 d: total increase by factor 40

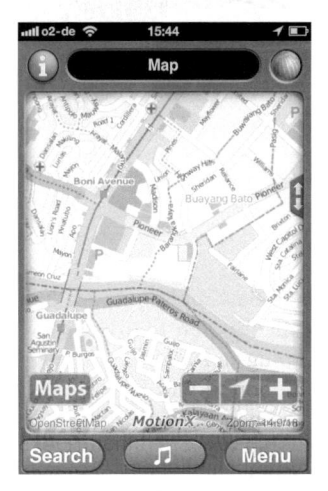

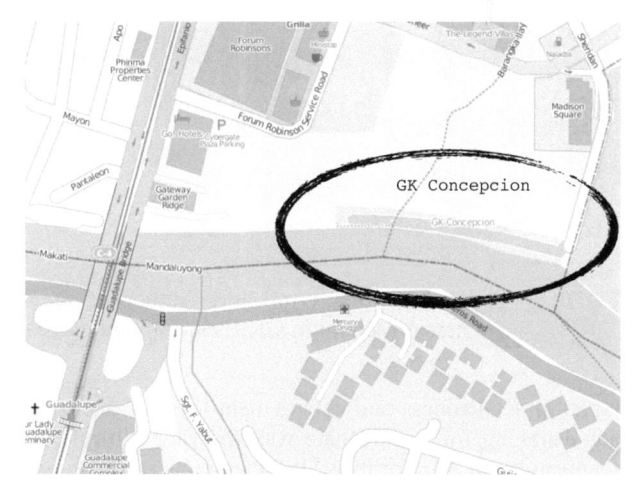

Fig. 5: Before (l.): Unmapped and untraceable GK Village Concepcion; after (r.): Uploaded mapping (screenshot osm.org, Feb 1, 2010) exclusive samples of house footprints (AUTHOR & AUTHOR)

Displayed at highest OSM enlargement level, we realise that the house samples are nearly to small for such maps, but they are visible. Telus Village doesn't have streetnames so far, but house numbers had been assigned recently and could be attributed in OSM. This final step would mean: Telus Village is findable, people can be visited, mail can be delivered – mission completed.

3 Research in Progress

Little by little one goes far. In 2010 we worked on GK village *Espiritu Santo*, one of our study sites "where the streets have no name" (AUTHOR & AUTHOR 2010). In the meantime streetnames had been assigned and we were able to update the data on OSM. No records of these new street names were available, only in the field we could check and document them (Fig. 6). Beside the street name signs, the photo shows the successful implementation of a simple green wall, proposed by one of our MLA students in 2010 (AUTHOR et al. 2010).

Fig. 6: Updated mapping (Author & Author) of street names, a new multi-purpose hall and new preschool in GK Espiritu Santo (screenshot osm.org, Feb 1, 2010)

Because detailedness and enlargement level of the used mapping platform are limited, we are searching for an adequate way to bring geo-referenced photography into play. There seems to be an App for that. '360 Panorama' allows – via an ingenious real-time *panorama painting technique* – to create fast and effective geo-referenced 360^0 views despite the ubiquitary hindrances and stumbling blocks in the informal environment, which render all too systematic and exact photo documentations virtually impossible. It had been a false assumption (AUTHOR & AUTHOR 2010) to think that a tripod could help to deliver better photo material, tripod positions could be calibrated, shooting angles be fixed and picture intervals be unitised. No chance for such finicky fiddling on squatter or ex-squatter territory. We tag our results as *Google Street View* surrogate (Fig. 7), a low-budget, self-made alternative to *Street View*, wildly handmade but sufficient and bridging the waiting time for the Google camera vehicle that never arrives.

Fig. 7: *Google Street View* surrogate. Panoramas of GK Telus Village and adjacent slum neighbourhoods, recorded with the 360 Panorama App (Author & Author)

This photographic material can be placed in Google Earth or Biosphere3D (PAAR & CLASEN 2007) – OSM doesn't provide this feature yet – communicating the reality and development of GK villages worldwide. As visible on the panoramas (if not displayed in black and white), colour plays a crucial role in the GK concept. Slums are dark, sometimes pitch-black. The GK movement tries to guide the poor back to the light and paints all new houses in bright colours. To document these colours, we used the App 'Color Set' (Fig. 8).

Fig. 8: Application of iPhone App 'Color Set' for documentation of the facade colours in GK Telus Village, Manila

'Color Set', a digital colour chart based on the German RAL colour systems 'RAL Classic' and 'RAL Design', can be used in the field to identify the exact tint of the GK house walls. Such detailedness constitutes a theoretical extreme of grassroots GIS, the finest graininess that we could generate. We think that this goes to far concerning our purposes, but we had to test it.

4 Wish List and Feedback Forum

4.1 Longed-for Tools

- 'MotionX GPS' is a powerful App, but the scattering of the integral iPhone GPS receiver is very high. The segmentation and narrowness of the informal city doesn't correspond to such a GPS signal variance and vice versa. An external GPS antenna could make the difference.
- Our Filipino OSM volunteers reported that they sometimes use such external antenna in the urban context, mounting it on a bamboo stick to receive GPS signals of higher precision. When mapping in areas of high housing density they recommended to set some reference points in open areas neighbouring these density zones, to have a possibility of averaging potential discrepancies. We consider to include this into our future fieldwork.
- When using iPhone App '360 Panorama' we repeatedly thought that an App with a comparable comfort and simplicity for the creation of linear (long and flat) panoramas would be extremely helpful for analytic work in the context of informal settlements. The '360 Panorama, App is one of the most simple and efficient fieldwork tools we found so far, but it should feature a better exposure compensation for strong light contrast in outdoor situations. And it would be good – especially for landscape architects – if the App could record a bigger visible portion of sky and ground.
- In addition to geo coordinates and compass data, the camera pitch should be saved as metadata in the EXIF standard, which is not the case in the tested iPhone 'ProCamera' and 'Camera' Apps.
- Tightness and density of informal settlements afford an ultra-wide-angle lens to allow photographing of complete facades – maybe unaccomplishable for an iPhone.
- 'GyroSurveyor' ought to measure distances simply on a flat level surface but we couldn't get it work properly. An inbuilt laser would be a useful sensor for future iPhones.
- An App that combines and synchronizes photography with a voice recorder would be expedient.
- OSM map services and also MotionX currently provide a zoom scale of maximal 18. This is very limiting if building footprints of about 18 m^2 – common in most GK villages – shall be displayed. We would like to map and display even tiny 'square foot gardens', which contribute to the aimed self-sufficiency of GK villages.
- Why not finishing this chapter with some utopianism: Megaurbanity necessitates an *Urban Chaos App*, able to record the overwhelming mixture of multiple noise, stench of excrements and waste, extreme air pollution, numbing exhaust fumes and other 'abstract environmental pollution' (SPEED & SOUTHERN 2010). Under this muser category also falls an App that could deliver image-based 3D reconstructions,

generated in real-time during fieldwork walk-throughs. It is the time-consuming data post-processing, which breeds such hallucinations.

4.2 Sisyphean Struggle

It remains an unanswered question of principle, how much information and how many data should be recorded for Grassroots GIS. Comparable with the side effect of digital photography – too many pictures to be viewed – our grassroots approach seems to demand resolute self-censorship and healthy braving the gap. Getting lost in detail would be ill-advised. As long as digital data gathering and processing takes significantly more time than detailed pencil sketching in the field, proper scepticism concerning our digital tools and methods persists. Yet a basic catalogue of data that we need seems to become apparent. The iPhone is a smart shenanigan and the flood of new Apps will lead to new ideas of using them. But although the overall number is overwhelming, not all too many Apps seem to be suitable for mapping activities and spatial design work preparation. This is reassuring at least.

References

Apple Store (2010), Accessible under http://www.apple.com/iphone/apps-for-iphone/.
Author & Author (2010), *Grassroots GIS – Digital Outdoor Designing Where the Streets Have No Name*. In: Buhmann, E., Pietsch, M. & Kretzler, E. (Eds.), Peer Reviewed Proc. of Digital Landscape Architecture 2010 at Anhalt University of Applied Sciences, Berlin/Offenbach, Wichmann, 69-78.
Author et al. (2010), *Logbook Landscape Architecture no.02*. Berlin: epubli.
Burdett, R. & Sudjic, D. (Eds.) (2007), *The endless city: the Urban Age Project by the London School of Economics and Deutsche Bank's Alfred Herrhausen Society*. New York, Phaidon Press.
Davis, M. (2007), *Planet of slums*. London:,Verso.
Gawad K. (2010), *Building Communities to End Poverty*. http:// www.gk1world.com.
Goodchild, M. F. (2007), *Citizens as Sensors: The World of Volunteered Geography*. In: GeoJournal, 69, 211-221.
Ingold, T. (2004), *Culture on the Ground: The World Perceived Through the Feet*. In: Journal of Material Culture, 9 (3), 315-340.
Over, M., Schilling, A., Neubauer, S., Lanig, S. & Zipf, A. (2009), *Virtuelle 3D Stadt- und Landschaftsmodelle auf Basis freier Geodaten*. AGIT 2009. Salzburg, Austria.
Paar, P. & Clasen, M. (2007), *Earth, Landscape, Biotope, Plant. Interactive visualisation with Biosphere3D*. Proc. at CORP, 12th International Conference on Urban Planning & Regional Development in the Information Society, Vienna, 207-214.
Papanek, V. (1985/2000), *Design for the real world. Human ecology and social change*. Chicago, Academy Chicago Publishers.
Speed, C. & Southern, J. (2010), *Handscapes – Reflecting upon the Use of Locative Media to Explore Landscapes*. In: Buhmann, E., Pietsch, M. & Kretzler, E. (Eds.), Peer Reviewed Proc. of Digital Landscape Architecture 2010 at Anhalt University of Applied Sciences. Berlin/Offenbach, Wichmann, 164-172.

Teaching Digital Methods in Landscape Planning – Design, Content and Experiences with a Course for Postgraduate Education

Roman LENZ, Werner ROLF and Christian TILK

1 Introduction

In the last years at Nürtingen-Geislingen University a course for further training and professional education in landscape and environmental planning was developed. This course is taught in German and called „Geodatenmanager Umwelt". It's part of the advanced education programme „U3 – Umweltinformatik Unterricht für Umweltplaner" which focuses on the topics of Geographical Information Systems (GIS), Data Management Systems, and Visualisation (LENZ & ROLF 2003; ROLF & LENZ 2005).

The course introduces into relevant software, but apply and teach them from a landscape planning point of view. The course has two weeks of joint training, with partly online-guided self study phases of four weeks. Clients are mainly professionals from governmental organizations, as well as freelancers. In addition, we often have some graduate students as tutors, who are interested in problem solving approaches of professionals as well as in more sophisticated GIS applications and other digital methods. The professionals not only benefit from updated methods and tools, but also from the skills and knowledge of the students. Interestingly, the course becomes more and more an update of what was offered already during the diploma or bachelor studies in Landscape Planning at our University, and, vice versa, helps us to improve our curriculum for Bachelor and Master nowadays (e. g. LENZ 2004). Hence, we believe, the course contains as well as continuously updates somehow the most important digital methods and tools in Landscape Planning, strongly related to current and ongoing changes of professional needs. In other words, courses for professional education, together with academic curricula, support each others.

In the following we describe the course development over the last 10 years, and the current competence focus as well as different learning methods and tools that can be seen as an selection of the best established methods of the programmes evolution.

After more than a dozen courses we also have had many evaluations, which show the feedback of students and professionals that will be presented besides some proven applications useful for teaching applied GIS methods.

Because new developments took place in the most recent years this is not an update of a description published earlier (ROLF & LENZ 2005), but will also lead to new conclusions. Therefore the outlook will contain hints that may inspire other programmes on how digital methods can be taught rather advanced and "with all senses" in universities curricula of landscape planning.

2 Programme Description

2.1 Course Development

This professional education programme evolved as a consequence of a course that was originally developed for students with whom a problem orientated approach was trained.

From 1996 until 2000, within the project "ECCEI – European Canadian Curriculum on Environmental Informatics" international students from Germany, Italy and Canada, coming from faculties of different disciplines (informatics and environmental management) were trained to find IT based solutions for environmental issues (LENZ 2000). Core element of ECCEI was the so called "Short Alpine Course on Environmental Informatics – SACEI" whereas the Soelk Valley in the central alps (Austria) was the training field for environmentally relevant issues that had to be solved within a one week summer school like workshop.

Although the concept originally was thought to educate students, in the year 2001 the methodological-didactical approach was tested for first time as a training course for professionals named "U3 – Environmental Informatics Education for Environmental Planners". To apply this approach to the needs of professionals further presence modules and online modules for self study phases and training were developed and combined. The outcome was an in-service training programme of four month duration. The applied use of GIS, database and visualisation and internet technologies was the focus.

After 4 courses with 77 participants the programme became "U3plus". The concept changed in the way that now professionals and students were now together course participants. The idea was to offer students the opportunity to learn more about the practical needs and applied IT use and to deepen their interest in environmental informatics beyond the universities curriculum. On the other hand the professionals should benefit from the advanced GIS skills and knowledge of the students, and could fill out a tutor like role. Within the next years three courses with 10 students and 44 professionals took place.

In 2007 the concept was updated again. The complete programme was divided into single specific issues and the modules got a more stringent structure. The training programme "U3" was therefore separated into different parts. The main course is now named "U3-Geodata-Manager Environment". It's an in-service training lasting three month and does have a very strong GIS focus. Other course contents (like database, internet technologies etc.) were extracted into single outlook course modules (each of a weekend duration) called "U3-Extensions". Last ones could be optionally chosen on single topics offered. However the problem orientated approach with a one week workshop like situation still remains as a characteristic of the methodological-didactical concept in form of a single one-week workshop-like module as part of the "U3-Geodata-Manager Environment" training programme.

Since that time another 77 participants, almost every tenth of them were students, summing the number of participants up to 208 in the past ten years, were visiting the "U3" programme.

2.2 Competence Focus – Learning Goals

Learning goal is the use of Information Technologies (IT) within the field of environmental planning and resource management, with a focus on Geographical Information Systems (GIS). Within the course usually ArcGIS in its most actual version is being used as an example, but other software solutions, especially Open Source GIS, are being used as well. The participants should get a practice relevant overview of the most important methods and tools and gain ability to break down planning issues for IT based solutions. Herewith participants should be enabled to perform basic applications on their own but also achieve ability to discuss more complex issues with experts. This involves to estimate quality aspects, validity of models as well as management abilities. The relevant issues are split into several different modules (Fig. 1).

Module	Content	Type & Workload
Introduction Seminar	• Introduction in GIS within the context of environmental informatics and its use for environmental planners	*Online Module* Workload: 10 h
	• Exemplary exercises of different project phases to deepen comprehension of GIS use in practise for „green jobs"	*Presence Module* Workload: 6 h
Basic Seminar	• Overview and practical use of GIS methods and tools (data management, capturing, analysis, presentation)	*Presence Module* Workload: 24 h
	• Training and exercise of the tools in use	*Online Module* Workload 20 h
Practical Seminar	• Learning an appropriate and efficient use of GIS methods and tools within an environmental project (incl. data and project management as well as integrated use of complementary IT solutions)	*Presence Module* Workload: 40 h
Outlook Seminars/ Extensions	• Overview and practical use of optionally choosen additional and complementary IT tools and methods (CAD, internet and database technologies, mobile services,..)	*Presence Module* Workload: 12 h

Fig. 1: Different modules with their main learning goals

The learning goals can be associated to different aspects as follows.

Appropriate and efficient use of GIS tools and methods
Participants get an overview of practice relevant GIS tools and methods for data management, data capturing, data analysis and presentation. The aspect of data capturing includes quality aspects of geo data, quality check and quality improvement. Data analysis contains attribute and spatial analysis of vector and raster data as well as data derivation due to geoprocessing and data modelling. This includes documentation aspects for comprehensibility of manipulated GIS data and to keep transparency of data analysis and modelling for further decision making processes.

Data Management
This means participants learn how an efficient data management is being organised, taking into account different conditions, i.e. integrated use of geo data server for multiple user etc..

Besides that they get introduced into relevant environmental geo data from different sources and in different scales and learn how to handle a combined use. Data documentation with metadata can be associated to this topic as well.

Project Management
This involves to learn how to approach and transform environmental planning issues for IT processing and the implementation of the appropriate methods and tools. Therefore it is important to know how to design a GIS project, and how to structure and document it. Within the course participants become aware of potentials and limitations of GIS software and learn appropriate complementary other IT tools like interactive visualisation systems.

2.3 Teaching Approach – Learning Methods and Tools

The course is set up as a blended learning concept (SCHMIDT 2005). Several online and presence modules complete the programme (Fig 1.). Hence, we believe that transfer of knowledge can be organized more efficiently while separating lessons with teaching attendance from exercises and study parts were the personal needs of the participants may differ regarding time requirements. Besides that it offers more flexibility in timetabling of course work, which is an essential requirement of an in-service course programme and accommodates to the needs of participants that cannot be absent from business for a longer time (WIEPKE 2006; SAUTER & SAUTER 2002). Therefore three different main methods can be distinguished:

1) Teacher-centred lectures with tutor supported exercise blocks
Particularly lessons in which basic knowledge will be introduced are preferred to be taught in teacher-centred lectures (GUDJONS 2007). Exemplary, short, tutored exercise blocks usually go along with them. In this way participants learn the principles and get familiar with the software use as well. In this combination this method is used as a key to access prior knowledge. Like a colourful flower bouquet being presented it offers the possibility to negotiate the wide range of methods and tools but the lecturer still has the opportunity to respond on specific interests of the participants, depending on their field of practise.

During the exercise blocks students support the course participants as tutors. As they are usually advanced in the use of GIS tools because of the universities curricula, they can help the professional participants by the software use. In preparation on this the student tutors coincidentally have to self-reflect their knowledge to identify own knowledge gaps and finally to resolve them. In this way this methods helps them as well to enlarge their own knowledge.

2) Problem orientated approach, project work within working groups
This, more or less, is one of the main characteristics of the U3 programme and in particular of the so called "practical seminar". Here the participants are faced with a real world problem and the task is to "solve this problem" with the help of IT i.e. GIS (BLÖTZ 2008). As a first step groups are build perspective the level of profession and/or regarding the interest of deepening certain methods. This will lead to a process that we call an internal differentiation among the heterogeneous participants, as within more homogenous teams learning goals can be achieved more efficient (SCHITTKO 1991, KLEIN-LANDECK 2004). The only preset is that the groups have to be mixed with students and professional participants together, so they can both learn from each other during the project work.

During the project work the different teams will get individual support by the lecturers, who now have the task of supervisors. Just in case one or more groups end up in a situation were new fundamental knowledge on specific methods or tools are needed theoretical parts are taught with short information blocks as "Lectures on the fly".

After this module of each working group a study is expected that will be presented by the end of the practical seminar. Besides the approach and solution being presented the assignment has to contain as brief description of the used methods and tools as well as possible traps, problems and workarounds.

As case studies several planning tasks are given as choice, that are prepared already in a way that they offer the use of a wide range of methods and tools. Nevertheless own project ideas usually are being developed by the participants as well.

3) Online-guided self study phases

Besides the presence modules several online modules are implemented, which are strongly e-learning supported. The main tool for our online modules is the so called U3-Learning-Management-Platform (U3-lmp), a database once designed for the course needs and realized by programming with PHP and mySQL. The U3-lmp can be accessed via in the internet. Participants have a personalised user account and will guide after login through an user friendly front-end. An additional mailing list, realised with Majordomo, supports the communication among the participants. Via the backend an administrator is able to organize the participants in different courses and modules, i.e. users, roles and courses can be managed quite easily.

Within the modules different themes can be created where the lessons and tasks can be uploaded by the lecturers assigned to the theme. The participants again can upload their results topic associated. The working results can be shared with other co-authors, so that even group work can be done. Furthermore a course blackboard is included besides simple true and false self-tests. Last ones do not have any automated analyses functions integrated, as they are used mainly for self-evaluation through the participants themselves.

Some of the tools can be described similar to common e-learning platform functions of adding, editing, organising resources, learning materials and assignments (BETT & WEDEKIND 2003, GRÜNWALD 2008). Such as discussion forums, chat rooms and web-conferences or grade items are not supported. Even though being aware that other e-learning software packages on the market already offer far more sophisticated options and additional tools; the focus of our work was set on the preparation of suitable learning contents. They exist out of learning materials, tasks, sample data and step-by-step solutions. The material is conceived to prepare course participants to a defined common knowledge background in preparation to an up-coming presence module as well as to recapitulate and exercise lessons learned during the presence modules as a post-process.

Anyhow, because the U3-lmp has, from the technical point of view, not been significantly developed since implementation in 2001 – which is one year before the first Moodle version 1.0 was placed on the marked, in August 2002 (HOEKSEMA & KUHN 2008) – it's most likely to replace it in near future, probably as soon as at our university a university wide e-learning platform is widely accepted and used.

3 Programme Results

3.1 Outcome of Project Work

Within the last years many different projects were worked out. At best, the issue meets the interest of the participant:

- from the thematic point of view (f.i. because of the professional specification).
- from the technical point of view (f.i. because of the intention to learn specific methods and tools)

Table 1 contains a compilation that gives an idea about the thematic spectrum and the used methods as well as being used in the past courses.

Project title/Thematic issue	Used GIS methods and complementary components*					
	VD	RD	MG	WG	3D	DB
Analysis of a biotope connectivity system in for different habitat types	x	x				
Analysis of landscape structures to support habitat connectivity for moor lands	x	x				
Analysis and models to detect potential sites for wind power plants	x	x			x	
Analysis to indentify different thematic landscape scenery tours for touristm	x	x			x	
Development of an internet based information system for biking routes	x	x		x	x	
Development of an geo data service for the administration	x	x		x		
Development of a GIS supported field mapping method for a biotope register	x		x			
Development of a mobile touring guide for landscape exploration	x		x			
Development of a hiking information system for different user groups	x					x
Design of an information system for biotope management	x					x

Table 1: Overview of different projects and their thematic spectrum
(* VD= Vector Data GIS methods; RD= Raster Data GIS methods; MG= Mobile GIS components; WG= Web-GIS components; 3D=3D-visulisation technologies, DB= with integrated database interface)

3.2 Evaluation Results

During the years the courses a internal evaluation was done by the end of each course. Around Three aspects of this evaluation will be presented:

	Disagree ←――――→ Agree
a) The Contents improve my professional qualification as an environmental planner	
b) Contents could been transferred into professional practice during course already	
c) The Practice seminar is very useful and should, from the conceptual design, kept	
d) To reach learning goals the conditions at university would be adequate enough	
e) Mixed groups, students+professionals, contribute to make the course a success	

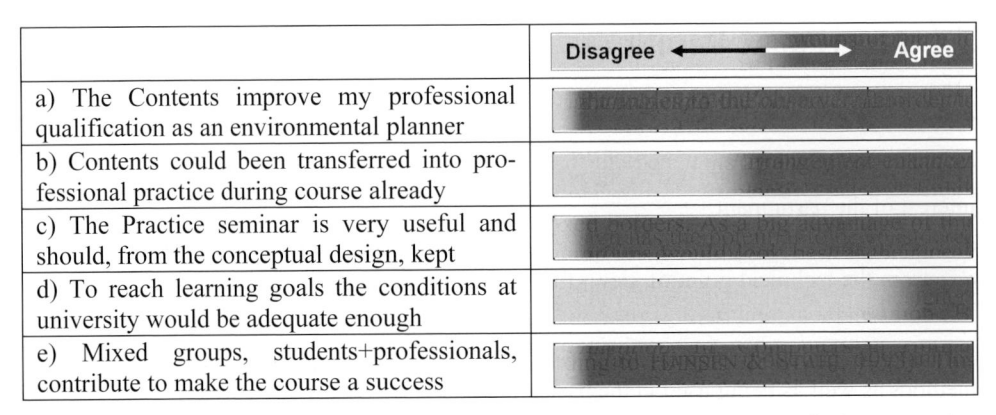

Fig. 2: Evaluation sheets tallied (2a-2d: 123 evaluated participants out of 10 courses; 2e: 69 evaluated participants out of 5 courses)

1) The practical relevance of the course contents

Because the programmes alignment is profession orientated and the target group of environmental planners is clearly defined, it is not very surprising that almost none denied that this course improves the professional qualification as an environmental planner (Fig. 2a). More interesting seems the feedback to the question if the learned topics could contribute to perform tasks in the daily business of profession, while the course is still running (Fig. 2b). More than half of the participants could transfer the lessons learned right away.

2) The methodological-didactic approach of the practical seminar – module 2

More than 90 % of the enumerated questionnaires agree that this module should be kept as it is (Fig. 2c). Furthermore about 70 % doubt that the same learning goals could have been reached with conditions at the university (Fig. 2d). Here the question focused on the situation of working together for one week in teams within a seminar building – away but on site – instead of a regular seminar situation on campus. Here participants often mention that the flexibility of the schedule for the working groups is one practical point to make the learning process productive. Besides that, the intense atmosphere and the option to compare the digital data models with the real world during a field trip also helps to get a better impression of data significance and validity.

Furthermore the working progress was evaluated during the seminars using a moodbarometer, which is supposed to document the atmosphere within the different working groups. Every participant was asked to mark two times a day (around noon and in the evening) his personal satisfaction with a point on a scale. In the context of the working group these marks are used by the supervisors as an indication about the progress within the different groups. A look at them show that it is fairly uncommon that there is a steady increasing satisfaction among group members. Its more or less an up and down, which is documenting success and achievement as well as setbacks. In any case, if at the end of the course the line shows up again it's most likely that satisfied project result and therefore learning goals have been achieved (Fig. 3).

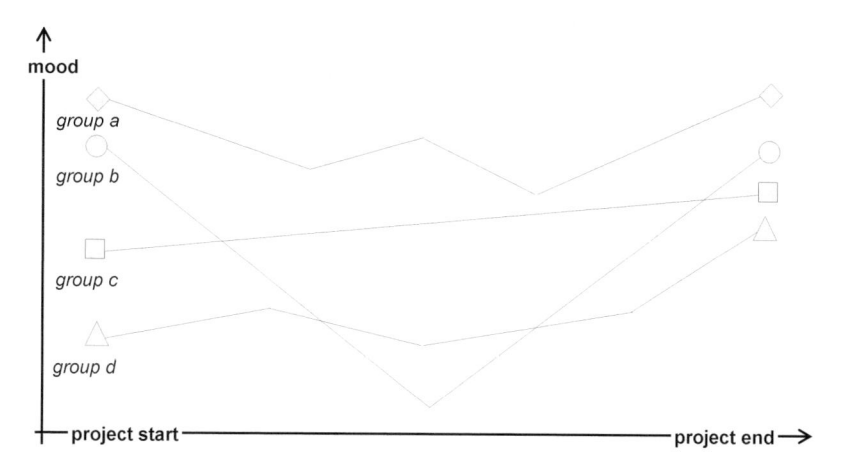

Fig. 3: Schematised figure of evaluated mood-barometers, documented during the practical seminar, may indicate the progress of the working group and achievement of the learning goals.

3) The effect of students and professionals working/learning together
Although just half of the courses could be evaluated from this aspect (see course development at 2.1), the majority of the participants clearly agrees to the idea of mixed groups and to the fact that quality of the course would benefit due to this (Fig. 2e). Students often mentioned to appreciate to learn from the structural approach of the professionals and also to get a better impression of practice relevance of GIS. Professionals on the other hand find the time with students not just refreshing for the atmosphere but also benefit from their experience and skills in the use of GIS.

4) The efficiency of the e-learning module
Although this aspect was never structurally evaluated by the questionnaire the results and outcomes as well as the debriefing of the online module allow some conclusions. Therefore professionals appreciate the option to recapitulate lessons from the presence modules and to take time to exercise regarding their personal needs. The contents were evaluated as suitable and effective particularly for post-processed recapitulation of lessons learned and for exercising.

On the other hand the discussions have shown, that it needs a lot of self-discipline and therefore online modules are sometimes neglected, in particular during work intense times as lots of participants are freelancers. Because of this we discussed with participants again and again if a more compulsory assignments would enforce attendance. But usually those ideas are rejected as being too school-like. One option to increase motivation during online modules is seen in more interactive face-to-face communication tools (like virtual team rooms) and periodic meetings.

5) Benefits to the educational quality for students of the landscape architecture faculty
Especially during the first years synergy effects between the programme development and the development of the faculties GIS curriculum took place and indeed still do. This is owed to the fact that the assistants of the faculties GIS laboratory were involved in the programme development and therefore could test new approaches and teaching methods.

Later, after GIS introduction was a integral part of the general studies the introduction module was not being considered as a course to teach students basic elements rather an option to deepen them. In the first view years this effect could been observed quite well. Besides the majority of the students may got introduction on GIS basics always a few used the offer to get familiar with more sophisticated GIS applications. But particularly since transition form diploma to bachelor programme this positive effect fails to appear. From personal discussion we know that this occurs because of inconvenient time frames and the tight studies timetable. Therefore the willingness to attend at courses outside classes seem to decrease significantly.

4 Conclusions and Outlook

Especially the central part of the programme of joint training, the so called "Practical Seminar", is well appreciated. But this is not just because we run a "real world project" but also because our class room for this week is on a remote place of the Swabian Alb, where we live and work together – very often until late in the evening. This "summer school like" situation – mainly in the winter season – communicates in an excellent way the various ways of learning, and often brings group members together, who have similar interests and knowledge backgrounds. This so called internal differentiation, combined with the general blended learning concept during the whole course, proved to be quite successful and effective.

10 years of experiences with e-learning methods again and again make us learn that even well evaluated and accepted learning materials are half worth of it, if learning motivation will not constantly raised. This is even more important for an in-service training programme. Therefore interactive tools, like virtual team rooms and regular meetings seem to be essential.

Regarding the students decreasing willingness to attend at courses outside classes due to new time conditions new stimuli needs to be thought about to increase interest again for more sophisticated GIS applications that go beyond the basic GIS curriculum of the faculties. Those can be seen for instance in new corporation forms between the programme and the faculty, i.e. if project and study area of the practical seminar relate to students semester project. Especially in combination with professionals working together at the practical seminars would be mutually beneficial.

References

Bett, K. & Wedekind, J. (2003), *Lernplattformen in der Praxis*. Münster, Waxmann, 248.
Blötz, U. (2008), *Planspiele in der beruflichen Bildung*. Bielefeld, Bertelsmann, 271.
Gudjons, H. (2007), *Frontalunterricht – neu entdeckt: Integration in offene Unterrichtsformen*. Stuttgart, UTB, 227.
Grünwald, S. (2008), *Learning Management Systeme im universitären Betrieb*. Lulu, Deutschland, 224.
Hoekseman, K. & Kuhn, M. (2008), *Unterrichten mit Moodle. Praktische Einführung in das E-Teaching*. 1. Aufl. München, Open Source Press, 229.

Klein-Landeck, M. (2004), *Differenzierung und Individualisierung beim offenen Arbeiten. Beispiel: Englischunterricht.* In: Pädagogik, 56 (12), 30-33.

Lenz, R. J. M. (2000), *Project Overview European-Canadian-Curriculum on Environmental Informatics (ECCEI).* Proceedings International Transdisciplinary Conference, Zürich.

Lenz, R. (2004), *The IMLA study program: how to strengthen methodology in Landscape Architecture.* In: A Critical Light on Landscape Architecture, Proceedings ECLAS Conference 2004, 16. – 19.09.2004, As, Norwegen.

Lenz, R. & Rolf, W. (2003), *U3 – Umweltinformatikunterricht für Umweltplaner – oder: Lernen mit allen Sinnen.* In: Studienkommission für Hochschuldidaktik an Fachhochschulen in Baden-Württemberg (Hrsg.), Beiträge zum 5. Tag der Lehre, Fachhochschule Nürtingen. Karlsruhe, 174-176

Rolf, W. & Lenz, R. (2005), *U3-Umweltinformatik-Unterricht für Umweltplaner – Ein Fort- und Weiterbildungskonzept an der Hochschule für Wirtschaft und Umwelt Nürtingen-Geislingen.* In: GeoForschungsZentrum Potsdam (Hrsg.), Innovationen in der Aus- und Weiterbildung mit GIS, 2./3. Juni 2005 in Potsdam, Tagungsband, CD.

Sauter A. M. & Sauter, W. (2002), *Blended Learning. Effiziente Integration von E-Learning und Präsenztraining.* Neuwied, Luchterhand, 344.

Schittko, K. (1991), *Differenzierung in Schule und Unterricht. Ziele – Konzepte – Beispiele.* München, Ehrenwirth, 202.

Schmidt, I. (2005), *Blended E-Learning: Strategie, Konzeption, Praxis.* Diploma Thesis. HS Bonn-Rhein-Sieg. Publ. Examicus, 105.

Wiepke, C. (2006), *Computergestützte Lernkonzepte und deren Evaluation in der Weiterbildung - Blended Learning zur Förderung von Gender Mainstreaming.* Studien zur Erwachsenenbildung, 23. Hamburg, Kovac, 342.

Nature by Culture – From „Romantic" to „Organic"

Rainer SCHMIDT

Keynote DLA Conference

1 Introduction

The confrontation of „nature" and „culture" within the title provokes a professional and personal consideration of the values of nature and culture. It seizes the theses that "nature" might be to be made or renewed by "culture". The attribute "from romantic to organic" provides an outlook onto the enfolding of a position about philosophy and history, which overcomes the history of the human desire for a counterbalance of burdens through "nature" since the Romantic period and which finds a contemporary approach through the making of open spaces by means of design. This approach is called "organic". A design method "organic" is herewith introduced. The name indicates its origins on organic forms as well as its generation by form-giving. Open spaces build the context as a resource for the establishment and the transformation of "nature".

We know and we feel that our highly technological and dynamic societies with daily increasing demands for the use of energy exhaust any kind of "nature", i.e. what has been left from it after many centuries of cultivating. We know that this "nature" needs increasingly to be strengthened, preventatively and also counterbalancing the already caused damages, in contrast to the more and more intensive attacks of "civilization" – as a basis for the survival of all of us.

We guess that – only by means of a new culture – we can recognize, feel and enrich the old conditions of "nature" in a way which allows that we protect relics of "nature" for our survival and, furthermore, can initiate new perspectives for "nature" in the social space of our divided everyday lives by means of collective action. Which leading ideas may help, and what this might mean today and practically for landscape architecture and for the role of design … we want to ask.

2 Responsiveness towards Sensual Perception

The task of landscape design seizes the understanding of the framing conditions of „nature" in society and in the economy and looks, building up on this, for an answer to those people for whom we form landscape and "nature" in rural and urban spaces for the everyday life of living and working.

Open spaces are resources for designing landscape and "nature". They offer places, which may become cells of gaining value and which can be connected as systems of "nature". They are – in place and time – the source for settling, the medium for action in society, in the economy and in nature and are, consequently, the expression of what might characterize the production of spaces by design: Quality in use, image and concept.

Long- and short-term action of those ones who produce spaces, searches for answers to regional and local conditions for a user-friendly and pictorial organisation of space-building elements, – an organisation which can create atmosphere, an organisation which can – more than fulfilling function – activate all the senses and can make spaces become places of experiencing meetings or niches of retracing from everyday life – an organisation which provides spatially precisely appropriate offers for new perspectives in perceiving and acting towards the ways of producing space and life in place and region.

Human perception is the exchange between the external nature of the physical world and its rules about renewing working and living and the internal nature of a person. The senses of human beings complete each other concerning the perception of conditions and signs. They lead with a clearer orientation than any reflection towards feeling well in or refusing spaces. The responsibility in designing spaces towards sensual perception is, accordingly, comprehensive and far reaching fort the tasks of conceptualizing landscape fort the benefit of renewing „nature".

Environments of everyday life, mostly distant to "nature", dominate in appearing to mankind as a hardly over-lookable layering of different elements, be it material, be it sound, smell or taste, competing with each other for attention. In order to prevent any resulting irritation of the senses, in order to counterbalance and to avoid disturbances, and in order to orient the human perception constructively onto positive effects, comprehensive concepts for designing open spaces are needed.

Those ones can define the opportunities to perceive space and its conditions in a new way by a theme of design and by creating atmosphere through structure, texture and materiality, and can make "nature" in a new way accessible by artificial means.

Time makes human action readable in space. Culture becomes evident in the reflections of human influences on structure, texture and material of space. Open spaces are the "business card of a town or region". They are the proof of a culture which determines space, uses it and feels it – physically, mentally and socio-economically. Valid is: "Space is a socio-technological biotope of culture".

Thus, use, image and concept of open spaces are – in their perception by the human senses – always results of collective or individual adaptation to forms and materials of spatial offers, i.e. to the framework conditions for perception and action.

The wider the offers are defined, the more space is provided for processes of adaptation. "Nature" can be created for this artificially, can be reminded in its meaning or can be made accessible by new means in reality.

Action is carried out in relationship to the offers within the reach of framework conditions, by realizing the offers and incorporating them intuitively and rationally as options for action. Intuition plays a big role herewith and determines finally the orientation in behaviour and their patterns in space, concerning offers and restrictions.

Offers cause encouragement, are strengthening, support the harmony of mankind with themselves and with the environment. They support also and above all the active searching of people for being unified with nature. They support, this way, "nature" in mankind and also in the material surroundings. This happens on public and on private ground. Action serves for enfolding different private and public interests: concerning the use of open spaces as a resource for "nature" in the short-term perspective as well as the renewal of space and "nature" in open spaces through cultures of the production of spaces and "nature", in the long-term perspective:

Offers for the human perception of space do always meet short-term activities in space as well as long-term effecting elements of „nature" in space. These ones are. Individually as well as in their working together an expression of historically meaningful identities, bound to

- Geographical origins: "Heimat"/family
- Territory: Area within administrative boundaries
- Ways of production: pre-industrial, industrial, late industrial
- Ways of communication: Dialogue societies – to date[1].

A responsibility towards human sensual perception for the production of spaces can be called to be fulfilled, if the intervention by design allows the expression of these continuously emerging and each other overlaying identities in a way that the over-layerings in space may always find new expression through action.

Open space is also considered as being subordinate to the principle of responsibility towards the human senses, and is furthermore, independently from the options for the expression of identity in time and space, artificially generated or renewed for the experience of "nature".

Open space is artificially maintained open towards the rules of land markets and built as the social ground for different realities of the economy which are supposed to become enfolded innovatively.

This is referring to the range of social and economic realities from individual entrepreneurs towards the public welfare in neighbourhoods, communities, regions …

This orientation towards economic interests relates in its range equally to the addressees for the production of space for perceiving „nature" through culture. Production and reproduction find their cultural synthesis specifically in designing open spaces, as the space which has been maintained open reflects the cultural understanding of "resource" in the respective society and economy.

3   Role of Design: „Control"? … Through Offers for Perception

Since the culture of the public spaces of late medieval times we are, relative to the different period and direction of design differently, used to experience the open space, specifically the "urban public" space and its design as an expression of "control".

This refers to the public spaces of the „Founder-time", to those of the time after WWI, and – in a specifically restricting way of course also to the space of the Third German Regime.

---

[1] This distinction refers to a lecture of Patrick Schumacher at Bauhaus Foundation, Dessau, in the beginning of the 2000ies.

The tasks of designing open spaces served – until at about the time of the break after the industrial crisis (1985 in West-Germany/1995 in East-Germany) – only within this framework. i.e. by means of subdividing areas and accessibilities, for offers of sensual perception and for their expression by action.

This was accepted as long as the industrial ways of production needed the public space as "controlled space" on one side, and – on the other side also allowed, as families and neighbourhoods still provided for sufficient differentiation in different spaces. This is different today. We are living in a society of individualization, not at least because of new ways of production and their technologies. The open and public space is immediately confronted to the individual privacies. Late industrial ways of production and technologies cause, more intensively than the industrial ways of production, long-term irritation through masses of visual/ audial impacts on the human senses. They increase the individualization and generate equally demands for its dissolution, for endurance and continuity of healthy living conditions in/ with „nature". The desire of mankind for counterbalancing disturbances through experiencing nature corresponds in principle to the desire during the period of "Romantic" , however, is much more intensively and in a multi-layered way burdened by disturbances of the senses. Landscape architecture provides, under these conditions, for design the role, to oblige different environments with their dimensions in time and space to "nature", and equally to mediate towards mankind new perspectives of perception, of use and – tackling space – of offers for the experience of "nature". This mediation takes place by bringing together philosophy, arts, technology, planting knowledge, knowledge about materiality. – fields of knowledge and of making, which only by working together allow for a comprehensive landscape design with varied approaches and good results for different topographies.

We know: The more efficient a framework for action in open space has been established for feeling and using the open space, its elements of "nature" and the effects of "nature" in people, the less this framework causes restrictions, the more the "control" by design speaks to the senses, affects innovatively by enriching, calming, counterbalancing, orienting, finally being productive... This insight makes us question planning in its well known sense of guiding processes by restricting perception and action in principle and, above all, fort the aims of designing open landscape and "nature". Consequently, non-planning is the idea, to support perception and action in space by determining the „columns for freedom" in space spatially, socially and economically by appropriate concepts and their appropriate implementation in everyday life. The appropriateness for implementation and their materiality build the proof fort the quality of the concepts. Means of „control" by design are also and especially existing in informally established spaces (s. informal settlements from different periods with rich gardens, including fruit and vegetable cultivation, in urban areas in Istanbul). We take these ones for illustrating the widely defined frameworks of "control" through "non-planning". Herewith, control" is understood as a framework of socio-aesthetical and material qualities for perception and for individual/ collective action, which in their orientation and frequency reflect the spirit of time and the meanings given by society to place. Today dominate, in consequence of ongoing private and professional situations of individualization, needs for meeting others as well as for a dialogue with oneself through contemplation and through incorporation as well as through taking a deep breath in the open spaces and in "nature". The active relationship with „nature" experiences also and specifically in urban conditions a renaissance of the demand, mainly where the forces of "civilization" become evident most intensively, where economic and social conditions look in most intensively for a counterbalance of personal burdens.

There, garden- work and related recreation of the forces of body, mind and senses turn very suddenly and intensively into the centre of needs for perceiving space. These needs again mark the demands on the land-markets. Residences with immediate access to open space are traditionally the „runners" of demands. Have these demands in times after WWII in Germany, East like West, mainly been related to the single family houses in one side and on the allotments of the big settlements on the other, thus, you can find these demands today in modified forms in all urban locations.

Roof terraces, terraces and balconies are expression of the search for contact with "nature" on available areas; "urban agriculture" is an idiom of our time which even puts the growing of food into the context of "nature" by culture in highly densified spaces. The open spaces reflect the remaining resources of "nature" in spaces of densification and of demands for nature there, on private and on public ground. The social awareness and the emotions of people for values and resources of (re-)production come, this way, back to being closely and productively in touch with establishing families of plants and caring for them, for the soil, for its appropriateness and its materiality for uses and its compatibility for rain water retention and energy production. Landscape and open spaces can mediate multi-facet images for these values. The affect three-dimensionally through structures, textures, materials ... and build sentences of a space-become language – which is to be read, to be spoken and to be danced. What is essential? Space – to be felt and to be used. Qualitatively high-value means and concepts dissolve „control" accordingly completely as a restriction, and initialize – instead – experience, contemplation, movement, meeting …This impulse is by means of designing the open spaces, via the perception of messages and towards their implementation, in a multi-layer way, sent to the senses and in a multiple way reflected there. Feeling of happiness and fulfilling can ideally be results. This would relate to the old wisdom that people are happy when they can harvest, – when they can carry home at night or in the end of the week or at specific times the fruits of their work, i.e. when care and economic use or cultivation of their worked-on "fields" show materially fruit …

4 From „Romantic" to „Organic"

The desire of mankind, that the forces of „nature" might counterbalance and heal the socio-economic and technological progress by culture, has so far been strongly expressed by the philosophies of the "Romantic". This desire has generated during the period "Romantic" the imagery and the establishment of landscape parks. Such a desire has the social and economic power, to generate in principle new design approaches. Since Vitruv, and in the New Age since the Enlightenment, there has been a comprehensive philosophical approach from the "making professions" to envisage the human scale…

The „Romantic" has become since a philosophy which had the practical aim to re-establish conditions for sensual experiences, establishing in a new way through setting-into-scene by artificial means "nature" within the remaining nature. The Modern Movement continues the theme "nature" within its three phases beyond new premises. Nature was reconsidered and determined, relative to the orientation of arts, structure and/ or ornament.

Roots of Modern Movement (from 1800-1910): Structure and ornament were inseparably linked by the structure. Nature was the theme if units in materiality and space-building. Built form was looked at as part of "nature".

Flourishing of Modern Movement (from 1918-1930): According to the complexity of the emerging technologies and the serial production, answers arose, which the spatial organisation of conditions separated functionally. Structure and ornament became opposites in the game of forces about power, which has been – in the basic ideas about the blood-and soil-ideology from the 1920ies – still comprehensively conceptualized. Styles arose, f.e. "youth-style", "Heimat-style", which expressed the separation of functions.

Re-establishing the Modern Movement (1950-1973/ 1973-to date): This period continues. It was, with influences like New sculpturalism/ Post-Romantic-Urbanism, still for a long time in the field of tension under the functional requests of mass production. The structure was freed again for incorporating the ornament, but the structures were differently sculptural, geometrically abstract or metaphor-like. A synthesis of structure and ornament became again mature with structural offers for perception and action.

A direction of science, called „Bionic", founded the application of „structural" patterns in architecture, landscape design, street art. It led during the 1990ies by means of computer-aided design-methods (Parametric design) to a new synthesis between ornament and structure in building space. The functionality of the Flourishing of Modern Movement was replaced by contents, like telling stories, and their values for opening up through design new angels of view. Patterns of communication and structures of building space, three-dimensionally building up on these patterns, set the columns of concepts. They are exposed in materiality and planting and are made perceivable as "living environments of nature". This method fulfils – with individual compositions – the phenomena of architecture as auto-poietic systems. It prepares for action, processes, communication by open frameworks for occupying spaces structurally.

Organic forms became only in the period after the industrial crisis and its break (in West-Germany around 1985) new „racing horses" of comprehensive approaches of the "making professions". Structure and ornament gained impulses from organic forms and their artificial materiality in architecture and building space. The relationship with "nature" was the new search for findings about the stream of forces in the structures of nature (flora, fauna, topography). This went along with applying the found patterns in architecture, landscape design, street arts and building space and led there during the 1990ies the conflict of the Modern Movement towards a new synthesis („The ornament is the structure"). In my work, a new design-method emerged: „Organic".

The „organic" has been evoked as an updated meaning and materiality of the desire for being close to nature during the „Romantic" period, – counterbalancing the forces of re-(production) according to its time –, by holistic concepts of design, which can be experienced as relief from the burdens of living and working. The „organic" plays, this way, a role for creating structural offers for perception which make „ratio" merge with „intuitio".

This position comes from architecture and is, in its time, not totally uncritically introduced: „There is a global convergence in recent avant-garde architecture that justifies a new style Parametricism. ... It is a style rooted in digital animation techniques, its latest refinement based on advanced parametric design systems and scripting methods. ... Parametricism finally brings to an end the transitional phase of uncertainty engendered by the crisis of Modernism and marked by a series of episodes like Post-modernism, Minimalism, Deconstructivism ..."[2]

This statement indicates that the quality f applying methods and their implementation determine whether a new episode and a new style might be replaced potentially by real qualities of design. Style is then positively understood as direction. Zaha Hadid comments on the tasks of producing qualities in the space of different topographies: "I have always believed that a formal repertoire is critical in urbanism. I am particularly interested in shaping the ground plane by carving, imploding and exploding; not just as a formal gesture, but as a way of dealing with the complexity of the programme – the social component in architecture. We have to go back to the ground, study it, learn how to programme it as an event space. It is not just a formal issue but a programmatic one. Form and programme cannot be separated from each other; topography brings them together." ...[3]

5 Parametricism and nature

Let us summarize the central statement/question for the method: The open and public space is immediately confronted to the individual privacies. The design of open spaces has to integrate under these conditions the uses of individual people, to incorporate them into different dimensions of space and time and equally to oblige them to nature in a new way. Aims of the "organic" are to mediate new perspectives of perception for experiencing "nature" to people and, thus, to establish "nature" in a new way. For Landscape design remains the question: How is the style of design Parametricism, to be applied conceptually for supporting "nature"?

The design method „organic" finds – as an answer to the question – impulses in architecture concerning the three dimensions of the essential patterns for structures, for building cells and connections, for implementing concepts for the use and the image of open spaces – this is based on the understanding of "landscape urbanism" (s. a. Charles Waldheim).

Nature is often misunderstood and inadequately simplified as „enduring", differently to contemporary ways of living, to communicating, to realize changes of structure, texture, materiality ... The new design approaches "organic" apply organic forms for elements of structure and texture. They enrich, this way, functional spaces with meaning and equally with the structural and textural materiality of the "in-scripted nature" of environments (people/ their movement in spaces, flora and fauna).

In order not only to reflect but also to renew „nature" , design needs to know the rules of "nature" as far as being able to make sure that, by "in-scripting methods" of the software of

---

[2] Schumacher, P. (2009), *Parametricism. A new global style for Architecture and Urban Design*. In: Zaha Hadid Architects, ILEK Institut für Leichtbau und Entwerfen Universität Stuttgart, 18-23, John Wiley + Sons.

[3] Schumacher, P. (2009), *Parametricism. A new global style for Architecture and Urban Design*. In: Zaha Hadid Architects, ILEK Institut für Leichtbau und Entwerfen Universität Stuttgart, 18-23, John Wiley + Sons.

information technologies, the design of vernacular transformation and equally for global progress can be seized in an appropriate way, calculated and applied as parameter.

Herewith, the ethnical differences of people who experience and understand „nature" are to be considered. We distinguish occidental and oriental positions relative to the understanding of "nature" [4]

5.1 Nature is changeless

a. Nature as a snap-shot of an organism with vitalistic functions and teleological purpose
b. Nature as a snap-shot of a machine with mechanistic and non-teleological purpose.

5.2 Nature as a constantly evolving whole

a. Nature reflects Western philosophical and cosmological thinking in science
b. Nature fulfils some form of teleological design – as a universe evolving between a finite beginning and a finite end.

5.3 Nature is god-given

Synthesizing natural and supra-natural realities and distinguishing the different fields of influence on nature relative to the understanding "God is nature, nature is god" finds:

- Unities of creation, of humanity, of truth and knowledge, of life;
- Spiritual worlds (roh, arab.), physical worlds (dunyia, arab.);
- Intermediate worlds (barzakh, arab.), after worlds (akhirat, arab.).

This understanding is as comprehensive as to frame any holistic approach to the Organic!

Valuating the approaches to the understanding of nature means to take the oriental approach as a starting point for the widely defined approach of „control" in designing open spaces and for supporting „nature" by culture: It is about, to give space to people and cultures for bringing together different meanings and realities of "nature" in space, physically and spiritually. This needs the holistic appliance of Parametric tools, in order to lead starting points for this, f.i. vernacular ornaments of locally and regionally specific topographies, towards open frameworks for use and image of space.

6 „Organic" as Method

How is „nature" by means of culture to be advanced to a new culture? The answer on this question is based on many levels of professional experience, of the generation of creativity and of the management of resources in the offices. The practical results are always witnesses of all these levels and their working together. Essential, however, is: Without qualified resources of labour and without sufficient financial means for implementing the concepts, the path towards the result is difficult. These difficult paths, however, are the

---

[4] Loo, S. P. (2007), *School of Educational Studies, The two cultures of Science: On Language – Culture Incommensurability concerning ‚nature' and ‚observation'*. In: Higher Educational Policy, 20, 97-116, Palgrave Journals.

ground on which many tasks for projects operate. The temporary positioning about the location is consequently closely linked with the temporary position to the own working situation and to the framework for performance of the client, to the cooperation with colleagues and employees and – last not least – to the own anchoring in private life, family and friendship. And: both the sides have a long-term effect. The real quality of the establishment and of the daily usability of created gardens and landscapes are, thus, the expression of position towards work, implementing the idea in the context of working and living and finally the implementation of the idea in form of structure, texture and material.

The success of the own projects starts with the classical appliance of the rules of proportionality and materiality of spaces to be felt and used – above all by the three-dimensionality of the set effects of hedges, shrubs, trees and completing artificial elements in modelling topographies in garden and landscape and their view connections in near and far perspectives: The "Romantic" was the biggest challenge for this. Because the movement towards the experience of "nature" in space was contemporarily the pre-step for the development, to synthesize "nature" with the tools of forms and colours of the Modern Movement.

„Organic" leads together the heritages of „Romantic" in a new way. Informal patterns of natural growth are made transparent for informal processes of the space-building occupancy of spaces.

Fig. 1: Theme Gardens, Parkstadt-Schwabing, Munich

The heritages of „Romantic" are f.i. reminded by setting into scene regionally typical landscape. In the Park-city Schwabing, Munich, this happens in Form of a strictly arranged park, established with view connection from the framing buildings towards the Alps, and enriched by "organic" niches (theme gardens) beside "pavillons" (s. Fig. 1): The planting creates the spatial effect which invites rot he individual/ collective occupancy of the spaces within the „pavillons". "Organic" uses here the form-giving language of contrasting and builds sculptures by means of structures and planting. Large-spatial theme gardens are object of „Organic" at the BUGA 2005, Munich. Places are interwoven with the network of paths as "cells" of an over-dimensional horizon of experiencing "nature" and become sensually perceivable by staying in the "cells" (s. Fig. 2).

Fig. 2: Theme Gardens, BUGA 2005, Munich

Also small-spatial parks, like the garden of the Leonardo Hotel, Munich, are taken by „organic" to new dimensions of „nature". The arrangement of space-building planting leads via paths towards places for staying.

Built form arrangement becomes a self-standing place in a suburban location. The arrangement and the movement of grasses in wind tell stories about the origins of this place and its offers for contemplation (s. Fig. 3). "Organic" generates implantations by means of materiality and planting within the existing "nature". The user is demanded by staying in space to be sensually active. The design has the maturity of arts and equally allows openness for the experiment of perception. The forms become an invitation to dive into a new world of use, image and concept – as deeply as the third dimension of the open spaces is perceivable.

Fig. 3: Park area, Hotel Leonardo, Munich

Fig. 4: Killesberg-Park, Stuttgart

This relates in specific ways to the park „Killesberg", Stuttgart. There, by means of elevating the surface of the area and crossing paths in there, the visitors are left with the fact that they are only visible between the artificial hills. They appear to be smaller in the new landscape, as only their upper bodies can be seen from path to path (s. Fig. 4).

The widely arranged new creation of organic forms for city and landscape expresses „organic" in Arabic and Asian countries, thus, with the masterplanning Hangzhou, where the landscape builds spaces (s. Fig. 5). The interchange between materiality and perception replaces the open space as "object" of the restricting design by the open space as „medium" of a challenging design for contemplation, dialogues, settings into scene.

Fig. 5: Masterplanning, Hangzhou

Main contributions to arts and space building consist in total of
- Sculpturing of locations within the specific context of tpography and climate, i.e. enriching "nature" by nivelling, levelling, contouring (f.i. Killesberg, Stuttgart)
- Over-dimensioning of symbols of „nature" as „cells" of contemplation within the context of interlinked paths and fields for exploring new experiences of "nature" (f.i. BUGA, Munich)
- Contrasting of built form and open spaces for expending the variety of experiences within the limited space of a plot (Hotel Leonardo, Munich)
- Complementing heights by depths in the arrangement of high rise towers over a surface of a water pool serving for energy production (Gateway City RAK)
- Pointing of the locations of a building in a park (Garden of the Villa Kranz, Munich, Park St. Gilgen Österreich …)

In all these works, organic forms as artificial implantations have gained importance for the new experience of "nature" by means of building space, lighting, planting, etc. ... "Control" by design goes along with the widely set offer for perception and occupancy of newly created spaces.

„Control" does not contradict to the organic structures, but these ones become, in the abstraction of artificial elements, an invitation to dive into a new world of use, image and equally clearly recognizable concepts – as deeply as the third dimension of the open spaces is perceivable.

7 Conclusion

Most of the content of this lecture has been taken from the very many lectures and publications of the author. The theory of Jacques Ranciere about "The division of the sensual. Politics of arts and their paradoxies" was considered, discussed and carefully put in relationship to the creative work of the author.

The result of the consideration is the position:

Open spaces are of high quality, when they

- contribute to the spirit of time in place counterbalancing the effects of technological progress by means of therefore appropriate concepts (from "Romantic" to "Organic" – by culture)
- are daily used by people who value highly the range of options for movement in space and for the perception of space for action,
- carry responsiveness towards the purpose to enjoy a healthy living (within the framework of "control" – for the benefit of "nature").

Sources of literature

Haase, A. (2004), *Kultur der Anlage und Nutzung von Raum*. In: Dreyer, C., Führ, E. & Hauser, S. (Hrsg.), *Baukultur*. Stadtkultur, Lebenskultur, 8. Jg, Heft 2, März, BTU, Cottbus.

Lefebvre, H. (1991), *The Production of Space*. Oxford, Blackwell.

Loo, S. P. (2007), *School of Educational Studies, The two cultures of Science: On Language – Culture Incommensurability concerning ‚nature' and ‚observation'*. In: Higher Educational Policy, 20, 97-116, Palgrave Journals.

Ranciere, J. (2006), *Die Aufteilung des Sinnlichen. Die Politik der Kunst und ihre Paradoxien*. Berlin, PoLYpeN.

Schumacher, P. T. (2009), *Parametricism. A new global style for Architecture and Urban Design*. In: Zaha Hadid Architects, ILEK Institut für Leichtbau und Entwerfen Universität Stuttgart, p. 18-23, John Wiley + Sons.

Schumacher, P. T. (2011), *The Autopoiesis of Architecture, A new framework for architecture*. Vol. 1. Wiley.

Waldheim, C. (2006), *The Landscape Urbanism Reader*. Princeton Architectural Press.

LandscapingSMART

Peter PETSCHEK

Keynote: 27 May 2011

1 Introduction

In 2010, most Landscape Architecture offices were using standard graphic oriented computer programs. Presentation graphics are a major concern among professionals. Companies even require students who are looking for an internship to have a good understanding of the software [1]. This development was confirmed by a questionnaire prepared as part of a master thesis last year. The student asked what software Landscape Architecture offices are using. Twenty seven out of one hundred and nine offices answered and named standard products like AutCAD, Vectorworks, SketchUp, ArcGIS, Photoshop and InDesign [8]. Looking at today's education scene in Landscape Architecture one can also state that students put a lot of emphasis on computer plan graphics as part of their Information Technologies (IT) education. This interest of professionals and students in presentation plan graphics is quite understandable as design studio work with a lot of credit points, and competitions with the chance to build are evaluated based on paper plans. But it is necessary to remind students and colleagues that IT in Landscape Architecture deals not only with the production of good looking plans. Therefore the term landscapingSMART has been introduced. It describes efficient ways to use IT-tools in the design and the construction process of Landscape Architecture. LandscapingSMART is an overview and as long as IT tools and software are further developed it never will be complete. The process consists of four main parts:

Data Acquisition
Terrain data and aerial photos today are provided by government agencies or private firms. If data on site is needed, either local surveyors or Landscape Architects use Tachymeter/GPS technology to build digital terrain models (DTM) out of existing terrain and locate existing vegetation.

Data Output
Digital terrain models are transferred via USB stick to machine guidance systems. No stake out plans and measurement on site are necessary. Today construction companies demand digital terrain data as the usage of machine guidance systems becomes more and more popular and saves money.

Data Manipulation
Based on DTM data and via flat bed cutter or CNC milling systems, analogue models can be built. The tradition of model building comes from the field of architecture where it is common to use models in order to test ideas. Models are also very popular among Landscape Architects. Analogue Models demonstrate definitive advantages when it comes to free forms like terrain. The haptic, intuitive aspect using model building sand for the form finding process cannot be simulated with a computer program. Nevertheless in the

further design and construction process, digital data are necessary for volume calculations and machine guidance. How can the transfer from analogue to digital take place?

Often colleagues in professional practice think that DTMs are only necessary for large reclamation and grading projects like golf courses. The opposite is true. Digital terrain modelling should be used for every site design project as grading and above/below ground drainage is more precise using the DTM technology. In addition the DTM data can be used directly by a dozer or digger.

Data Presentation
Successful planning needs public acceptance. Recent discussions of large infrastructure projects in Europe proof the importance of informing the people. Plans are not very accessible for everyday person. They use abbreviations and symbols which not everybody understands. Usually plan exhibitions of projects with public interest take place in city halls which people do not visit in their spare time. Analog models, perfect for design development, also demand a high level of abstraction by the viewer. Convincing project communication is extremely important. The well known Sender Receiver Model is most critical in the communication process. Decoding of information differs between a professional and a layperson not being used to read plans. The recipient of the information has to be taken more into consideration. In brief when at home most kids play with high end video game consoles which have best 3D graphic quality then it is also for the parents no longer enough to look at plans as part of a PowerPoint presentation or at the wall in a city hall [6].

2 Data Manipulation – CRP Fills a Gap

Photogrammetry is a process of deriving metric information of an object via measurement on photographs of this object [4]. The photogrammetric technique involves reproducing the trajectory of a ray running from the lens of a camera to a point. The position of the point in 3-D space is obtained by a crossing ray running to the same point but from another camera station. The principle of photogrammetry in its broadest sense lies in the usage of stereo pairs of images to convert flat 2-dimensional images into 3-dimensional models.

Aerial photogrammetry or sometimes called airborne photogrammetry, typically refers to oblique or vertical images acquired from distances that are greater than 300m [3]. Aerial photogrammetry delivers the most accurate and reliable earth observation data.

On the other side, close-range photogrammetry (CRP), in which an object-to-camera distance of less than 300m is involved. The ability of CRP to produce real-time measurement and to use sequences of images has led to many new applications [4]. Presently they include precise measurement of buildings for architectural historic preservation, surveillance of nature dynamics and biometric applications. Comparing close range photogrammetry and close range laser scanning CLR has the major advantages of fas. production, low purchase and operating costs [5].

As previously stated physical models play an important role in the landscape architectural digital design process. The big question is how to transfer the analogue data back to the digital world. CRP is one the way to handle the transfer. Taking PhotoModeler Scanner of EOS Systems Inc. as an example the workflow consists of: camera calibration, data import, image processing and data export. With a calibrated camera (simple automated process) the data creation takes no longer than an hour. A point cloud created with CRP can be post-processed with regular DTM software. The new digital terrain is ready for volume calculations, site analysis (slopes, water run-off, exposition, etc.) profiles and finally a machine guidance system on site. Therefore CRP fills an important gap in the process chain of landscapingSMART and transforms an analogue model, modified by using model sand into a digital terrain model.

3 Data Presentation – 3D PDF for the Public

High Tech Graphic Cards are the standard in all personal computers. The graphic card calculates processes on hardware basis, which otherwise have to be done by the software or the CPU. With the help of these cards real time is possible. Real time means 25 images per second are shown on the screen. The very tedious rendering time of animations is no longer necessary. Not many years ago real time visualizations were only possible on high end computer systems. The driving force behind the development of fast graphics is the computer game industry and the huge consumer market asking for speed and realism. Real time applications outside the game sector profit from this development. Also Landscape Architecture can use real time. The goal of real time planning is to make a project understandable for all participants of a planning process via a 3D context. The acceptance of interactive models is high. Studies proof that people are asking for it [6]. Different methods exist: Digital Globe Systems, software applications like Quest3D and 3D PDF[9].

An interesting tool for real time based 3D plan communication is the PDF format of Adobe Systems. The Portable Document Format (PDF) was developed by Adobe Systems in order to view documents with text, images and 2D vector grafics independent of hardware, operating system and software. Now 3D visualizations can be also integrated in standard PDFs. Users can rotate and zoom in to reveal hidden detail. Not like in a computer game environment, where the viewer is positioned inside the model, in a 3D PDF you look at the model from outside. One either has to navigate to a position inside or one is guided in the model to predefined locations via buttons.

Based on a previous research project, which looked into the demands different user groups like computer kids, parents, planners using a real time environment, certain adaptation to the PDF platform were made [7]. The two most important are: simplicity (button amount, size and design), ease of orientation (button for reset back to the original position).

4 Conclusion

LandscapingSMART is not a revolutionary new process. It just tries to shift the focus to the overall process of using IT in Landscape Architecture. More and more technology will be developed. Ubicomp (Ubiquitous Computing) is more and more influencing our daily life. It is an interdisciplinary field of research and development that utilizes and integrates pervasive, wireless, embedded, wearable and/or mobile technologies to bridge the gaps between the digital and physical worlds. Not all makes sense for landscaping, but a critical / positive attitude has to be developed among students. This is the task of IT education in Landscape Architecture. It is more than digital plan graphics.

Fig. 1: Data Output with a GPS Dozer. Machine control systems can be efficiently used also in landscape construction projects

Fig. 2: The haptic, intuitive aspect using model building sand for the form finding process in Landscape Architecture

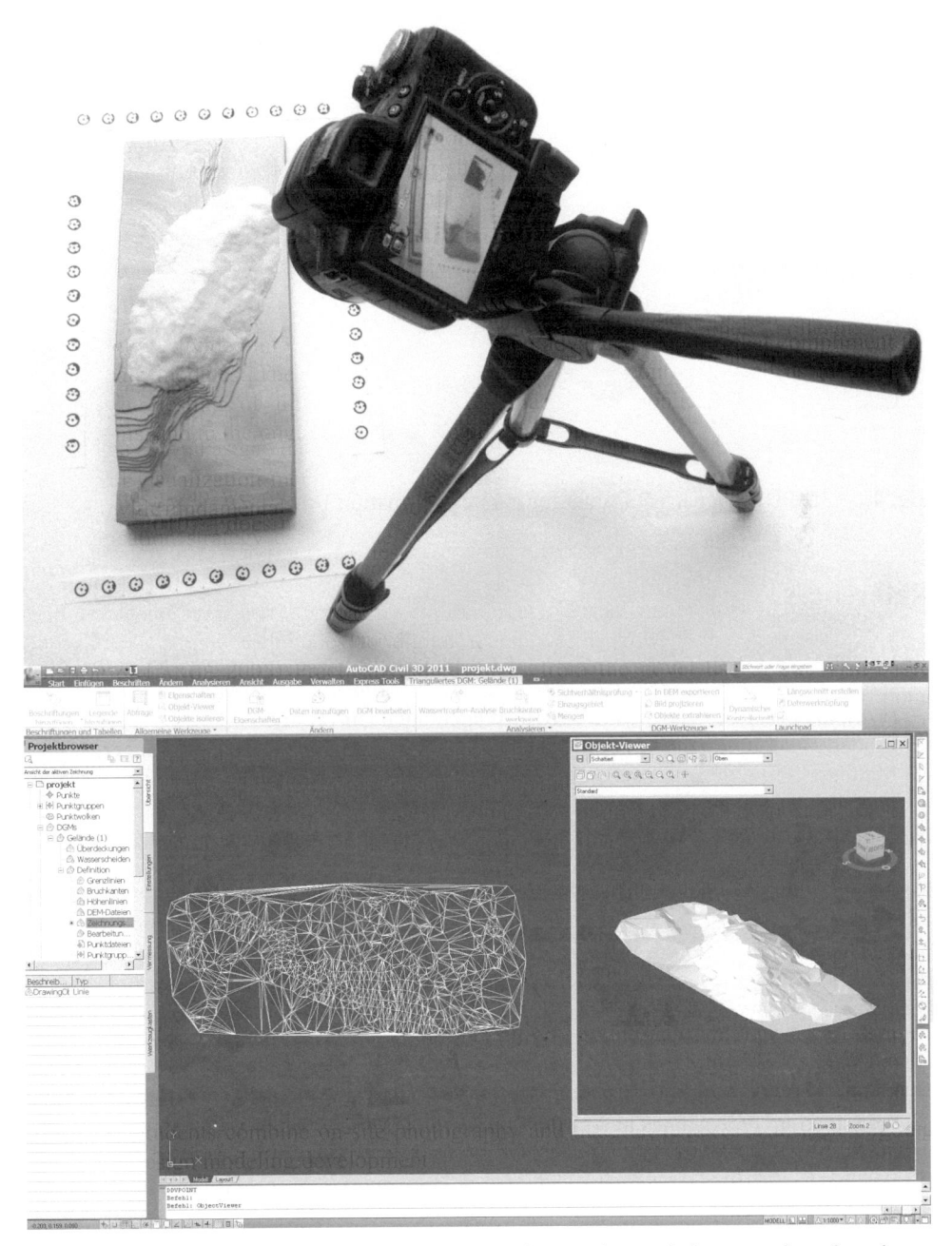

Fig. 3: Close Range Photogrammetry allows the precise and fast transfer of analogue model data to digital terrain models

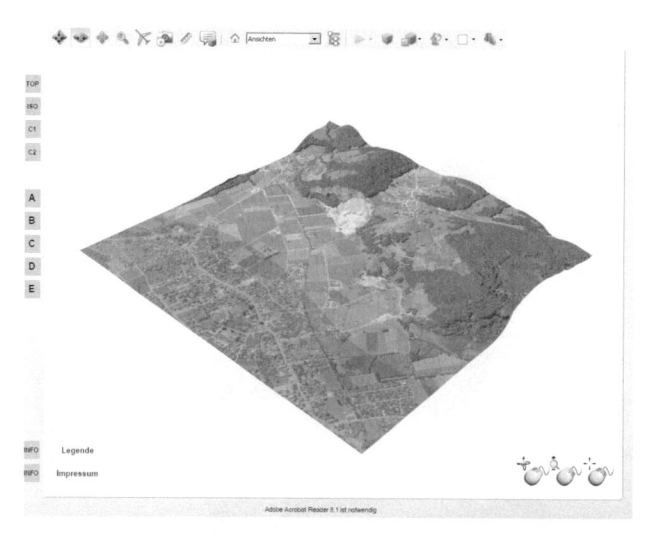

Fig. 4: 3D PDF is a perfect format for public information

References

[1] Calabria, A. (2010), *The Art and Science of Design Communication Media in Landscape Architecture Today: Changes in Curricula and the Profession.* In: Carsjens, G. (Ed.), Landscape Legacy. CELA 2010, Wageningen University.
[2] Linder, W. (2006), *Digital photogrammetry, a practical course.* Berlin, Springer.
[3] Matthews, N. (2005), *Close-range photogrammetry.* US National Science and Technology Center, Denver.
[4] Mikhai, E., Bethel J. & McGlone, C. (2001), *Introduction of Modern photogrammetry.* Wiley & Sons.
[5] Gu, Q. (2010), *The application of close range photogrammetry in digital terrain modeling for landscape architecture design and construction – Comparison and test of data acquisition methods.* IMLA Master Thesis, Rapperswil.
[6] Petschek, P. & Lange, E. (2004), *Planung des öffentlichen Raumes – der Einsatz von neuen Medien und 3D Visualisierungen am Beispiel des Entwicklungsgebietes Zürich-Leutschenbach.* CD CORP-04 Proceedings, Wien.
[7] Petschek, P. (2005), *Terrain Modeling with GPS an Real-Time in Landscape Architecture.* In: Buhmann et al. (Eds.), Trends in Real-Time Landscape Visualization and Participation. Proceedings at Anhalt University of Applied Sciences. Heidelberg, Wichmann.
[8] Wolfer, L. (2010), *Darstellungsmethoden in der Landschaftsarchitektur.* IMLA Master Thesis, Rapperswil.
[9] Zeile, P. (2010), *Echtzeitplanung. Die Fortentwicklung der Simulations- und Visualisierungsmethoden für die städtebauliche Gestaltungsplanung.* Dissertation, Fachbereich Architektur/Raum- und Umweltplanung/Bauingenieurwesen, Universität Kaiserslautern, Kaiserslautern.

Computer-Supported Collaborative Learning with Wikis and Virtual Classrooms across Institutional Boundaries – Potentials for Landscape Architecture Education

Ellen FETZER and Heike KAISER

1 Preliminary Developments

The collaborative online seminars presented in this paper build on two principal foundations. One is given by LE:NOTRE, the European Thematic Network in Landscape Architecture[1], which has created a new forum for academic collaboration among landscape architecture scholars. Without the new and fruitful connections made within this network the setting up of collaborative teaching units with inputs from different European countries would hardly have been possible. The second foundation is the IMLA[2], a landscape architecture Master's course in which distance learning with virtual classrooms has been practiced already since 2002.[3] Based on the expertise gained in this programme the idea came up to open virtual classroom education to the member schools of LE:NOTRE. Since 2007 the Universities of Nürtingen-Geislingen[4] and Kassel[5], both Germany, have taken up a coordinating role in this context. Up to now, seven online seminars have taken place. They have been attended by over 150 landscape architecture students from as many as 40 different countries.[6]

2 Motivations for Developing and Offering Online Seminars

The motivation for implementing such a rather complex model of academic education is based on a broad set of expected benefits for students, teachers and even for the institutions involved. From the beginning it was assumed that these benefits would compensate for some of the common drawbacks of online education (such as technical problems and the limitation of communication channels compared to face-to-face contacts). The internet serves as a tool for bridging the boundaries between academic institutions. It enables teachers to offer a course collaboratively while continuing their daily business on campus. In doing so, they will not only distribute the required workload but also gain fruitful new

---

[1] LE:NOTRE: Three EU-funded subsequent thematic networks involving among one hundred European landscape architecture schools in the period from 2002 – 2013. http://www.le-notre.org.
[2] http://imla-campus.eu.
[3] Virtual Campus IMLA (2002/2003), a project by Prof. Dr. Ulrich Kias and Dipl.-Ing, (FH) Michael Ditsch funded by the Virtual University of Bavaria, Germany.
[4] HfWU Nürtingen-Geislingen is a member of the IMLA consortium.
[5] Department of Landscape Planning, Prof. Dr. Diedrich Bruns.
[6] All seminar websites can be accessed via this wiki http://landscape-diary.net/wiki/index.php?title=Case_Studies (Wiki author: Ellen Fetzer).

knowledge though the exchange with international colleagues and students. This happens of course also during summer schools, international workshops and exchange semesters in a face-to-face context. But the difference made with international online seminars is that they are included in the daily rhythm of studies. In addition to lectures, seminars and studio work on campus the students also have their regular weekly meeting in the virtual classroom. They learn how a specific topic is handled in other European countries and collaborate in international small groups. Given the increasing number of landscape architecture practises operating on a global level it becomes more important that graduates are able to work in an intercultural and partially virtual team. But first and foremost stands the idea that the internet opens new options for a student-centred learning approach based on the learning theory of constructivism.

3 Building on Educational Constructivism

Constructivist learning methods are very common in landscape architecture education. According to JEAN PIAGET (1977) an individual will create knew knowledge through an interaction between his/her previous conception and new experiences. If this new experience is in line with a previous concept this process is called *accommodation*, which is basically an extension of the knowledge scope. In contrast, *assimilation* takes place if new knowledge contradicts with previous ideas, finally leading to a change of the prior concept. This happens for example when we learn through failure. Furthermore, constructivism calls for an active learning process that is rich in authentic experiences for the learner. In this context, the role of the teacher becomes the one of a *facilitator*, motivating and accompanying an independent learner in his/her process of dealing with reality. Also, learning in groups through interaction is principal. We find this model in landscape architecture design studios where students develop solutions for real life problems accompanied by an active dialogue with their teachers (BRUNS et al. 2010). Ideally, both teachers and learners are aware that there is no unique answer to the problems being dealt with. Unfortunately, this paper does not allow for elaborating further on constructivist learning[7]. But the case study described further down will show how the seminar process and the activities included have been designed according to the principles of constructivism.

4 E-learning Tools for Student-centred Learning

As seminars embrace various communication and information flows a mix of technologies is required for translating these flows into a virtual learning environment. From the beginning it was clear that the possibilities for students to be actively involved would be essential for creating an authentic learning environment. Therefore, the tools applied had to support self-directed activities of the students but also allow for authentic communication. The structural backbone of the seminars is given by weekly synchronous meetings in the virtual classroom (90 minutes). Virtual classrooms allow for synchronous communication of larger groups and usually have additional functions such as application sharing, file

---

[7] See for example Reich (2007) for a broad overview of constructivism as a learning theory.

exchange, collaborative editing, breakout rooms and interactive tools like participants' arrows or text fields.[8] Virtual classrooms make a high level of interactivity among the participants possible which is a big asset. However, they are often only used for transmitting lectures without further learning activities. Unfortunately, most applications support this approach and do not represent the group adequately. Therefore, VITERO was used for the seminars. The advantage of this application is that the group is represented like in a roundtable discussion. In addition, many interactive tools supporting collaborative learning are available such as card clustering for moderated discussion.[9]

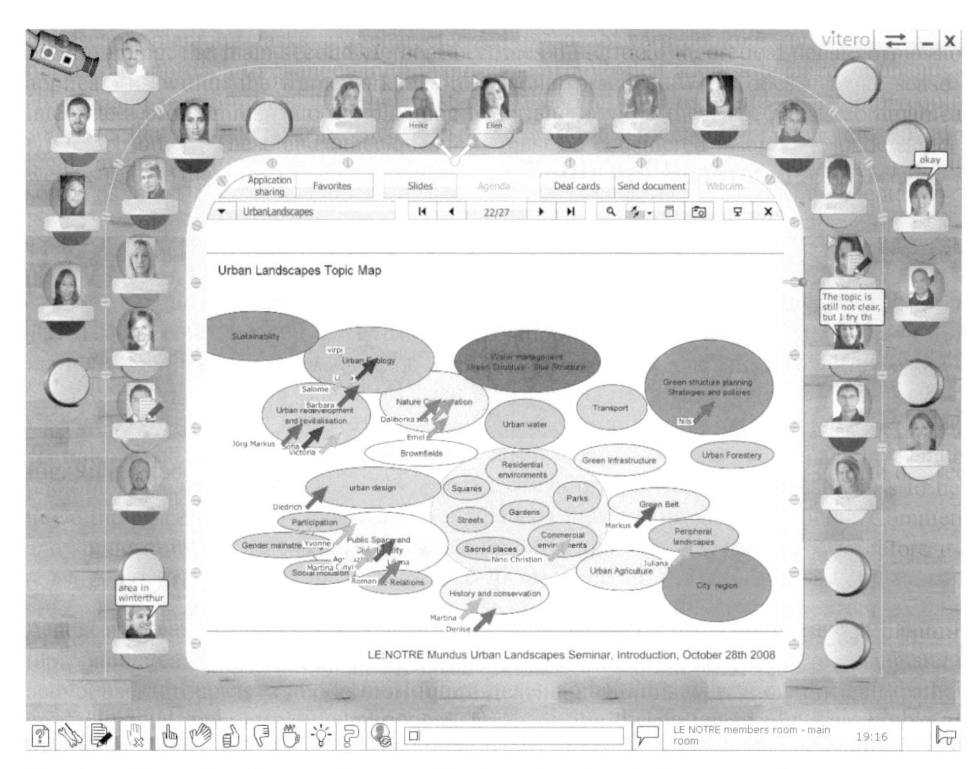

Fig. 1: Urban Landscapes Seminar: Students localise their case studies on a topic map with participants' arrows. Working groups have been formed on this basis.

In addition, it was necessary to have a platform where students would be able to publish and discuss their seminar work. A wiki was found to be the ideal solution for this[10]. The strength of a wiki lies in its horizontal, non-hierarchical structure that leaves much freedom to learners for structuring contributions. However, this freedom may turn also into

---

[9] For more information about how to apply these tools with learning groups: http://www.uni-kassel.de/hrz/db4/extern/collaboration/wiki/index.php?title=Virtual_Classroom (author: Ellen Fetzer, 2009).

[10] The wiki software applied is Mediawiki, also used for Wikipedia. This server software is open source. http://www.mediawiki.org.

confusion. Therefore, templates where widely used for pre-structuring the students' pages. A wiki is a very powerful tool for collaborative writing (CRESS & KIMMERLE 2008). It records all stages of a text so that the text production can be entirely tracked back or even restored if problems or abuse occurs. This is ideal for providing freedom for the learner on the one hand while being able to identify authorship for assessing the results. The wiki was used for externalising knowledge (when students document their case studies), collaborative writing (discussing and documenting group tasks), aggregation of knowledge (as wikis allow for collecting dispersed information) and documentation (as a basis for assessment). Furthermore, the wiki is used for managing the course organisation such as publishing schedules, minutes and recordings and giving access to additional resources.[11]

5 Case Study "Public Participation in Landscape Architecture and Open Space Planning"

The aim of this seminar was to offer an introduction to consultation and to the scope communicative forms of planning may take in landscape architecture. It took place from October 2009 - January 2010 and was attended by 22 students coming from 14 countries and 8 study programmes. The activities where organised on two levels: regular online meetings in the virtual classrooms and group tasks which had to be documented on the seminar wiki. The synchronous sessions where used for expert lectures, discussions and student presentations. At the beginning, the students where asked to read subject-specific articles and to visualise these as concept maps. These where then presented to the plenary. In the second phase each student documented a case study on the wiki which had to include a process model indicating all relevant participation and consultation steps of the project described. These cases where then clustered thematically as a basis for the working groups. After externalising knowledge in the form of a case study the students where asked to compare these within the thematic group. This was done with help of a set of core questions. The students were further asked to identify qualities and drawbacks between their cases, to discuss these in the group and to derive a concluding synthesis. The result of this group work was again presented and discussed in the plenary session.

The procedure described above is mainly based on the model of learning being a process of reconstructing, constructing and deconstructing which is described by REICH (1998) in his model of interaction-based constructivism. Students start with reconstructing knowledge when they identify a potential case study out of their previous experience. They then start to construct knowledge by documenting the case and presenting it to the group. At last, they enter into a process of deconstruction as they are asked to compare the characteristics of their case with other examples, mostly from different cultural contexts. In doing so, they will experience both *accumulation* (if the comparison confirms previous assumptions) and *assimilation* (if previous assumptions are contradicted).

---

[11] The seminar wiki can be publicly accessed: http://draco.hfwu.de/~wikienfk5/index.php/Main_Page.

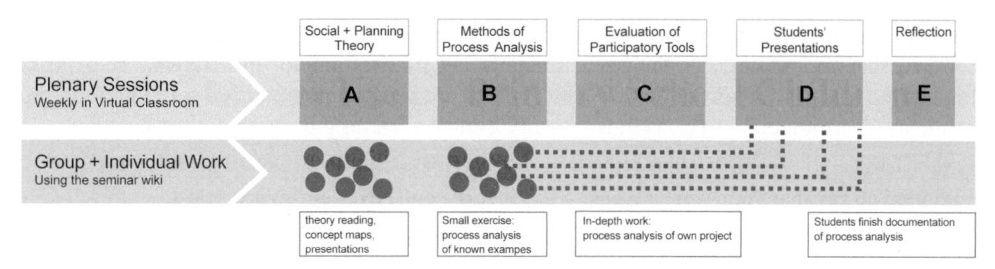

Fig. 2: Process model of the seminar "Public participation in landscape architecture"

6 Some Selected Evaluation Results in Comparison

Since fall 2008 the seminars have been evaluated with a standard set of questions. In the following, some results are presented showing the last five seminars in comparison.[12]

The content of the meetings was interesting and I gained new knowledge.					
Seminar Title	Urban Landsc.	Rural Landsc.	Public Particip.	Cultural Landsc.	Landsc. Concept
Year	2008	2009	2009	2010	2011
Answers	20	16	21	20	30
Yes, absolutely	45%	56,25%	42,86%	86,67%	26,67%
This was mostly the case.	35%	31,25%	47,62%	6,67%	40%
…was of average interest	15%	12,5%	9,52%	6,67%	30%
…was not interesting	5%	0%	0%	0%	6,67%

These figures give important information with regard to the motivation of the learners. Their profound interest is very relevant for the overall performance of the seminar as it strengthens the participants' will to overcome potential technical obstacles and to face the challenges of virtual communication. In general, the vast majority of the participants stated that the contents were of absolute interest or that this was mostly the case.

Did you find it easy to express your thoughts in the VITERO meeting room?					
Seminar Title	Urban Landsc.	Rural Landsc.	Public Particip.	Cultural Landsc.	Landsc. Concept
Year	2008	2009	2009	2010	2011
Answers	20	16	21	20	30
Yes absolutely	35%	37,5%	42,86%	33,33%	33,33%
mostly	25%	50%	47,62%	53,33%	33,33%
sometimes	40%	12,5%	9,52%	13,33%	30%

This question tries to illustrate in how far the participants felt comfortable in the virtual team room as a basis for participating actively in the discussions. As the seminar concept builds on active participation and involvement as a principal learning method it is crucial to

---

[12] See http://landscape-diary.net/wiki/index.php?title=Category:Seminar_Evaluation (30.01.2011).

have a virtual environment in which students can easily be stimulated. Except for the first seminar the vast majority of the participants found it absolutely or mostly easy to express his/her thoughts in the virtual team room. This is a principal precondition for achieving the seminars' learning objectives as these mainly build on learning through dialogue.

Would you say that your working group has met the objectives of this seminar?					
Seminar Title	Urban Landsc.	Rural Landsc.	Public Particip.	Cultural Landsc.	Landsc. Concept
Year	2008	2009	2009	2010	2011
Answers	20	16	20	15	30
Yes absolutely	10%	6,25%	23,81%	40%	23,33%
mostly	70%	56,25%	61,9%	53,33%	66,67%
Minimum	10%	25%	9,52%	0%	6,67%
No, we did not	10%	12,5%	4,76%	6,67%	3,33%

Working group processes are dependent on many factors that are likely to be very different not only in an international learning group. Besides the actual task, the seminar groups had to cope with additional study commitments, diverse learning and working styles and unequal language capacities. In addition, most students were confronted with different concepts of the subjects being dealt with due to the variety of national backgrounds. In this delicate context, some students achieved great results mostly if they happened to be in a more or less homogenous group. Others sometimes failed to meet the demands even if they were known for performing well in other learning situations. There was a risk factor throughout the seminars with the principal assignment being designed as a group task. However, despite all likely dangers the majority of the participants thought that they performed absolutely or at least mostly well as a group.

7 Competences for Seminar Coordinators

Teaching in an online environment requires a specific combination of competences moving beyond traditional education. As defined in the European Qualifications Framework (2008) "competence means the proven ability to use knowledge, skills and personal, social and/or methodological abilities, in work or study situations and in professional and personal development". Although the principles of designing, organising and executing both online and traditional classroom courses seem quite similar, coordinators of online seminars "need training and support to…adopt this new teaching paradigm [and] need to be cognizant of how the details of their course will be implemented in the new environment" (LEVY, 2003).

First of all, a key competence deemed necessary prior to the actual start of the online seminar is the ability to create a consistent course concept. This includes specifying course requirements, communicating its aims and expectations and defining grading criteria. Invited lecturers need to be confirmed and introduced to using a virtual classroom well ahead of the time of their lecture. Furthermore, additional resources, such as expert literature, are to be collected and made accessible. The course schedule ideally provides a general structure while allowing for flexibility and negotiation. All of this as well as further information should be communicated in plain English at a level that does not overwhelm students who are new to this study method. Coordinators will handle virtual classroom software including all interactive tools and further devices because they take over the moderator's role during seminar sessions. In general these types of software can be handled with basic IT skills after short practise. However, the moderator needs to react with a sense of improvisation in case of technological problems. Additional software for concept mapping[13] also needs to be mastered beforehand. Apart from that, coordinators will edit and structure wiki pages and may also need to become acquainted with a learning management system such as Moodle.[14]

As the online course begins, the coordinator takes over his/her role as *facilitator*, focusing not only on course contents by providing appealing and informative presentation slides but mainly on the development of the learning community. Student-student and student-coordinator interaction are essential for a fruitful learning environment. But both need to be accomplished without overwhelming students who may be endangered by cognitive overload. Moderators should use best practices for encouraging the students' active participation in online discussions and for getting them to respect working phases and due-dates. He or she will further provide adequate minutes (and recordings), give prompt and objective feedback on assignments, and, when appropriate, use humour for breaking up potential communication gaps in the virtual room. As mentioned already earlier a learner-centred approach is essential. Based on this approach an online seminar coordinator will take the different curricular and cultural backgrounds of the participants into account, will accompany the delicate collaboration process of internationally mixed working groups and will show patience regarding the completion of deliverables with respect to concurring exam periods or holidays. In addition, reflection times are required during which students can identify their strengths and weaknesses and develop critical thinking. He or she will further stimulate the participants' prior knowledge in the best way and encourage them to bring real-life examples from their cultural backgrounds into the online classroom.

At all times, coordinators must maintain the momentum of the course. This may require taking actions that might not be needed in a traditional face-to-face setting, such as stimulating participation and redirecting discussions if headed into the wrong direction. Coordinators will need to contact students individually (typically by email) if they face problems, are not participating as expected or even are disruptive. However, it always needs to be remembered that there are real people attached to the profile pictures in the virtual seminar room. After all, the ultimate priority of an online course should be to establish a vivid and fruitful community of learners.

Although the above list of competences may seem complex, it attempts to illustrate "[different] levels of competence - entry/novice, experienced, specialist-rather than a once for all attainment. Interpreted broadly, competence is not trained behavior but thoughtful capabilities and a developmental process" (KERKA, 1998). Thus, rather than dissecting and focusing on individual competences, it is suggested that, for online seminar coordinators, true competence mandates: mastery of all the individual competences in complex

---

[13] During the seminars presented here two open source applications for digital concept mapping have been applied: VUE (Visual Understanding Environment, http://vue.tufts.edu) and Cmaps (http://cmap.ihmc.us). Cmap even allows for synchronous collaboration on shared concept maps.
[14] In this case Moodle was only used for sharing copyright protected literature during the seminar offering a restricted area to students.

combinations; employment of a variety of knowledge, skills, attitudes, and values; and a standard of excellence that practitioners will obtain and continuously demonstrate through a process of active research. It goes without saying that coordinators must also master the subject being taught.

8 Outlook

We hope that the long list of competences discussed above will not discourage anybody who is reading this paper not only with the intention of becoming familiar with online seminars, but also with the aim of possibly organising or taking part in such seminars in the future. Most of the competences mentioned are naturally well developed among experienced educators. Basically, these abilities need to be transferred to the new context and combined with knowledge of the key e-learning tools. The generation of digital natives is currently entering university education, so students easily adapt to the virtual environment. However, achieving the seminar objectives is in no way only a matter of mastering a digital tool. After entering the virtual room and editing the first wiki page the students' success will strongly depend on their communicative, organisational and analytical skills. And like in face-to-face education some will provide better results than others. Coming back to the first chapter of this article it remains to be said that cooperation of higher education institutions through online seminars is still at the beginning of exploiting its whole potential. For landscape architecture academia the further development of the LE:NOTRE project will be crucial for developing this idea further.

References

Bruns, D., de Vries, J. et. al. (2010), *Tuning Landscape Architecture Education in Europe.* Vs. 26, 21.09.2010, published on http://www.le-notre.org, 37.

Cress, U. & Kimmerle, J. (2008), *A systemic and cognitive view on collaborative knowledge building with wikis.* In: International Journal of Computer-supported Collaborative Learning, 3 (2), 105-122.

European Parliament and Council (2008), *Recommendation of the European Parliament and of the Council on the establishment of the European Qualifications Framework for lifelong learning,* 4.

Kerka, S. (1998), *Competency-based education and training: Myths and realities.* Retrieved December 03, 2011: http://www.eric.ed.gov/PDFS/ED415430.pdf, p 4.

Levy, S. (2003), *Six factors to consider when planning online distance learning programs in higher education.* In: Online Journal of Distance Learning, Spring 2003, VI (1).

Piaget, J. (1977b), *Problems of equilibration.* In: Appel, M. H. & Goldberg, L. S. (Eds.), Topics in cognitive development, 1, 3-14, New York, Plenum.

Reich, K. (1998), *Die Ordnung der Blicke. Band 2: Beziehungen und Lebenswelt.* Neuwied, Luchterhand, p. 41 ff.

Reich, K. (2006), *Konstruktivistische Didaktik, Lehr und Studienbuch mit Methodenpool.* Weinheim/Basel, Beltz.

Development of Randomly Mixed Perennial Plantings and Application Approaches for Planting Design

Wolfram KIRCHER, Uwe MESSER, Jessica FENZL,
Marcel HEINS and Nigel DUNNETT

1 Introduction

The planning of diverse and species rich perennial plantings is very time consuming. To increase the efficiency of the planning process the concept of randomly mixed plantings was developed in a series of trials at Anhalt University of Applied Sciences, Bernburg, Germany. Compared with plantings based on a graphically depicted planting plan the visual quality of a mixed model was assessed equally ranking.

1.1 Planning of Perennial Plantings – Common Approaches of Planning Perennial Plantings

Monoplanting: One simple and popular strategy in public green spaces is planting with a single species. Among the typical perennials used in monoplantings are competitive species of *Geranium, Salvia, Lavandula*, etc. Monoplantings are less cost intensive to planning, construction of the planting and maintenance as long as hardy and long lived perennials are used. Monoplantings are monotonous, since blooming occurs only for a certain period or even not at all. Evergreen shrubs such as *Cotoneaster* and *Lonicera pileata* do not show the seasonal changes typical to perennials and deciduous trees and shrubs.

Planting in groups or blocks (Fig. 1), usually of more than two different species, are another way of creating public green spaces. Block planting is essentially a more complex version of the above and is perhaps the most common approach to landscape planting (DUNNETT, KIRCHER & KINGSBURY 2004, p. 246).

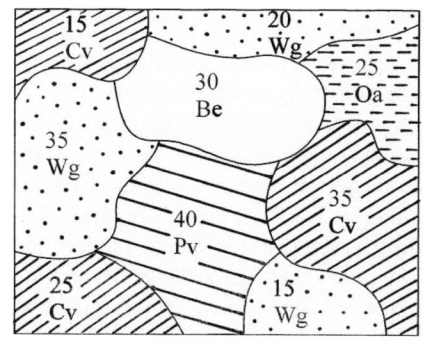

Fig. 1: Planting Plan arranged in groups

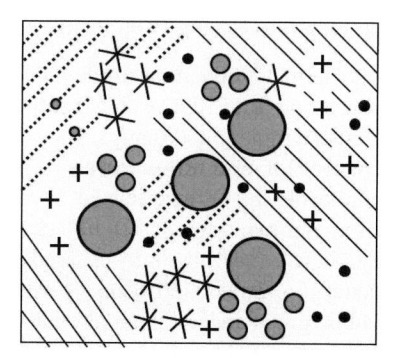

Fig. 2: Planting Plan arranged according to sociability levels

Block plantings consist of several different species planted for effect in groups of three to five or even more than one hundred. Block plantings can be subdivided into so-called drifts, which are strips, usually linear and at tendentially right angles to the observer, in order to provide perspective. Drifts contain groups of plants arranged in extremely narrow rows running more or less parallel to the main direction of the bed. This arrangement enhances the depth effect, but is more expensive to plan and maintain. Drift planting was used with great skill by Gertrude Jekyll in herbaceous and mixed borders. As a big advantage of this method ROBINSON (1998) mentions that the narrow groups would look best at their peak but also that they would not detract from the border after fading.

Planting according to sociability levels (Fig. 2): A very naturalistic approach is to arrange plants according to sociability levels I to V (according to HANSEN & STAHL 1993). This planting strategy is applicable to perennials with a more "wildlife" effect. Plants of low sociability levels (groups I and II) are set individually or in small groups of three to nine. Plants of higher sociability levels (groups III to V) are set in groups of 10 to 20 or more and arranged loosely around those of groups I and II (HANSEN & STAHL 1993).

1.2 Randomly Mixed Planting

This approach aims to completely abstain from a drawn planting plan. Species and variety are carefully selected according to their habitat, competitiveness, flowering, height, and reproductive behaviour. The amount of each type of plant to be used is recorded according to these criteria in a list. By laying out the plants are distributed as evenly as possible over the entire area, starting with the species present in the lowest amounts. The exact position of every plant is not predetermined in a planting plan, but determined by chance. This planting strategy gives a natural effect. The idea behind this strategy is to create a plant community in an ecologically sound, competitive balance, comprising the ideal type of vegetation for public green spaces. Ideally, species showing various striking aspects, forms, heights and propagation strategies complement each other to form a self-regulating system. Within this dynamic model the survival of the entire planting under extensive maintenance is more important than survival of individual plants. The conditions at individual sites result in different competitive conditions and vegetation patterns despite identical plant components. Moreover, there are possibilities for introducing structure such as dominant visual elements or rather theme plants (especially "dominant species" – see table 1; definition following BORCHARDT, 2006, complemented)

The term "Staudenmischpflanzung" (mixed perennial planting) was coined in 1994 by KOLB and KIRCHER at the Institute "Landesanstalt für Weinbau und Gartenbau" (LWG), Veitshöchheim, Germany. They were seeking to develop a simplified version of the concept of sociability levels (HANSEN & STAHL 1993), which would be practically applicable by inexperienced workers, who had never before worked with herbaceous perennial plantings. It is not necessary to have a planting plan with a list of the prescribed numbers of the plants to be used (KIRCHER 2000). In times of limited public funding the concept of mixed herbaceous perennial planting is a reasonable way nevertheless to provide public green spaces as an alternative to costlier approaches with intricate planting plans or rather seasonal bedding.

2 Material and Methods

2.1 The Research Project "Mixed Herbaceous Perennial Planting" at Anhalt University, Bernburg

Experiments to create suitable perennial mixtures as well as establishing and maintaining methods are carried out since 1999 at Anhalt University of Applied Sciences in Bernburg. This dry region provides an annual precipitation of only 470 mm in average. The trials were supported by the German Federal Ministry for Education and Research (BMBF), the German Perennial Nurseries Association (BdS), and the German Research Foundation (DFG). Till 2010 around 30 mixtures have been developed; fifteen have been optimized on the basis of knowledge and assessments gained in the project and are now published as recommendations (http://www.prof-kircher.de; FENZL & KIRCHER 2009). Methods and results from the assessments are recorded by MESSER (2008), who elaborated many aspects of the research project in his PhD-Thesis, assumed by the University of Sheffield.

2.2 Perennial Mixtures from Further Research Institutions

Additionally to the research at Anhalt University a remarkable amount of mixtures with promotional "trade names" were tested, assessed and optimized since the end of the 1990's especially by these institutions:

- Bayerische Landesanstalt für Weinbau und Gartenbau (LWG), Veitshöchheim (P. SCHÖNFELD)
- Schau- und Sichtungsgarten Hermannshof, Weinheim (C. SCHMIDT)
- Lehr- und Versuchsanstalt für Gartenbau/Fachhochschule, Erfurt (C. PACALEI, W. BORCHARDT)
- Zürcher Hochschule für Angewandte Wissenschaften Institut Umwelt und Natürliche Ressourcen, Wädenswil, Switzerland (D. TAUSENDPFUND, A. HEINRICH)

Together with Anhalt University these protagonists currently are publishing 29 recommended plant mixtures in http://www.stauden.de/cms/staudenverwendung/mischpflanzungen/mischungen_alphabetisch.php?navid=87.

The most widespread planting is "Silbersommer", a concept of 36 taxa, created by the Arbeitskreis Pflanzenverwendung (Committee of Planting Design; see http://www.stauden.de/cms/staudenverwendung/mischpflanzungen/forschung/ak_verwendung.php?navid=93).

2.3 Creating Applications of Randomly Mixed Perennial Plantings

Small plots can be designed aesthetically satisfying with pure randomly mixtures. On bigger areas more predictability of the resulting planting is desired. Thus at Anhalt University some variants of mixed plantings were developed to enhance the quality of the designed vegetation's appearance. Nevertheless they should not effect significantly more time consumption than working with the pure randomly mixing strategy. FENZL & KIRCHER (2009) published six application variants for practitioners. In chapter 3 these variants shall be introduced and explained by their theoretical background.

3 Results

3.1 Preconditions for Well Performing Mixed Plantings

Plant species have to be selected and arranged by taking into account the following trait (MESSER 2008):

- Choice of suitable site/habitat conforming (HANSEN & STAHL 1993)
- Thematic focus of the planting (i.e. blue-yellow contrast)
- Growth rhythm (short-term dynamics, annual aspects, height in various seasons, long-term dynamics)
- Life expectancy of the plants (biennials, short- and long-lived perennials)
- Plant sociability (according to HANSEN & STAHL 1993)
- Reproduction and rate of propagation
- Population biological strategies (runners, rhizomes) (GRIME, HODGSON & HUNT 1986)
- Aesthetic criteria (layering, color combinations, texture)

To guarantee a visually pleasing and sustainable relief within the planting, it is recommended to distinguish between 5 different plant categories according to height and long and short term space requirements (table 1).

The diversity of most mixtures is guided by around 12 to 20 species/cultivars per theme for sunny habitats. "Silbersommer" is the richest mixture with 36 taxa. Under shady conditions it is better to reduce the diversity appropriately. Concrete recommendations for mixtures in woodland habitats are currently investigated by tests of the Committee of Planting Design.

The diversity of most mixtures is guided by around 12 to 20 species/cultivars per theme for sunny habitats. "Silbersommer" is the richest mixture with 36 taxa. Under shady conditions it is better to reduce the diversity appropriately. Concrete recommendations for mixtures in woodland habitats are currently investigated by tests of the Committee of Planting Design.

Category	Definition	Recommended proportion of plants
Dominant species: structure plants, framework plants	Forming the structural framework of the planting, e.g. grasses (*Miscanthus sinensis, Cortaderia selloana*), large-leaved perennials (e.g. *Rodgersia*) or upright plants (e.g. *Veronica longifolia*); mainly C-, C-S or S- Strategists	5 – 15%
Companion plants	Recurring, stabilizing elements (e.g. *Salvia nemorosa, Hemerocallis lilioasphodelus*) which define the visual character of the planting and emphasize the structure plants. Long lived plants; mainly C-, C-S or S- Strategists	30 – 40%
Ground cover plants	Usually small perennials of up to 30 cm height which must be used in larger numbers, usually as a carpet between gaps between plants of the first two categories, i.e. *Geranium x cantabrigense, Omphalodes verna, Waldsteinia geoides;* mainly C-, C-S or S- Strategists	≥ 50%
Filler plants	Short lived plants, responsible for a quick cover and visual display in the first one to three years. Quick in growth and spreading generatively, but weak in competition, declining whilst substituted by the dominant, companion and ground cover plants (e.g. *Linum perenne, Aquilegia canadensis, Digitalis purpurea);* R-, R-S, or C-R-Strategists	5 – 10%
Scattered plants	Plants with a short growth period that do not require much space. However, these are very showy and dominant when in bloom, such as flowering bulbs (e.g. *Allium sphaerocephalon, Anemone blanda, Narcissus* 'Hawera') or very slim perennials (e.g., *Codonopsis clematidea, Campanula persicifolia*)	added additionally in great amounts: 20 to 50 bulbs/m²

Table 1: Classification of perennials (MESSER 2008, definitions following BORCHARDT 1998, supplemented by FENZL & KIRCHER 2009; C-, R-, S- Strategy see GRIME, HODGSON & HUNT 1986

3.2 Applications of Randomly Mixed Perennial Plantings

The planning strategy of mixed perennial plantings can be recommended for small to medium sized beds in public and private green spaces, if a natural display is desired. Considering the proportions and amounts of plant- categories recommended in table 1 is essential for an attractive relief as mentioned above. Bigger scaled planting areas can be arranged in a meadowy style, allowing more intricate and intertwining plant structures: the proportions of taller species may be increased. Also combinations between sown vegetation and planted perennials are conceivable (see Fig. 8/Variant 6).

Suitable for tenders with mixed plantings are for example the following themes:
- Planted meadows, eventually in combination with sowing
- Planted prairies
- Mille-Fleur-plantings (seasonal bedding with small inflorescences, in well balanced color combinations)
- Woodland underplanting (mixes with lower diversity in species)
- Marginal plantings in ponds (reed forming plants shall be combined with only filler plants or rather shade tolerant ground coverers)
- Traffic islands and roundabouts as well as small plots between asphalt sealed surfaces
- Narrow beds along fences, walls and buildings
- Rock gardens and dry stone walls (plant lists divided into differentiated habitats like „sunny gaps", „shady gaps", "mural crown" etc.)
- Extensive roof gardens
- Plantings between pavement crevices

Particularly on bigger planting beds it might be beneficial not to dedicate the complete plant distribution to chance. Six adaptation possibilities are introduced hereafter.

Fig. 3:
Variant 1 – Pure mixed planting list with names and quantities of plants (species, genus, variety, cultivars). Plants are arranged in similar distances to each other. The image on the right hand shows a randomly planted plot of "Bernburger Blütenschleier".

plant list example	planting plan example	distribution pattern on the plot
30 *Aster amellus* `Rudolf Goethe´ 25 *Buphthalmum salicifolium* 50 *Carex Montana* 100 *Thymus serpyllum* `Album´ etc.	no depicted plan is necessary	

Variant 2 – List with names and quantities of plants plus additional remarks about positioning or grouping
The grouping of certain species allows to include areas providing a calming appeal. Groups in rows or streams are effective for grasses or vertically growing perennials like *ampanula persicifolia* or *Verbascum* spp. "Filler plants" do not leave an ugly hole after vanishing if

arranged in narrow groups. It is also possible to prevent tall plants (dominant species) from being placed nearby the bed's edge by only adding an appropriate written remark into the plant list. On vast areas the observer's eye is assisted by a consistent use of all species in small groups, causing a "zooming" effect communicating with the extension of the complete planting bed's size.

plant list example	planting plan example	distribution pattern on the plot
30 *Aster amellus* `Rudolf Goethe´ in small groups of 3 25 *Buphthalmum salicifolium* randomly distributed 50 *Carex Montana* In linear streams of 5 100 *Thymus serpyllum* `Album´ in groups of 10 etc.	no depicted plan is necessary	

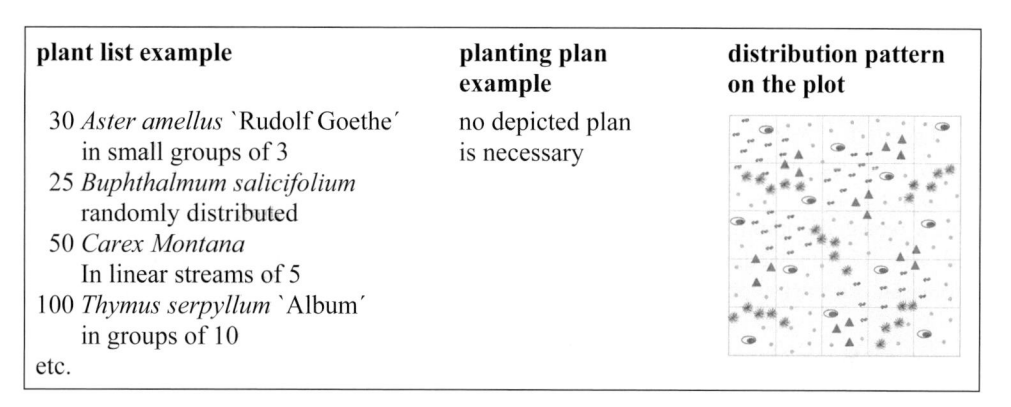

Fig. 4:
Variant 3 – Illustration of tall, dominant plants in the plan, accompanied by randomly mixed lower perennials
On middle and bigger sized areas it can be necessary to include taller Plants into the vegetation pattern to provide visual leading. This can be realized by adequate perennials, subshrubs or even shrubs arranged in sufficient distances to each other. To create naturalistic distribution patterns these species should be depicted into a graphical plan or sketch. The shallower species can be added randomly as in variant 1 or added by additional remarks about positioning or grouping as in variant 2.

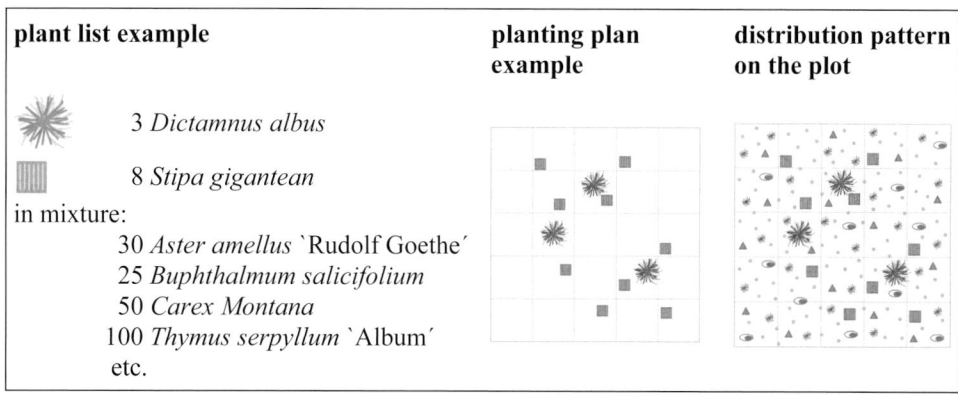

plant list example	planting plan example	distribution pattern on the plot
3 *Dictamnus albus* 8 *Stipa gigantean* in mixture: 　30 *Aster amellus* `Rudolf Goethe´ 　25 *Buphthalmum salicifolium* 　50 *Carex Montana* 　100 *Thymus serpyllum* `Album´ 　etc.		

Fig. 5:
Variant 4 – Illustration of clear defined recurring core groups accompanied by additional perennials in randomly mixture

Core group planting = "Kerngruppenpflanzung" is defined by BORCHARDT (2006). To be sure that desired combinations of particular partners will be realized within the planting clearly specified groups can be depicted detailed (left alternative in box below). The same effect will result by using one symbol as placeholder for the specific species combination per core group. The rest of the designated perennials can be added randomly as in variant 1 or 2. The planting on the right hand shows repeated combinations of *Gypsophila* 'Rosenschleier' and Iris Barbata-Media, the former covering reliably the unpleasant summer appeal of the faded latter ones.

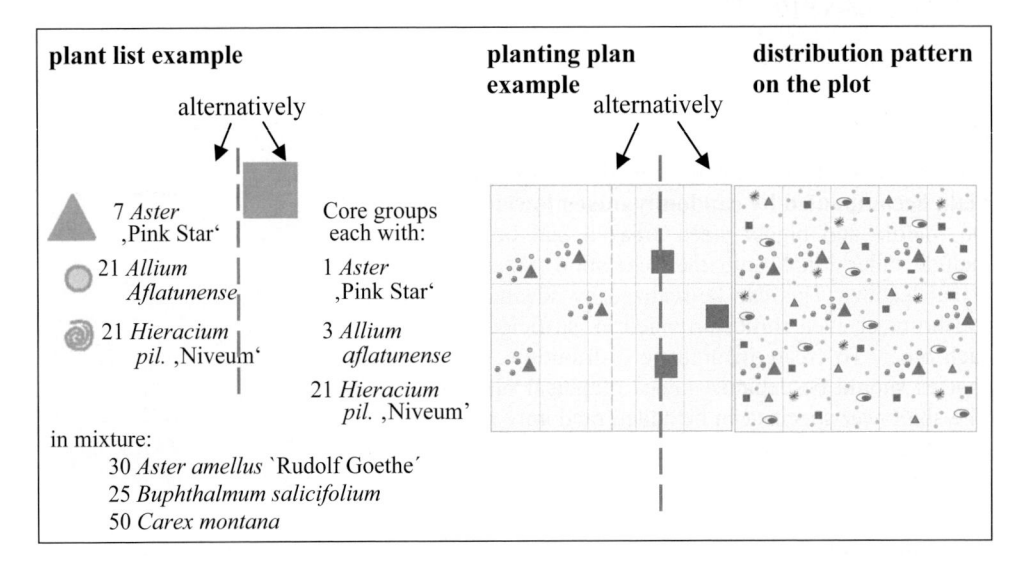

plant list example

alternatively

▲ 7 *Aster* 'Pink Star'

● 21 *Allium Aflatunense*

◉ 21 *Hieracium pil.* 'Niveum'

■ Core groups each with:
1 *Aster* 'Pink Star'
3 *Allium aflatunense*
21 *Hieracium pil.* 'Niveum'

in mixture:
30 *Aster amellus* 'Rudolf Goethe'
25 *Buphthalmum salicifolium*
50 *Carex montana*

planting plan example alternatively

distribution pattern on the plot

Fig. 6:
Variant 5 – subdividing the planting bed into smaller parts with different mixtures or with the same mixture in different multitudes per species

Extensive planting areas, for example several hundred square meters, may look too uniform if covered consistently with one mixture only. If partitioning into dequate sized plots it should be paid attention to form transitions from plot to plot or rather to refrain from colliding completely different plant combinations against each other. Neighbouring plots should contain similar ranges of species but arranged in different amounts respectively proportions. Of course this approach can be combined with all variants mentioned above inside of one single plot. The image shows a "Garigue" planting at the Botanical Garden of Würzburg: Different mixtures can be recognized, but some species such as *Asphodeline lutea* are skipping over the borders.

plant list example	planting plan example	distribution pattern on the plot
area A: 100 *Aster amellus* `Rudolf Goethe´ 50 *Allium sphaerocephalon* 60 *Asphodeline lutea* etc.  area B: 30 *Aster amellus* `Rudolf Goethe´ 300 *Allium sphaerocephalon* 24 *Dictamnus albus* etc. area C: 100 *Allium sphaerocephalon* 30 *Asphodeline lutea* 8 *Festuca gigantea*	area A area B area C	

Fig. 7:
Variant 6 – Combination of mixed plantings with seeding or rather spontaneous vegetation
The most naturalistic approach is to combine existing vegetation or seed mixes with a mixed perennial planting. To include pot plants or young plants from trays is rewarding for vegetatively propagated perennials (i.e. *Hemerocallis* cv.) as well as for species with a weak or unreliable germination (*Salvia* species). Also plants with a very slow development (*Dictamnus albus*) are worth to be planted as bigger specimen.

plant list example	planting plan example	distribution pattern on the plot

30 *Iris* `Libellula´
20 *Salvia officinalis*
15 *Verbascum olympicum*
etc.
 	
plus sowing mixture (1g/m²):
  5 % *Festuca pallens*
  2 % *Dianthus carthusianorum*
  1 % *Linum perenne*
etc.
 	
or scattered between existing spontaneous vegetation
	no depicted plan isnecessary	

4   Conclusions and Outlook

4.1   Contemporary Strategies Related to Mixed Perennial Plantings

Precursors of the mixed planting strategy were already performed in artificial prairie plantings such as the first prairie restoration project, the Curtis Prairie, in Madison, Wisconsin, USA. A combination of sowing, planting and transplanting sod (DIECKELMANN & SCHUSTER 2002; WASOWSKI 2002) was distributed in mixture. The "Matrix Planting" approach of Peter Thompson (THOMPSON 1997) is also based on mixed plantings. Oudolf combines a few mainly tall growing species in drifts or blocks which are neighboured to each other in borders, as can be seen at his famous "Wisley Border" in RHS Wisley Gardens (GERRITSEN & OUDOLF 2000). The latter project can be characterized as an ornamental type of "variant 5" (Fig 7). Further examples of planting methods referring to the randomly mixed strategy see in MESSER (2008).

Fig. 8: Piet Oudolf's Border at RHS Wisley Gardens is a combination of diagonally running streams with mixed tall perennials

4.2 Mixed Plantings in Academic Education

It is to be discussed which level of knowledge and skills shall be taught in academic landscape architecture courses based on four steps of intricacy:

Step 1: Coming to know only the mixed strategy as one possibility beside monoplanting, planting in groups and habitat planting with sociability figures. To experience a synoptic view on existing mixtures from diverse research institutions and their implementation into the planning process.
This step should be a fundamental target in undergraduate Landscape Architecture courses.

Step 2: Understanding basically principles of functioning sustainable combinations such as habitat-conformance, well balanced strategy types and heights as well as a precise aesthetical concept.
At least undergraduate students with emphasis in planting design should reach this level.

Step 3: Ability to change existing mixtures in case of differing site conditions or aesthetical demands by substituting particular species according to their individual role within the planting. For this step it is important to mediate a deeper insight into plant traits concerning ecological as well as aesthetical features. Postgraduate courses should target at least this step.

Step 4: Ability to create individual mixtures. Profound plant knowledge is necessary as well as a grounded understanding of ecological traits and site demands of plants. Design skills and a well developed know how in aesthetical enhanced plant composition are to be educated intensively. This can be aimed in specific postgraduate courses focusing on planting design

4.3 Applications of Mixed Perennial Plantings in Academic Education

Undergraduate courses should include at least variant 1 – 4 (see 3.2 Fig. 3-6) of mixed plantings to implement these methods into planning for public as well as for private gardens. Variant 5 (subdividing the planting bed into smaller parts with different mixtures or with the same mixture in different multitudes per species) demands more skills from the planner: to guarantee fluent transitions between neighbored plots mixtures have to be carefully changed in their species ranges (Step 3 in 2.2). Also a combination between mixed planting and sown or spontaneous vegetation (variant 6; Fig 8) requires this skills. So variants 5 and 6 can be targets for postgraduate courses for an adequate insight into site analyses, aesthetical and ecological principles are necessary as well as an adequate knowledge of plant ranges.

Table 2 summarizes 7 levels combining the application variants from chapter 3.2 and the intricacy steps from the above chapter 2.2.

Variant	Step of intricacy	Required knowledge				Proposed for academic course	Time demand in planning
		Knowledge in Site analysis	Knowledge in aesthetic & design	Knowledge in vegetation ecology	Knowledge in species		
1	1	low	low	–	–	Undergraduate	very low
1	2	low	middle	low	low	Undergraduate*	low
2	1 – 2	middle	middle	low	low	Undergraduate	low
3/4	1 – 2	middle	middle	low	low	Undergraduate	middle
1 – 4	3	high	middle	middle	middle	Postgraduate	high
5/6	3	high	high	high	middle	Postgraduate	high
1 – 6	4	high	high	very high	high	Postgraduate*	very high

Table 2: Matrix for estimated required knowledge/skills, proposed level in academic Landscape Architecture courses and predicted time demand in the planning process. Knowledge levels: – = no knowledge required; low = basic knowledge; middle = advanced knowledge; high = professional knowledge; * = for specific courses or studios focussing planting design.

References

Borchardt, W. (1998), *Pflanzenkompositionen [Plant compositions]*. Stuttgart, E. Ulmer.
Borchardt, W.(2006), *Planungsstrategien für Staudenpflanzungen – Nachdenken lohnt sich*. Deutscher Gartenbau, 25.
Dieckelmann, J. & Schuster, R. (2002), *Natural Landscaping – designing with native plant communities*. 2nd Ed. Madison, WI, USA, The University of Wisconsin Press.
Dunnett, N., Kircher, W. & Kingsbury, N. (2004), *Communicating ecological plantings*. In: Dunnett, N. & Hitchmough, J. (Eds.), The Dynamic Landscape. London/New York, Spon Press.
Fenzl, J. & Kircher, W. (2009), *Bernburger Staudenmix – Ein Forschungsprojekt der Hochschule Anhalt*. Bernburg, HS Anhalt (FH).
Gerritsen, H. & Oudolf, P. (2000), *Dream Plants for the Natural Garden*. Frances Lincoln.
Grime, J. P., Hodgson, J. G. & Hunt, R. (1986), *Comparative Plant Ecology*. London, Unwin Hyman.
Hansen, R. & Stahl, F. (1993), *Perennials and Their Garden Habitats*. Cambridge University Press.

Kircher, W. (2000), *Zufällig gemischte Staudenpflanzungen [Randomly mixed perennial plantings]*. Deutscher Gartenbau, 32.

Kircher, W. (2004), *Planung von Staudenpflanzungen*. Grünforum, LA 6.

Kircher, W. & Trunk, R. (1995), *Wenn der Zufall regiert – Naturnahe Bodendeckerverwendung*. Deutscher Gartenbau, 21.

Messer, U. (2004), *Stauden im Schatten der Bäume*. Grünforum LA 3

Messer, U. (2008), *Studies on the development and assessment of perennial planting mixtures*. Research Thesis for PhD, University of Sheffield, Department of Landscape, Sheffield, United Kingdom.

Robinson, N. (1998), *The Planting Design Handbook*. Aldershot, England, Goner Publishing Company.

Thompson, P. (1997), *The Self-Sustaining Garden*. London, Batsford.

Wasowski, S. (2002), *Gardening with prairie plants*. Minneapolis, MN, USA, University of Minnesota Press.

The Diagrammatic Landscape

Ross MCLEAN

1 Introduction

This paper reflects on an ongoing research led project with students of landscape architecture at Edinburgh College of Art with the purpose to enhance understanding of visual reasoning, in particular the engagement of spatio-temporal dynamics. The living quality of a visual field is generated by the tension between the spatial forces acting within it (KEPES 1969), providing capacities to elicit sensations, effects and motivated gestures. For landscape architecture to invest visual technique with agency stems from the capacity to visually configure spatio-temporal dynamics, visually translating dynamic interactions through the interplay of abstract gestures and effects. This describes the dynamic potentials of the visual field, but one that requires critically framing the conceptual motivations that direct these complex configurations. The diagrammatic landscape presents this framing, as a performance imperative that seeks to explore how signs are motivated toward particular effects to strategically engage the landscape.

Fig. 1: Example of the composite qualities of montage

2 Key Concepts

This project developed from research that synthetically framed visual concepts from contemporary practice, then presenting these to students within a workshop structure where in turn they could experiment with abstract material and provide a reflective basis to refine the conceptual framework. This process sought to explicate the nature of visual configuration, where the diagrammatic landscape is a synthetic framing to enhance knowledge of visual performance. From this we can summarise some of the key concepts that were effective in advancing understanding on visual reasoning.

Abstracting & Relating: The relational aspect describes a process that allows us to manipulate the pliability of landscape through an art of relationships that critically engage contemporary circumstances (CORNER 1999). Signs consist of elements that act as substitutes to what they reference in reality, which combines to visually display the nature of a dynamic interaction through relational conditions (VIDLER 1999), elucidating spatial dynamics with a view to the transformation of a given context.

Fig. 2: Example of the range of visual elements, from abstract to pictorial

Hybridising & Interrelating: Two different modalities are particular to the visualisation of landscape; 'vertical representations to horizontal signs' (KWINTER 1992). By identifying this planar modality a comprehensive visualisation of the landscape can be formed, where operational overview is dialectically interrelated with scenic perspective. Hybridising can work across planes of modality, juxtaposing visual information by integrating both scenic and operational modalities within a single dynamic visual field.

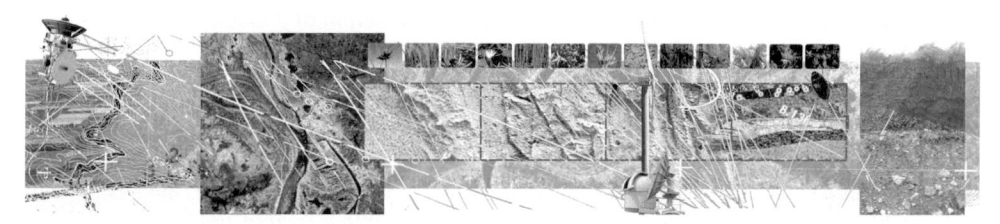

Fig. 3: Example of the diversity of elements and scales within synthesised configuration

Synthesising & Constructing: the visual field can interrelate across scales, modalities and planes of expression, where varying scales, scopes and types of data can be brought into expressive interplay, shifting between the general and specific, individual and collective, fluid and fixed, as a dynamic overview where relationships between patterns, process and scale, events, movements and space can be examined.

Compressing & Intensifying: through a process of elimination and reduction, the isolation of specific aspects can create precise statements of expression, as a progressive compression of detail that retains information relevant for a particular purpose (HOFSTADTER 1979). Attention to the signifying act of compressing relates to the motivation of producing an intensification of effect, free from superfluous expression.

Dismantling & Re-connecting: The process of layering makes complex situations more manageable through dismantling proceeding through a set of criterion, which act to rationalise and filter the landscape. The process of layering provides a strategy of revealing through a dismantling that aims for dynamic interplay through re-connection, as a complex interplay which combines to give heightened intensity when fused.

Fig. 4: Example of the configuration of layered and phased visual orders

Phasing & Adapting: Phasing provides a strategy to make distinctions on the temporal performance of landscape, sequencing predictive phases as a calculated projection over key stages of developmental increments. Phasing also offers the potential to give up the assumption of long term prediction (CZERNIAK 2001), recognising that economic, social and ecological patterns require adaptive sequencing, where phases visually determine resources to catalyse new phases of development.

Aligning & Enabling: Aligning evokes processes, systems and structures caught in supple fusion, of interactions emerging, self organising, adapting and shifting, configuring more dynamic processes of evolutionary change. This anticipatory framework eables phenomena to emerge, expand and proliferate as an orchestrated simulation of dynamic behaviours.

Differentiating & Synchronising: At a more advanced level visualisation can involve multi-ordered lines of configuration to configure a co-evolving visual trajectories that correlate differing spatio-temporal timelines with their own internal logic of programme, structure and process. Synchronisation occurs through linear and lateral parallelism that help to monitor multi-variant processes and emergent structures, where timeline has its own nature and pattern of growth as a complex of signs, set within a field of fields.

3 Conclusion

Within each key concept presented here there are many more considerations to enhance understanding of visual performance, but this summary outlines primary ideas for students to regard when exploring the compositional versatility of visual language. An important aspect of this understanding is how the process of configuration has strategic implications for engaging the spatio-temporal condition of the landscape, where a more conscious engagement of the signifying acts that determine visual configuration enhances understanding of how visual performance connects strategically to those of the landscape; of structures, systems, and processes. This is an important issue, as understanding the visual alone can lead to superfluous statements, but when a greater sense of the visual capacity to connect to the strategic implications of constructing landscape is grasped it enhances the operational potentials of visual performance.

Ferdinand de Saussure identified that the conception of meaning in sign systems was structural and relational rather than referential, proposing that no sign makes sense on its own but only in relation to other signs (SAUSSURE 1983). This establishes that the material practice of landscape architecture lies in signs, symbols and associations, which condition a compositional approach to design. The diagrammatic idea places emphasis on the relational, as both the power of composite interplay, but also the relational performance of abstract material to be conceptually connected to a sense of fabricating the landscape, where to determine the sign is ultimately to determine the landscape. This emphasises the productive, as much as the representational, qualities of the visual as a signifying process that seeks to interconnect thought with production.

What is important in this research is that often exploration into visual performance tends to focus on the result rather than the process of visual reasoning, where signs are put into types, rather than a process to explore the generative qualities of visual material to elucidate spatio-temporal performance. This project sought to enhance the eloquence of students' visual fluency, to further question the strategic and pragmatic implications of a signifying process to advance their basis of operating.

References

Blackwell, A. (2001), *Thinking with Diagrams*. Dordrecht, Kluwer Academic pub.
Corner, J. (1999), *The Agency of Mapping: Speculation, Critique and Invention*. In: Cosgrove, D. (1999), *Mappings*. London, Reaktion Books, 213.
Czerniak, J, (2001), *Downsview Park Toronto*. Harvard Design School, Prestel.
Eco, U. (1976), *A Theory of Semiotics*. Bloomington, Indiana Uni. Press.
Hofstadter, D. (1979), *Gödel, Escher, Bach,* NY, Basic Books.
Howett, C. (2002), *Systems, Signs and Sensibilities.* In. Swaffield, S. (2002), *Theory in Landscape Architecture: A Reader*. Uni. of Penn. press, 108.
Kepes, G. (1969), *The Language of Vision.* NY, Dover.
Kwinter, S. (2003), *Who's Afraid of Formalism*, In: Zaera-Polo, A. & Moussavi, F. (2003), *Phylogenesis*: *FOA's Ark.* Barcelona, Actar pub., 96.
Saussure, F. (1983), *Course in General Linguistics,* London, Duckworth.
Vidler, A., Zegher, C. & Wigley, M. (1999), *Diagrams of Utopia: the Activist Drawing* London, MIT Press.
Weller, R. (2006), *An Art of Instrumentality: Thinking Through Landscape Urbanism.* In: Waldheim, C. (2006), *The Landscape Urbanism Reader.* NY, Princeton Archi. Press, 69.

From Reality to Virtuality and Back Again – Teaching Experience within a Postgraduate Study Program in Landscape Architecture

Pia FRICKER, Christophe GIROT, James MELSOM and Pascal WEMER

1 Introduction

The central theme of this paper is to illustrate new didactic concepts for implementing visualization techniques within a one-year postgraduate program (MAS LA). Within the broad categorization of visualization in the thematic of Landscape Architecture, we are principally interested in the envisioning and development of landscape design concepts. Given the breadth of possible visualization techniques and applications, the postgraduate students are encouraged to both learn specific workflows and actively propose alternative methods of working and visualizing space. It is not our goal to train specialists in CAD software but rather give the students an overview of current information technologies relevant to their field. The paper reports on concrete examples of the use of information technologies in teaching and their related potentials as well as challenges.

If one asks leading Swiss landscape architecture offices about the visualization skills of university graduates, the answer may be anywhere from sobering to alarming. Based on the current poor state of the educational system for landscape architects within Switzerland, most graduates are not able to represent large-scale landscape architecture projects in a convincing manner. Since the term landscape architect is not protected in Switzerland and there are no masters-level programs, the need for action is acute. The general focus of the Master of Advanced Studies in Landscape Architecture (MAS LA) of Professor Girot's Chair of Landscape Architecture at ETH Zurich is the integration of cutting-edge modeling and visualization technologies as design tools within the field of large-scale landscape architecture. The concept presented here is directed towards bringing graduates of planning disciplines as well as people with extensive professional experience up to date in terms of planning and analytical tools.

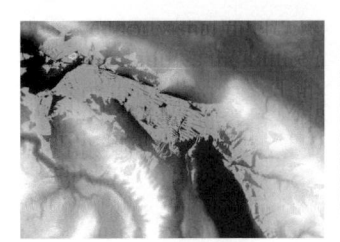

Fig. 1: Student work: Viewshed analysis using SAGA and ArcGIS (Integration of swiss topo data, terrain analysis – analytical hillshading, manipulation of point cloud data)

2 How to Teach 'New Design and Visualization Tools'?

At present, information technologies are an essential component of design and construction. Contemporary architecture and landscape architecture as designed by top offices would be unthinkable without them. Without computer-assisted manufacturing and logistics, modern form language and structural solutions would hardly be realizable. The current trend in designing spectacular and complex forms and structures is ongoing. However landscape architects often tend to be skeptical of or helpless in using these technologies. This is why reflections as well as questions of method and theory stand at the forefront of our efforts.

The development of a digital chain poses an essential element to this end. Subsequently, we would like to take a look at the exercises we have conceived to supplement the learning process. The development of new design and visualization tools and their linking with each other stands at the forefront of the MAS LA. We do not limit the term 'visualization' to the photorealistic depiction of a design, but rather see it as a highly differentiated playground with application possibilities from the beginning of a design process until the final implementation phase. Elements like using GIS information, visualization tools, and programmed tools should be applied professionally and adjusted according to varying requirements.

2.1 We program

What are the tools that adequately serve current landscape architectural trends and how can they be conveyed? The past years have shown us that programming within architecture has become as commonplace as CAD drafting (MITCHELL 1990). Within urban planning but also building construction, for example, parametric designs are often the only solution to dealing with complex form language. To this end we would like to use the computer to realize landscape architectural projects that would not be possible using conventional methods (BOHNACKER, GROSS, LAUB & LAZZERONI 2009). This requires the further development of digital tools, which allow for the subsequent design and working with the extracted information.

2.2 Applications within teaching

The communication of such new ways of thinking poses a challenge, especially in a postgraduate program with students with different academic backgrounds and interests. Through simple, manageable exercise sequences, we attempt to give insight into the many possibilities of programming and parametric landscape design (LEINONEN 2010). The resulting skills may then be expanded in their final synthesis project and used in conjunction with a design. The course of studies is intended to be a part-time program over two semesters and is divided into 7 themed modules (Landscape Modeling, Landscape Visualization, Research and Analysis, Landscape Programming, Landscape Scanning, Working with GIS Data, Landscape Video and Photography) and 1 concluding synthesis module. The modules focus on the practice-oriented use of current CAAD/CAAM (computer-aided architectural design/computer-aided architectural manufacturing) technologies in the area of landscape architecture. Each module begins with a phase where new techniques are learned. In this phase, individual exercises are connected to current issues in landscape architecture. In the second part of the module, participants grapple with complex problems, which will be discussed during a concluding presentation within the framework of a panel discussion with a group of experts.

Three modules of the current MAS LA program will now be described that are exemplary of the program's technical and didactic implementation:

3 Basis Data for Landscape Architectural Design

In this module the students learn where and how they can glean relevant topographic basic data for the development and visualization of their design. In addition to learning how to integrate 3D data from national land surveying offices in various GIS applications, the main focus is on capturing the topographic characteristics of a landscape space using photography, video and landscape scanners on-site.

The possibility for acquiring up-to-date planning data by using a high-resolution long-range terrestrial scanner gives us the ability to control the digital chain from the outset. As opposed to spatial planning, the use of GIS data in landscape architecture is still somewhat new. A large didactic challenge involves the overcoming of the optical inhibition threshold. Conventional GIS processing programs used by spatial planners have a unique visual language, which many landscape architects find rather unappealing. The indisputable potential of these programs in landscape architecture is tested in a customized module.

3.1 Method

In order to document the specific topographic characteristics of a place, the students learn techniques from the area of photogrammetry, among others. A 3D model can be created using image recognition software and a collection of photos or video stills of the project area. The use of a mini-drone (UAV) offers an alternative method for modeling and visualizing territories of several square kilometers (i.e. a floodplain or river delta) in three dimensions.

An additional method for gathering 3D information on a part of the landscape is the use of terrestrial 3D laser scanners. In recent time, this tool from the field of geodesy has shrunk from the size of a small truck to that of a football and is simple enough for students to use quickly and efficiently.

The sampling of the landscape delivered with 125,000 light pulses per second can generate – in combination with photos – a photorealistic point cloud that can, in turn, be combined with less detailed (i.e. aerial) data from national land surveying offices in order to create a complete 3D model of the project area.

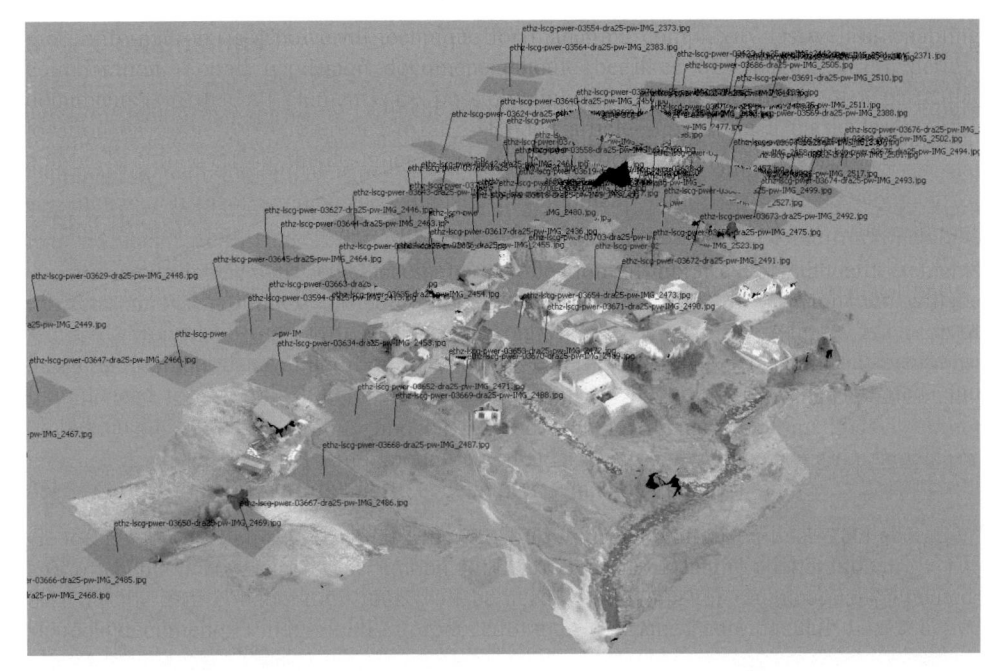

Fig. 2: Airborne image-based 3D modeling of a student project site

Fig. 3: Students working with a terrestrial laser scanner (TLS) near Lago Lucendro (Gotthard)

3.2 Challenge

One of the greatest challenges with regard to the use of 3D GIS data is not only the handling of 3D data in general but also the sheer numbers of polygons, points, or pixels involved. If the relevant data is found on an official databank or gathered on-site, these have to be in the right format, in an appropriate resolution, as well as uploaded, managed, combined, filtered, and visualized in coordinate system that is compatible with already existing information. In addition, the integration of data gathered in this manner with modeled project designs puts relatively heavy demands on the spatial imagination of the students for the navigation and their ability to work on the models in the 3D environment. This, on the other hand, is connected to a certain learning curve. An additional hurdle is the repeated import and export as well as transformation of the GIS data in a large variety of other applications and data formats as well as types.

Fig. 4: Combination of point cloud data from airborne and terrestrial laser scanners, Gotthard tunnel entry Airolo

3.3 Advantage

In the teaching context, students can profit from more favorable conditions in acquiring official data and the use of the still relatively costly laser scanners. However, most of the students of the course will most likely never have access to their own laser scanner with which they can collect the required GIS data on-site. Still one will confront data originating from aerial or terrestrial laser scanners more and more often, and it will become increasingly important for project development in landscape architecture. Thus a proficiency in handling this data will become more and more significant.

Regarding the use of photogrammetric software, this remains – at least in the consumer sector – a quick, low-cost and relatively precise method in order to generate one's own 3D data of the topography of a specific place. Especially for self-employed recent graduates or those who return after the course as an employee to their office, these new software applications offer a good alternative to the often less-detailed data from official sources.

4 Module Landscape Visualization

4.1 Structure

The module is divided into separate foundation and advanced modules that compliment the intermediate modules. These take place both early in the course as well as in the later stages. The testing of the techniques takes place with application to both discreet assignments, and to the ongoing yearlong Synthesis project.

The initial visualization module takes place early in the course in order to familiarize the students with fundamental software and workflow techniques, as well as establish a strong communication strategy early in the synthesis project, which can also serve them in the later modules (MERTENS 2010).

Various techniques and approaches to landscape visualization are introduced, from traditional analogue techniques to digital techniques borrowed from other disciplines. (BISHOP & LANGE 2005) Practically, this ranges from the novel use of site visit-based photography and sketching, to the generation, modeling and manipulation of colored point clouds.

Fig. 5: Students combine on-site photography and material textures with their ongoing design modeling development

The main exercise is a site with an established and restricted problem-set, allowing broad experimentation, yet limiting the risk of focusing on design issues alone. Students are encouraged to recognize their own preferred techniques, working preferences, and aesthetic choices, which are supported in the development of new directions and manners of working.

4.2 First module – immersion in landscape visualization Structure

Within a four-week structure, the students are introduced to various processes, software combinations, and workflows. A single site provides the basis for the individual application of the various techniques, which are then combined intuitively by the students based on their proposals for the site.

The choice of project site stems from its specific landscape themes and clear problem-set. This broad categorization has lead to sites ranging from the reconfiguration of a site through the structured placement of excess tunnel excavation material in a suburban/agricultural context, to the redevelopment of an evolving urban transport node. Key to the choice is easy access for the students, as initial spatial impressions and photographic site data are key to the process and techniques. Additional data beyond simple CAD and GIS sources is deliberately limited and focused, requiring the students to observe, record and generate their own impressions, observations and additional data. This approach emphasizes the visual and sensual aspects of the site, and leads the students to develop their own personal skills of site-analysis and development. During the module, there is a clear emphasis on the process, over the design outcome, in terms of documentation and experimentation. The design results, however, often benefit greatly from this means vs. goals approach.

This focused approach also leads to the use of a limited number of software, namely Rhinoceros and related plug-ins such as VRay, RhinoTerrain, and various custom-made plug-ins. Since the software is not specifically landscape oriented in their application, specific workflows and tools, rather than generalist knowledge is the focus. The content and visual approach of the final project presentation is similarly directed towards landscape design outcomes, rather than requiring a specific content or format. The result combines abstract data, photography, and rendering with plan, section, and perspective in a freeform and often experimental manner.

4.3 Second module – generative algorithm workflows in landscape visualization

The second visualization module begins with a workshop in which the various possibilities of generative modeling and visualization are explored. The Grasshopper plug-ins and related additional extensions lend Rhinoceros a visually simple method of approaching flexible and extensible visualization workflows (HIGHT 2008). Various data inputs from Rhinoceros can be combined with external inputs such as GIS data, image sources, environment data, and connected to other software for added functionality.

The project applications shift from brief example problems to the ongoing Synthesis project, allowing the students to directly integrate their findings and developments into their thesis project. This generative mode of working also allows the integration of much of the production from the intervening teaching modules, where programming, photography, point cloud, GIS and abstract analog sources may be combined into discrete landscape design workflows.

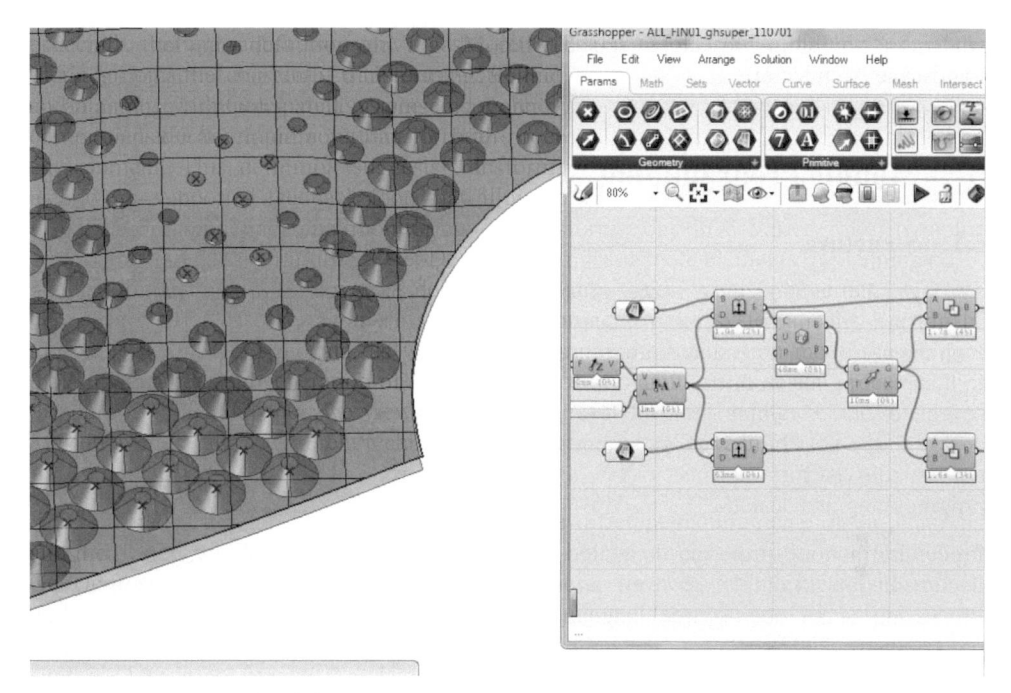

Fig. 6: An example of the large-scale implementation of design parameters, generating volume, surface grading information, and areas

Fig. 7: Project experimentation with point cloud data, combining existing and proposed landscapes in one interactive model

4.4 Challenges within the teaching process

Over the years of course development, the key challenges for the teaching program have greatly re-shaped the method of teaching. Of particular importance are the short timeframes allocated to each module, requiring a well structured and specialized approach. The other aspect that has most likely influenced the course structure the most is the varied background of the students, whether in discipline, methodical approach, culture, or education (HAGAN 2008). This particular challenge has been the impetus to directly focus on the development of design workflows, based on each individual. The key benefit for the students is the ability to further specialize their own interests within the broad field of Landscape Architecture.

5 Module Programming

5.1 Structure

Through the use of customized programmed tools we see a further possibility for supporting design and planning processes in an innovative way. For this, we experiment with the use of Processing as a programming language (REAS & FRY 2010, WANNER 2010). In this manner, the students were introduced to the principles of parametric design. Through several exercises, which build on each other in content and a lecture series with guest speakers, the possibilities of 'programmed design' are learned and discussed. The goal of the module is for students to be able to recognize and define starting points for programming in a design.

Processing is our preferred language of choice due to its simplicity, the excellent documentation available, as well as the ambition of its academic user community. Processing is an open-source language developed by Ben Fry at MIT's Media Lab (GREENBERG 2007). It has the advantage of offering a very simple development environment. In addition, its procedural introduction makes it easy to learn, and we have observed that the learning curve is very good.

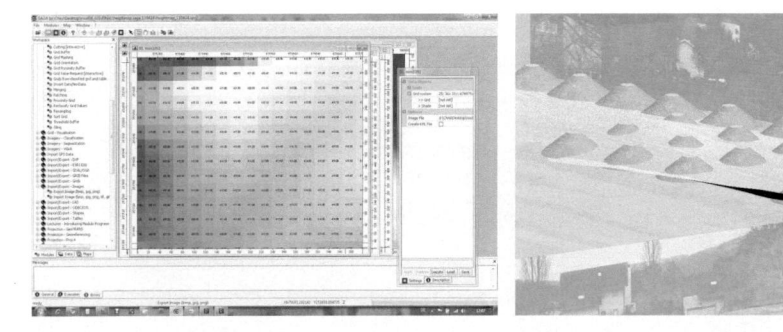

Fig. 8: Student project: Generation and control of a design project by a customized program and verification through 3D output

5.2 Method

The didactic approach poses the greatest challenge in the teaching of a programming language. Landscape architects are visually influenced people, who for the most part feel helpless and at odds towards the mathematic and prosaic structure of a program. This first hurdle is quickly overcome in comparison to other programming languages with the learning of Processing. Already after the first three lectures, students are capable to create scripts with variables and loops and test their scripts in a visual output window. The five-week module concentrates essentially on the learning of the basic principles of modular programming based on a series of exercises built on each other. The relatively short timeframe yields significantly better results in comparison with semester courses, since a beginner at programming can reach his or her goals more quickly in a concentrated period of time as opposed to divided up and distributed over an entire semester. Through the use of sketchpad, the students can exchange scripts and deepen their individual interests. Since we focus on applied programming, we teach the efficient use of existing tutorials including scripts from the Internet. Through each exercise, the script becomes more complex in content and the output adjusted to the individual's demands. The versatile output possibilities of Processing allow the programmed result to be shown, i.e. as a printout, or real or lasered 3D model (BURMEISTER 2006). These different visualization possibilities provide a new level of communication as a design-supportive medium. In addition to the programming, the main second element becomes the critical theoretical debate of possible applications within the framework of professional practice. When does it make sense to integrate a programmer into the planning team, and how do these specialists communicate with each other? These and other questions must be considered in order to understand as well as realize the potential of parametric design.

Fig. 9: Student project: Application of programmed tool to simulate the growth of green pockets

5.3 Next Steps

A next step would be to network with other universities that teach and research in similar topics and areas. The example of the open-source programming language Processing and its active user group is a model for future teaching. In this way, the difficult path of implementing existing knowledge and skills in built landscape architecture can be accelerated and qualitatively improved. Of course not every landscape architect needs to be a programmer. However, we tend to believe that programming constitutes an important role within landscape architecture and therefore a basic understanding must be conveyed within the course of studies. In so doing we place great emphasis on using programming languages that are simple to learn and also offer interfaces for further applications.

6 Summary

Within the framework of a concluding thesis project, students are encouraged to combine the tools and test new forms of working method. We challenge the students to go beyond the boundaries of conventional domains and test the tools in analysis, design, and visualization. The programs and different CAAD/CAAM techniques, which the students have become acquainted with in the different modules, complement each other and are to be applied and recombined to explore new design methodologies in the final project (KOLAREVIC 2003). The concluding module acts as a test case for the questions or agenda defined throughout the teaching year.

During the MAS LA a number of discussion/presentation deadlines serve to test one's individual focus or agenda and make them more precise. At each deadline the students are asked to prepare a meaningful presentation that is complemented by new examples of use. During the final working phase they are supervised during one-on-one weekly meetings. The scope of the final project is defined during interim presentations.

7 Acknowledgements

The research and teaching work described in this paper are the result of a very committed teaching team at the Chair of Christophe Girot, special thanks to all of them..

The authors would like to thank the MAS LA students of 2010-2011: Aemmer Karin, Baumgartner Christine, Figna Lorenzo, Goula Ioulia, Kuratli Salome, Manolis Dimitrios, Stavrotheodorou Vasiliki, Ankita Thaker and 2011-2012: Tom Beterams, Qingzhao Hu, Elpiniki Kekelou, Alexander Kochan, Eleni Nodara, Pascal Werner, for their dedicated participation in the program.

References

Bishop, I. D. & Lange, E. (2005), *Visualization in Landscape and Environmental Planning, Technology and Application.* New York, Taylor & Francis, London
Bohnacker, H., Gross, B., Laub, J. & Lazzeroni, C. (2009), *Generative Gestaltung.* Mainz, H. Schmidt.
Burmeister, K. (2006), *Vom Personal Computer zum Personal Fabricator.* Hamburg, Murmann.
Greenberg, I. (2007), *Processing. Creative Coding and Computational Art.* Friends of Ed.
Hagan, S. (2008), *digitalia.* Routledge, N.Y.
Hight, C. H. (2008), *Architectural Principles in the Age of Cybernetics.* N.Y., Routledge.
Kolarevic, B. (2003), *Architecture in the Digital Age.* Design and Manufacturing. N.Y., Spon Press.
Leinonen, T. (2010), *Designing Learning Tools.* Aalto University, FIN.
Mertens, E. (2010), *Landschaftsarchitektur visualisieren.* Basel/Boston/Berlin, Birkhäuser.
Mitchell, W. (1990), *The Logic of Architecture.* MIT, USA.
Reas, C. & Fry, B. (2010), *Getting Started with Processing.* O`Reilly.
Wanner, A. (2010), *Processing.* New York, Taylor & Francis, London.

A Gender-based Typology of Determinants of Video Games Use by Primary School Children

Evangelia POLYZOU, Nikolaos HASANAGAS and Konstantia TAMOUTSELI

1 Introduction

1.1 Aim

This research examines factors which make children video games users, either incidentally (namely just preferring video games to outdoor play) or systematically (long duration). A gender-based typology of determinants is proposed. The results are expected to contribute to a more accurate understanding of conditions stimulating children to become video game users. They also suggest an understanding of the factors affecting the different reasons video games are used, either as a substitute for outdoor reality of outdoor play or as a complementary pedagogic tool. According to research findings the role of video games varies depending on factors including gender, the educational level of the family, the degree of socialization and individual integration into outdoor play activities.

1.2 Literature review

Extensive research data GEE (2003), GENTILE & GENTILE (2008), SANFORD & MADILL (2007), SQUIRE (2006), LIVINGSTONE & BOVILL (1999), LUCAS & SHERRY (2004), PROVENZO (1992), and SELNOW (1984) refer to the educational and social role of video games, with very little concern for the examination of gender relevance to the perception and use of video games. Gender as a factor affecting preference of place for play is examined in the research of FROST & KLEIN (1979) as well as PELLEGRINI & SMITH (1998). However their studies were not focused on digital landscapes.

FREITAS & NEUMANN (2009) have laid emphasis on the visual learning experiences but not on the pedagogic influence of the interactive virtual landscape of video games which may also offer such experiences. SCARFO (2007), HAMALAINEN (2008), MORENO-GER et al. (2008), VIRVOU & KATSIONIS (2008), ÜZÜN et al. (2009), KRIPPNER & MEDER (2011), VANGSNES at al. (in press) have pointed out the didactical usability of virtual reality and computer games. HERRINGTON & STUDTMANN (1999) and JANSSON (2010) have analyzed the pedagogic function of landscape but they did not focus on the pedagogic function of video games which are also based on digital landscapes. ORLAND et al. (2001) have emphasized the possibility of using virtual worlds as tools for perceiving landscapes but without pointing out the pedagogic framework of digital landscapes.

1.3 Innovation and applicability in Landscape Architecture

Research results concerning factors which affect the attitude of young people toward video games have not been developed until now. Thus, such a research can be considered as purposeful and innovative. In this research it is intended to develop a gender-based typology of determinants and user groups.

Landscape Architects are not only involved in design of real playgrounds. Some them are also interested in designing virtual landscape used in video games (ADAMS 2002, PACYGA 2011). Therefore, Landscape Architects should be aware of the free-time behavior and preferences of children regarding playing not only in real landscapes but also in video games. They should also be aware of certain determinants which have proven to exert important influence on this behavior and preferences such as gender, family and social environment. The knowledge of such factors can be useful for designing more attractive digital landscapes with more desirable or effective pedagogic and social influence depending on gender and the social and family environment of the users. The paper is only going to point out that users of different gender present different preferences and behavior as video game users. The exact elements they prefer on the digital landscape does not pertain to the aim of this paper but it may constitute a question for further research. Certain landscape features whose perception and preference depends on the gender of the observer have been proposed (DOWLER et al. 2005, HASANAGAS et al. 2007, LEVI-STRAUSS 1969): „private" (feminine) vs „public" (masculine), „exotic" (feminine) vs „European/ western" (masculine), water (feminine) vs land (masculine), „nature" (feminine) vs „culture" (masculine), while it is also research-worthy whether „wilderness" is perceived as a feminine or a masculine property of a digital (and also real) landscape. More specific features whose possible gender-related perception may be examined in future research projects could be situations of danger or risk, colors and color-related values, landscape diversity, biotic and abiotic elements, categories of fauna or flora (e.g. meso- or megafauna). Moreover, preferable combinations of these features and thereby landscape types can be proposed by future research. The applicability of findings related to digital landscape in real landscapes is also a question for future research.

This paper has basic rather than applied research orientation. Moreover, an important education-related function of this paper is to show to Landscape Architecture students that the gender as well as social and family characteristics should be taken into account in the design of digital landscape of video games as well as of real landscapes which are expected to exert a pedagogic influence as playgrounds and this topic constitutes a challenging research issue.

2 Material and Methods

Two hundred fifty two (252) fifth and sixth-grade students constitute the sample in this study. The participants belong to different primary school classes from various geographical areas in Greece (Athens, Thessaloniki, Kastoria and Crete). The balanced synthesis that has been achieved corresponds to analytic statistics (correlation between variables) that has been used to extract results. It was a judgement sample achieving the best possible balance of basic charcteristics (variety of places ranging from rural to urban areas etc). This is the most appropriate sample for analytic statistics, as it makes correlations more valid.

Standardized questionnaires were distributed to the pupils and were filled out with the help of teachers, when necessary. Pupils were asked to answer questions about their satisfaction with outdoor play time, the duration of their play time with video games and their preferred landscape elements in virtual world. In-depth interviews with pupils were conducted first to formulate the questionnaires and to interpret the quantitative results. The correlations were extracted by using the Pearson test, after normality control (HASANAGAS 2009 and 2010). The gender-based typology was constructed considering the moderating effect of gender on the correlations.

3 Results - Determinants of Susceptibility to Playing Video Games

3.1 General determinants

Gender (Table 1) seems to be a basic factor of susceptibility to acquiring a preference for the digital experience of the imaginary world of video games (-0,226). Children who are not well integrated into groups (0,172) or are not quite satisfied with the time available to them for outdoor play (0,127) tended to use video games as an alternative. On the contrary, children who feel that they have enough time to play outdoors tended to avoid using video games (-0,210).

252 total	Play (play outside =1, play video games=2)
Gender (boy=1,.girl=2)	-,226(**)
	,000
Exclusion from outdoor play activities	,172(**)
	,006
Inadequacy of outdoor play time	,127(*)
	,044
Adequacy of outdoor play time	-,210(**)
	,001

** Correlation is significant at the 0.01 level (2-tailed).
* Correlation is significant at the 0.05 level (2-tailed).

Table 1: General determinants of susceptibility to playing video games

3.2 Determinants in the case of boys

The lack of outdoor play integration (0,176) (Table 2) and the inadequacy of available outdoor play time (0,186) appear to be the main factors of susceptibility to digital experience for boys. This can be attributed to the aggressiveness typically displayed in boys' behaviour in everyday life which often leads to adventurous activity which can often make integration into play groups more difficult (GENTILE & GENTILE 2008, LUCAS & SHERRY 2004).

149 boys	Play (play outside =1, play video games=2)	Duration of playing video games (no=1, 1-2h/w=2, 3-14h/w-3, up to 14h/w=4)
Exclusion from outdoor play activities	**,176(*)**	,153
	,031	,063
Inadequacy of outdoor play time	**,186(*)**	,032
	,023	,697
Adequacy of outdoor play time	**-,286(**)**	-,096
	,000	,243

** Correlation is significant at the 0.01 level (2-tailed).
* Correlation is significant at the 0.05 level (2-tailed).

Table 2: Determinants in the case of boys

3.3 Determinants in the case of girls

Inadequate integration or the lack of outdoor play time does not affect girls' video game experience. Their experience with video games is typically related to the educational level of their parents (0,203, 0,261) and is a factor related both to the time they spent with video games and the type of video games chosen. Girls use videogames as a pedagogic tool rather than as an escape from a social reality. This could be explained based on research that views entertainment video games as learning experiences that offer excellent educational design with challenge, motivation, and learning principles that may even be used to improve formal educational approaches (GEE 2003, GENTILE & GENTILE 2008, SANFORD & MADILL 2007, SQUIRE 2006).

103 girls	Play (play outside =1, play video games=2)	Duration of playing video games (no=1, 1-2h/w=2, 3-14h/w-3, up to 14h/w=4)
Father's studies (no=1, yes=2)	,004	**,203(*)**
	,966	**,040**
Mother's studies (no=1, yes=2)	,064	**,261(**)**
	,524	**,008**

** Correlation is significant at the 0.01 level (2-tailed).
* Correlation is significant at the 0.05 level (2-tailed).

Table 3: Determinants in the case of girls

4 Conclusions and Outlook

Gender has a moderating effect on the variables and thus can be a sound basis of typology. In the case of boys, socialization factors, particularly the social game environment outside the home, seem to be the driving forces for their becoming mainly incidental users of video games, while girls tend to become systematic users of video games due to family-related factors. Thus, video games tend to be a means of escape for boys with socialization deficits, whereas for girls video games become an educational tool under the guidance of parents.
Challenges for future research include the collection of a larger sample from a wider geographical area and a possible typology according to the place of origin or of residence. A basic question for future research is also what types and characteristics of digital landscapes the users prefer and what the determinants of this preference are.

References

Adams, E. (2002), *The Role of Architecture in Video Games*. Gamasutra (Gamasutra.com).
Dowler, L., Carubia, J. & Szczygiel, B. (2005), *Introduction. Gender and Landscape. Renegoatiating morality and space.* 1-15. In: Gender and landscape: renegotiating morality and space. Ed. by Dowler, L. et al. Oxon and NY, Routledge.
Freitas, S. de & Neumann, T. (2009), *The use of 'exploratory learning' for supporting immersive learning in virtual environments.* In: Computers & Education, 52 (2), 343-352.
Frost, J. L. & Klein, B. L. (1979), *Children's Play and Playgrounds*. Boston, MA, Allyn & Bacon
Gee, J. P. (2003), *What video games have to teach us about learning and literacy.* New York, NY, Palgrave Macmillan.
Gentile, D. A. & Gentile, J. R. (2008), *Violent video games as exemplary teachers: A conceptual analysis.* In: Journal of Youth Adolescence, 37, 127-141.
Hamalainen, R. (2008), *Designing and evaluating collaboration in a virtual game environment for vocational learning.* In: Computers & Education, 50 (1), 98-109.
Hasanagas, N. D., Styliadis, A. D., Papadopoulou, E. I. & Birtsas, P. K. (2010), *Land Policy & Socio-Spatial Impacts in a burned forest: The case of Chalkidiki, Greece (2006). RevCAD.* Journal of Geodesy and Cadastre.
Hasanagas, N. D., Birtsas, P. & Kyprou, S. (2007), *Perception of Environment by the students of Environment-related Schools.* Proceedings 2921-2926, CEMEPE/ SECOTOX. Skiathos island, Hellas.
Hasanagas, N. D., Charalampous, A. & Moutsou, G. (2009), *Restoration alternatives and attitudes of local population towards Kavala quarry – Northern Greece.* Ed. by Kungolos, A. et al. Proceedings of the 2nd International CEMEPE & SECOTOX Conference, Mykonos, Greece, 1591-1597.
Herrington S. & Studtmann, K. (1999), *Landscape interventions: new directions for the design of children's outdoor play environments.* In: Landscape and Urban Planning, 42 (2-4), 191-205.
Jansson, M. (2010), *Attractive Playgrounds: Some Factors Affecting User Interest and Visiting Patterns.* In: Landscape Research, 35 (1), 63-81.

Krippner, U. & Meder, I. (2011), *Cultivating, Designing, and Teaching: Jewish Women in Modern Viennese Garden Architecture.* In: Landscape Research, 36 (6), 657-668.

Levi-Strauss, C. (1969), *The Raw and the Cooked: Mythologiques, Volume 1 (Raw & the Cooked).* University of Chicago Press.

Livingstone, S. & Bovill, M. (1999), *Young people, new media: Report of the research project: Children, young people and the changing media environment.* London, London School of Economics and Political Science.

Lucas, K. & Sherry, J. L. (2004), *Sex differences in video game play: A communication-based explanation.* In: Communication Research, 31(5), 499-523.

Moreno-Ger, P., Burgos D., Martínez-Ortiz, I., Luis Sierra, J. & Fernández-Manjón, B. (2008), *Educational game design for online education.* In: Computers in Human Behavior, 24 (6), 2530-2540.

Orland, B., Budthimedhee, K. & Uusitalo, J. (2001), *Considering virtual worlds as representations of landscape realities and as tools for landscape planning.* In: Landscape and Urban Planning, 54 (1-4), 139-148.

Pacyga, J. (2011), Landscape architects and designers as video game designers. SketchUp for landscapes (http://sketchupland.posterous.com/landscape-architects-and-designers-as-video-g, last visit: 2-4-2012).

Pellegrini, A. D. & Smith, P. K. (1998), *The development of play during childhood: forms and possible functions.* In: Child Psychology & Psychiatry, 3 (2), 51-57.

Provenzo, E. F. (1992), *The video generation.* In: American School Board Journal, 179 (3), 29-32.

Sanford, K. & Madill, L. (2007), *Understanding the power of new literacies through video game play and design.* In: Canadian Journal of Education, 30 (2), 432-455.

Scarfo, B. (2007), *A working paper on the structuring of environmental design education. In:* Landscape Research, 13 (3), 3-5.

Selnow, G. (1984), *Playing videogames: The electronic friend.* In: Journal of Communication, 34, 148-156.

Squire, K. (2006), *From content to context: Videogames as designed experiences.* In: Educational Researcher, 35 (8), 19-29.

Tüzün, H., Yılmaz-Soylu, M., Karakuş, T., İnal, Y. & Kızılkaya, G. (2009), *The effects of computer games on primary school students' achievement and motivation in geography learning.* In: Computers & Education, 52 (1), 68-77.

Vangsnes, V., Økland, N. T. G. & Krumsvik, R. (in press), *Computer games in pre-school settings Didactical challenges when commercial educational computer games are implemented in kindergartens.* Computers & Education.

Virvou M. & Katsionis, G. (2008), *On the usability and likeability of virtual reality games for education: The case of VR-ENGAGE.* In: Computers & Education, 50 (1), 154-178.

Materials and Digital Representation

Anna M. THURMAYR

1 Introduction

The ability to make use of digital technologies has become second nature to young adults today. Unlike their middle-aged teachers, most recent students remember doodling on the computer as part of their kindergarten experience. These students, familiar with the digital techniques of drawing, erasing, and painting since childhood, can all explore new software playfully, quickly and with great ease.

This is in strong contrast to the poor knowledge of regulatory issues and construction techniques among students in undergraduate courses. Very few students have carried out activities such as working with concrete in their families' back yards and standard construction formulae well known to earlier generations of students, such as the 1:2:3 ratio mix of cement, sand and gravel, are unfamiliar to them.

In an attempt to address this situation the author – teaching in the traditional field of "Construction Materials and their Application in Landscape Design" at a Canadian university – is experimenting with basic concepts of both construction detailing and computer aided design. Student reflection on the inherent sensuous and vibrant qualities of materials and their digital representation plays a key role in this experiment.

2 Material Collection

When working at the interface between conceptual design and detailed design it is essential to be responsive to materials, textures, colours, light etc. "Over the past decade ... many universities have built up material collections to provide students with access to material samples" (JOST 2011, 126). Students "can touch materials, see how they react to light, examine their surfaces, and learn how certain materials might work in combination" (JOST 2011, 124). Of course there is nothing like having the physical materials on hand, but as long as there are limitations on the establishment of a significant collection of real samples a digital material library can act as an effective substitute for hands-on learning to a significant degree.

3 Field Trips as Source

The author's course starts with research in the field. Trips to nurseries, quarries, lumber yards, and urban sites support lectures about materials in a demonstrative way and are themed "walk, observe and understand" (Fig. 1). These trips serve as a source of knowledge for students and also allow them to start developing an appreciation of the range of materials used in construction today, such as natural stone, concrete, brick, asphalt, wood, plants, metal and water.

Fig. 1: Students attending field trip "Lawn and Meadow"

Fig. 2: Field trip journal "Lawn and Meadow" by student Nathan McLeod

4 Digital Material Library

When asked to collate impressions students do not carry back samples from the field but create image-based, archetypical textures by hand or by camera and store them digitally. The graphic work of the book "Miniature and Panorama" (VOGT 2006) serves as the inspiration for setting up and laying out the collected pictures. Field trip journals (Fig. 2), a written summary of the main characteristics of each material and a collection of precedents showing best practice are appended to the digital images. Students can then begin to build up a personalized library showing a wide range of materials, their manufacture and application in landscape design. By starting a digital material library in the early stages of their careers students can create an indispensable archive, which can be added to and developed as they go through their professional lives. This library will ultimately be as unique as a student's own handwriting (Fig. 3).

Fig. 3: Stone samples of the Material Library by student Bing Wang;
Lawn, plant, soil samples of the Material Library by student Bret Mack

5 Material Rendering

The CAD software product Vectorworks, together with the fully integrated Renderworks developed by Nemetschek AG, allows easy interaction of CAD drawing with image-based textures from file formats such as TIF, JPG, GIF and BMP. This system is well known among professionals in Europe for its presentation capabilities in all phases of the design process, but is less developed in other parts of the world. Imported bitmap images can be

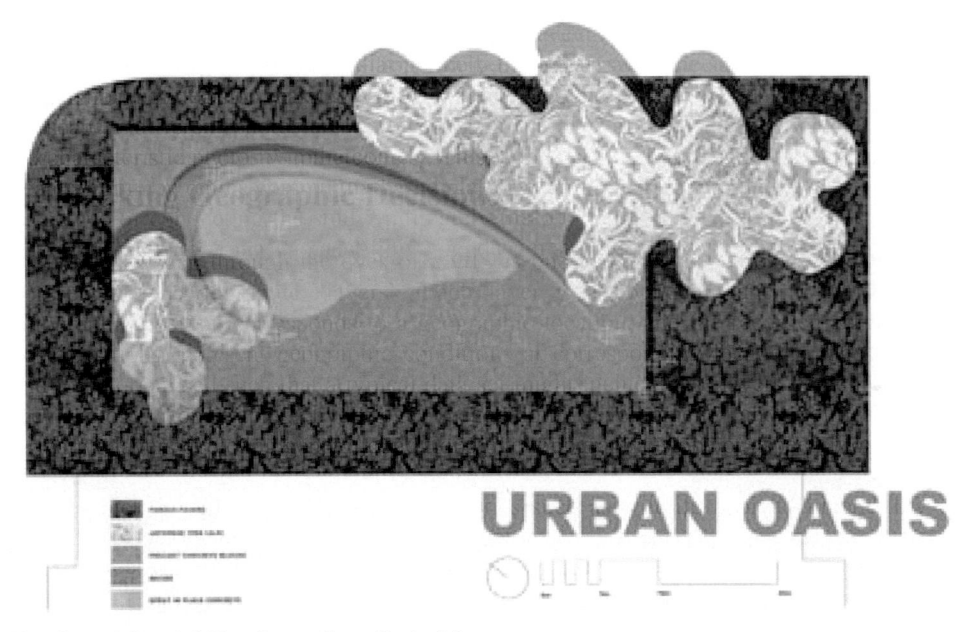

Fig. 4: Material Plan by student Carly Moore,
based on a plaza design of Tedder Timmermanns, landscape architects

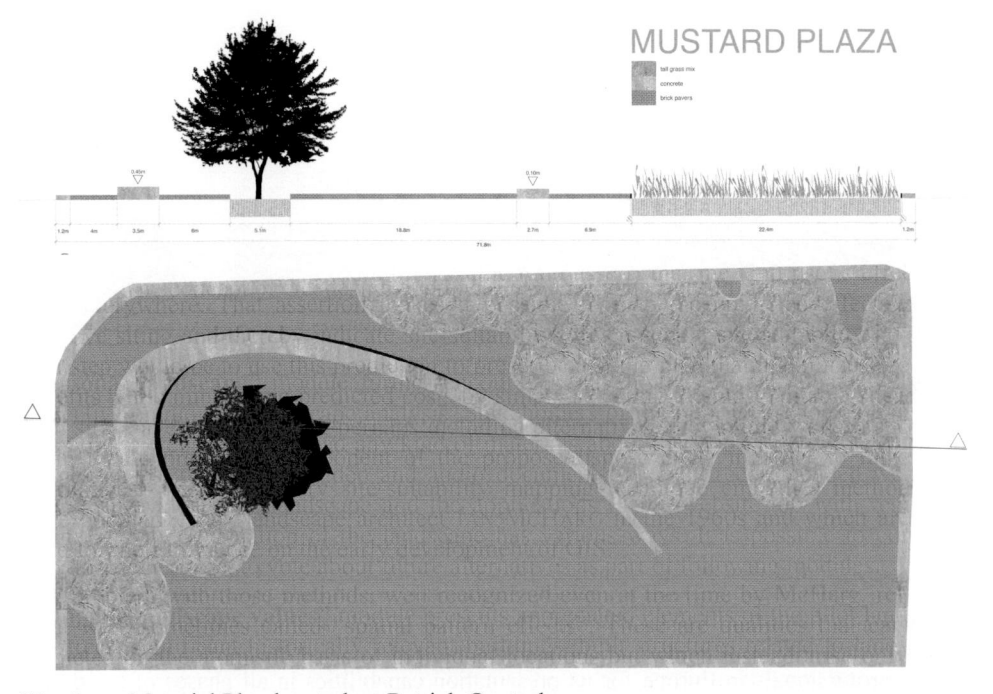

Fig. 5: Material Plan by student Patrick Oystryk,
based on a plaza design of Tedder Timmermanns, landscape architects

applied to vector based graphics and subsequently repeated, reshaped and rendered in real time. The 3D modelling program SketchUp supports similar rendering possibilities and has found the way into many design offices worldwide. Anyone who ever spent time in the pre-digital era placing Letratone sheets by hand on manual drawings will understand the importance of these revolutionary software developments. Switching back and forth between basic CAD layouts and rendered drawings is now easily done so it is remarkable that Vectorworks unlike SketchUp is still playing a secondary role within professional CAD applications in Canada.

Rather than comparing and evaluating programmes this paper aims to discuss teaching experience in depth from an expert's view. The author has been using Vectorworks for more than 13 years. But no hindsight is required either to appreciate the high value of an organized digital library using Vectorworks or to learn the application of materials. Following the first assignment students are asked to develop a rendered Vectorworks plan using their digital material collections. Since more than 30 students register in this course, it is not possible to explore detailing issues throughout a design process. Therefore a sketch design done by professionals is placed at the student's disposal for reference purposes. On the basis of this provided concept the students work out a material proposal (Fig. 5) and some construction details (Fig. 6). The final assignment in the course enhances knowledge about materials and their application and activates graphical experiments. This procedure also reflects a situation that is common in design offices as in most cases the responsibility for design and construction detailing is distributed amongst several people.

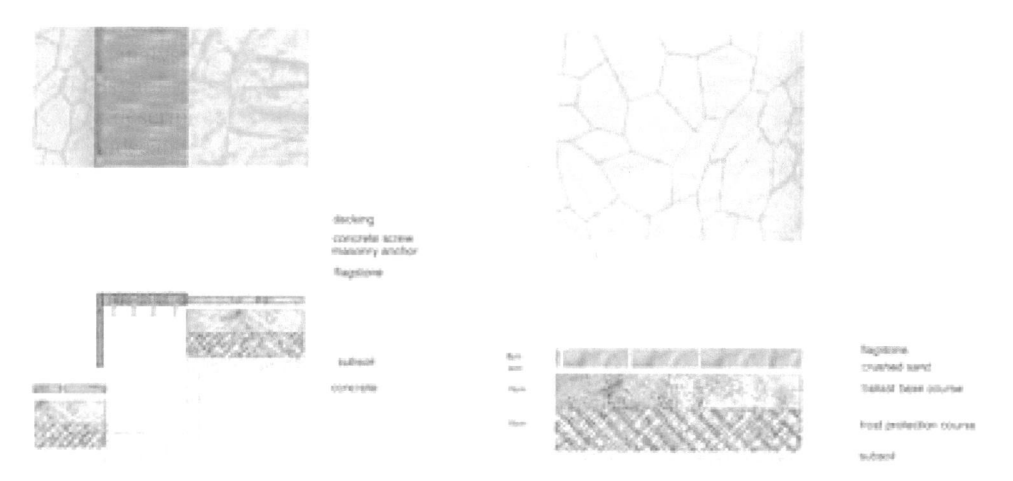

Fig. 6: Material Details by student Trent Workman

6 Results

When the author initially introduced this software into her teaching in a Canadian program of Landscape and Urbanism (three years ago) she received scepticism from both teaching colleagues and students. The author's colleagues were not convinced that the software was relevant to their own subject matter and students were unsettled as AutoCad is the predominant software in Canadian practice. But exceptional work has been submitted over the last three years and more and more students articulate their interest in Vectorworks today. Students who went through the course and who are now attending graduate courses also value their training in Vectorworks components and libraries, since "it has translated to other programs and been very helpful" (Carly Moore). These results and continuing feedback are encouraging the author to pursue this experiment and to try to establish it further.

As it is imbedded in a Vectorworks file the digital material library allows endless rendering possibilities, which are not only to scale but can also be used for the development and representation of further design projects. Thus, a construction material course that includes the application of this software functions as an excellent foundation for design studios and/or seminars (Fig. 7).

The sense that the author tries to instil into her teaching is that the art is to mediate between the greater whole of the original idea and the specific material. As the architect Peter Zumthor puts it when describing the music of Johann Sebastian Bach: "Its construction seems clear and transparent. It is possible to pursue the details of the melodic, harmonic, and rhythmical elements without losing the feeling for the composition as a whole – the whole makes sense of the details" (ZUMTHOR 2010). It applies to music as well as to landscape design that the knowledge of notes and accordingly materials does not necessarily result in successful work. The question is how to compose, how to realize an idea. If we succeed in this, projects can be made to shine and vibrate.

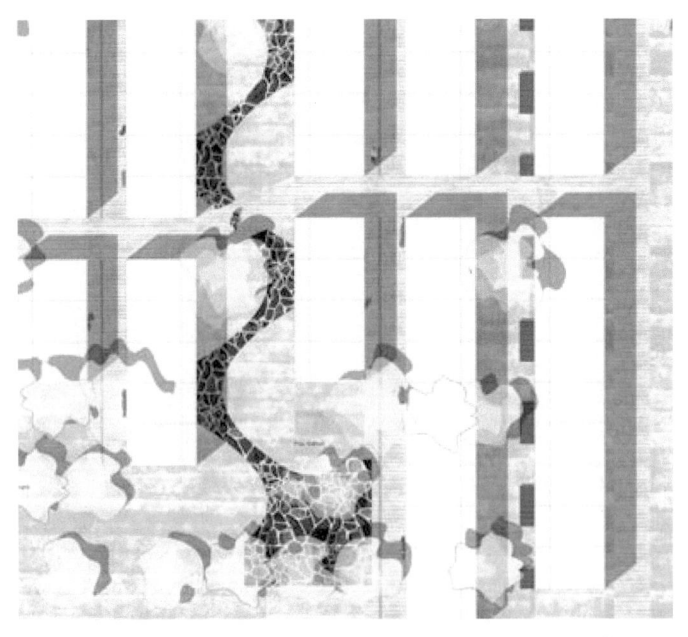

Fig. 7: Studio design: Site plan by students Yuanchenxi Gao and Bing Wang

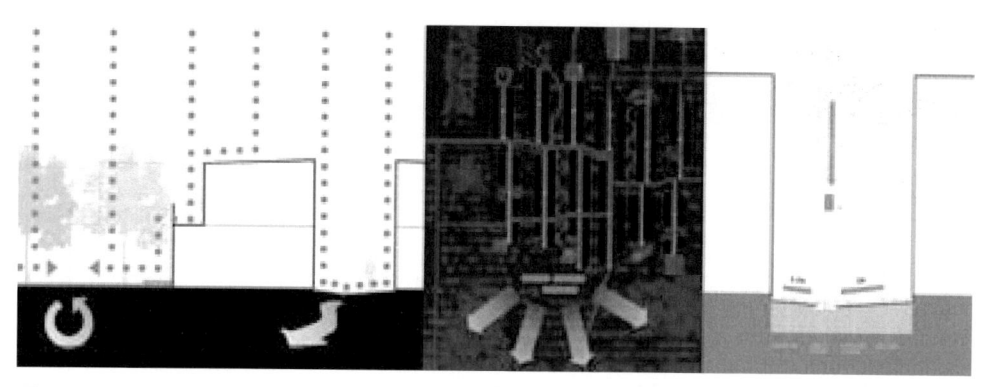

Fig. 8: Studio design: Idea and Detail by students Yuanchenxi Gao and Bing Wang

Fig. 9: Studio design: Perspective by students Yuanchenxi Gao and Bing Wang

The work of the students Yuanchenxi Gao and Bing Wang exemplifies how both the concept and the details can be developed concurrently and consistently throughout the project. In an undergraduate design studio these two students were asked to develop student housing and its layout in a rural setting. The concept of living in an orchard was the starting point (Fig. 8). The drainage and irrigation details were a logical consequence of the concept and subsequently, the overall site plan became a meaningful composition of details, illustrated by perspectives (Fig. 9).

Reflection on materials and the study of construction details help students to both build ideas and illustrate their own designs. The beauty of this method of study is that whilst keeping track of the original idea they can consider what is possible and focus on what makes sense in relation to a specific design goal. They learn that the original design idea influences the selection of construction materials as much as the material selection can influences the design.

7 Conclusions and Outlook

This paper illustrates, with reference to students' work, that the construction of a design idea and the principles of Bach's music composition are following similar rules in order to achieve convincing results. Placing as much value on the reflection of details and materials as one does on design ideas and representation results in convincing projects due to the inherent coherence that results.

Applying digital drawing techniques to material proposals and construction drawings in a Canadian context not only matches what is common practice in other places, it also embraces the distinct willingness and exceptional ability of students to learn by and with computers. Digital sketching within CAD software is an upcoming research field and the more skilled software users become the more potential they see in using these programs as a sketching type tool. The profession needs software developments supporting this demand thus allowing advanced doodling, the easy and smart interaction of bitmap and vector images in 2D and 3D.

References

Jost, D. (2011), *Material Evidence.* In: Landscape Architecture Magazine, 8/2011, 124-134.
Vogt, G. (2006), *Miniatur und Panorama: Vogt Landschaftsarchitekten Arbeiten 2000-2006.* Baden, L. Mueller Publisher.
Zumthor, P. (2010), *Thinking Architecture.* Basel/Boston/Berlin, Birkhäuser.

Managing the Visual Resource of the Mediterranean Island of Gozo, Malta for Tourists – A Studio Approach for International Conversion Students Bridging Different Levels of English

Erich BUHMANN, James PALMER and Matthias PIETSCH

This contribution to the DLA Conference 2011 proceddings appeared in gis.SCIENCE – Zeitschrift für Geoinformation, 24. Jahrgang, 4/2011, pages 123-137, published by Wichmann, VDE Verlag Berlin/Offenbach.

Please find the full online contribution at
http://www.kolleg.loel.hs-anhalt.de/landschaftsinformatik/435.html.

Abstract

The visual quality of the Mediterranean Island of Gozo is one of the major selling points of the island. Which landscape elements do potential tourists prefer and which landscapes are preferred by Gozitans and Maltese? The concept and results of an academic studio on visual perception, carried out by international landscape architecture students are described below. The article introduces the method used in selecting representative images, their evaluation and indicates the need for further implementation strategies.

The study shows that applying methods of a visual landscape inventory and a landscape perception study can be analyzed based on the results of the analysis work as a studio approach for international conversion students, bridging different levels of English. The use of predominantly visual methods could be a catalyst for teaching basic research methods to students with different cultural backgrounds and who have different skill levels in academic reading and writing in English. Dealing with visual aspects such as field work, evaluation, analysis and design of suggestions addresses the design skills which are often very well developed and not limited by the level of knowledge of the teaching language.

This paper documents the evaluation of the survey and the comparison with the geographic approach by MEPA and some recommendations based on favorable landscape elements and composition. In cooperation with the University of Malta we have now also interviewed Gozitans using the same set of photos. Further analysis and translation into implementation on Gozo will be discussed by the local authorities and the public.

Keywords: Visual Resource Mangement, Mediterranean, Tourism, Landscape Architecture Studio

Interdisciplinary Studio for Teaching 3D Landscape Visualization – Lessons from the LVML

Ulrike WISSEN-HAYEK, James MELSOM, Noemi NEUENSCHWADER, Christophe GIROT and Adrienne GRÊT-REGAMEY

This contribution to the DLA Conference 2011 proceedings appeared in gis.SCIENCE – Zeitschrift für Geoinformation, 24. Jahrgang, 4/2011, pages 117-122, published by Wichmann, VDE Verlag, Berlin/Offenbach.

Please find the full online contribution at
http://www.kolleg.loel.hs-anhalt.de/landschaftsinformatik/435.html.

Abstract

There is an urgent necessity of accredited education and training programs for 3D landscape visualization at various levels of user. Experts have to gain the competence for assessing those techniques in terms of their options and limits in supporting communication in the planning process. This paper presents an interdisciplinary approach for teaching studios to improve students' awareness of principles in 3D landscape visualization and application in participatory landscape planning and design. The studio was conducted in the LVML – Landscape Visualization and Modeling Lab at the ETH Zurich. Conducting studios that are interdisciplinary with regard to both the lab leaders as well as the students foster knowledge transfer from research findings to education and thus to praxis and seem to accelerate the speed of learning.

Keywords: LVML – Landscape Visualization and Modeling Lab, interdisciplinary lab, education, 3D landscape visualizations, planning workshops

GeoDesign Concepts and Applications

A System for GeoDesign

Stephen M. ERVIN

Keynote: 27 May 2011

Abstract

GeoDesign enhances traditional Environmental planning and Design activities with the power of modern computing, communications, and collaboration technologies, providing on-demand simulations and impact analyses to provide more effective and more responsible integration of scientific knowledge and societal values into the design of alternative futures. For practitioners, as well as students and researchers, this requires integration of many kinds of information, and a number of software tools, together in a comprehensive system, as no single familiar software product or approach (CAD, GIS, BIM, etc.) will suffice. The elements of such a idealized system are outlined and described in this paper.

1 Introduction

'GeoDesign' is the current manifestation of an age-old practice – planning, designing, implementing and evaluating changes to our built & physical environment – transformed by modern tools including digital databases, representational and analysis software tools (CAD, GIS, BIM, et al.), and modern communications technologies, practices, interfaces and approaches, including embedded sensors, multi-media feedback, web-based interactions, group decision making, mobile devices, social networks, and others – and given new urgency by scientific advances in understanding and analysis of Earth's natural systems [GEODESIGN.ORG].

A hallmark of geodesign activities is that they are usually multidisciplinary, across a range of domain areas, and that they feature a relatively tight coupling between ideation (design) and evaluation (including scientific, economic, social, et al.) or 'generate' and 'test' in Simon's immortal paradigm (SIMON 1981) – that provide rich and complex interactions between design practices and scientific methods.

Geodesign projects leverage the powers of digital computing and communications technologies to foster information-based design and provide timely feedback (sometimes 'real-time') about implications of proposed designs, often including impacts and evaluations covering a larger area, greater complexity, or longer time-frame than the immediate design proposal (e.g. the impacts over time on watershed-scale hydrological processes of a proposed new dam, or the aggregate carbon footprint of many individual building component /system decisions.)

To date, the technical infrastructure for geodesign has been achieved by using the tools of existing GIS, CAD and BIM systems, coupled with spreadsheets, databases and emergent web techniques. This ad-hoc approach has not been altogether satisfactory, as there are still many 'interoperability' issues to resolve; and, perhaps more critical, there is still considerable debate about whether and how 'design' is supported and enabled (or thwarted) by these existing tools. Many designers report that the ideation phases of design – creating new representations in some medium to express, explore and refine concepts and relationships – are hampered by the relative clumsiness of using a mouse and software, especially compared to the fluidity of pencil on paper. Some report that 'tablet-based' interfaces such as using a 'Wacom tablet', improve the experience. And yet others, mostly (younger) 'digital natives', are immune to these frustrations and report that pencil on paper seems an impoverished medium to them. A new era of software design and delivery emphasizes multiple inter-operating 'apps' that perform specific tasks, with access to a rich array of databases and emergent interface methods, that may be 'mashed- up' to create new synergies. In that spirit, the following proposal is for a system of component parts, with more emphasis now on the parts, than on their systemic integration. That will have to come from practice.

2 System Components

A hybrid GeoDesign System (GDS) will require a combination of the best of various legacy tools (CAD, GIS, BIM, e.g.), and some new and best-practice techniques, such as object-oriented diagrams and key-indicator'dashboards`. These will be obtained not by perfecting any one software product, but by leveraging interoperability and providing essential modularity and flexibility in component design.

The ideal system presented here has no current manifestation, and is not formulated within the constraints of any existing software products. The integration points required between the several components presented below are complex, but are glossed over here for the purpose of presenting the broad essential set of components required for a 'geodesign system`.

This proposed system will entail at least sixteen essential, interrelated, components, as identified below with simple descriptive names. In the next section, each is described in more detail.

GeoDesign System Components:
1. Environment/Context-Base
2. Configuartion
3. Elements (Objects, Classes, Properties, Methods)
4. Constraints
5. Analyses
6. Simulations
7. Dashboards
8. Version Manager
9. Time/Dynamics Manager
10. Level Of Abstraction (LOA) Manager
11. Diagram Manager
12. Algorithmic Interface
13. Text/Media (Hyper-Annotations)
14. Library

15. Collaboration Tools
16. Design Methods Coach

GEODESIGN SYSTEM : TOOLS & HELPERS

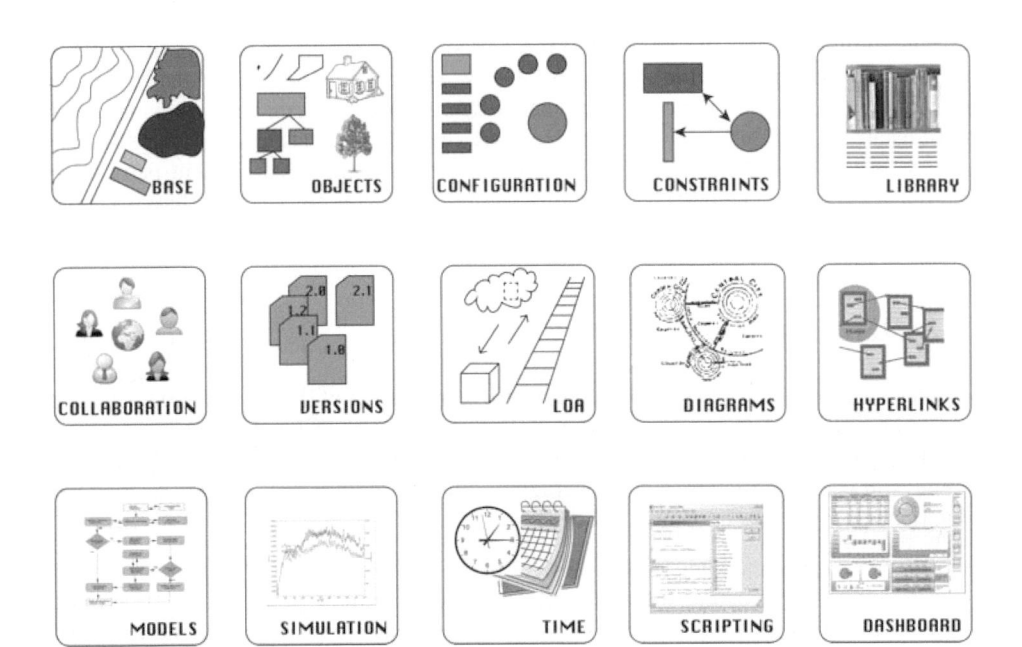

Fig.1: GeoDesign System:Tools and Helpers

3 System Components: Details and Interrelationships

3.1 Environment/Context – Base

Most geodesign projects happen within a specific geographic context, or area. While some digital information for that area may be available as 'layers' in a GIS or other database, sometimes all that is available is an air photograph, survey map, contour-/hypsometry plan, or other 'base' information, which sets the stage, and the frame, for the project. It's important that this base environment information extend a good amount beyond the working boundaries of the project, so that edge effects can be most effectively considered. Information in this "Environment/Context-Base" category (often, one or more separate display layer(s) includes that which is not expected to change or be changed for the purposes of the design project, and which serves as a 'reference' for visual orientation. Other elements of base information may well change over the course of a project, as conditions change, or other better or different information becomes available. Linkages from the active design layers and elements back to this reference become constraints that can be generalized in diagrams for describing the design ("the treatment plant should be located near the stream, on a flat slope".) This component is a multi-media, hyper-linked, multi-layer, geo-referenced repository – not just a 'background image'.

3.2 Configuration (the 'plan')

The 'Configuration' at any instance is a record of all the elements specified by the designer, including their attributes, spatial layout, and other logical relationships with each other, as well as to base or other contextual information. Relationships such as 'house face parallel to sidewalk' or 30 maple trees at 12' on center, are aspects of a configuration, as are arbitrarily complex polyline shapes that cannot easily be described in words. The configuration may be a fully 3D detailed and dimensioned set of construction documents; or may be a simple 2D concept diagram. Keeping track of the configuration at different levels of abstraction, and its evolution and states over time, along with supporting evidence including inspirational notes and impact evaluations, is the true function of the GDS.

3.3 Elements (Features, Objects, Classes, Properties, Methods)

The geodesigner will mostly work, in the 'Configuration' described above, with 'Elements' – the atomic and molecular particles of the design. These may be broad and aggregate, such as a forest or neighborhood, or specific and detailed, such as a doorknob or a single tree. Since these will also be created, refined, and manipulated at various levels of abstraction, and with kinship relations (e.g., a 'water park' is a special kind of 'park', etc.), the 'object oriented' paradigm of modern computer programming will be essential.

Objects & Properties
Object oriented programming (BUDD 2002) and related approaches such as object oriented databases, have a number of advantages in the production and maintenance of software that have made them standard approaches for most big modern software systems. Two essential aspects of this approach stand out: 1.) 'Objects' are bundles of attributes such as height, color, cost, along with 'behaviors', embedded in 'methods', such as 'calculate cost of installation' or 'calculate new water level', etc. This makes these objects stand for, and often behave like, real-world objects, rather than just primitive computer graphics elements such as points, lines, polygons, tables, etc. (Even just the nomenclature helps designers, who benefit cognitively from feeling like they are 'placing a door', or planting a tree, rather than 'drawing a rectangle', or a circle.)

2.) Objects exist within a 'taxonomy', or 'hierarchy', such that some objects are 'classes, (such as 'trees'), while other objects are 'specialized instantiations' of the class, such as 'maple trees', which inherit all the attributes (including methods) of their class, and may also have some more detailed and possibly additional attributes beyond those they inherit; and there may be deeply nested class-subclass-object relationships (such as 'this particular maple tree', that has an age, location, height, maintenance record, etc.)

One efficiency that results is that attributes and methods that all individual instances share can be defined at the class level, and applied everywhere, while individual objects can also have additional local, or idiosyncratic, attributes and methods as needed. In software engineering this means that shared libraries of common methods and procedures and bundles of attributes can be developed and shared as 'class libraries', to be re-used and adapted as needed. Object-oriented is widely accepted as an efficiency in modern day programming. Geodesigners could benefit from the same efficiencies and expressiveness with access to libraries of object-orient geodesign elements (OOGDE's).

3.4 Constraints (Relationships)

The relationships between elements in the design include simple position and adjacency and proximity – the most elemental geo-relationships – as well as other more complex or dynamic relationships, such as different wells tapping a single aquifer, or a sound absorbing wall whose height is a function of the width of an adjacent roadway. Some of these relationships are simple, linear and algebraic, like the last one; others are complex, dynamic, and possibly heuristic and incalculable (the behavior of neighborhood children as a function of the size and design of a neighborhood park and availability of water...) The formalism of 'constraints', which have been shown to be very powerful in computational simulations of real world objects and systems, (BORNING 1979), must be managed by the GDS, and are essential inputs into the 'Analysis and Simulation' functions below.

3.5 Analyses (Models)

Familiar to many GIS projects, which use geoprocessing models to extract A System for Geodesign information of create new, analysis models are essential to the GDS. The tight coupling between analytic models, simple or complex, and design moves, manual or algorithmic, is a hallmark of "GeoDesign". Regularly comparing design proposals to their predicted impacts, in addition to using automatic algorithmic approaches to design, are distinguishing characteristics of geodesign projects, that leverage the power of computing.

Analysis models may be purpose-built in some suitable interface (e.g. ArcInfo's "model builder") (ALLEN 2011), may be taken from a library of off-the-shelf analyses, or may be performed essentially outside of the GeoDesign System, by exchanging parameters extracted from the design configuration, running complex models synchronously or asynchronously, with results possibly reimported into the design, or reported on a 'Dashboard' (see below).

3.6 Simulations

'Simulations', like analytic models, are performed to learn more about the properties or behavior of the design, over time or under varying conditions. These usually step forward in time, in some controlled and granular way, and may produce animations that are observed for visual inference, or may generate other quantitative outputs, just like a simpler model.

Simulation toolkits are available for many particular design domains, e.g. transportation, hydrological and structural performance, an others. A particularly effective approach is 'agent-based' models, in which objects designed to simulate certain kinds of behaviors (those of animal species, or human shoppers, e.g.) are 'let lose' in a virtual environment, and their behavior and interactions recorded and analyzed. This is particularly effective in a distributed design infrastructure in which one team may be designing an environment, another designing suitable agents, and yet a third overseeing simulations.

3.7 Dashboards

'Dashboards' are element of modern information systems that have gained popularity for managing complexity, and engaging human-computer interaction (FEW 2006). Simply defined as 'a visual display of key performance indicators', there is a presumption that the dashboard display is concise, and taken in at a glance, using cognitive features such as color ('red' for a warning or danger situation, e.g.), as well as instantly updated, or at least current and timely, so that they can help guide the design processes. Dashboards can be used to monitor goals and thresholds by simple counting (such as 'need to provide 1000 parking spaces; 497 provided, 503 to go' or 'cut and fill not balanced', etc.) as well as warnings based on more complex simulations and analyses such as 'danger of downstream lake eutrophication' or 'increased public health risks over time'. Real-time distributed processing and feedback of results, in a suitably high-level/aggregate form to help inform evolving design decisions, is a must for highly effective geodesign. The specification of a suitable dashboard is a design problem in itself, inasmuch as it depends on situation-dependent determination of key performance indicators, some knowledge of both generic visual interface principles as well as user-specific customizations, and integration with analytic and other routines. Generic geodesign dashboard 'templates' are a likely place to start, based on real-world experience and community-approved defaults, that will doubtless be specialized over time in various applications and use-case scenarios.

3.8 Version Manager

Geodesign projects, like all design projects, may generate multiple variants and states over time. Managing these, with a consistent naming convention for retrieving different versions or demand, with sufficient modularity so that individual elements or arrangements can be copied and incorporated into yet more versions, is a demanding task. Most human designers or design teams depend on basic file naming and time-stamping to achieve this; some have access to more sophisticated version control systems such as the 'VCS' used by software programmers. When geodesign projects feature teams of collaborative designers, managing versions across time and teams becomes even more complicated – and essential. Stored versions need not just elements and relations, but 'metadata' describing the conditions, goals, special considerations, etc. so that they can be recovered in the future and interrogated for 'why is this?'.

3.9 Time Manager

'Simulations', 'Analyses' and 'Versions' all need information relating to time, identifying special properties of certain moments in time (groundbreaking, construction calendar, future generations) as well as the properties and impacts of dynamic processes in the environment (flooding, growth, social change, e.g.) [COSTANZA & RUTH 1998] The time manager may be considered a cross-cutting set of properties attached to various elements, combinations, or arrangements; required since time based analyses, simulations, projections, and impacts are so important in many GeoDesign projects.

3.10 LOA Manager

'Levels of Abstraction (LOA) ' are critically important to many design processes (JAQUES 1978). Designs often start at a high level of abstraction (verbal goals, e.g., or abstract ideas like symmetry), proceed through a diagrammatic phase in which basic elements and relationships are identified, and finally move to more resolved, concrete specifications of materials and locations. Managing the relationships between high-level abstraction elements ("barrier"), and subsequent refinements ("18" thick concrete wall") is the job of the LOA manager, as well as keeping track of the current state, and enabling switching between LOAs (extracting diagrams from configurations, for example.)

Level-of-Detail (or LOD) is a closely related concept. In many digital imaging and rendering, and cartographic presentations, the idea is to present different levels of detail depending on distance away from the subject. A map at 1:100,000 shows different content (less detail) than one at 1:25000; a tree in the foreground of a 3D rendering may have leaves with veins, whereas a tree in the distance has only a fuzzy green shape. In general, the greater the LOA, the lesser the LOD, but since one has to do with cognitive distance (LOA), and the other geometric distance (LOD) the two are not exactly inverse, or comparable. Determining appropriate relationships between them, both in principle and in practice, is an area for continued research.

3.11 Diagram Manager

A special and central case of managing levels of abstraction is understanding diagrams. In diagrams, deceptively simple graphics are imbued with rich meaning [ERVIN 2010]. Circles, lines and arrows encode objects, relationships and constraints; and become instantiated with more concrete elements as the design is refined. The 'grammar' of diagrams – conventional elements and relationships, both graphic and conceptual – deserves study all on its own, and the development of an interface to software tools for diagram management are just as important as more freeform, expressive, 'sketching' tools.

3.12 Algorithmic (Scripting) Interface

Some forms of geodesign will require not just graphical manipulation of objects and interfaces, but it is possible that some 'Scripting' opportunities will arise, as well. Both for automation of routine or repeated tasks, and for more complex algorithmic approaches, such as optimization or rule-based allocations, a programming language need to be provided. It should include at a minimum the three basic programming/scripting capabilities: named variables and procedures, repetition, and conditional branching [TERZIDIS 2006].

Whether this is simply a full-featured programming environment such as Java or .NET, along with an API provided for all other components; or a custom-provided scripting capability, is a detail for implementation. Many geodesigners will likely be able to get along with the later, while determined algorithmic designers will require the former. Agent-based modeling, mentioned above, is an example of the use of algorithmic tools for simulation and design.

3.13 Text/Media (Hyper-Annotations)

Both elements in a configuration, and steps in a process, can benefit from annotation, providing context, additional information, motivations, etc. The web has shown that 'hypertext annotations' – the ability to make any element of a document 'hot', and linked to more information, is a powerful technique for managing complexity, as well as enabling collaboration. So all geodesign documents should be like 'web pages', or 'hypertext documents', in that all elements are possibly linked to others (a movie of a particular location on a site, or a reference to an extended article supporting a particular analytic approach, or a note from the designer saying why a particular color was chosen…).

3.14 Library

As the geodesigner will need access to a collection of libraries of standard elements (objects, analysis routines, past designs and precedents, etc.), a library function will be required. More than just a typical file system, it will need a layer of finding aids on top of it, accessing meta-data, in addition to full text search and file-/folder-naming conventions. It will need to integrate both local repositories and collections, as well as shared, and global, ones.

3.15 Collaboration Tools

Geodesign projects are increasingly marked by requiring multidisciplinary collaboration, and often public (at least some amount of) participation [DesignCollaboration.org]. The best of emergent collaboration tools will be needed, allowing for (automatically updated) shared documents, individually produced and (perhaps automatically) reconciled contributions, as well as shared decision making techniques such as Delphi methods, questionnaires, etc. In the management of any project, the results of collaboration may become hyperlinks, as well as stored in a library for future reference.

3.16 Design Methods Coach

There are many ways of approaching geodesign problems, varying from individual to individual, and from case to case, depending on a myriad of considerations. Nonetheless, some generalizations about design methods have been made. In particular, Steinitz has proposed a taxonomy of kinds of design problems and design methods (STEINITZ 2010). His component of the GDS is closely related to the library, containing access to precedents and other geodesign projects; but can also be more, providing a virtual coach, with tools to help diagnose problem characteristics, suggest appropriate design methods, and help structure methods in the process. This module could record and classify design processes, and could 'learn' over time, over many different geodesigners & projects, and become more and more valuable, linked up to the library described above.

Much more research is needed on design methods, their details, and their utility and effectiveness in different circumstances, but this component of the GDS could be a valuable adjunct, for both novice and Expert designers alike.

4 Discussion and Summary

In the above idealized description, no real reference has been made to many important 'implementation' and 'interface' questions: how many screens? How big? Mouse, keyboard, or other controller? Sound? Immersive displays? And many others... The evolution of computing devices, interfaces, and experiences is well along on its own exponential curve, and the options are seemingly endless. Some manifestations of the GeoDesign System may be all-cell-phone or tablet-computer- based; others will require a roomful of equipment and high-end multi-millions-of-pixels display devices; others will be entirely in 'the cloud'. None of the functional requirements described here are dependent upon those choices. In a System for Geodesign some combination of ergonomically convenient hand-held device for in-field work, coupled with multi-screen desktop environment, supported by asynchronous on-line access to some more massive computing power and databases (the web? the cloud?) will doubtless be common. Details of performance, as well as cognitive affordance, may well vary depending upon the form of the interface, inputs and outputs. That multiplicity of experience will be essential. There will certainly be discipline- and application-specific variants among the wide range of geodesign applications.

Whether 'sketching' in such a system is performed with graphite on paper, and scanned into digital form, or performed on a touchscreen device, also doesn't matter much. Right now, in 2011, sketching via computer software is generally a frustrating experience. Much work needs to be done on interfaces to support easy, expressive, gestural input, to match the sketchbooks and tracing paper of yore. When those developments are further along, the ability of the digital GDS to also record metadata, annotations, diagrammatic intent, contextual and environmental data, etc., will result in a powerful combination.

At its core, GeoDesign is not a new enterprise. Landscape architects, planners, architects, transportation engineers, and other allied professionals have long been engaged in this noble enterprise. What's new, and deserves a new term, is the rich deep and broad support given by modern computing and communications. Whether from sophisticated scientific models or inclusive participatory techniques, the scope and style of planning for better alternative futures (however defined) is radically affected by these technological supports and affordances. To take maximum advantage of these developments will require some shifts in working styles – away from 'single-tool' solutions, towards 'mash-ups', just as geodesigners have already evolved away from 'main-frame' or 'personal-computer' towards 'networked' approaches that may include aspects of both. Its in that spirit and direction that these outlines of essential components have been offered. Some years of experience and validation, adjustment and refinement are now required, as the practice, art, and science of geodesign co-mature.

References

Allen, D. W. (2011), *Getting to Know Arcgis Modelbuilder*. ESRI Press.
Borning, A. H. (1979), *Thinglab – a constraint oriented simulation laboratory*. Stanford University.
Budd, T. (2002), *An Introduction to Object-OrientedProgramming* ,Addison Wesley Longman.
Costanza, R, & Ruth, M. (1998), *Using Dynamic Modeling to Scope Environmental Problems and Build Consensus*. In: Environmental Management, 22 (2), 183-195.
DesignCollaboration.org, http://www.designcollaboration.org/.
Ervin, S. M. (2006), *On the Necessity of Diagrams*, http://video.esri.com/watch/55/2010-geodesign-summit-stephen-ervin-on-the-necessity-of-diagrams.

Few, S. (2006), *Information Dashboard Design*. O'Reilly Media.
GeoDesign Summit (2010), *http://www.geodesignsummit.com/*.
Jacques, E. (1978), *Levels of Abstraction in Logic and Human Interactions.* Cason Hall & Co.
Simon, H. A. (1981), *The Sciences of the Artificial.* 2nd edition, MIT Press.
Steinitz, C. (2010), *Ways of Designing*, http://video.esri.com/watch/54/2010-geodesign-summit-carl-steinitz-ways-of-designing.
Terzidis, K, (2006), *Algorithmic architecture*. Elsevier.

The Potential of GeoDesign for Linking Landscape Planning and Design

Barty WARREN-KRETZSCHMAR, Christina v. HAAREN,
Roland HACHMANN and Christian ALBERT

Keynote: 1 June 2012

1 Introduction

In many European countries, landscape planning and landscape design have developed in different directions. It appears that the more environmental planning is legally codified and elaborated, the more the prototypes of landscape planning and landscape design differ from each other. This development is especially evident in Germany. Furthermore, the legal implementation of the European Landscape Convention may well increase this dichotomy.

Landscape planning and design emphasize different values, such as creativity or transparency, and they typically use different methodological approaches. For example, landscape design relies more on intuition and creativity to generate design ideas and concepts. On the other hand, landscape planning is often restricted by legal and implementation considerations and required to use transparent approaches such as standardised procedures and GIS analysis. In order to explore this relationship, landscape planning and design prototypes can be used to characterise hybrid approaches in practice and to identify opportunities for integrating design approaches into landscape planning. (V. HAAREN, submitted 2012)

Without doubt, linking the diverging approaches is a desirable goal. Landscape design ideas and concepts can help to communicate and even implement landscape planning objectives. Convincing landscape designs hold the power to engage and persuade citizens, politicians and land users. Even conservation objectives may be more easily implemented in the wake of a fetching design idea.

A prerequisite for better integration of design into landscape planning is the definition of situations where design approaches are feasible and preferable in the context of landscape planning. Conversely, situations need to be identified in which planning approaches can support the design context. The underlying hypothesis is that the design or planning approach has evolved based on requirements of the different tasks and situations where it is applied. However, there is rarely a situation or homogenous context that demands the application of only one approach, either a design or planning type. Growing recognition of the need for hybrid design-planning theory is acknowledged in the increasing spectrum and systemization of approaches (STEINITZ 2010). A concept for linking the approaches must consider the different applications, ethics and values to be implemented as well as the methods and processes of planning and design.

Furthermore, the complex integration of planning and design approaches can be supported by GeoDesign. In this paper, the understanding of GeoDesign follows Flaxman (2009) and especially the wider definition of VargasMoreno (2010): GeoDesign is defined "as the act of integrating the constantly transforming techniques, concepts and approaches in design and planning with GeoSpatial systems and technologies. GeoDesign's aim is to create more comprehensive and rapid solutions for both processes and forms of the built and natural environment. This definition suggests transforming design and planning practices to a technology-supported feedback loop that allows the rapid conceptualization, articulation, visualization, modeling, and monitoring of transformations in a variety of geographic environments". This definition of GeoDesign encompasses both elements of planning (GIS, geodata, impact analysis) as well as design (feed back loops). We use the term GeoDesign here in a restricted sense: as an IT supported approach, which systematically uses an IT toolset for combining geodata, as well as analyzing and visualization models/techniques that support the rapid generation of spatial scenarios in planning and design processes. For the purposes of this paper, we view the GeoDesign toolbox as a complete and integrated toolset, even though the present technology may not yet be able to achieve this.

In Vargas-Morenos' definition, the planning and design processes are not specifically differentiated. Therefore, we need to examine the different requirements of the planning and the design processes in the European context, before we can identify opportunities for GeoDesign to support their integration. Furthermore, we need to investigate the role of available GeoDesign tools for integrating the planning and design processes.

2   Objectives and Approach

In order to understand how GeoDesign can strengthen the relationship of landscape planning and design processes, the following questions present themselves:

1. How can we classify and describe an integrated planning and design procedure?
2. To what extent can we support or replace intuition by automated analysis and generation of solutions?
3. In which context is the digital dashboard approach – multiple loop approach - the most efficient way to plan or design, and when is a more linear approach more efficient?
4. Which role can visualization and collaborative planning approaches play in the different hybrid situations?

The paper addresses these questions (with the exception of the second problem, which requires basic research) with the objective to define the design and planning context and requirements for using GeoDesign technologies, and thus, focus on contributions to its application and conception (in a European setting).

In a first step, the preconditions and requirements of planning as well as hybrid planning-design situations were summarized in a framework based on an existing review of the literature. Second, we integrate theoretical potentials of GeoDesign into this framework and discuss consequences for the approaches. Finally, we examine the application context of different visualization methods and of participatory GIS technology.

3 Landscape Planning and Design Prototypes

3.1 What are typical tasks of the landscape planning and design processes?

Landscape planning is an analytical process based on a legal framework that uses science-based methodologies to evaluate landscape functions and inventories. The legally binding nature of the landscape planning decisions requires that the process be transparent. Therefore, landscape planning relies on clearly derived planning objectives, reliable methods and comprehensive site analysis. The content of the core tasks and phases of landscape planning are closely linked (see Fig. 1):

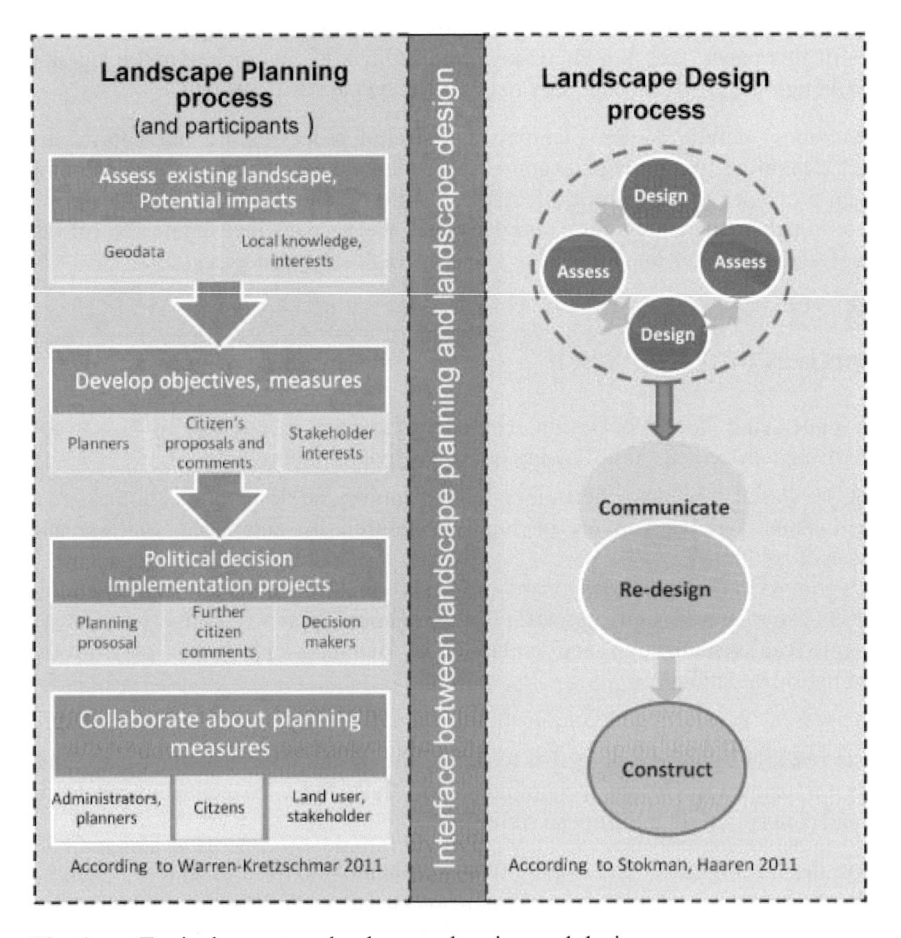

Fig. 1: Typical processes landscape planning and design

Inventory and evaluation: Geodata and local knowledge of citizens provide the basis for the inventory and assessment of the existing landscape condition and help to identify development opportunities of the natural resources and landscape functions, as well as potential impacts of existing and planned uses.

Planning objectives and concepts for development: The development of objectives and measures is based on the planner's analysis of the landscape and includes the comments of citizens and stakeholders. Scenarios are often used to formulate the objectives and possible alternative for the remediation, conservation, and development of nature and landscape.

Proposed planning measures: Finally, the requirements and measures for implementing the objectives are formulated in concrete planning proposals. This may include descriptions of alternatives that solve conflicts and information about their implementation. In participatory situations, visualization and scenarios can be helpful to prioritise concrete measures for preserving the most valuable landscapes.

Landscape planning tasks are characterized by a forward looking, precautionary and normative approach to changes in the landscape. They often address legal requirements or spatial land use conflicts. For the most part, existing land uses and ownership requires solutions that go beyond a physical or spatial solutions, for example economic, social or legal measures for safeguarding the environment. Furthermore, landscape planning solutions must be accurate, convincing and legally defensible. The implementation of landscape planning objectives becomes an important consideration in the development of solutions, and it depends on active participation, environmental awareness and the use of diverse legal and planning instruments.

The design process, on the other hand, is a combination of knowledge and intuition in an iterative, non-linear process. It includes discovery, research, analysis, testing, presentation, implementation, evaluation in an ongoing process of applying and adapting. (STOKMAN & V. HAAREN 2011) The process of reducing complex information into coherent designs is a central part of the process which designers perform. Visualization tools, such as sketches, photomontages and models, play an important role not only in the design process to help move the process forward, but also to communicate design proposals to clients. Comments and suggestions from clients are then incorporated into the re-design solution before it is constructed (see Fig. 1)

Landscape design tasks are characterised by the freedom to create something new on a site where change is expected by the client. Landscape designers consider economic, social and technical requirements when envisioning and shaping the landscape in order to create environmental and aesthetic experiences. Furthermore, the design results are tangible and visible as well as aesthetically pleasing and often stimulate public interest.

3.2 How do landscape planning and design tasks differ?

Landscape design is an experimental approach in contrast to the deductive approach of planning. The design mandate encourages a creative approach based on a strong design idea that produces an aesthetically pleasing solution. Transparency of the design process is not prerequisite for the acceptance of the results. This allows a non-linear, iterative, creative process. The planning mandate, on the other hand, is not necessarily to create something new in the landscape, but rather to safeguard or re-establish landscape functions or ecosystem services. Because the implementation of the planning result requires economic and legal measures, the planning process must be transparent, documented and reproducible. This requires a thorough analysis of the landscape and its impairments and projected changes and reflects a more linear process. Monitoring and participation add iterative loops to the process, but they remain traceable and reproducible.

The most distinct differences between the tasks and application of landscape planning and landscape design lie in the fact that landscape planning is restricted by governmental directives that have precautionary objectives. Landscape design tasks, on the other hand, focus on creating or improving the environment so it is aesthetically pleasing to everyone. There may not be a sharp divide between planning and design tasks, rather a different corset of requirements or conditions, in which the client relationships, legal requirements, potential for fundamental change of the existing site differ. Landscape planning and design share a common set of values, e.g. creativity, transparency, participation, however the emphasis and freedom to interpret the individual values vary when carrying out the tasks. Furthermore, the tasks not only overlap, but hybrid tasks have also developed, such as designs for urban brownfields.

3.3 Where are opportunities at the interface between planning and design?

The interface between landscape planning and design offers the chance to draw on information and applications from both. The synergy of approaches can improve the quality and acceptance of the respective planning and design solutions. Fig. 2 shows potential ways in which landscape planning and design approaches and applications can support each other.

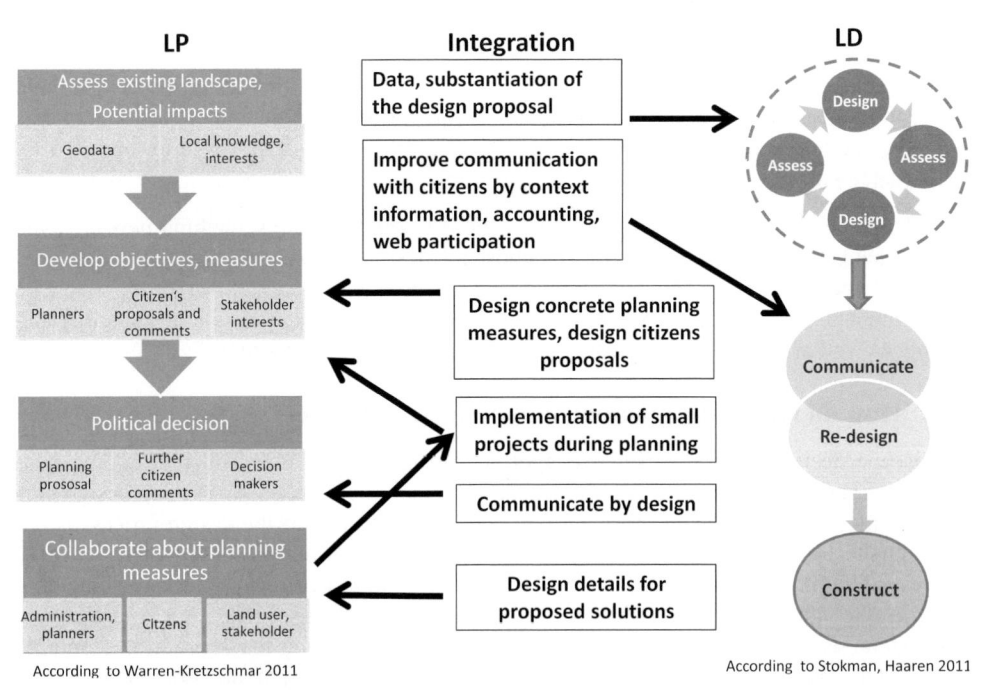

Fig. 2: Opportunities for integration of planning and design

A framework has been proposed (in V. HAAREN 2012 submitted) in order to identify situations in landscape planning where a design approach may support or improve for communication and implementation.

3.4 How can landscape design improve landscape planning?

In landscape planning, decisions are made about the management of the landscape that may not create a visible change in the landscape. Instead the landscape planning measures may protect a resource that is invisible to the observer, i.e. water quality or biodiversity. The importance and tangibility of such planning proposals is difficult to communicate to the public. Consequently, it is difficult to illicit support from politicians or local enterprises because there is nothing to show. Planners are forced to consider methods and means to obtain the public's acceptance for such projects, even when citizens and stakeholders may have to relinquish some of their privileges. Landscape design may offer opportunities to solve such landscape planning dilemmas.

Landscape designers have the creative freedom to develop design ideas that convey non-visible landscape concepts. Furthermore, designers are involved in the "creation of images, ideas and concepts" as well as "places that have meaning and that mirror the interests of the residents." (LYNCH & HACK 1984/2005) Such an approach may be exactly what landscape planning needs to make their "invisible" objectives and solutions to environmental problems visible to the public. A design that brings the issues of the landscape to life, communicates the underlying issues of a landscape plan, could be an answer to the Achilles heal of landscape planning. Creative designs can also emotionally engage citizens, politicians and land users and raise awareness about land degradation problems or reconcile the functional and aesthetic requirements of the landscape. Finally, designers can give planning measures a concrete expression or transform citizen proposals into a design solution. Such physical expression of planning objectives in small projects, which can be implemented during the planning phases, can move the abstract planning objectives to understandable, visible solutions for the citizens.

3.5 How can landscape planning support design?

Landscape planning can provide landscape design with a wealth of information about the landscape that can be used to reframe the design problem. Designers can draw on the planners' landscape analyses as well as methods and models. Environmental information can be used not only to substantiate the design proposals but also to provide an important understanding of the site context. As SWAFIELD (2002) points out that the design action may be at the site scale, but design thinking must embrace other relevant scales. Finally, landscape planning information can help to distinguish the limits of design ideas in terms of legal restrictions i.e. by species protection which is also relevant in urban areas. Finally, ecological and economic information and experience in (web) participation processes may also improve communication about design ideas.

The opportunities to combine planning and design (Fig. 2) stress the possibilities to improve transparency and communication in both processes. In the following, we look at the capacities of GeoDesign (GIS, scenario technique and visualizations) to support these common objectives, especially in a participation or collaborative planning process.

4 The Capacities of GeoDesign for Supporting Collaborative Planning

4.1 Requirements of Participation

For participation it is important that the range of services of the administration is organized so that citizens are able to actively participate and communicate. This includes direct contact through emails or forms, involvement of citizens in decision making processes, or the possibility to voice opinions about public issues. Transparency is guaranteed primarily when comprehensive and current information about legislative processes are made available, but also when responses to inquiries are documented. For this, the administrations use not only web-based GIS but also other solutions in the appropriate participation phases, such as systems for: content management, document and work flow management.

In addition to the current technical standards and the legal requirements that are involved in the development of an e-participation system or an integrated GeoDesign system, the needs of the involved parties must be considered first. This is the only way can the administration efficiently provide citizens and interest groups with useful participation tools. Assuming an intuitive interface, new ideas can be developed with these tools and alternatives, e.g. with respect to environmental impacts, can be publicly discussed or in a forum that use a limited access concept. Furthermore, the plausibility control of interactive scenarios can be integrated, which provide the initial spatial analyses of the input on-the-fly.

It must be possible at any time during the participation process to add or edit written background information, to integrate further geodata, to change participation deadlines, or to include other stakeholders. Furthermore, the individual interest groups should be informed and extensive access authorisation should be given to the respective parties, before the participation process starts, and all changes need to be documented, i.e. protocol of versions. A statistic about how the participation proceeding was used should summarize the number of visits and hits on different pages. Following an initial review and evaluation of the submitted comments, a simple classification and internal assessment is undertaken before it is presented to the political bodies.

4.2 GeoDesign in participation

Visualization supports citizen participation in the decision making process by responding to participants' needs for understanding the planning situation. Participants require orientation, the ability to picture planning changes and to understand the effects of planning proposals. Above all, they look for interactivity or the chance to interact with the information and to try out their own ideas and proposals. Interactivity of the visualization plays an important role in engaging citizens' interest as well as promoting credibility of the visualization (WARREN-KRETZSCHMAR 2011). GeoDesign responds to many of these requirements for successful participation (see Fig. 3). It offers the possibility to visualize the results with many different output types, e.g. maps, diagrams, models. This variety allows viewers with different spatial understanding capabilities to choose the visualization type that they understand, e.g. 2D map for overview or a 3D model that shows spatial relationships or both. Although GeoDesign may not yet satisfy the participants need for a "picture" or quickly rendered image of the landscape change, it does provide participants with the opportunity to interact and exercise control over the planning information and proposals.

GeoDesign responds to visualization requirements of participant

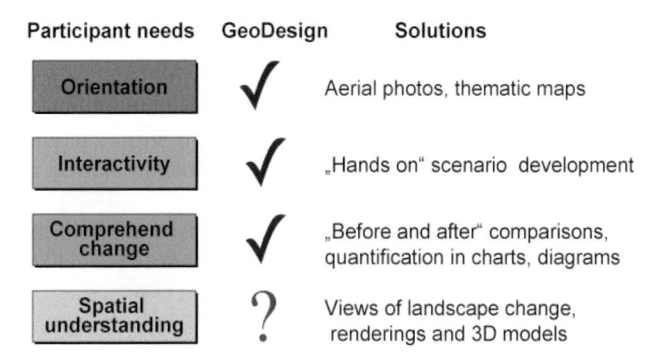

Participant needs	GeoDesign	Solutions
Orientation	✓	Aerial photos, thematic maps
Interactivity	✓	„Hands on" scenario development
Comprehend change	✓	„Before and after" comparisons, quantification in charts, diagrams
Spatial understanding	?	Views of landscape change, renderings and 3D models

Fig. 3: GeoDesign supports the visualization requirements of participants

5 GeoDesign at the Interface between Landscape Planning and Design

The workflow and software capabilities of GeoDesign can support the tasks of both landscape planning and design during different phases of the processes (see Fig. 4).

Landscape planning already takes advantage of GIS and models. The first steps of the ideal GeoDesign workflow (FLAXMANN 2009), which starts with the study area and existing digital information about the site, are identical with that of landscape planning. An array of site information as well as models or analysis methods, which can be adjusted to the requirements of the planner or designer, are also available. In addition, IT tools that support citizen participation are used in landscape planning practise. However in public participation, the ability to quickly portray landscape change, design ideas or citizen proposals remain unsatisfactory. Visualisation (supported by GIS data) is used, for the most part, to show an overview of the landscape, while detailed or realistic visualisation of small projects "on the fly" remain the exception. In addition, impact modelling, which uses GIS exclusively, is not fast enough for the face-to-face participation process.

Integrated visualisation components of GeoDesign, which encompass tools for visualizing large complex areas as well as smaller project sites could play an important role at the interface between planning and design. Such components offer the ability to bridge the scale change often associated with the different tasks of planning and design. For example, with GIS based visualization software, e.g. VNS from 3DNature, it is possible to frame the information for the design and planning question according to scale and point of view and thus to overcome the CAD fixation of design (see Fig. 5). The hybrid tasks of landscape planning and design may range from landscape scale analysis down to site design decisions.

For example, decisions are required at both the landscape and site scale when a landscape planning proposal for a habitat network is incorporated into the site design of urban open spaces that supports both biodiversity and recreational use. While the data base remains the same, the flexibility of the view and scale of the visualization support the transition from landscape analysis to detailed, site specific design.

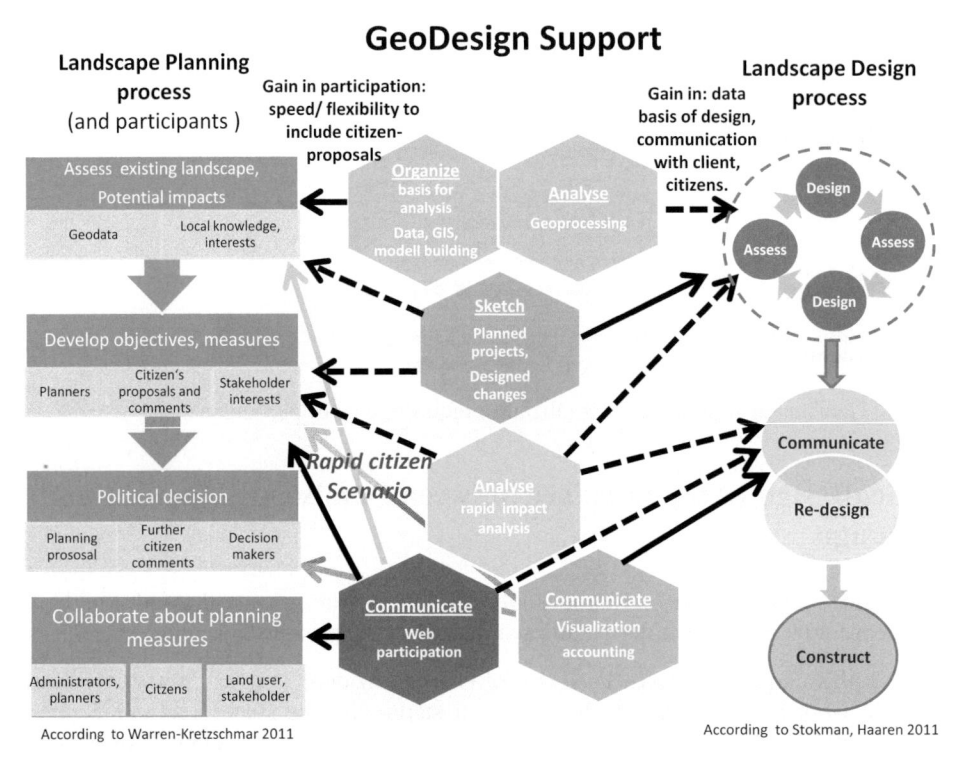

Fig. 4: GeoDesign supports the different phases of the landscape planning and design processes

Fig. 5: Flexible scale and view point of visualization support hybrid tasks and bridges scale differences of landscape planning and design. (VNS renderings created by Anne Meise)

However, the greatest innovation of GeoDesign for landscape planning lies in the sketch component and the related rapid assessment options. This is very important for improving both the participation and political decision making processes. It can be used to visualize citizen proposals and analyse their environmental impact very rapidly. GeoDesign not only makes the planning process more intuitive and accessible to participants; it also allows citizens to participate at the beginning of the planning process, by visioning their ideas. Furthermore, the sketch function can be used not only to design planning and citizen proposals but also to complement the planning proposal with respective design projects.

The sketch function supports an intuitive approach and encourages a more informal design attitude. Subsequently, "sketching" may stimulate ideas for small planning projects that can be implemented during the landscape planning process. Studies show that such fast and tangible outcomes of the planning increase citizen's acceptance of the planning measures (OPPERMANN et. al 1997). The combination of sketching capabilities and impact analysis together with a broad variety of visualisation options (including software not supported by GIS) can help to focus planning. These capabilities integrated with web participation tools would be a desideratum for technical development that offers an integrated approach to online participation.

On the other hand, when GeoDesign tools and workflow introduce planning elements into landscape design process, some principal changes can be expected. In the initial design step, a digital data base (GIS) about the existing situation and context can substantiate site information. Furthermore, the environmental or social impacts of the subsequent design proposals can be rapidly assessed, thereby supporting communication with the client or public participation. Although, as mentioned, landscape planning can profit from detailed visualisations of planning measures, landscape design might profit from GIS based visualisations that show the landscape context of a project site. In addition, web based participation tools could open design projects to public participation. At present, the use of visual media to creatively illustrate design concepts prevails in design. A GIS based program could expand these communication tools with new opportunities to communicate non-visible site factors as well as project features and impacts with easily processed charts of information.

Clearly, the strength of GeoDesign to quickly analyse and test ideas along the way is beneficial for both landscape planning and design. GeoDesign workflow and infrastructure can be integrated into the different planning and design approaches. This is important in both large scale planning tasks, but also in design projects However, while we see advantages for integrating GeoDesign tools into landscape planning in general, there may be less of an advantage (unfavourable cost benefit ratio) in using GeoDesign in small landscape design projects. The investment for preparing the GIS data base and implementing the models for analysis must be considered.

Furthermore, some shortcomings of using currently available technology for GeoDesign in planning practice were revealed when tested in a landscape planning case study in the region of Hanover, Germany. The case study found that further technological improvements are needed in order to enhance the tools' capacities for conducting complex analyses quickly, and to improve software usability by non-experts. Harnessing the power and usability of web-based GIS could alleviate this problem. Furthermore, discussions with planning practitioners revealed that the automated impact assessments of GeoDesign were insufficient in cases that required local interpretation and case-sensitive expert judgment in addition to rule-based analyses.

6 Conclusions

In conclusion, the technical support system of GeoDesign has the potential to be considered a "natural" interface between planning and design approaches, because it supports hybrid tasks of both processes. GeoDesign tools could help both landscape planners and designers to incorporate new approaches. Geodesign reflects the existing workflow and approaches of landscape planning. It may support participation by fast, simple modeling of planning alternatives which are proposed by citizens, especially if the the tools' capacities for conducting complex analyses quickly will be enhanced in the future. The use web-based GIS could alleviate the present problems.

The capacity of Geodesign for speedy modeling may also support typical elements of the design process (iterative loops) by fast assessment of design alternatives. In addition, for landscape design, Geodesign offers opportunities for an improved information base and context information by linking design to GIS. However, it may not yet reflect the requirements of the creative design process sufficiently. As ERVIN (2011) points out, more information about the different design approaches used by designers and planners is needed. The designer aims to reduce complex information into a strong design idea by using analytical knowledge and creative abilities (compare SCHWARZ-V.RAUMER & STOKMAN 2011). When the creative process can be understood better and models developed for the process, then GeoDesign moves closer to not only supporting the analytical processes but also the creativity of the designer. An understanding of the different methods and decision processes in different situations could be translated into GeoDesign. Finally, the user interface may be the key to the intuitive thinking of the designer. If GeoDesign can develop an interface that supports the way the designers develop ideas, it then offers the intuitive designer (and planner) a treasure of information to support design decisions.

GeoDesign can encourage designers to see GIS as a useful tool, and planners, who engage in Geodesign, may be able to use design methods to give form to planning measures. Nevertheless, an efficient adoption of different GeoDesign tools and components must reflect the different tasks and applications of landscape planning and design, how they can complement each other and which modeling detail is needed for the respective planning or design task. Finally, GeoDesign now offers a combination of sketching capabilities and impact analysis. When detailed visualization options and web participation tools are integrated into GeoDesign, it will not only become an indispensable part of participatory planning, but it may also smooth the gap between planning and design applications.

References

Ervin, S. (2011), *Object Oriented GeoDesign*. Redlands, California: 2011, GeoDesign Summit, 06-07 January 2011.http://video.esri.com/watch/195/object-oriented-geodesign.

Flaxman, M. (2009), *Fundamental Issues in GeoDesign*. Digital Landscape Architecture, Keynote Lecture.

v. Haaren, C. (submitted 2012), *Ethics and Aesthetics in Landscape Design and Landscape Planning: Conflicting concepts or potential for integration? A theoretical approach from the perspective of landscape planning*. Submitted to JOLA, 1/2012.

Lynch, K. & Hack, G. (1984, 2005), *Site Planning*. MIT Press.

Oppermann, B., Luz, F. & Kaule, G. (1997), *Der 'Runde Tisch' als Mittel zur Umsetzung der Landschaftsplanung*. Angewandte Landschaftsökologie, 11.

Luz, F. (1994), *Zur Akzeptanz landschaftsplanerischer Projekte*. Frankfurt am Main, P. Lang.

Schwarz-v.Raumer, H. G. & Stokman, A. (2010), *GeoDesign – Approximations of a catchphrase*. In: Buhmann, E. et al. (Eds.), Digital Landscape Architecture: Teaching & Learning with Digital Methods & Tools. Anhalt University of Applied Sciences.

Steinitz, C. (2010), *Ways of Designing*. Redlands, California 2010, GeoDesign Summit, 06-08 January 2010. http://video.esri.com/watch/54/2010-geodesign-summit-carl-steinitz-ways-of-designing

Stokman, A. & von Haaren, C. (2010), *Integrating Science and Creativity for Landscape Planning and Design of Urban Areas*. In: Weiland, U. & Richter, M. (Eds.), Urban Ecology – a global Framework. Oxford, Blackwell Publishing.

Swafield, S. (2002), *Theory in Landscape Architecture*: A Reader. (Penn Studies in Landscape Architecture). University of Pennsylvania Press.

Terzidis, K (2006), *Algorithmic architecture*. Elsevier.

Vargas-Moreno, J. C. (2010), *GeoDesign: The Emergence of a Tight-coupling Approach in GIS and Spatial Planning*. In: Planning and Technology Today, 2010.

Warren-Kretzschmar, B. (2011), *Choosing appropriate visualization methods for public participation*. Dissertation, Leibniz University Hannover.

Speaking of GeoDesign

C. Dana TOMLIN

Keynote: 27 May 2011

1 Coming to Terms

Juggling, pole vaulting, making love, operating a backhoe ... To be sure, there are a number of things in this life for which analytical scrutiny may be antithetical to performance at any given moment. Even in those cases, however, an appreciation for fundamentals does tend to improve performance over time.

This is certainly true of that peculiar form of decision making called "design." In academic programs that attempt to teach students how to be creative in fields such as Architecture, Landscape Architecture, Urban Design, or Regional Planning, the moment at which this actually happens is often one that ultimately remains both magical and mysterious. And perhaps this is as it should be. Nonetheless, the likelihood of achieving such moments is undoubtedly affected by a number of less magical and less mysterious but no less useful concepts, conventions, and capabilities that can indeed be taught in an academic setting and embraced in professional practice.

The term "geodesign" is one that has recently – and rightly – been promoted in reference to geographic decision making by way of that combination of rational and non-rational thought that is more typically associated with things like designing a building, sculpting a landscape, composing a song, or crafting a work of art. While the notion is certainly not a new one, the deliberate attempt to give it a name and place that name in the spotlight (both of which have been done quite successfully) presents an opportunity to articulate the notion in ways that can have real consequence.

This is particularly true in terms of the media to be employed. The current focus on geodesign is largely due to JACK DANGERMOND, a landscape architect by training but also founder and president of the company responsible for one of the world's most prominent geographic information systems (GIS). Indeed, much of the current discussion about geodesign has been oriented toward the ways in which it can be supported through the use of GIS and – perhaps more importantly – the ways in which GIS might be extended or refined to better assume that role.

This focus on GIS as the primary medium for geodesign has enormous implications, not only in terms of particular data-processing capabilities, but also in terms of the manner in which those capabilities are combined into more sophisticated procedures. In effect, it establishes the fundamental vocabulary by which geodesign strategies are depicted and deployed. Importantly, it also establishes the broader language through which such strategies are initially derived, gradually developed, and ultimately deliberated over time.

To appreciate just how significant that can be, consider the role of formal language in other creative fields of endeavor, from the composition of music or the choreography of movement to the engineering of materials or the forging of legislation.

The language of GIS has already affected a number of fields in which this technology has been applied. The question at hand is how this language should affect, and be affected by, efforts to articulate geodesign.

2 Making Decisions

To address that question, it may be helpful to start by stepping back. Geodesign is a form of decision making and, as such, it shares structural characteristics with other types of prescriptive activity, from optimizing delivery schedules to determining what's for dinner. It can be argued that all such activities encompass an interplay among three major types of information and two major types of information-processing activity.

At the risk of sounding more formal than is either necessary or appropriate, those three types of information might be cast as "predictions", "descriptions", and "prescriptions". Predictions predict *what could be*. They articulate the desired result(s) of a decision-making exercise and anticipate its potential effect(s) in both positive and negative terms. As such, predictions can be equated with what might otherwise be referred to as "objectives", "goals", "purposes", "outcomes", "consequences", "effects", "results", "aspirations", "intentions", "evaluations", or "assessments". Descriptions, on the other hand, simply describe *what is*. These are what might otherwise be termed "observations", "conditions", "constraints" or "parameters". They are selected aspects of the *status quo* that must be taken into account but are not themselves seen as subject to alteration. In contrast to these are prescriptions, which deal specifically with such alterations – particularly those intended to achieve positive benefits without suffering negative consequences. Prescriptions prescribe *what should be* and, as such, are much like "solutions", "treatments", "recommendations", "suggestions", "propositions", "plans", "designs" or "decisions".

The two types of information-processing activity that relate predictions, descriptions, and prescriptions to one another can be expressed by way of the terms "proposal" and "disposal". To propose is to put forth prescriptions that attempt to perform well in regard to effects that have been predicted and conditions that have been described. To dispose is to predict the effects of given prescriptions under described conditions. As illustrated in Figure 1, the output of each of these two activities becomes input to the other to form a cycle of iterative refinement that begins and ends with predictions.

Fig. 1: The Propose/Dispose Cycle

This cycle also implies that, for certain types of decision at least, prescriptive strategies can and should be informed by their predictive counterparts. After all, it only makes sense to predict how a game will be played before one attempts to prescribe a game-winning strategy. In some fields, in fact, it is possible to automatically invert the predictive statement of a problem into a prescription for its solution. A variety of mathematical optimization techniques are remarkably effecting in doing just that.

Indeed, formal optimization techniques may well have a role to play in geodesign. There is much about geodesign, however, that is likely limit that role. One such concern is the fact that this form of decision making is inescapably *geographic* in nature.

3 Making Geographic Decisions

The world as seen through the eyes of a GIS is one in which geographic conditions are represented as cartographic "layers". Each layer is a bounded plane in which every cartographic position corresponds to a geographic location, and the recorded attributes of those positions represent geographic conditions at corresponding locations. These layers tend to be organized such that a typical data set might include base layers depicting things like geology, topography, hydrology, vegetation, development, and political units; derivative layers depicting general conditions like topographic slope or aspect, proximity to the nearest road, or the total amount of open space per county; and additional derivative layers depicting more specialized characteristics such as wildlife habitat vulnerability or alternative sites proposed for future development.

Such derivative layers are usually generated by way of GIS operations that perform tasks such as classifying attributes, superimposing other layers, measuring distances and directions, characterizing sizes and shapes, simulating geospatial processes, and so on. Significantly, most of these operations accept input and generate output in a common format: the cartographic layer. As such, they can easily and flexibly be combined by using the output of one as input to another. Thus, no one operation need be overly complex or highly specialized. Complexity and specialization are instead achieved as primitive operations are combined into cartographic "models".

The construction of a geographic decision-making model begins with its predictive component: a set of layer-generating operations that attempt to predict the selected effects of potential prescriptions, given a particular set of site descriptions. Consider, for example the generation of a site-suitability layer for a new land use. Since no new land use has yet been sited, the initial presumption is that this land use could be sited just about anywhere. For the moment, at least, we therefore assume (and assert) that the land use is effectively sited everywhere. That assertion makes it possible to map the predicted effects of this tentative siting and thereby indicate site suitability. Given such a site suitability layer, the next step would be to use this prediction to prescribe a siting plan that maximizes predicted benefits while minimizing predicted costs by simply selecting the most suitable sites. Once such sites have been selected, perhaps including alternatives, they might well be subjected again to the predictive component of the propose/dispose cycle in order to rank those alternatives. This elementary site-suitability mapping technique embodies methods that were popularized by landscape architect IAN MCHARG in the 1960s and which are often cited for their influence on the early development of GIS.

One problem with those methods, well recognized even at the time by McHarg, relates to what were sometimes called "spatial pattern effects". These are qualities that cannot be predicted on the atomistic basis of individual locations but which instead arise only when multiple locations are considered in a holistic manner. Consider, for example, the simple requirement that a proposed land use occupy at least ten hectares. Now consider the less-simple requirement that this land use occupy *exactly* ten hectares or the decidedly-unsimple requirement that those ten hectares be in the shape of the letter "G". How about a truly complex requirement that this "G" be sited in meaningful juxtaposition to two other yet-to-be-sited land uses: one in the shape of the letter "I" and the other shaped like an "S"? In each of these cases, it would be quite possible to develop a GIS model that predicts the degree to which alternative land use plans satisfy the stated criteria. What is not so clear is whether a GIS could also be programmed to generate such plans.

And yet, we know that a human decision maker could generate such plans with ease. So how does a human decision maker actually make such decisions? Most would probably not do so by following the definitive steps of an algorithm but, rather, by following the more experimental steps of a heuristic. Whereas an algorithm is essentially a linear recipe, a heuristic is an iterative and exploratory procedure. In a typical heuristic, each iteration attempts to make progress through trial and error by

- posing an interim solution that improves on previous ones, and
- generating useful knowledge for future iterations.

Consider, for example, the kind of decision making that takes place when a group of strangers mulling about is asked to quickly "line up". What are the thought processes involved in attempting to unravel a tangle of twine? Or for that matter, what goes through you own mind as you solve the following problem in reference to Figure 2? *Without raising your pencil, draw exactly four straight lines connecting all nine dots.*

The cyclical structure of the propose/dispose cycle lends itself well to such iterative processing. Indeed, there are a number of effective heuristics for the allocation of geographic space that have gained well-deserved acceptance in fields ranging from political redistricting and retail siting to complex routing, military tactics, and timber harvest scheduling. Experience has shown, however, that both the geometric and the social dimensions of geographic problems are often such that it is well worthwhile to keep humans in the loop. In fact, it is also often a good idea to engage those humans in *design*.

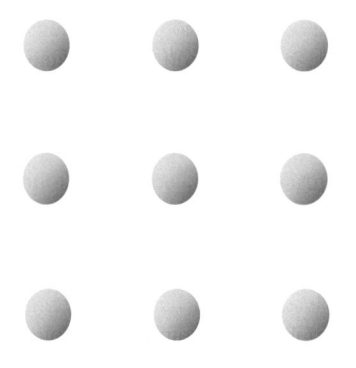

Fig. 2: Connecting the Dots

4 Making Geographic Decisions by Design

So how does design differ from other forms of decision making like planning or engineering? It is a familiar question and one whose answer would seem to be straightforward but which is nonetheless seldom articulated. This is true even in academic and professional settings where disciplines like urban design, regional planning, and environmental engineering reside just down the hall from one another. In reference to that propose/dispose cycle, what seems to distinguish design-flavored variations of the cycle most is – simply and perhaps surprisingly – the fluidity of predictions. In particular, it is the choice of those selected qualities that initially motivate the design exercise and ultimately measure its success. In planning or engineering problems, these qualities tend to be well articulated from the outset and seldom subject to change. In design problems, on the other hand, they tend to be much more fluid.

One way to deal with this is reflected in ever-so-slightly-satirical quips like the following.

> *As an accomplished artist, I am certainly able to tell you precisely what it is that I am looking for. In fact, I can often do so within just a day or two of having found it.*
>
> *To sculpt a masterpiece, just start with your block of stone and chip away everything that doesn't look like one.*
>
> *Depending on your field of endeavor, prescription without prediction is either inspired design ... or malpractice.*

A more constructive way to deal with that fluidity is to acknowledge, accommodate, and take full advantage of emergent design criteria. These are sought-after qualities that are well able to be predicted but may not even have been considered at the start of a decision-making process. Rather, they come into focus and assume significance only as tentative decisions are iteratively generated, evaluated, and refined. Consider, for example, an exercise in rearranging a set of bedroom furniture. In an attempt to achieve better cross-ventilation, you move the bed to a corner between two windows. Having done so, a certain pleasing balance between the bed and the dresser begins to emerge. While this criterion was not there at the outset, it soon assumes even more importance than cross-ventilation.

The question at hand, then, is fairly straightforward. How can the job of recognizing, recording, and revising emergent geodesign criteria be accomplished *better* through the use of GIS?

5 Making Better Geographic Decisions by Design

What is it that purportedly makes the back of an envelope such an attractive medium for certain types of design? On the one hand, it is certainly quick, inexpensive, easy to use, readily available, and decidedly flexible. On the other hand, it is really not very good at evaluating or even recording – let alone generating – new ideas without relying almost entirely on the fertile imagination of that person at the other end of the pencil.

So how about GIS? Sure, it would always be nice to enjoy higher speeds, lower costs, easier operation, greater availability, and increased flexibility, but none of those niceties is particular to geodesign. What seem to be most critical to the use of this technology in a design setting are the degrees to which it is able to assist its user in doing three things: generating, recording, and evaluating refinements as part of the propose/dispose cycle.

5.1 Generating Geodesign Proposals

In terms of generating new proposals, there are two conspicuous ways in which GIS might be better able to play a more constructive role in support of geodesign.

The first is not especially exotic but is nonetheless likely to have real and immediate impact. It is simply to help in cataloguing examples of previous solutions. In a world where web searching, knowledge tagging, geopositioning, aerial imaging, social networking, and mobile computing have almost instantly become routine, our ability to record geodesign precedents is already well established. What remains for GIS is to deliver those precedents to the geodesigner not just in terms of pretty pictures but, rather, in terms of things like dimensional standards, spatial templates, legal citations, price lists, and environmental impact assessments that are keyed to particular components of the design project at hand.

The second is much more exotic but also much narrower in likely impact: to improve on formal techniques for geospatial optimization. As indicated earlier, this is already an active and productive field but one whose more direct impact on geodesign may call for greater interaction between spatial allocation heuristics and the designer's intuition. In Figures 3 and 4 is an example. Given the "landscape" of site conditions depicted in Figure 3, suppose your task is to connect the upper right and lower left corners of this landscape with a path that minimizes the total amount of darkness along the way. If you will spend a few moments attempting to do so, you will soon come to appreciate the path shown in white. This path solves the same problem but does so for the other two corners. It was "designed" by way of a standard technique for generating minimum-cost routes.

To the geodesigner, however, it may not always be sufficient to simply accept such an optimal solution. Indeed, it may be much more helpful to show how that solution compares to sub-optimal alternatives. Just as a negotiator might search for "wiggle room", the geodesigner might well want to know how – and where – an optimal path might be jostled just

Fig. 3: The Optimal Path

a bit in order to accommodate additional siting factors without unduly increasing its overall cost. Figure 4 responds to that concern by placing the minimum-cost path, as still shown in white, within a context of more and more costly alternatives that are represented by increasingly-darker shades of grey. The technique used to generate this output is no more recent or complex than that which was used to generate the optimal path, and yet this output may well be of much greater use to the geodesigner.

Fig. 4: Sub-optimal Paths

5.2 Recording Geodesign Proposals

In terms of recording new proposals, it can be said that trying to sketch with a GIS is no more difficult and no less effective than trying to console a crying baby ... by issuing instructions to its baby sitter ... over a phone line ... via text messages ... while driving ... a backhoe ... in heavy traffic. It is clear that GIS could benefit enormously from the kind of intuitive input gestures that the SketchUp application has introduced to the world of architectural modeling. In contrast to those of SketchUp, however, each gesture here should be guided not so much by what has already been drawn as by the landscape beneath that drawing. One should be able to reposition, reorient, resize, and reshape in ways that immediately and automatically interact with geographic conditions. A point drawn on a topographic surface, for example, should be able to initiate a downstream path, to follow a contour, to grow its upstream watershed, or to delineate its viewshed. And ideally, each of these should be immediately updated as that point is repositioned.

In order to be effective in the particular context of geodesign, it is also (and perhaps even more) important that the GIS user be able to record proposals at varying degrees of both spatial and thematic precision. Spatially, this would suggest that proposals be cast not in terms of crisply-bounded features (or at least not necessarily so) but as continuous fields of presence/absence on which transitions may be either gradual or abrupt. Thematically, it would suggest that proposals be cast not in terms of crisply-articulated conditions (or at least not necessarily so) but as qualities defined at higher levels of abstraction. Consider, for example, a back-of-the-envelope scribble. The degree to which this is to be taken as literal or figurative is entirely a matter of its scribbler's intent. And very importantly, that intent must be allowed to change again and again as certain portions of the scribble are anchored while others are permitted (and indeed encouraged) to float in a sea of additional opportunities and constraints. In short, what is needed is something akin to the blob diagrams that have long been employed in traditional landscape design. Note how the map layer shown in Figure 5, for example, depicts three different spatial phenomena: the specific location of an uppercase "G", the more general location of an "I-like" form, and the vague notion of some sort of squiggle.

Fig. 5: Blob Diagrams

5.3 Evaluating Geodesign Proposals

Arguably, the greatest opportunities to empower geodesign through GIS lie not in the generation or the recording of design proposals but, rather, in their evaluation: the ability to dispose of such proposals in a manner that leads to their subsequent evolution.

Perhaps the most straightforward way in which GIS can assist here is to enable geodesign proposals to be taken, quite literally, into the field. While we have done well over the past several decades in bringing more reality into the lab, we have only recently begun to establish the technology necessary to move in the other direction and effectively take more of our virtual worlds back into the real one. As we consider how best to deliver pertinent information to that geodesigner who is currently standing in a very-real world of sights and sounds, temperature and humidity, sunlight, winds, risks, recollections, and so on, it is clear that GIS has much to offer in effectively organizing and presenting stored information.

The second most straightforward way in which GIS can assist in the evaluation of geodesign proposals may be to make them more readily available to the public. The same digital media that hold such promise for placing precedents within reach of the geodesigner have also spawned interests and expectations in online rating mechanisms that portend a fundamental shift in the geodesigner's role. At the risk of endorsing "design by committee" in any traditional sense, suffice it to say this. While individuals are still just as likely to be better at proposing, disposing will still almost always be better handled by committee – particularly when something like GIS is available to present options, compile reactions, and return results to the public in meaningful terms.

Closely related to this is the prospect of making design-evaluation models available and easy to invoke. Recent developments geographic data access, model sharing, and web-service computing make it easy to envision a time when off-the-shelf capabilities will make it possible to routinely upload ones data and plans in order to assess anticipated impacts. Even more encouraging along these lines are developments in what has come to be referred to as "parametric design." The fundamental notion here is that the component elements of a design proposal can be made sufficiently sensitive to their proposed environments (and each other) that they can self-adjust in order to achieve a better fit. Not only can that proposed pine say "Ouch!" but it can also walk away if necessary or even metamorphose into a hemlock. Though good geospatial examples of this are not yet common beyond the worlds of infrastructural networks, an ability to combine the geometric capabilities of GIS with the bookkeeping magic of relational database management systems holds enormous promise for the construction of such "smart" digital landscapes.

Much of this promise relates to the possibility of endowing digital trees, lakes, roads, buildings, and other elements of intelligent maps with sophisticated behaviors. Given such a set of elements, the geodesigner need not be a forester, hydrologist, transportation engineer, or an expert specializing in any one of their particular behaviors in order to effectively work with those elements in a game-like design environment. In fact, it may well be the game-like nature of that environment that is ultimately more important than the sophistication of those behaviors. By enabling that geodesigning gamer to tap into human powers of perception and choreography that are more visceral than cerebral, it is often possible to detect patterns, to capture opportunities, and to "cultivate serendipity" in ways that would otherwise never happen.

With or without the benefit of a game-like environment, another way in which GIS could certainly improve on its use as a design tool is simply to present its evaluations in more evocative form. How about levels of sound that rise or fall in pitch and/or volume as evaluations change? A mild electric shock might even be used to convey the implications of an ill-advised decision (a technique that actually works quite well with my dog).

The most productive critic is always one whose criticism, regardless of its content, is delivered in a format that lends itself to immediate ideation. In evaluating geographic decisions, however, this is often made difficult by the fact that causes and their effects may well occur at different locations. While it is one thing to note, for example, that a limited number of allowable units per district has just been exceeded by the last one proposed, it is quite another to determine just where that problem should be noted in order to be corrected.

This is also true of visual impacts, which may well be felt in the eyes of the beholder but are often more helpfully cited at the location of whatever it is hat has been beheld. And note how solutions to hydrological problems almost always seem to lie upstream? In short, the ease with which one can damn the darkness may often belie the difficulty of turning on a light, particularly when the light switch is located elsewhere.

Less difficult but perhaps just as important is the prospect of developing evaluation models at different levels of abstraction. Half-baked ideas should be taste-tested as such, with more general criteria and greater tolerance in earlier iterations of the propose/dispose cycle. What might initially be evaluated as "open space", for example, might later be assessed "outdoor recreation" and only after that identified and judged as "a nine-hole golf course".

This idea of employing different models of evaluation at different points in the design process is also central to one final recommendation. In recognition of the fact that what most distinguishes geodesign from other forms of geographic decision making is the role of emergent criteria, what would happen if evaluation models themselves could be selected, weighted, adapted, and even reinvented as a consequence of their own earlier findings? In other words, what if the propose/dispose cycle could be cast as one in which predictions learn from prescriptions as much as the other way around? It is an engaging question and one that could prove to be fundamental to the broader issue at hand. It is also a question that will never be answered or even well posed, however, without an appreciation and an aptitude for design.

GeoDesign – Approximations of a Catchphrase

Hans-Georg SCHWARZ-v.RAUMER and Antje STOKMAN

Abstract

The paper discusses methodologically the term ‚Geodesign'. The history of the term is illustrated by three approaches which can be considered as complementarily covering the broad meaning of the term. A classification scheme for the description of geodesign outcomes leads to the description of product-lines according to the purpose and function of Geodesign. The authors suggest to consider ‚Geodesign' a collaboration and convergence program between geo-science and the design of spatial visions.

1 Introduction

„'We've been doing GeoDesign for years' was a statement commonly overheard at the first Geodesign Summit held in January 2010" (Artz 2010). Our contribution tries to balance what is new and what is old and what is worth to follow up as a perspective concerning the catch-term „GeoDesign". We try to come to a conclusion about what fits and what does not fit under a broad definition of the term. But this contribution also simply bundles up some own examples in order to define main „product-lines" 'GeoDesign' could have. Finally an outlook will be given into the potential future of 'GeoDesign' as a bridge between landscape planning and landscape design.

2 Approaches

The nineties. Not later than 1993 KUNZMANN (1993) uses the term 'Geodesign' to discuss opportunities and threats related to illustrative sketches communicating ideas of spatial structures like the "European Banana". His fear is that simple iconic map like representations of ideas on spatial development replace the time consuming activity of reading texts as well as reanalysing numbers and reinterpreting complex maps which underlay the text. In the meantime for example DÜHR (2007) has clarified that the Banana was an extreme simplification on the way to balance out the degree of generalisation and to maintain the idea of a territory-specific concept. What remains is that we have to accept that maps are „socially produced and discursively embedded within broader contexts of social action" and that conceptualisation and developing graphical images help – and often must help – to communicate spatial relations in a consumable language. This affords that the graphical and the linguistic structure of a cartographic representation collaboratively addresses its intentional meaning. All in all 'Geodesign' coming from the nineties was strongly related to non-binding planning cartography which was working on spatial scenarios and visions. Dühr (2007, 58) summarizes different types and functions of 'geodesigned futures' which are conceptualized in Fig. 1 – a very helpful scheme to classify

'Geodesign' case studies in general. BBSR (2011) shows and PÜTZ et al. (2009) discusses the actual state of the Art of the "German branch" of 'Geodesign'.

Ecological Design. Looking back to the roots of ecological design, IAN MCHARG (1969) in his book "Design with Nature" put forward a system of analyzing the layers of a site such as the history, hydrology, topography, vegetation, etc. with the aim of compiling a complete understanding of the qualitative attributes of a place as a set framework for planning. His system is considered to be the foundation of today's Geographic Information Systems (GIS) and therefore also for GeoDesign. McHarg´s approach has been further developed into an institutionalized system of ecological planning (not design) which is basically an analytical process. Landscape objectives are drawn from both scientifically based landscape analysis and from normative democratically legitimised goals (as stipulated in laws).

However from the point of view of ecological design in the 21st century, the approach of "nature showing the way" is criticized by NINA-MARIE LISTER (2007) as a too deterministic model of nature: good design does not mean that the correct reading of the landscape would necessarily prescribe appropriate design. Whereas in landscape planning science is perceived as a deterministic imperative for design, the landscape designers call for a more open, process-oriented and flexible design process. That also means that the designers themselves have a more active role as "creative agents" who consider their interpretations of ecological realities not as solutions but as choices and trade-offs within an evolving, open landscape system.

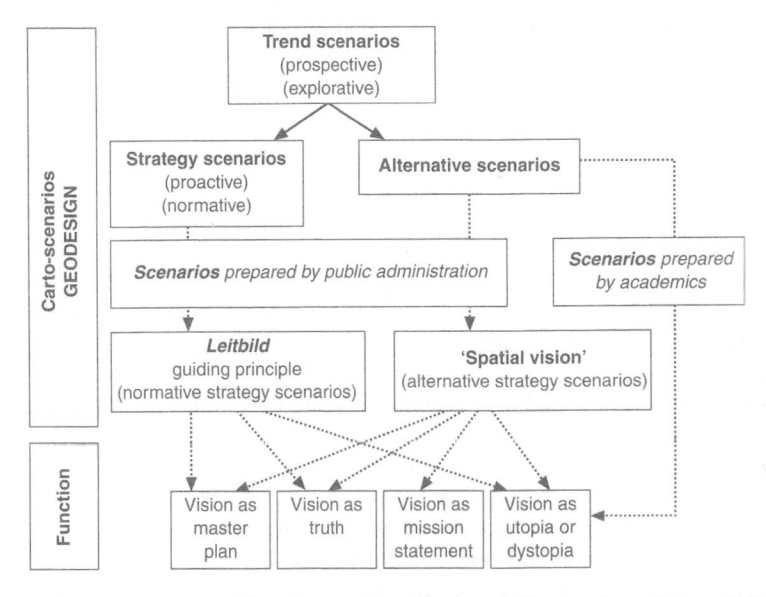

Fig. 1: Types and functions of 'geodesigned futures from DÜHR (2007, 58)

The Geodesign Summits. Since the First Geodesign Summit 2010 the term 'Geodesign' gets popular in the GIS community. The term now concerns the gap between *GIS* and visual communication on spatial ideas. From this starting point ESRI publishes ArcSketch™ but the discussion goes more in depth, it

- Enlarges the role of GIS as a backbone of data driven spatial reasoning
- Includes the role of participative and collaborative planning approaches
- Considers spatial visualisation as a key in such approaches
- Includes modelling as a tool to generate visualisations of spatial structures and
- Tries to re-establish a rational view on spatial planning.

The summits in 2010 (http://www.geodesignsummit.com/pdf/agenda.pdf) and 2011 (http://www.geodesignsummit.com) tried to gather the 'who is who' contributing to the topic in the United States and completed the list by some flagships from e.g. Canada and Europe. The intention of the summits is to combine the presentation of existing and innovative technologies, experiences and results with creative and innovative reasoning in 'idea labs'. As a focal methodological approach the use of geo-information directly or processed by spatial analysis and statistics, simulations, geo-processing models and Multi Criteria Analysis is used to develop 'plans' which are optimized by adaptive loops through impact analysis and participatory communications.

3 Terms, and Towards a Unifying Definition

Despite having highly ranked promoters, the term is not yet clearly defined. 'Geodesign' must be specified somewhere between mapping, modelling, sketching, visioning, planning, generating, constructing and engineering. But the term must also be specified according to the common understanding of 'geo' and 'design'.

Geo[1]. All approaches agree the convention of 'geo' coming from geospatial information, which can be descriptive, analytical or – as a modelling result – conjectural, which covers bio-, geo- and landscape-ecological as well as social, socioeconomic, economic and socio-cultural aspects, which includes their physical spatial manifestations of physical human land-use structures and infrastructure facility patterns, and which provides insights in spatial coincidence, patterns and processes. When using geo-information technologies like GIS, spatial modelling and 3D-visualisation (including the use of ingenious VR-devices) or when bethinking expert knowledge, in both cases doing Geo-design means to be "real world related" (FISHER 2010) or "evidence based" (TANZER 2010) or "consequence anticipation guided" (GOODCHILD 2010). It intents to include natural, semi-natural and man-made environments and mechanisms (system process), geo-pasts and geo-futures into what we consider as a design process.

Design. The task of design is the purposeful manipulation of an object or arrangement (e.g. physical space and its organization) in terms of information, function and systems with regard to functional performance, aesthetic qualities and social affects. Designers use the creative design process as a fundamental tool for synthesizing complex factors into cohesive designs. Landscape designers consider landscape "less [as] a quantifiable object than an idea, a cultural way of seeing, and as such it remains open to interpretation, design

---

[1] In general we know three meanings of 'geo': 'abiotic', 'earth' or 'spatial'.

and transformation" (CORNER 1999). Von SEGGERN et al. (2008) describe spatial design in the context of designing urban landscapes as "searching for interpretations, for development possibilities and a spatial Gestalt which is able to unite multiple requirements – from function, to process, aesthetics, construction, material, symbolism and ageing – in a convincing whole" (von SEGGERN et al. 2008). To express findings and ideas about the landscape, the visual representation tools of landscape design are of particular importance as they do not only represent an abstract system of colour-codes but mainly operate as a mechanism for the new interpretation of landscape. It is only through representation and picturing that landscapes become culturally visible and meaningful – "the world is visually prefabricated through its potential for being seen" (WALDHEIM 1999). Using a variety of tools from pencil to computer, designs can be represented and communicated in the form of drawings, photo-montages, plans, diagrams, models, films and texts.

Geodesign. Our suggestion now is not to define sharp conditions for a belonging to the subject, but we suggest three dimensions which could help to classify and to discern an approach being 'Geodesign'.

(1) The first classification aspect can be labelled *technology*. Besides the fundamental separation of doing Geodesign analogously or digitally the used technology refers to the dimensionality of space considered. Depending on the technology involved we can do Geodesign in 2, 3 or 4 dimensions and have to accept the limitations coming from the capability, usability and suitability – as the case may be – of relevant software products[2], devices[3] and, related to both, limitations with regard to grain and/or scale.

(2) As a second dimension for characterizing Geodesign product-lines we have to look for the *role of geo-information* in the design process. E.g. Multicriteria Evaluation results can be transferred directly into decisions, but they also can be considered as an input from a 'side event' into a planning process, which allows a certain or efforts a huge degree of freedom and which leads to a result that is not consequently following the rational suggestions of geo-information processing. And the workflow can (and in the ambitioned thinking of the recent promoters of Geodesign should) go beyond a one-way workflow. Given a high degree of interactivity between man/group and machine ideally an iterative loop can be established by a multiple walk through the rational planning concept "evaluation of conditions and options –> decision/design –> impact evaluation".

(3) The third criterion considers the *function* of 'Geodesign' products according to its binding character between vision and prescription (refer to Fig. 1). Depending on the function of the result we have to use less or more precise Geo-data and have to be less or more accurate in modelling, accepting evaluation procedures or representation detail. And we have also – besides an appropriate lay out of the visual representation – to select an appropriate lay out of the collaboration between the actors themselves and between the actors an the GeoIT-facilities.

---

[2] ArcSketch is a very ingenious take off, but it has to be improved; the discussion about geodesign will enforce efforts the integration of CAD and GIS.
[3] We just like to paraphrase the limitations of devices by the similarities of the words 'cave' and 'cage'.

4 Product-lines and Examples

As we can see there is a wide range of combinations of inputs and outputs, methods and techniques, targets and target groups which are touched by the brand 'Geodesign'. Now we go more in depth concerning the *functions* of 'Geodesign' refering to the concept of Moll (1992), who first separates external from internal purposes when using maps in the planning process, and who then separates basic analytical maps from cartographic representations for participatory purposes and from representations of binding objectives laid down in a unifying or prescriptive form. Some own examples can be assigned to each of this 'Geodesign' product-lines and use cases.

Representations for analytical purposes. Here 'Geodesign' products are addressing decision-makers – if used internally – or they try to give evidence or justification if used externally. A lot of examples exist where the label 'Geodesign' is used for applying Multi Criteria Evaluation (MCE) methods to propose patterns for landscape or urban development. MCE-methods or other analytical applications of geo-processing models per se do just weakly meet the term Design. The results can closely be used as an input of a design process, but usually an additional creative revision will be necessary. Supplementing this kind of geo-information processing based 'Geodesign' SCHWARZ-V.RAUMER & SADEK (2008) suggest to use spatial disaggregation/allocation procedures to create hypothetical land use patterns according to assumptions on strategically preferred spatial structures in urban planning. The scenarios can be used to find strategies for minimizing forest cover decline and related loss of retention capacity. This design approach goes beyond drawing a geo-data based sketch of new development areas as an input for impact analysis (DANGERMOND & ESRI 2010, FLAXMAN 2010). Here geo-processing model parameters are used to implement strategies into a complex land-use change model which then creates referential spatial structures (Fig. 2).

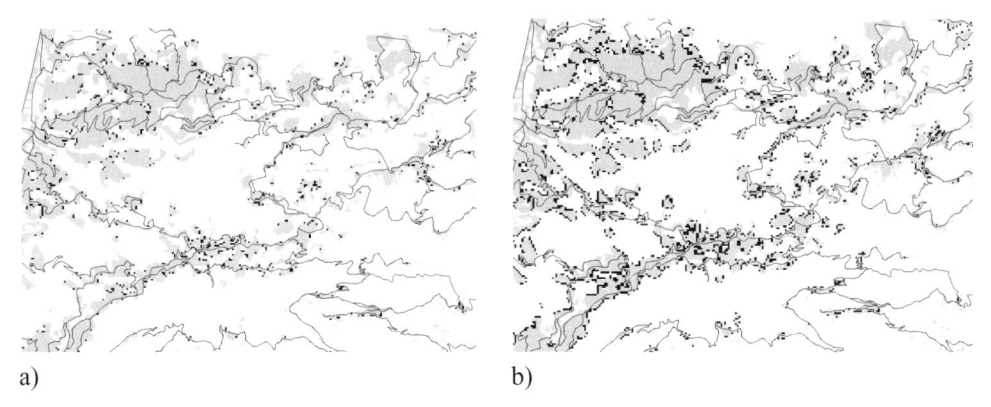

a)   b)

Fig. 2: Forest decline (= black raster cells) according to two future urbanisation scenario patterns, designed using a geo-processing model: a) unbiased urban development and b) development preferably in high density urban areas (urban areas 0 grey raster cells) (SCHWARZ-V.RAUMER AND SADEK 2008)

Representations for participation purposes. Here we must separate 'Geodesign' *done by* from that *done for* an addressed target group. Examples for the 'done by' group are reported by von HAAREN (2010) and CARLOS (2010) who interactively work with planning actors and clients to adaptively develop plans and thus inherently improve their acceptance. To discuss the 'done for' group we refer to AGGENS (1983, cit. in OPPERMANN 2001), who separates in an "Orbit-model" different groups participating in the planning process according to the degree of being involved by the scale 'unsurprised apathetics – observers – reviewers – advisors – creators – decision-makers'. He compares the degree of being involved with an energetic level. Graphic representations of plans here play not only the role of being a medium for communication. Maps, CAD-drawings and other 'Geodesign' products sent by a visual channel are able to 'energize' a person's involvement und to push his/her role more close to the kernel of the planning process. For that purpose again GIS can be helpful, particularly when geo-processing models build up spatial future visions. And the activating effect of such visualisations can be amplified when spatial scenario representations are used to demonstrate the consequences of per se non-spatial alterations. For example SCHWARZ-V.RAUMER et al. (2007) (also refer to KAULE & SCHWARZ-V.RAUMER (2008)) did that when visualising the difference of future land-use patterns in the northern Benin according to a different birth rate (Fig. 3). The visualisations have been successfully used in a stakeholder discussion about the necessity of the implementation of watershed management structures.

Representations for normative purposes. The target of spatial representations for normative purposes is to lay down and to socially unify interpretations of the present or of ideas, plans and visions concerning the future. Depending on the degree of being binding they are to be considered as a suggestion, an intention or a prescription. The European banana for example tries to unify the perception of the European spatio-economic structure and so to facilitate the internal and external communication about future strategies. The professional work of landscape architects on the other hand offer concepts for to get accepted as a binding design of geographical space. If – in the sense of the Ecological Design approach – the concept strongly incorporates landscape and geo-ecological conditions and processes, landscape architects do a normative 'Geodesign' job. The essence of these ideas is reduced in complexity by making use of the human capability to reduce complex information into coherent designs at the interface of analytical and creative knowledge. The complexity of the design needs to be condensed into a strong design idea that is easy to communicate and appeals to the client and public. Comprehensive information that might have been accumulated within the iterative process of analysis and design recedes into the background and is aggregated into a comprehensive and convincing idea that signifies the future and gains support. This idea is communicated through a spatial concept that is expressed in drawings and texts that communicate its essence and form a clear basis for more specific design decisions. Carefully selected, meaningful metaphors and titles are used to explain spatial concepts, like the vision of the "Tidal City Hamburg" (STOKMAN 2010) developed by the office osp urbanelandschaften presented in Fig. 4. The spatial vision for giving more space to the flood is expressed by the two landscape types of tidal lakes and tidal islands within the low-lying marshlands of the Elbe river in Hamburg.

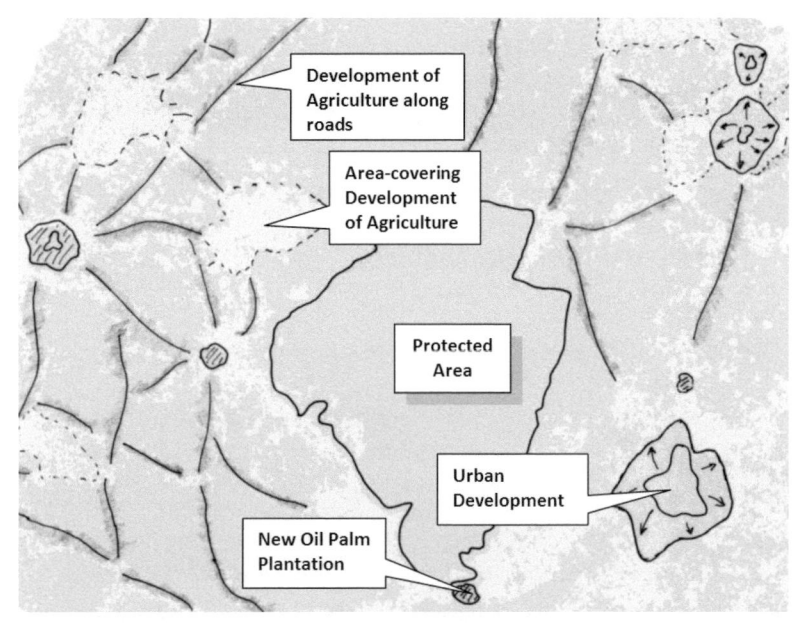

a) birth rate = 3.2

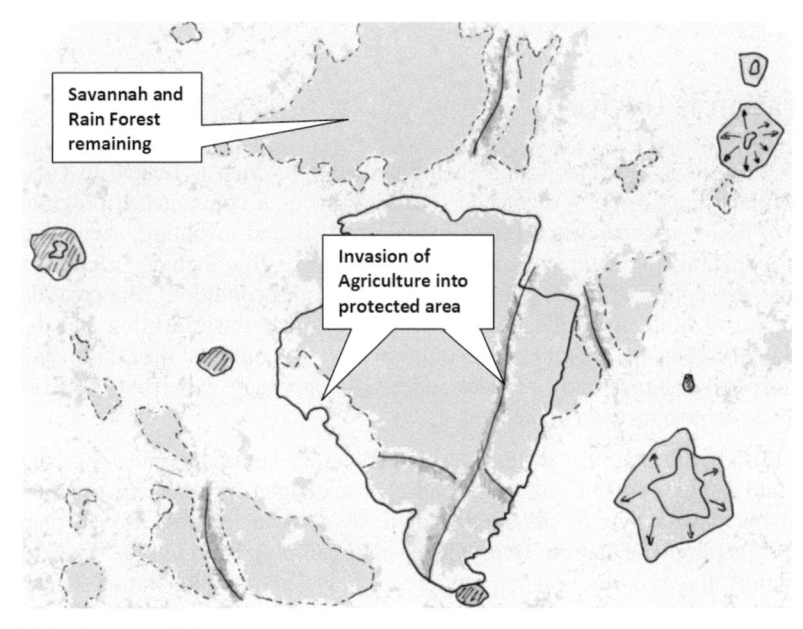

b) birth rate = 3.5

Fig. 3a/b: Two sketches visualising different future land-use patterns and processes in the north of Benin and overlaying the result of a 30year land-use change simulation in the background. According to SCHWARZ-V.RAUMER et al. (2007).

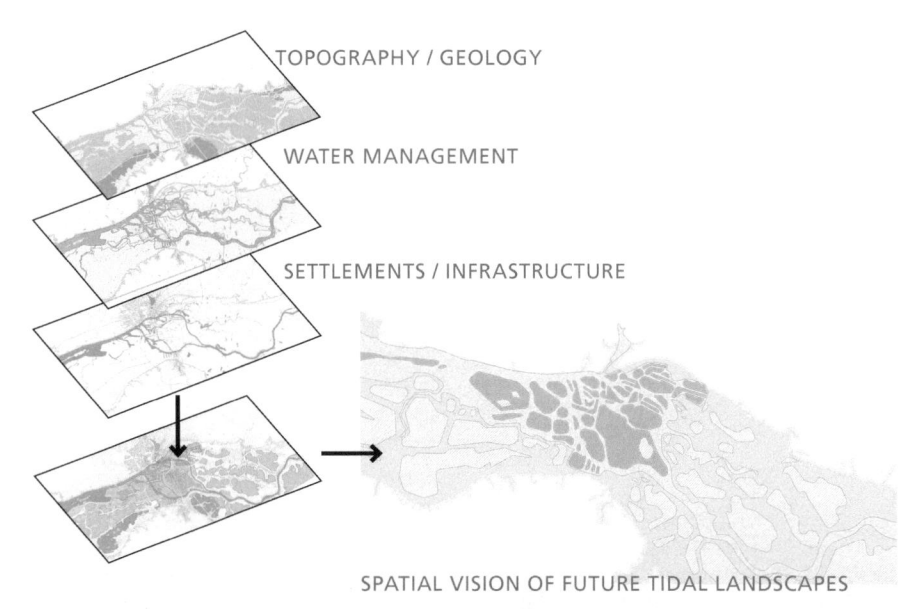

Fig. 4: A process of reducing complex environmental factors into a cohesive spatial concept as demonstrated by the Hamburg Tidal Elbe concept as a vision for the Elbe valley of the future (STOKMAN 2010)

5 Collaboration Is the Key

We suggest to not only focus on 'Geodesign' being something which is related to GIS-application and geodata-processing. We consider 'Geodesign' as a convergence program for bridging the gap between geo-science based spatial analysis and inventing, sketching, communicating and shaping the future of spatial environments. This includes a circular relation between the perception of given geospatial structures and conditions, the creation of ideas, and the control of their implications. It is possible that this workflow can be established individually, but in general we would prefer to combine the skills and knowledge of landscape designer and architects, GIS/GI-technology experts as well as landscape researchers, ecologists and engineers.

STOKMAN et al. (2010) analyse the different approaches of landscape planning and landscape design and suggest to link adaptive landscape management and experimental landscape design by new ways of interaction and towards a process-driven plan development and project implementation. Here 'Geodesign' can play the role of a key link, but there still remains the task to find appropriate new ways of collaboration which optimally include GI-technologies into adapted terms of interaction specified for the group of actors involved. Here it is worth to learn from science-policy facilitation. For example the "Joint Fact Finding (JFF) process" suggested by KARL et al. (2007) can be adapted. However there also exists a wide range of experimental fields to develop collaborative work modes and to replace an „Inform and Ignore"-practice. Work modes which can be tested in and should lead to improvements of teaching and learning in landscape planning, architecture and design.

6 Conclusions

The *first* conclusion we draw from the above considerations is a suggestion for a future research agenda (and perhaps for the next Geodesign Summit): it should not focus on technical or methodological aspects, but should emphasize the importance (1) of visual communication theory and visual language development[4], and (2) of the development of collaborative design settings which goes beyond simple workflow descriptions (DANGERMOND 2010) and yinyang ideas (GOODCHILD 2010). This would help to prevent a fall back to the idea of Wegener's planning machines (cit. in SCHWARZ-V.RAUMER 1999).

Secondly: the long lasting discussions about the rational planning approach[5] and about applying quantitative methods (MCE, Modelling) in planning should not be overheard due to the positivistic proclamations of the good deeds the new 'Geodesign' concept promises. Using an advanced technology and being stuck in the old usage of methods does not increase evidence of MCE and modelling analysis – and evidence is a fundamentally necessary prerequisite when using science in policy making and in planning.

Thirdly: 'Geodesign' could serve as an important contribution to bring back the designers type of intuition, creativity and emotion into the process of data analysis and representation, while at the same time it contributes to a new designer's perspective which is driven by a better understanding of man-environment interactions and which is a result of the take over of new collaborative design settings.

References

Artz, M. (2010), *Geodesign: Changing Geography by Design.* http://www.directionsmag.com/ article.php?article_id=3435.
BBSR (2011), http://www.bbsr..bund.de/cln_016/nn600826/BBSR/DE/ Raumentwicklung/ RaumentwicklungDeutschland/ Leitbilder/Konzepte/Fachbeitraege/NeueLeitbilder/ RaumentwicklungLeitbilder.html.
Carlos (2010), http://www.geodesignsummit.com/videos/day-one.html.
Chapin, F. S. & Kaiser, E. J. (1979), *Urban Land use Planning.* 3rd. edition.
Dangermond, J. (2010), http://www.geodesignsummit.com/videos/day-one.html.
Dangermond, J., ESRI (2010), *GeoDesign and GIS – Designing Our Futures.* In: Buhmann, E. et al. (Eds.), Peer Reviewed Proceedings of Digital Landscape Architecture 2010 at Anhalt University of Applied Science. Berlin/Offenbach, Wichmann, 502-514
Dühr, S. (2007), *The visual language of spatial planning: exploring cartographic representations for spatial planning in Europe.* Abingdon, UK.
Flaxman, M. (2010), *Fundamentals of Geodesign.* In: Buhmann, E et al. (Eds.), Peer Reviewed Proceedings of Digital Landscape Architecture 2010 at Anhalt University of Applied Science. Berlin/Offenbach, Wichmann, 28-41.
Fisher (2010), http://www.geodesignsummit.com/videos/day-one.html.

---

[4] Here ArcSketch must be improved and should be oriented to a unified symbolic language like the chorèmes suggested by Brunet (cit. in DÜHR 2007).
[5] "The rational model has come under fire […] no one follows the pure rational model […]. Instead, it is claimed that decisions are made by "muddling through" […]." (CHAPIN & KAISER 1979).

Goodchild (2010), http://www.geodesignsummit.com/videos/day-one.html.

Karl, H. A., Susskind, L. E. & Wallace, K. H. (2007), *A Dialogue, not a Diatribe. Effective Integration of Science and Policy through Joint Fact Finding.* In: Environment, 49 (1), 20-34.

Kaule, G. & Schwarz-v. Raumer, H.-G. (2008), *Bridging the gap between knowledge and policy action: Land use is the key – confidence is the condition.* In: River Basins – From Hydrological Science to Water Management at Kovacs Colloquium UNESCO Paris, June 2008. IAHS Publication, 323.

Kunzmann, K. R. (1993), *Geodesign: Chance oder Gefahr?* In: Planungskartographie und Geodesign. Hrsg.: Bundesforschungsanstalt für Landeskunde und Raumordnung. Informationen zur Raumentwicklung, 7.1993, 389-396.

Lister, N. (2007), *Sustainable Large Parks: Ecological Design or Designer Ecology?* In: Czerniak, J. & Hargreaves, G. (Eds.), Large Parks. New York, Princeton Architectural Press, 35-57.

McHarg, I. (1969), *Design with Nature.* New York, Natural History Press.

Moll, P. (1992), *Einsatz thematischer Karten in der öffentlichen Verwaltung: Werbung, Information und Bindung.* In: Festschrift Günter Hake zum 70. Geburtstag. Wissenschaftliche Arbeiten der Fachrichtung Vermessungswesen der Universität Hannover, 180, Hannover, 77-92.

Oppermann, B. (2001), *Die Katalysatorfunktionen partizipativer Planung im Umweltschutz – kooperative und partizipative Projekte als neue Instrumente einer umsetzungsorientierten Umwelt- und Landschaftsplanung.* Dissertation an der Universität Stuttgart. (http://elib.uni-stuttgart.de/opus/volltexte/2001/784/).

Pütz, T. & Schmidt-Seiwert, V. (2009), *Kartographie versus Geodesign? Visualisierungsbeispiele aus dem BBSR.* Informationen zur Raumentwicklung, 10/11.2009, 727-739.

Schwarz-v.Raumer, H.-G. (1999), *Bewertungsverfahren: Bedeutung in der raumbezogenen Planung, Methodik und GIS-Einsatz.* In: Kilchenmann, A. & Schwarz-v.Raumer, H.-G. (Hrsg.), GIS in der Stadtentwicklung. Methodik und Fallbeispiele. Berlin u. a., 35-63.

Schwarz-v.Raumer, H.-G., Printz, A. & Gaiser, T. (2007), *Ein "Spatial Scenario Design Model" zur strategischen Unterstützung der Landnutzungspolitik im Ouémé-Einzugsgebiet (Benin).* In: Strobl, J. et al. (Hrsg.), Beiträge zum 19. AGIT-Symposium. Salzburg. Heidelberg, Wichmann, 725-730.

Schwarz-v.Raumer, H.-G. & Sadek, M. (2008), *A GIS based scenario design modelling framework for ecological impact analyses of land use change: applied in the case of urbanization in the Lebanon.* ILPÖ-Workingpaper.

Stokman, A. & v. Haaren, C. (2010), *Integrating Science and Creativity for Landscape Planning and Design of Urban Areas.* In: Weiland, U. & Richter, M. (Eds.), Urban Ecology – a global Framework. Oxford, Blackwell Publishing.

Stokman, A. (2010), *Tidal City Concept Hamburg. Poster Theme 4: Climate change and climate proofing urban areas.* Conference "Deltas in Times of Climate Change", Rotterdam, Netherlands. 29 September – 1 October 2010.

Tanzer (2010), http://www.geodesignsummit.com/videos/day-one.html.

Von Seggern, H., Werner, J. & Grosse-Bächle, L. (2008), *Creating Knowledge.* Innovation Strategies for Designing Urban Landscapes.

Waldheim, C. (1999), *Aerial representation and the recovery of landscape.* In: Corner, J. (Ed.), Recovering Landscape: Essays in Contemporary Landscape Architecture. New York, Princeton Architectural Press, 121-139.

Foot Soldiers of GeoDesign

Joerg REKITTKE, Philip PAAR and Yazid NINSALAM

1 Introduction

Considering the growing variety of electronic tools being deployed by geo-oriented landscape designers and researchers in fieldwork operations, we are able to distinguish two main groups of users. The first group we refer to as 'aviators'. Namely, colleagues who profit directly from satellite data as well as all sorts of sophisticated aerial imagery. Aviators' GeoDesign fieldwork is fueled by an approaching armada of flying devices like drones and all sorts of helicopters, carrying high-capacity cameras and scanners that deliver image data as well as geo data for precise 2D imagery, mapping, and 3D modeling.

The second group we refer to as 'foot soldiers'. Being a foot soldier – of GeoDesign (FLAXMAN 2010) – does not mean that the chosen equipment is not related to the mother technology of the aviators – satellites, radio, and data networks – rather, it has to be light, portable and inconspicuous. An ideal stage of operation, for the GeoDesign foot soldiers, is the informal city, amidst unclear and labyrinthine urban grounds, street sites or waterscapes. Their mission is to provide for terrain data and details that cannot be gathered by the aviators (Fig. 1). Even with all the available remote sensing technology, in order to build complete and highly detailed models of complex terrain and urban territory, the direct contact to ground and detail will remain indispensable.

Fig. 1: More than 200 people room under this vast street bridge – only detectable by the foot soldier after climbing down the ladder (Photo: Rekittke)

Compared to the sophisticated machines and efficient methods in remote sensing, the craft of the GeoDesign foot soldier appears to be more intricate and laborious – but it can deliver unique results. In this paper, we pick such a result as the central theme. We will describe a method of on-site data and image gathering, which allows the processing of highly detailed and widely correct 3D models of labyrinthine spaces. The necessary equipment used is comparatively low cost; the software and online services, free.

2 Methodological and Geographical Context

2.1 Low-key, low-cost

This paper describes the last of a series of three fieldwork studies in the context of the research project "Grassroots GIS", financed by the School of Design and Environment, Department of Architecture, National University of Singapore (NUS). The method at hand is the sequel to fieldtrips and works by the authors in 2010, 2011 and 2012, published within the framework of the Digital Landscape Architecture conferences (REKITTKE & PAAR 2010, REKITTKE & PAAR 2011). The research is conducted in combination with special landscape design studios in the context of the NUS Master of Landscape Architecture programme. In 2012, we embedded our NUS based research into an interdisciplinary research program entitled Future Cities Laboratory (FCL), under the aegis of the Singapore-ETH Centre for Global Environmental Sustainability (SEC). The Centre serves as an intellectual platform for research, scholarship, entrepreneurship, and postgraduate/postdoctoral training, aiming at the provision of innovative scientific methods, instruments, and product research. We operate in the research module "Landscape Ecology" (GIROT & REKITTKE 2011).

Grassroots GIS relies on the bottom-up principle and integrates 2D and 3D geospatial data and tools for the purpose of landscape design activities. We developed a toolbox and user-generated geospatial content process that supports mapping, storing, staging of interactive design, disseminating, and generating interactive visualisations of landscape architectural interventions in the context of urban informal settlements. Grassroots GIS implies easy and free access to applied tools, geodata, and georeferenced design data. This premise calls for attention to open-source, open standard and cost-free or low-cost tools and data storage possibilities (REKITTKE & PAAR 2010).

2.2 Jakarta and the Ciliwung River

The outline of the FCL research module "Landscape Ecology" has been published recently. The article pictures the extreme condition of Jakarta and the rivers which cross the city (GIROT & REKITTKE 2011): "The urban catchment of Greater Jakarta, this metropolitan Moloch, has reached a population of about 28 million today and is expected to reach 35 million by the year 2020. But the exact figures are irrelevant; one senses overpopulation and its consequences everywhere one looks. Jakarta is located in a delta plain where 13 rivers merge into the Java Sea; the Ciliwung River is only one of them. The city is quite literally a sinking ship, as some parts of Jakarta sag almost 25 centimetres per year. This extreme subsidence rate is due to abusive ground water extraction, with the ground giving-in to the city's sheer weight. Jakarta is also prone to cataclysmic flooding, which has become a major annual problem for the city. The worst flood in the history of the Indonesian capital happened in February 2007: it covered almost 60 percent of the total urban area. In Kampung Melayu, one of our areas of study, the water level of the Ciliwung River reached a height of more than 10 meters over the valley floor."

Fig. 2: Ciliwung River in Jakarta – a natural watercourse in the shape of an open-air sewer (Photo: Rekittke)

Scientific spot tests confirm that almost all residents of informal settlements along the riverbanks discharge all their waste and sewage directly into the rivers (TEXIER, 2008). These people and their culture of pollution are not the only reason for the status of the rivers – the whole of Jakarta distinguishes itself by a total lack of a sewage system citywide. The environmental consequences of negligence and mismanagement are enormous (Cochrane et al., 2009), not only for the realm of the Ciliwung River. An incredible amount of artificial sedimentation – plastic waste – makes up its riverbed in central Jakarta (Fig. 2). Are we able to generate any impetus towards the improvement of this worst-case situation on the ground of the Jakarta Megacity?

2.3 Fieldwork mission

Via our research we are trying to contribute to a potentially positive answer of this profound question. Any realistic landscape design intervention along the Ciliwung River will be intrinsically tied to the relief along the river. The difference between the harsh, comparatively "lank" slum layer along the immediate riverbanks, and the quasi-ideal, dense mixed-use urban idyll of the kampung – the Indonesian urban village – with occasionally high quality houses, is widely defined by altitude differences. The simple formula "lower location = bad housing / higher location = better housing" cannot be generalised but applies to most of the area along the river (Fig. 3). Analysis and understanding of the terrain logic necessitates precise cross sections and three-dimensional models, which – typical in the complexity of narrow informal worlds – have to be acquired in the cumbersome foot soldier way. One way to build such models is based on the handmade photo-to-point-cloud technique that we approach in this paper.

Fig. 3: Same kampung – different altitude (Kampung Melayu, Jakarta): Slum layer down at the riverbank, high-quality housing up on the hill (Photos: Rekittke)

Being regularly flooded or comfortably living with dry feet becomes a matter of decimetres or even centimetres. Precise measurements in the field sometimes shape up as being problematic, because in the informally founded urban environment precision is a widely disregarded concept. Hence precise and measurable models of the site can contribute to additional clarity concerning profound design decisions.

3 3D Models of Unseen Spaces

3.1 Photo-to-point cloud

Lacking an expensive and chunky 3D terrain scanner, the common handheld digital camera serves as quick and easy imaging device for the foot soldier of GeoDesign. Instead, we are experimenting with consumer based photogrammetric 3D reconstruction software. The free software and online service 123D Catch (formerly Project Photofly) from Autodesk Labs Technology (AUTODESK 2012) allows for the creation of 3D models from digital photographs, usable for accurate measurements. The software and related workflow has primarily been developed and optimised for single object photography and corresponding modeling (Fig. 4), rather than for the capture and modeling of complex spatial entities. For our landscape architectural purposes, creative experimentation as well as unprejudiced trial and error can lead to sophisticated results.

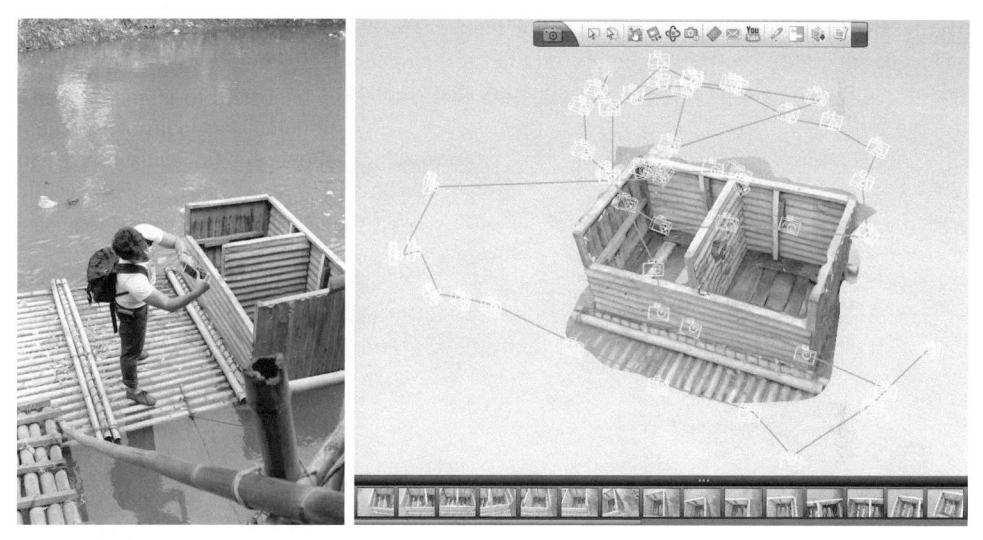

Fig. 4: Object centred photo 'catching' and 3D modeling: 'Slum W.C.' located on a floating bamboo raft on Ciliwung River. Recorded through circular photography from multiple viewpoints (Photo: Rekittke; 123D Catch sample: Ninsalam).

The standard workflow for 123D Catch models calls for the following steps: 1) Taking of multiple digital photos with any standard digital point-and-shoot camera; 2) Upload of photos in an image file format (jpg) via 123D Catch software – to a cloud server for stitching and processing. The software automatically downloads the results in a Photo Scene data file format (3dp). The cloud server mesh engine processes a high quality 3D model from the collection of overlapping photos. Provided that the system is fed with an extensive, well-taken set of photos, the resulting model will be spatially and dimensionally accurate. Via manual stitching – adding or removal of photos that had not automatically been included – the results can be adjusted; 3) Saving of the project as a video animation, or export of the underlying wire frame (model) into standard 3D formats (obj, dwg) for further editing in various 3D programmes.

3.2 From object to space

We are systematically trying to push the limits of *123D Catch*. Clearly laid out objects are easy to *catch* and the resulting models are good. It is evident that with a bigger and more complex spatial situation, the limitations of the software become clearer. It comes as no surprise that the 3D models of intricate spaces feature holes, due to the fact that sometimes even the most athletic foot soldier is not able to physically capture all necessary images from certain vantage points. When it lacks source data to fill in all regions of the generated 3D model, gaps are inevitable. Yet, there are several ways to optimise the results of works on larger complex spaces.

Shoebox method
Freestanding objects allow the photographer to revolve around them and take pictures from evenly distributed multiple viewpoints. For the catchment of an entire space this principle has to be inverted. The typical empty middle of the space has to be regarded as a virtual, see-through object that is scanned in a shoebox shaped manner (Fig. 5).

Fig. 5: Shoebox image capture method along a street at Manggarai Selatan, Jakarta. The small white camera icons indicate the location from which photographs were taken (Sample: Ninsalam).

Semi-circle method
Where we conduct our foot soldier handcraft, many streets especially alleys are too narrow for the application of the shoebox image capture method. In these cases we experiment with semi-circular camera movements around the alley mouth, to capture as many overlapping image instances as possible (Fig. 6). Generally, we have to try to achieve enough visual redundancy for satisfactory post-processing results. The described field photography

method does not allow us to control the adequateness of the image capturing on site – only with Internet access, e.g. in the hotel or a nearby *7-Eleven* store, we can examine our results. This bears the risk of coming home with fragmentary material.

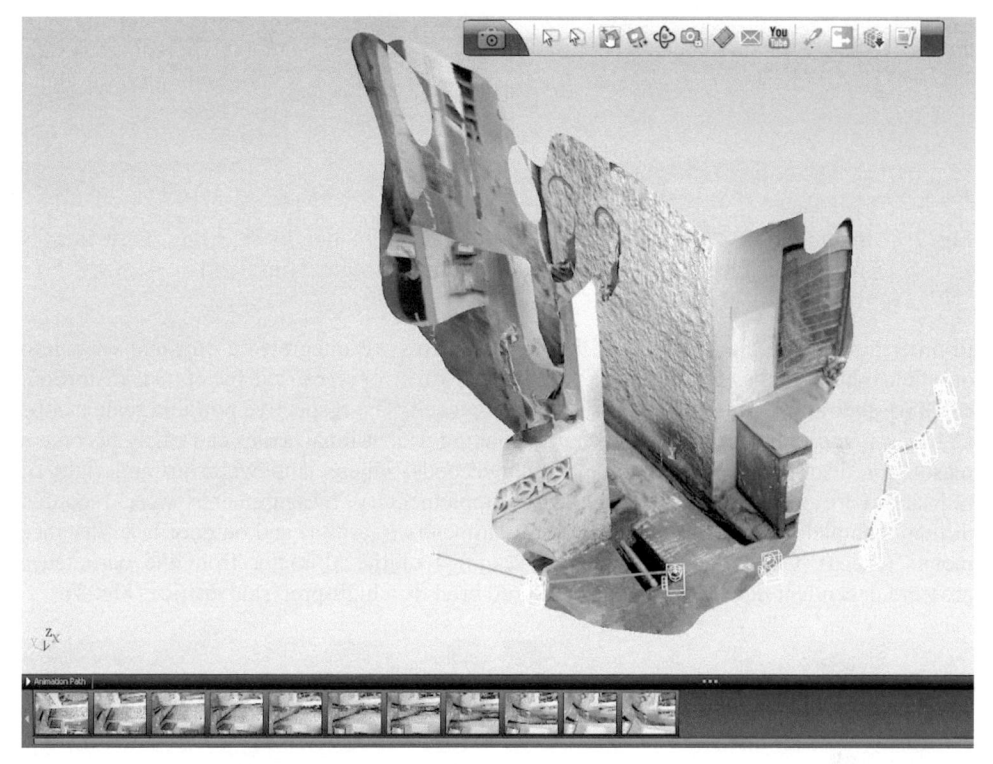

Fig. 6: Narrow alley situation. A semi-circular path was taken by the photographer to capture the space (Sample: Ninsalam).

We generate 3D working models for the purposes of measuring, design decision-making and participatory planning. Perfection and aesthetics of the models are subordinate. One very useful feature of *123D Catch*, which literally bridges the gaps in the generated spatial models, is the ability to fade in referenced original 2D images as backdrop to the 3D model, by selecting the respective images in the image bar of the programme window. This 3D/2D mixed depiction provides a better comprehensibility and reading of the model with regard to its spatial context (Fig. 7). This feature can also serve as an antetype for further experimentation with mixed media techniques of working models. Grassroots GIS and the foot soldier pattern of thought call for the exploration of all thinkable ways in uncommon terrain and city modeling.

Portable scale artifice
After having generated a *123D Catch* model, it is possible to do accurate measurements of any object in the 123D Catch editor scene. To extract reliable data, at least one reference dimension in the model must be known.

Fig. 7: Fading in of 2D images as backdrop in the 3D model, by selecting single images in the image bar of the programme window (Sample: Ninsalam)

In order to enhance the accuracy of the measurements we integrated a *portable scale* into our field photographs. By doing so, the margin of error – caused by optical distortion, stitching aberration, finite resolution – can be reduced. The respective portable scale can be anything found in the field. More technical looking professional scales can easily become a reason for distrust in informal city neighbourhoods, where landownership and right of utilisation are often not readily settled. Supplementary measurements were recorded through handheld GPS devices with in-built altimeters (Garmin) and outdoor laser distance meters (Leica). Compared to the erratic readings of the altimeter from the barometric pressure dependent devices, the well-measured brick is a high-precision artifice (Fig. 8).

Fig. 8: Left: Measuring a brick before positioning it as a *portable scale* in the field. Right: Supplementary measurements with a Garmin device (Photos: Rekittke).

3.3 Post-processing of models

There are multitudinous possibilities of subsequent post-processing and optimisation of 123D Catch models. In this paper, we limit ourselves to bring up three interesting options, with reference to our intention to use the 3D work in analysis, design activity, project communication, and public participation. The chosen aspects correspond to three important expectations in our works: (a) Comprehensible documentation and interactive visualisation of the site; (b) Exploitation for a georeferenced 3D model of the urban landscape; (c) Interactive modifiableness in the course of design processes.

Iterative integration of partial improvers
During the fieldtrip, after coming back to the hotel with first sets of field photography, the material can be uploaded via 123D Catch and inspected for the first time. Insufficiencies can be determined and potential improvers defined. Returning to the field, the specific sites (for improvers) are thoroughly photographed again and the material is uploaded, together with the initial set of photos. After the iterative integration of all necessary improvers, we can achieve a significantly refined model of comprehensible visual quality (Fig. 9).

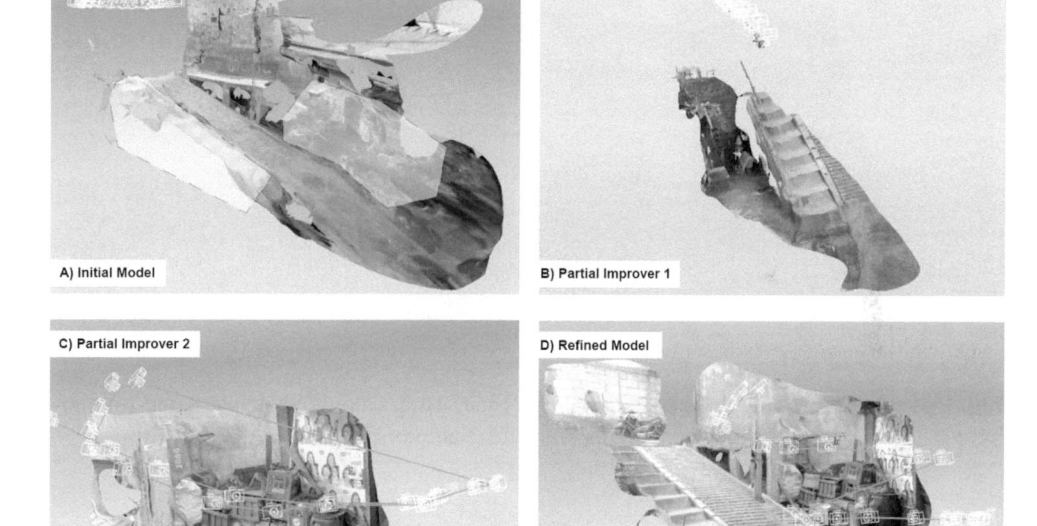

Fig. 9: Quality of the 3D model can be improved by adding localised picture of objects into the working model to generate refined model segments (Sample: Ninsalam)

The software 123D Catch demands discrete, logic spatial entities to be able to process usable models. The assembly of the single parts into an overall model of the entire spatial unit (e.g. a street or specific spatial cross section) cannot be processed within the programme. For this purpose we export the files into obj-format and compose them in Rhinoceros 4.0.

Embedding into the digital globe

Unfortunately, 123D Catch lacks geo-referencing. However, even in the toughest urban places, people have networked computers and know where they sit on the digital globe. Embedding 3D models into geo-virtual environments such as the digital globes Google Earth and Biosphere3D facilitates the intelligible discussion of planning and design decisions (SHEPPARD & CIZEK, 2008). This can be done via 3ds-format, using Google Sketchup 8 geo-referencing feature, which is saved in kmz-format (Fig. 10).

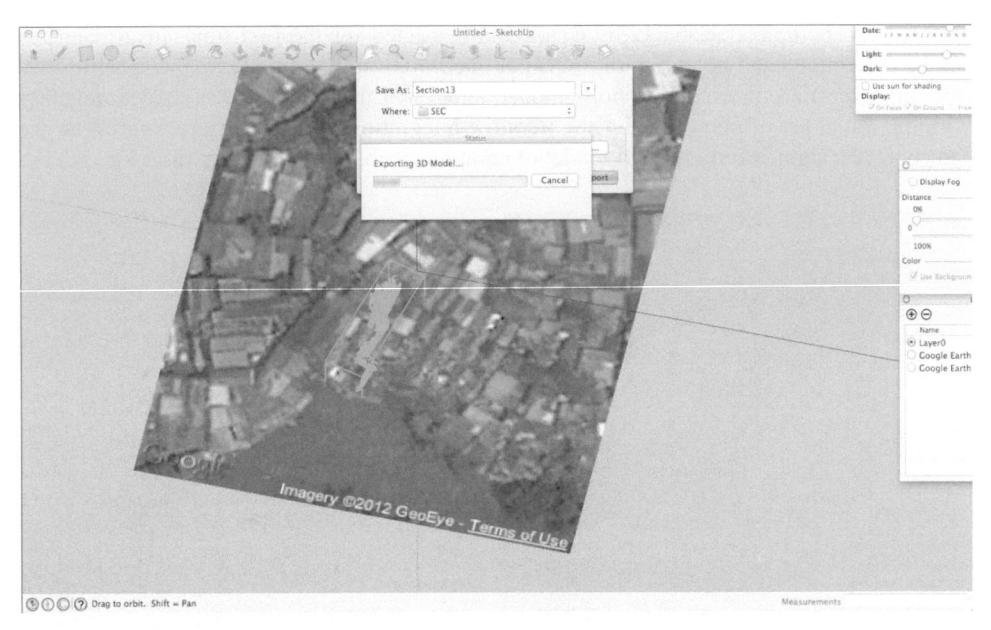

Fig. 10: Geo-positioning of a model in *3ds*-format, using Google Sketchup 8 geo-referencing feature which is saved as kmz-file (Sample: Ninsalam)

Interactive modifiableness

Supreme discipline of the GeoDesign foot soldier is the provision of sufficiently precise raw material from rough places, transforming it into a sophisticated 3D model – which then can be interactively modified, in order to display design alternatives. We started to export our Autodesk 123D Catch models as las-files into Autodesk Revit Architecture for future integration of 3D models of design proposals (Fig. 11). This is where we stand at the moment.

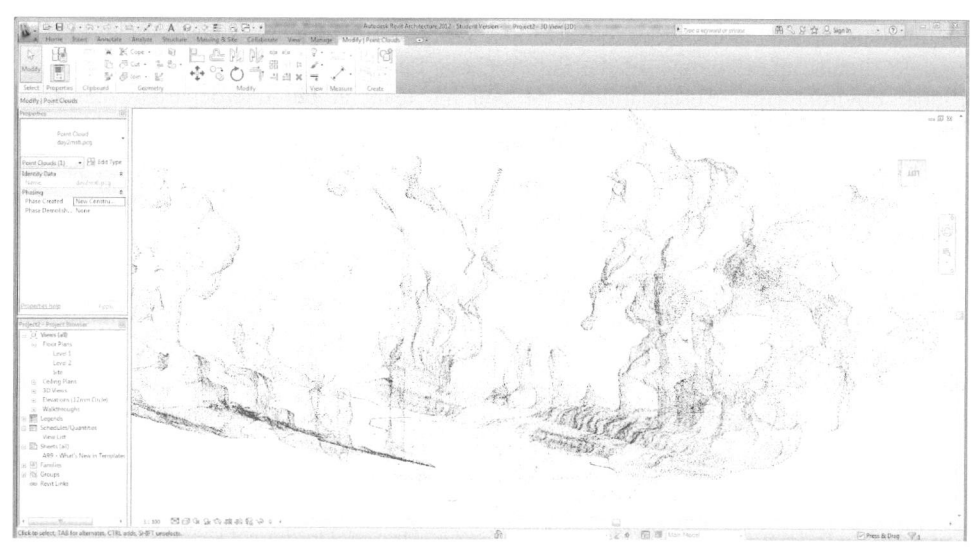

Fig. 11: Exported 3D information in *Autodesk Revit Architecture* (Sample: Ninsalam)

4 Writing on the Wall

We seek to remain moderate and grounded. We love muddling through the toughest urban places on earth. We advocate the utilisation of inexpensive, ordinary tools, and technology and we are convinced that the aviators will always need the foot soldiers and vice versa. But we are not ignoring the writing on the wall. There is no way around more expensive tools like short and long range 3D laser scanners for detailed terrain measurement and documentation – which are available on the market and can be readily employed by the landscape architectural guild now. At present, we are not certain if we can use these more expensive and flamboyant tools for Grassroots GIS operations on the informal city grounds. Such doubts do not have monetary justification. They derive from the fact that any informal city quarter is built on politically and judicially shaky ground. The people living in these parts of the city are wary and justifiably so. Another problem might be security issues concerning the laser beam. We do not speak of beautiful empty landscapes or deserted, suburban bourgeois living environments. We roam constantly crowded and super narrow city layouts. The mounting of exclusion zones and running around in hard hats plus high visibility vests are unthinkable.

5 Acknowledgements

The research and fieldwork described in this paper could not have taken place without the MLA students from the National University of Singapore. Especially through their contributions on top of being lion-hearted, weatherproof, and ever willing to travel together with us foot soldiers, many thanks to all of them. We would also like to thank the NUS Department of Architecture for the lasting support – in face of our out of the ordinary travel destinations. Special thanks goes to Christophe Girot, in his capacity as head of the research module "Landscape Ecology" of the Future Cities Laboratory, Singapore-ETH Centre for Global Environmental Sustainability. He and his team at the interdepartemental ETH Landscape Visualization and Modeling Lab are our favourite aviators.

References

Autodesk 123D Catch (2012), URL (Feb. 2012): http://www.123dapp.com/catch.

Cochrane, J. et al. (2009), *Water Worries*. Special Issue, Jakarta Globe Saturday/Sunday, July 25/26, 2009.

Flaxman, M. (2010), *Fundamentals of Geodesign*. In: Buhmann, E., Pietsch, M. & Kretzler, E. (Eds.), Reviewed Proc. of Digital Landscape Architecture 2010 at Anhalt University of Applied Sciences, Berlin/Offenbach, Wichmann, 28-41.

Girot, C. & Rekittke, J. (2011), *Daring Down the Plastic River in Jakarta*. In: Topos, 77, 55-59.

Rekittke, J. & Paar, P. (2010), *Grassroots GIS – Digital Outdoor Designing Where the Streets Have No Name*. In: Buhmann, E., Pietsch, M. & Kretzler, E. (Eds.), Reviewed Proc. of Digital Landscape Architecture 2010 at Anhalt University of Applied Sciences, Berlin/Offenbach, Wichmann, 69-78.

Rekittke, J. & Paar, P. (2011), *There is no App for that – Ardous fieldwork under mega urban conditions*. In: Buhmann, E. et al. (Eds.), Reviewed Proc. (online version) of Digital Landscape Architecture 2011 at Anhalt University of Applied Sciences, 26-36.

Sheppard, S. R. J. & Cizek, P. (2009), *The ethics of Google Earth: Crossing thresholds from spatial data to landscape visualisation*. In: Journal of Environmental Management, 90 (6), 2102-2117

Texier, P. (2008), Floods in Jakarta: when the extreme reveals daily structural constraints and mismanagement. In: Disaster Prevention and Management, 17 (3), 358-372.

GeoDesign Approach in Vital Landscapes Project

Sándor JOMBACH, László KOLLÁNYI, József László MOLNÁR,
Áron SZABÓ and Tádé Dániel TÓTH

Abstract

Vital Landscapes Project aims to interpret landscape management as an interactive, integrative, conscious and open procedure. Geodesign approach supports the frames of the research workflow and study area activities of the Hungarian project partner. It uses various combination of spatial data and software to interpret landscape in general and in the pilot area involving regional stakeholders. It integrates 3D visualisation, photography, landscape assessment, personal impressions, surveys of tangible landscape values and intangible heritage, future scenarios and decision support for a complex online landscape management. Geodesign approach has relevant mission in the project especially as the study site is probably the most controversial landscape of the country. This kind of exceptionally transformed landscape needs interpretation, awareness raising and vision for the future.

1 Introduction

Vital Landscapes Project is an international institutional cooperation of eight partners from seven Central European countries with the common aim of analysing different methods of landscape maintenance in the aspects of European landscape protection, environment and rural development policies. The general goal is to contribute to the sustainable development of cultural landscapes in Central Europe. Further aims are:
- Maintenance and protection of vital and attractive cultural landscapes,
- Enhancement of local's sensitivity towards landscape values,
- Involvement of inhabitants, farmers, media into the local decision-making process,
- Visualisation of landscapes and their changes,
- Exchange of experience and good practice of Central European landscape management methods.

Generally visual interpretation of landscapes is an essential part of raising awareness in the project. Landscape is considered in the project as European Landscape Convention defines. It is an area, whose character is the result of the action and interaction of natural and/or human factors. Most of the partners use huge amount of photograps or historical maps to interpret landscape changes. The geodesign approach, that has a wide range of photographic and GIS applications and is presented in this paper, mostly represents the Hungarian partner's application at the Department of Landscape Planning and Regional Development in Corvinus University of Budapest.

Application of innovative tools like 3D visualisation of landscapes and achieving active public participation in the project is a basic requirement. The idea of integrating GIS and planned or designed elements in visualisation brought the necessity of a complex geodesign application. Geodesign is hold as a combined "toolset" of future world developers in various fields from transport network planners to social science. Some think that geodesign is one form of decision making (TOMLIN 2011), some considers as a convergence-program for bridging the gap between spatial analysis, and shaping the future of spatial environments (SCHWARZ 2011). It is obvious that geodesign can involve efforts to analyse, plan or interpret alternative futures, describe landscape and its changes for stakeholders as many applications show (STEINITZ 2003, WOLF & MEYER 2010, LANGE & HEHL-LANGE 2010). It has relevance to be used for planning of complex systems where geographical, visual and design aspects do matter. From our landscape architect's point of view Geodesign is the combination of software from GIS to CAD, that integrates materials from photographs to digital maps, believes the purpose of informing locals and planners, and supports the activities from planning to implementation. Some believe that geodesign is not a new enterprise for landscape architects (ERVIN 2011) as only the supporting tools were strengthened by new computing technology and deeply renewed its concept. Definitely true. The task is to prove that we can analyse, assess and manage landscapes better and communicate this process more clever than without digital geodesign. Are we able to do it?

2   Study Area

Nagyberek Pilot Area in Southwest-Hungary used to be the largest swampy bay of Lake Balaton. This is one of the most transformed landscapes of Hungary with many contradictory characteristics. After the water regulations of the 19th century and the intensive agricultural use in the middle of 20th century it still plays an important role in the ecological system of the largest lake in Central Europe. This drained but still swampy region is mostly dominated by patches of forests, agricultural land, reeds, built up areas, network of channels and the international highway line between Ljubljana and Budapest or generally between Italy and Russia. The area of about 300 km2 is represented by water management, forestry, hunting, fishing activities, nature protection, extensive agriculture of pasturing, viticulture, fruit and honey production, highway and railway lines, traditional crafts, rural tourism and mass lakeshore tourism (Fig 1.).

Due to the changing environmental circumstances, the disadvantageous economic tendencies and peripheral location in the Lake Balaton touristic area, Nagyberek is bankrupt and has tremendous conflicts at the moment. Increasing unemployment, aging population, low level of infrastructure, opposition of nature protection, agriculture and forestry do all represent the site. Due to the poor soil quality and extreme water conditions agriculture has low productivity. The mass tourism at the Balaton lakeshore has very slight effect on local rural tourism of the whole area right now. Depopulation and ethnic conflicts make the situation even more difficult. Despite of dominant human influence and intensive agricultural utilisation in the past, the area still has relevant natural and cultural values. It seems like the conflicts with nature conservation of "Fehérvíz" marshland do not strengthen rural tourism, but decrease the breakout options in agriculture and forestry. In few settlements only extensive agriculture, activities related to heritage and unique landscape features and traditional land use can be beneficial activities for a limited number of farmers. There is a debate whether any of these could mean breakout choice for the study area.

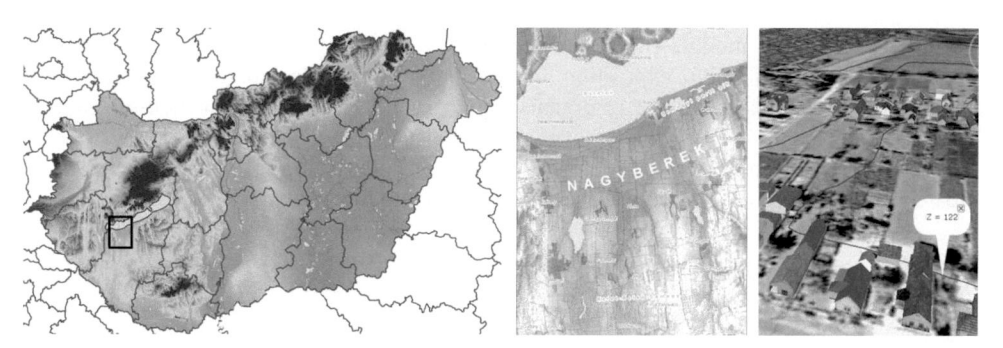

Fig. 1: Location of Nagyberek and "Visual Landscape" screenshot of a settlement

The various interests, functions, on-going political debates related to land ownership and land tenancy all affect landscape management in Nagyberek. It seems that the area's future depends on the awareness and activities of its stakeholders. What do they consider valuable in the landscape and how these should be maintained? This will determine the future of the area. This complex, but mostly plain and hidden landscape needs interpretation supported by Vital Landscapes geodesign approach to be understood by its stakeholders.

3 Method and Materials

Geodesign is used to explain the landscape changes and to illustrate the complexity of landscape development. The purpose is to make landscape characteristics visible, creating a website where all stakeholders can see and update the virtual landscape. This landscape is well introduced with all internal drivers of its development. The drivers are especially important for the Hungarians in the countryside, who think that landscape is not an image or a scenery, but a place where they live in, something they act for and fits their needs. The website (Vital Berek: e-berek.hu), under development, supports information exchange, maplike illustrations, photo upload and personal annotation options, platforms for landscape values and visualisations (Fig. 2). The basics of the website are Google applications: Google Maps and Google Earth. The modules of the application are based on GIS layers that contain different levels of landscape and offer different interpretation of landscape. Besides carthographic information there is a huge amount of photo, figure, diagram, table and visualisation to introduce the pilot landscape of the Hungarian partner.

"Info landscape" module presents statistical information of municipalities, micro-regions and micro-landscapes in the area. Written summaries of landscape characteristics are illustrated by diagrams, tables and figures. The future development of the module plans to make statistical information available on a digital vector map with attribute table.

"Time and Space" module provides information about the past, present and future of the landscape, interpreting historical maps, landscape transformation, present development tendencies, and outlook to the future. Besides the huge amount of cartographic information it is necessary to explain landscape changes and development process.

"**Photo Landscape**" module of the website allows registered users to upload landscape photos. The place of photo exposition can be marked with placemark on google maps. Occasionally photo competition is organised in various categories. The non-relevant or landscape indifferent photos are removed from the platform and the most adequate, nice and characteristic photos win the contest. This is the very first application of the website to create a common online community based platform about the pilot area.

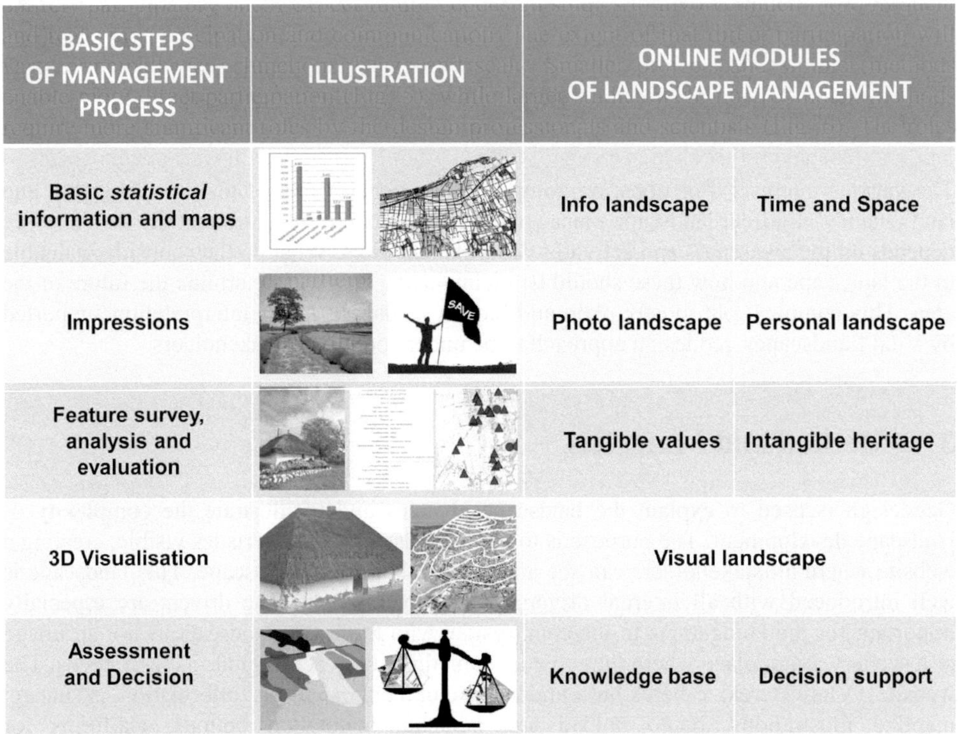

BASIC STEPS OF MANAGEMENT PROCESS	ILLUSTRATION	ONLINE MODULES OF LANDSCAPE MANAGEMENT	
Basic *statistical* information and maps		Info landscape	Time and Space
Impressions		Photo landscape	Personal landscape
Feature survey, analysis and evaluation		Tangible values	Intangible heritage
3D Visualisation		Visual landscape	
Assessment and Decision		Knowledge base	Decision support

Fig. 2: The concept of Geodesign approach in Vital Landscapes Project

"**Personal landscape**" module further encourages stakeholders to participate with their subjective opinions in landscape analysis. Stakeholders of the pilot landscape as registered users of the website can upload any kind of materials about the landscape e.g.: text, photo, images, maps, placemarks, video and audio documents, or even recommended weblinks. Users can evaluate the landscape and the applications of the website too with registered online activity. Evaluation of the landscape has various ways. It is possible to comment info, to rate heritage, to vote about future alternatives as part of following modules.

"**Tangible landscape values**" module presents appreciated elements of the pilot landscape. Old trees, significant historically relevant buildings, built structures related to land use or other various types of landscape values. The landscape elements are registered by users just as location, photos, name, description and categories of values. "Intangible heritage" module has a similar approach but deals with values of not exact location, not tangible objects, but spirit and history. It manages traditional characteristics of the landscape hold as values. Most of the heritage and landscape values are non-protected, and are in continuous danger thanks to inaccurate human activities or harmful natural processes. These are important part of our cultural heritage and determine landscape character as well.

"Visual landscape" module provides a 3D visualisation of landscape with help of Google Earth. Landscape characteristics and distinct landscape elements are visualised GIS layers in three dimensions. Various image and vector data is shown for the territory of the pilot landscape. Layers of Google Earth like road network, boundaries, panoramio photos, 3D buildings, sunlight etc. are integrated. Vital Landscape Project developed photo-realistic models of buildings in villages of pilot area, and made it visible on Google Earth.

"Knowledge base" module is a platform that refers on all other modules or integrates parts of them. It offers the visibility of additional landscape analysis and assessment. These all contribute to the final module where the users can decide about future development of the landscape. In **"Decision landscape"** module users can express their preferences concerning landscape development. These two modules enhance knowledge based landscape assessment and stakeholder oriented decision support for future scenario assessment.

4 Results

The most important result is the website that follows a kind of landscape planning and management process. It is based on the steps of landscape perception, data collection, survey, analysis, evaluation, visualisation, assessment, decision and planning. Most of the modules are continuously under development and are uploaded with materials, maps, photos, descriptions and GIS datasets. The website aims the stakeholders of the Hungarian pilot landscape, and the partnership is kept informed about the structure, techniques and the general development. The users of the website do mostly belong to the younger generation. They visit the photo contest and visual landscape applications. The photo contest organised on the website resulted in more than 150 photo uploads. This is promising in this extremely underdeveloped landscape. Two years of surveys in the pilot area resulted more than 200 landscape elements considered as relevant values. Students and landscape professionals do continuously work on further surveys. Another effort is to involve inhabitants and other stakeholders in heritage and landscape value surveys.

The website has multiple functions. Besides that it shows visualisation, use of GIS data, and historical images, which were general requirements in the project, it has basically a combined role of a community based knowledge management and GIS based decision support. Nevertheless it applies new visualisation channels to interpret landscape by integrating GoogleMaps and GoogleEarth applications and photorealistic visualisations of SketchUp or occasionally VNS3 software. Decision module offers alternative landscape futures and members of the community can express their opinion and make their decision in personal landscape module. The different alternatives can be commented and rated, the proposals can be liked or disliked, territorial developments can be supported or not supported. Rating landscape elements and future alternatives is a very important function, as on one hand it enhances awareness and activity of locals, and on the other hand professionals can gain information about values, way of thinking, and norms of locals.

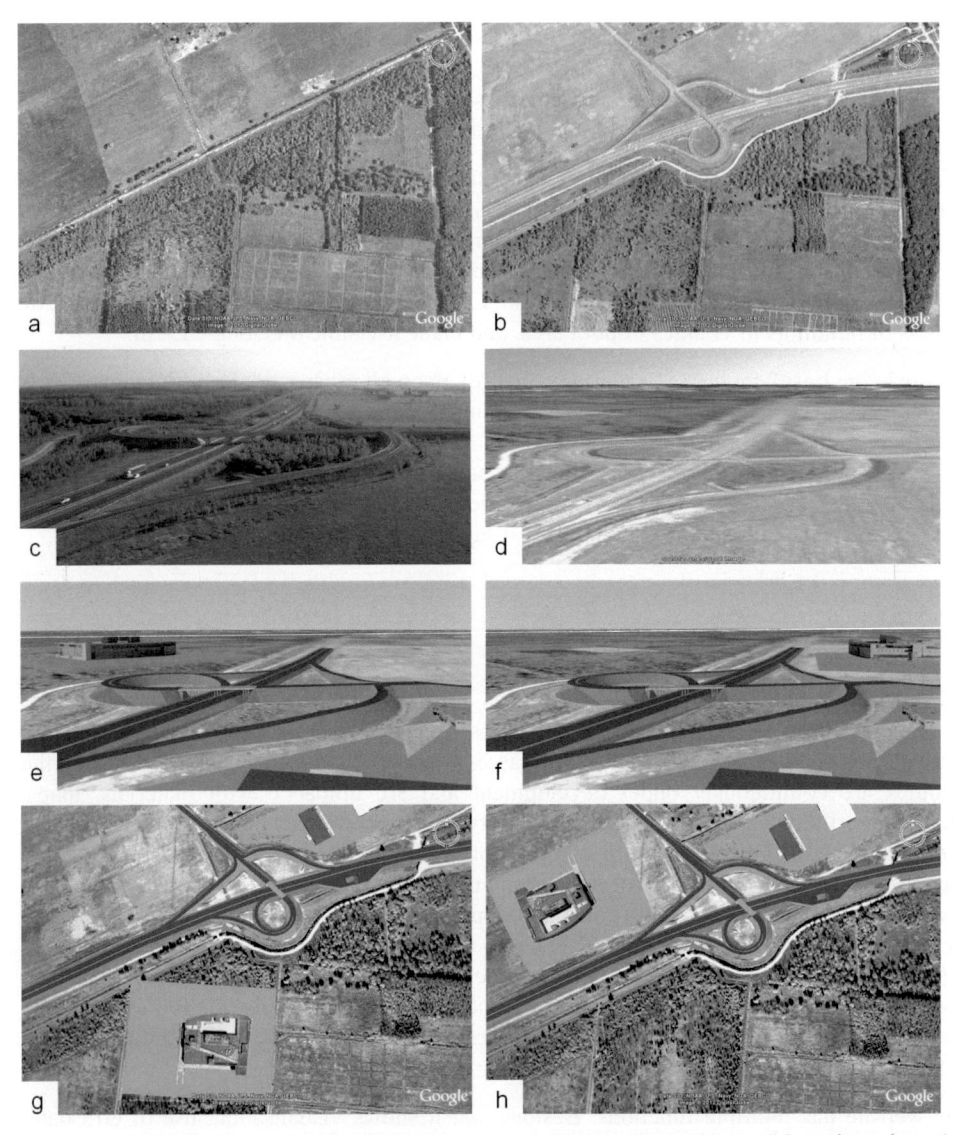

Fig. 3: Visualisation of past development and future alternatives of locating shopping mall. (a and b) Orthophoto 2000, and 2008; (c) Birds eye view aerial photo 2011, (d) Draped ortho based landscape model (e, f, g, h) 2D and 3D future alternatives

5 Discussion

The most challenging part of the approach is the concept that landscape interpretation and management process should be as online as possible. Google SketchUp and Earth provide simple, free available, easily understandable platform in a GIS browser application with sketching tools. Practically in kml or kmz format most relevant formats (tiff, shp, skp) can be transformed. Of course there are experiences involved in the approach how in general GIS based 3D visualisation is informative for local stakeholders and how important tool it is to address local communities with consultation about management (LEWIS & SHEPPARD 2006) or in case of experts how much of realism and detail matters in visualisation (APPLETON & LOWETT 2003). It is also well known how virtual globe systems such as Google Earth and related applications have rapidly growing popularity (SHEPPARD & CIZEK 2009). Besides visualisation the "know how" and values of active society is an important input for vital landscapes. We can find summaries about how much the knowledge of the past influences the preferences towards present values (HANLEY et al. 2009.), or how much the social background, ethnic origin does matter, or expert and lay-people preferences differ in evaluation (SWANWICK 2009, VOULIGNY et al. 2009).

Wide range of methods to predict future land use, based on actors, driving forces and various scenarios, are sprawling in landscape management. There are applications which combine modelling results with visualisation (DOCKERTY et al. 2006), researches concentrating on definition of key forces that are responsible for high rate of changes (SCHNEEBERGER et al. 2007). Land use modelling in our project applies CLUE-s model chosen from a wide range of spatially explicit land use change models developed over the last decade (VERBURG et al. 2006). The website ending up in decision landscape module could provide options for territorial assessment or in some case territorial impact assessment. Overview of current plans, programs or policies does also belong to the essence of territorial impact assessment. The goal is to avoid the possible negative impacts at the early planning period. In the pilot region Nagyberek the highway built recently will definitely have territorial impacts e.g. green field investments (Figure 3). These above are all relevant contributions to our Vital Landscapes approach. It aims to combine the best practices and to integrate various popular steps of planning and management. It supports locals to express their opinions and add further input of e.g. landscape values and to choose from different geographically referenced and visualised landscape development alternatives.

6   Conclusions, Outlook and Acknowledgement

Due to the holistic geodesign approach of Vital Landscapes Project the website seemed to be the proper platform for the communication of landscape management. It is expected that the approach and its related technical applications are reusable on other pilot areas, and rechargeable with other maps and data. Up to now from feedbacks it is clear that huge effort of NGO's and young professionals is necessary to address stakeholders, raise awareness with organising workshops, integrating the website introduction in student education. The future application should be based on more interactive participation of associations, organisations and municipal or regional landscape management. Thanks to Vital Landscapes partnership and leadership supporting activities. The content of the paper was prepared and the participation on the conference was supported by the Vital Landscapes Project (2CE 164P3) in Central Europe Operational Programme 2007-2013.

References

Appleton, K. & Lovett A. (2003), *GIS-based visualisation of rural landscapes: defining 'sufficient' realism for environmental decision-making.* In: Landscape and Urban Planning 65, 117-131.

Dockerty, T., Lovett, A., Appleton, K., Bone, A. & Sünnenberg G. (2006), *Developing scenarios and visualisations to illustrate potential policy and climatic influences on future agricultural landscapes.* In: Agriculture, Ecosystems and Environment 114, 103-120.

Ervin, S. (2011), *A System for GeoDesign.* In: Preliminary Proceedings Teaching Landscape Architecture, Anhalt University of Applied Sciences Bernburg & Dessau, 145-154

Hanley, N., Ready, R., Colombo, S., Watson, F., Stewart, M. & Bergmann, E. A. (2009), *The impacts of knowledge of the past on preferences for future landscape change.* In: Journal of Environmental Management, 90 (2009), 1404-1412.

Lange, E. & Hehl-Lange S. (2010), *Making visions visible for long-term landscape management.* In: Futures, 42 (2010), 693-699.

Lewis, J. L. & Sheppard, S. R. J. (2006), *Culture and communication: Can landscape visualization improve forest management consultation with indigenous communities?* In: Landscape and Urban Planning, 77, 291-313.

Schneeberger, N., Bürgi M., Hersperger, A. M. & Ewald, K. C. (2007), *Driving forces and rates of landscape change as a promising combination for landscape change research – An application on the northern fringe of the Swiss Alps. In:* Land Use Policy, 24, 349-361

Schwarz-v.Raumer, H.-G. & Stokman, A. (2011), *GeoDesign-Approximations of a catchphrase.* In: Preliminary Proceedings Teaching Landscape Architecture, Anhalt University of Applied Sciences Bernburg & Dessau, 106-115.

Sheppard, S. R. J. & Cizek, P. (2009), *The ethics of Google Earth: Crossing thresholds from spatial data to landscape visualisation.* In: Journal of Env. Management, 90, 2102-2117.

Steinitz, C., Rojo, H. M. A., Bassett, S., Flaxman, M., Goode, T., Maddock III, T., Mouat, D., Peiser, R. & Shearer, A. (Eds.) (2003), *Alternative Futures for Changing Landscapes.* Washington/Covelo/London, Island Press.

Swanwick, C. (2009), *Society's attitudes to and preferences for land and landscape.* In: Land Use Policy, 26S (2009) S62-S75.

Tomlin, D. (2011), *Speaking of GeoDesign,.* In: Preliminary Proceedings Teaching Landscape Architecture, Anhalt University of Applied Sciences Bernburg & Dessau. 136-144.

Verburg, P. H., Schulp, C. J. E., Witte, N. & Veldkamp A. (2006), *Downscaling of land use change scenarios to assess the dynamics of European landscapes.* In: Agriculture, Ecosystems and Environment, 114, 39-56.

Vouligny, É., Domon G. & Ruiz, J. (2009), *An assessment of ordinary landscapes by an expert and by its residents: Landscape values in areas of intensive agricultural use.* In: Land Use Policy, 26, 890-900.

Wolf, T. & Meyer, B. C. (2010), *Suburban scenario development based on multiple landscape assessments.* In: Ecological Indicators, 10, 74-86.

Testing GeoDesign in Landscape Planning – First Results

Christian ALBERT and Juan Carlos VARGAS-MORENO

1 Introduction

One of the fundamental aspects that is rapidly changing in design and planning is the process by which designs are generated and developed (VARGAS-MORENO 2008). In this context, computer-aided design tools and approaches are introduced that may provide planners and designers with immediate feedback on the potential impacts of their propositions, thus potentially enhancing design processes with real-time information on design implications and opportunities for greater involvement of non-experts.

GeoDesign is at the epicenter of these discussions. GeoDesign is "a design and planning method which tightly couples the creation of a design proposal with impact simulations informed by geographic context" (FLAXMAN 2010a, b). The GeoDesign concept has been proposed and fostered mainly through a series of international conferences organized by ESRI, the leading company for GIS software solutions, and has gained increasing attention over the last three years. GeoDesign conceptually builds upon the traditional design approach of sketching an idea, evaluating it, and redrawing the design. What makes contemporary approaches to GeoDesign unique however is the exploitation of nowadays available technology to provide rapid, model-based feedback on different design proposals.

Landscape planning offers a potentially very useful field of application of GeoDesign-approaches. Landscape planners often use scenarios as a basis for simulating and assessing possible future landscape configurations (alternative futures). A GeoDesign approach to landscape planning would enable planners to much more rapidly develop, alter and evaluate alternative futures. This could ease the participation of affected and interested parties in the planning process.

Research gaps exist on the one hand concerning the practical implementation of the GeoDesign-approach in landscape planning. On the other hand, the potentials and limitations of providing rapid, modeling-based feedback in participatory planning processes have not been thoroughly explored.

2 Research Objective and Methods

As a first steps towards exploring the usefulness of a GeoDesign-approach in landscape planning practice, one objective of this contribution is to explore how GeoDesign can be implemented in landscape planning within the ArcGIS software environment. Another objective is to discuss the opportunities and limitations of the currently available tools for using a GeoDesign approach to landscape planning in practice.

The research questions are:
1. Which steps does a GeoDesign process consist of and which tools for implementing them are already available in ArcGIS 10?
2. How can the identified GeoDesign tools be used to support landscape planning?
3. Which opportunities and limitations do current tools exhibit for application in planning practice?

Questions two and three will be explored on the basis of an exemplary test in a planning study for the region of Hanover, Germany. The region was selected due to the availability of detailed spatial data and the diversity of site conditions. The planning process focuses on the development of two scenarios of future landscape development, and exemplary evaluates their respective impacts on biodiversity. The two scenarios are understood as extreme options for landscape development that are both unlikely to occur as proposed, but nevertheless frame the development corridor within which landscape change will probably take place.

The research methods can be sorted to three parts: First, available literature and online resources are reviewed. This section results in the development of a framework of essential components of a Geo-Design process. Then, publications and online resources are used to identify tools and extensions currently available for ESRI's ArcGIS 10 for conducting these steps. The identified tools and extensions are then mapped to the components of a GeoDesign framework, resulting in an overview of available methods for each step.

Afterwards, the GeoDesign tools are applied in the landscape planning case study. The planning process consists of five steps:
1. Developing a conceptual model for assessing biodiversity
2. Translating the conceptual model into the model builder
3. Defining and simulating two extreme scenarios for landscape development
4. Assessing and reporting the impacts of the two scenarios on biodiversity

Finally, they application of the tools is critically discussed concerning its opportunities and limitations.

It is important to emphasize that the herein described study is only a first step for testing and evaluating the usefulness of a GeoDesign approach to landscape planning. More sophisticated simulations of land use changes in each scenario, as well as more elaborated and evidence-based assessment methods for evaluating biodiversity and other landscape functions would be needed in real planning applications. The case study therefore only presents preliminary results on opportunities and limitations of the use of contemporary GeoDesign tools in landscape planning, based on a critical evaluation from the planners' perspective.

3 Results

3.1 Basic components of GeoDesign

Interestingly, a search on the ISI web of science for the term "geodesign" as the topic or title reveals only two references, of which one is a two-page Croatian article from 2011, and the other an Austrian article from 1994. Therefore, the following review is based on other peer-reviewed publications, in particular DANGERMOND (2010) and FLAXMAN

(2010a) as well as grey literature (including ABUKHATER & WALKER 2010, DANGERMOND 2009, VARGAS MORENO 2010). Taken together, the literature argues that GeoDesign process includes at least the following three components:

- **The input process** *(a.k.a the design):* A sketching interface that allows for rapid generation of alternative designs consisting of spatial features with attributes in a geographic context.
- **The evaluation** *(a.k.a the impact):* A set of spatially informed models that assess the potential impacts of an entered design based on a series of evaluation parameters, and
- **The result:** (*a.k.a the report):* The instrument that communicates the outcomes of the impact evaluations in rapid, predetermined and understandable ways. The fast feedback then serves as input for another iterative cycle of sketching and evaluation.

3.2 Tools and methods for implementing GeoDesign with ArcGIS 10

The review of publications on the ArcGIS 10 software package (e.g. ALLEN 2011, HARDER et al. 2011, ORMSBY et al. 2010) shows that no comprehensive documentation is so far available of which tools can be used for implementing the entire GeoDesign process in a planning project.

The currently available methods for GeoDesign with ArcGIS 10 are summarized in table 1. For each component of the GeoDesign workflow, some tools are already existent. Other findings include:

- Sketching is implemented by a new feature of ArcGIS that was formerly known as the ArcSketch extension (but requires some level of customization).
- A useful approach for conducting rapid impact evaluations is to create a (simple) impact evaluation model in the model builder and defining it as a new tool. This tool is then readily available for impact evaluations. When started, it will prompt for the input layer to use, and then conduct the analysis.
- Reporting the results of the rapid impact evaluations can be accomplished in three ways: the display of the maps created by the impact models, tables that summarize the spatial extend of areas that fulfill specific criteria, and diagrams. In addition, a new ESRI prototype, called dynamic charting, has been recently released for rapidly creating and dynamically updating charts that summarize a given field of a data source (e.g. costs or area). A significant issue identified is the restriction to move from reports back to design or evaluation seeking iterative improvements. Other constraints will be discussed in full manuscript.

Component I: Sketching	Component II: Impact evaluation	Component III: Reporting
• Editable templates for ArcMap s included in mainstream ArcGIS 10 (formerly the ArcSketch extension)	• Creating an impact evaluation model in model builder	• Displaying maps that result from impact evaluation modeling • Displaying tables and charts that summarize the results

Table 1: Method components and tools for Geodesign currently available in ArcGIS 10

Taken together, the review has shown that tools already exist in ArcGIS to fulfil the core ideas of GeoDesign, but all require substantial customization or advanced knowledge of the software platform.

3.2 Test of the tools in the case study region

The region of Hanover has been established in 2001 in a unification of the former communities of the city and the surrounding county of Hanover. Hanover region encompasses about 230,000 hectares and is home to about 1.13 Million inhabitants. The region is at the transition between the North-German plain and the low mountain ranges of central Germany. The northern part of the region is a typical moor geest landscape dominated by moor or sandy soils, pine forests, grasslands and fields. The southern part of the region includes the fertile soils of the Hildesheimer Börde and parts of the Deister mountain range. The Leine is the main river, trespassing the region from south to north.

The land use data for the case study is derived from a detailed land use map based on CIR interpretation of aerial imagery. In order to make the land use map manageable, the originally large number of land use classes was synthesized into ten land use types.

3.2.1 Conceptual model for assessing biodiversity

The literature shows various modeling approaches for assessing biodiversity. For example, WALDHARD et al. (2004) developed a model that estimates and predicts plant species richness at the local to regional scale based on an empirical assessment. This study uses an approach proposed by von DRACHENFELS (2004) that employs the importance of biotope types for nature conservation as a proxy for biodiversity. Each biotope type is evaluated on a five point scale with reference to the situation in the state of Lower Saxony. Four evaluation criteria are used: (i) 'closeness to natural conditions', (ii) endangerment, (ii) rareness, and (iv) importance as habitats for plant and animal species. In order to make the von Drachenfels-method applicable to the land use maps of the case study, a raster scheme needed to be derived that connected the ten land use types with the respective value for biodiversity (see table 2).

Land use type	Biodiversity value	
Moor	5	Very high importance
Grassland	4	High importance
Forest	3	Medium importance
Moor (degenerated)	3	Medium importance
Fields	2	Low importance
Grassland (intensively used)	2	Low importance
Infrastructure	0	Not evaluated
Settlement area	0	Not evaluated
Waters	0	Not evaluated
Other land uses	0	Not evaluated

Table 2: Evaluation table of land use types and respective biodiversity value (summarizedand altered from DRACHENFELS (2004))

3.2.2 Translation of the conceptual model into the model builder

The ArcGIS model builder is used to develop a model for evaluating biodiversity (see figure 1) according to the evaluation rules defined in section 3.2.1. Input data for the evaluation of scenario impacts on biodiversity include (i) a shapefile with the land use configuration of the alternative future to be evaluated, (ii) an evaluation table (see table 1), and (iii) a symbology layer that defines the color code for illustrating the five levels of importance for biodiversity.

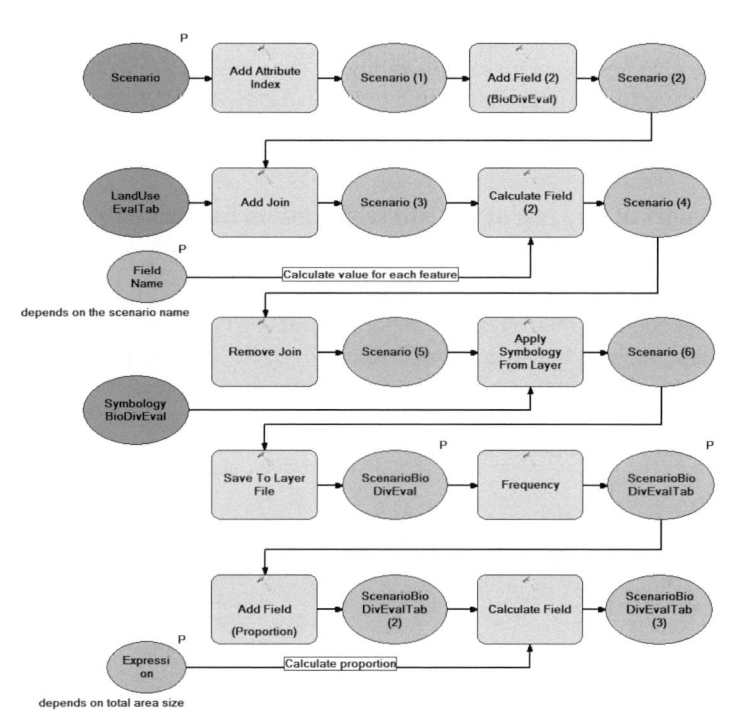

Fig. 1: Model structure for evaluating biodiversity

The developed model first joins the evaluation table with the attribute table of the alternative future land use map. Then, each feature in the shapefile will be attributed a biodiversity value according to the respective land use. The symbology is altered to no longer refer to the land use, but to the biodiversity value of each feature. In addition, a table is created that lists the area attributed to each level of importance. Fields in figure 1 that are indicated with a 'P' need to be checked and potentially altered by the user in each cycle, including the input files and the names and folders for the created files to be saved.

3.2.3 Definition and simulation of two extreme scenarios

Two scenarios are developed in the case study. The first scenario is termed "Bioenergy" and assumes that the production of biomass for bioenergy is strongly increased. The second scenario, "Climate mitigation", assumes that all arable land on organic and alluvial soils (moor and floodplains) is converted to grasslands (Table 3).

Land uses type in scenario "Status quo"	Land use type in scenario "Bioenergy"	Land use type in scenario "Climate mitigation"
Grassland	Fields	Grassland
Fields (on organic and alluvial soils)	Fields	Grassland
Fields (all other areas)	Fields	Fields

Table 3: Transition rules for the two scenarios to create alternative futures

The simulation of land use changes in each scenario followed the transition rules listed in table 3. Plots of land were selected that fulfilled the land use type criterion, or the combination of land use type and soils criteria. The attributes of the selected plots were subsequently changed to represent the land use type of the respective scenario.

3.2.4 Assessment and report of impacts of the two scenarios on biodiversity

To assess the impacts of the two scenarios, the respective alternative futures were used as input for the developed models in the ArcGIS model builder environment. The resulting evaluation maps were color coded for the respective importance of each polygon on the 1 to five scale. Furthermore, the respective distribution of areas for each level of importance was summarized in a table, and illustrated in a bar chart for comparison (figure 2).

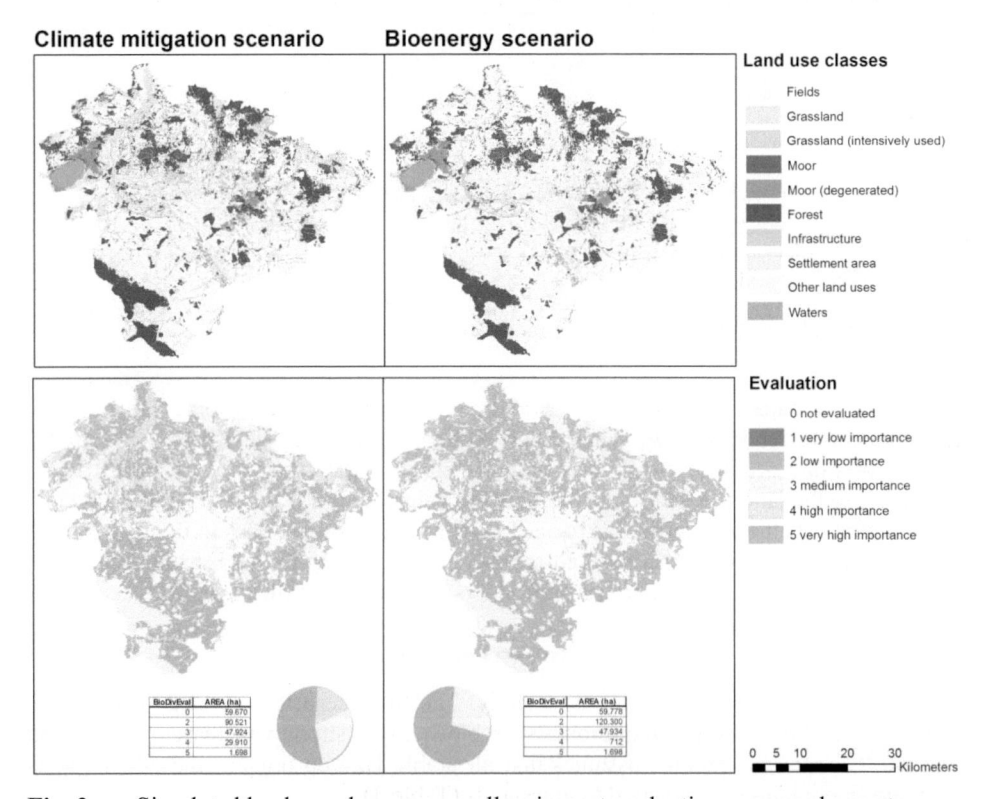

Fig. 2: Simulated land use changes as well as impact evaluation maps and reports

4 Discussion and Conclusions

The herein described study used a limited land use data set of only ten classes, two extreme scenarios, and only a simple impact evaluation model. The results therefore should not be understood as detailed recommendations for planning, but rather as stylized study examples for testing a procedure for implementing GeoDesign in landscape planning.

The study has shown a few opportunities and limitations. The following opportunities were identified:

- Transferring the impact evaluation concept into a model in the ArcGIS model builder was possible. The model builder environment allowed for detailed customization of the model to the specific planning needs. Developed models can be saved and used to evaluate different land use maps that follow a pre-defined land use coding.
- Translating the simple land use conversion rules of the scenarios into GIS was easily doable using simple feature selection and attribute change commands.

Limitations that became apparent relate mainly to the question of how useful the developed GeoDesign procedure would be in landscape planning practice:

- Customizations of models in the model builder can often not be conducted within a participatory planning session. Changing and pre-testing of models requires advanced knowledge and may take quite some time.
- While executing the evaluation model in the case study was possible within a few seconds, a test with a more complex model showed that the modeling time can strongly increase to about two or more minutes. Studies that employ large datasets and/or complex models should therefore consider using distributed computing power in the cloud.
- The need for tools to intuitively respond to the design-evaluation process in iterative ways has not yet been resolved. However, some discussions has started directing the characteristics of such systems (LEE 2010, VARGAS MORENO 2010).
- Most importantly, what is needed what is a way to develop a front-end appearance for ArcGIS that combines the sketching, impact evaluation and reporting components in a common interface that is intuitively usable for diverse audiences. This front-end appearance should be web-based to provide access also to users that do not have access to computing power and the ArcGIS software.

Issues for further research include (a) further developing the scenarios and models and adding additional impact evaluation models, (b) testing the approach with real stakeholders, (c) investigating the reactions and effects of the use of the GeoDesign approach in a participatory planning process on participants' perceptions and decision making (do they find it useful?), (d) further developing the software interface for non-experts, and (e) developing and evaluating options for transferring a front-end version of the GeoDesign planning process to the web.

5 Acknowledgements

The authors thank Johannes Hermes, Svenja Heitkämper, Rebekka Hofmann, Christiane Hörmeyer, Angelika Lischka, Felix Neuendorf, and Pia Wedell for support in conducting the analyses in the case study. Johannes Hermes also helped preparing the graphics. The study is related to the research project "Sustainable use of bioenergy: bridging climate protection, nature conservation and society", lead by the Interdisciplinary Centre for Sustainable Development at Georg-August-Universität Göttingen and, in part, the Institute of Environmental Planning at Leibniz Universität Hannover.

References

Abukhater, A. & Walker, D. (2010), *Making Smart Growth Smarter with GeoDesign.* In: *Directions Magazine, July 19.*
Allen, D. W. (2011), *Getting to Know ArcGIS ModelBuilder.* Redlands, CA, ESRI Press.
Dangermond, J. (2009), *GIS: Designing Our Future.* In: ArcNews, Summer issue.
Dangermond, J. (2010), *GeoDesign and GIS – Designing our Futures.* In: Buhmann, E., Pietsch, M. & Kretzler, E. (Eds.), Digital Landscape Architecture 2010 at Anhalt University of Applied Sciences. Berlin/Offenbach, Wichmann, 502-514.
Flaxman, M., (2010a), *Fundamentals of GeoDesign.* In: Buhmann, E., Pietsch, M. & Kretzler, E. (Eds.), Peer Reviewed Proceedings of Digital Landscape Architecture 2010 at Anhalt University of Applied Sciences. Berlin/Offenbach, Wichmann, 28-41.
Flaxman, M. (2010b), *Geodesign: Fundamentals and Routes Forward.* Presentation to the Geodesign Summit, January 6, 2010, Redlands, CA.
Harder, C., Orsmby, T. & Balstrom, T. (2011), *Understanding GIS: An ArcGIS Project Workbook.* Redlands, CA, ESRI Press.
Lee, B. (2010), *GeoDesign in Land-Use Planning.* Presentation to the Geodesign Summit, January 6, 2010, Redlands, CA.
Ormsby, T., Napoleon, E. J. & Burke, R. (2010), *Getting to Know ArcGIS Desktop.* ESRI Press, Redlands, CA.
Vargas-Moreno, J. C. (2008), *Participatory landscape planning using portable geospatial information systems and technologies: the case of the Osa region of Costa Rica.* In: Graduate School of Design, Harvard University, Cambridge, MA, USA.
Vargas-Moreno, J. C. (2010), *GeoDesign: The Emergence of a Tight-coupling Approach in GIS and Spatial Planning.* In: Planning & Technology Today.
Von Drachenfels, O. (2004), *Kartierschlüssel für Biotoptypen in Niedersachsen. Unter besonderer Berücksichtigung der nach § 28a und § 28b NNatG geschützten Biotope sowie der Lebensraumtypen von Anhang I der FFH-Richtlinie.* Niedersächsisches Landesamt für Ökologie, Abt. Naturschutz, Hildesheim, 240.
Waldhardt, R., Simmering, D. & Otte, A. (2004), *Estimation and prediction of plant species richness in a mosaic landscape.* In: Landscape Ecology, 19 (2), 211-226.

Visual Landscape Character
in the Approach of GeoDesign

Jan SZTEJN, Piotr ŁABĘDŹ and Paweł OZIMEK

1 Introduction

One of the important landscape indicators used in the field of GeoDesign are the visual aspects of the landscape, including any change resulting from the implementation of projects and spatial transformations. The authors of this paper perceive there still exists a gap in the assessment used for bulk data, with which designers have to do in a situation of large area of attractive landscape. They attempted to develop a monitoring tool to support landscape and design in the space using a combination of the best of various legacy tools (CAD, GIS, BIM, e.g.), by supplementing general information about the area of visual resource assessment. An important new approach is to gather information about the visual resources of the landscape as seen not from few individual viewpoints, or paths, but with the entire surface of the land being developed. It could be helpful for obtaining content of several components connected with Geodesign, which were distinguished by Stephen Ervin (ERVIN 2012). First of all, as a tool for monitoring, it could give us layers of specific information about environment. Moreover, these layers contain visual relationships between space elements and provide essential inputs into the analysis and simulation functions. Using the same tool we can establish relationships between existing and designed elements of landscape.

The study was conducted in based on the concept of the landscape character developed in the framework of LCA (Landscape Character Assessment), which is defined as: "a clear, recognizable, coherent system of elements in the landscape, which makes the landscape different from the others ..." (SWANWICK 2002). Mentioned elements can be interpreted in the context of the visual character of a landscape described as: "The visual expression of the spatial elements, structure and pattern in the landscape." The role of the researcher is the proper selection of these resources, which are visual indicators of a place. It can be interpreted as elements-per-view and degree of variation in Between-size view (ODE et al. 2008).

In order to illustrate the visual character of the landscape it was decided to present the results in the form of maps. Traditional maps shows the distribution of individual landscape elements in the orthogonal projection well, it may also show some phenomena in relation to the unit area, for example the number of inhabitants per square km. It can provide good performance of the physical elements of a landscape, however to illustrate the visual landscape character it is indispensable to show a panorama from particular place on the map. To demonstrate the phenomenon in evenly arrangement it is needed to analyse panoramas spaced in systematically intervals on the ground. In practice, due to the large area and a significant number of respondents panoramas it would be impossible, and therefore it was needed to build three-dimensional model of the landscape for the entire study area and simulate visibility on it (OZIMEK et al. 2009).

2 Methods

2.1 Case Study Area

The adopted classification of indicators was dictated by the specificity of the study area. The Investigation area covers the south western suburbs of Wroclaw and has a size of approximately 7 × 12 km. According to the micro-region division of Poland the area belongs to Wrocław Plain (318.53), a part of the Silesian Lowland (KONDRACKI 2002). The land here is almost entirely flat with small hills and the only distinctive haughtiness of Sleza mountain. The area is characterized by the agricultural landscape, where large surfaces agricultural crops are separated by small groups of green forests and mid-field. Traditional rural buildings are concentrated in small villages. The neighborhoods of Wroclaw promotes a dynamic suburbanisation phenomenon and the emergence of new buildings and new landscape elements.

2.2 Input Data

In the investigation numerical terrain model in the form of TIN was used and also orthophotomaps, which were gain from a local administrative unit responsible for collecting, processing and dissemination of geodetic and cartographic data. The maps of investigation area was built on a grid of squares TEMKART developed by Klimczak (KLIMCZAK 2001). This system consists of a geometric units, which have resulted from the division of Polish territory by polygons with dimensions of 125 × 125 meters. Sides of each unit are determined by the meridians and parallels. A digital code is assigned for each field representing its centroid. It allows to link a particular fields of TEMKART with the corresponding database record. The map was expanded by additional layers including buildings, roads, different types of greenery and water. These layers were produced from digital orthophotomap which was vectorised in order to give corresponding data. The cartographic elaboration and preparation of basis data for construction of the model was prepared in ArcGIS 9.3 in projected coordinate system PUWG_1992 and partly in the CAD environment.

2.3 Modelling Approach

The first step of the work was generation of three-dimensional digital terrain model with the use of AutoCAD Civil 3D. Achieved model included the area of 240 km^2 and with the Sleza Mountain taken into account it raised to 500 km^2. Additional layers were included and whole data were exported to Autodesk 3ds Max 2011, which environment is effective in the case of renderings. Such constructed digital model allows unrestricted selection of components that are important for analysing and distinguishing them in space.

In order to determine points for analysis, systematic sampling was performed. The frequency of points' distribution depends on the level of detail required for the analysis. It should be chosen properly to match morphological dynamics and land cover. As a basis to determine frequency of points' distribution, different maps can be used including: map of landform geodiversity containing information from map of landform energy, map of landform fragmentation or map of contemporary landform preservation (ZWOLIŃSKI). The highest geodiversity value on the map is treated as a determinant of points' distribution frequency. Thereby it is less probable to omit the important places in the spaces between samples. In the landscape attractive areas, the values of geodiversity are always high, therefore the number of samples is tremendous. Processing them requires an efficient computer system.

2.3.1 Visibility Map

The visibility map was created in Autodesk 3ds Max 2011. During that process a point light was placed in each of the achieved points. The assumption was made that the light rays trajectories, calculated according to raytracing algorithms, are straight lines. Thus, one can draw a conclusion that if the fragment of model is lit by the light it is also visible from point where the light is situated. It led us to creation of a visibility graph, which shows the parts of the terrain that could be seen from single point. The resolution of the graph were set in the way that 1 pixel in the image corresponds 25 square meters of terrain.

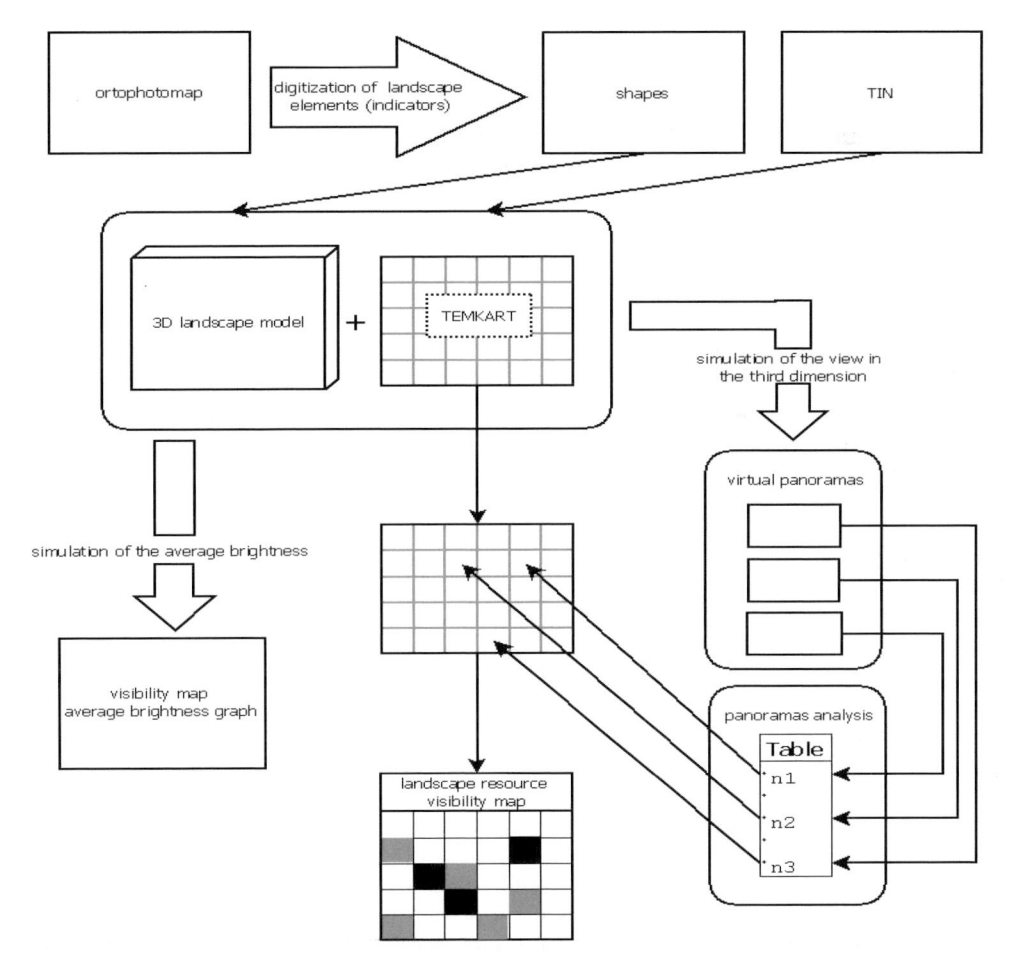

Fig. 1: The diagram showing steps of implementation of visibility map and landscape resource visibility map

After image binarization the visibility graphs from all points, achieved by systematic sampling, were composed in arithmetic mean to create the visibility map. Procedure of histogram stretching was applied in achieved map to improve its quality. In final image we obtain information about passive exposure, which reveals the degree of visibility of whole analysed area. The lighter the pixel in the picture is there are more places the corresponding area is seen from. In order to achieve map of active exposure one can compose samples located in the open space. Visibility graphs made for the huge number of points take into account all the items affecting the landscape. Visibility map contains accumulated information about the nature of the landscape of whole analysed area. That type of maps can be applied to perform analyses of changes in landscape visibility character caused by terrain transformations or changes in the way of its management. It is important during map creation process that the denominator of arithmetic mean for different design states was the same. Then, the differences in maps achieved for the states before and after performing simulated transformation indicated the degree of impact on visibility. Making such analyses on digital models, on which one can perform simulations, is a part of GeoDesign.

2.3.2 Virtual Panoramas

The second element is the virtual panoramas analysis. In each point of systematic sampling, a virtual wide-angle camera was placed at the height of 1,8 meter above the terrain to simulate human eye-level view. Its focal length was set to 18 mm, equivalent of 90 degrees field-of-view. A series of four renderings was performed from each point, to present a view in each side of the world. Model elements were simplified and marked easy recognizable colors from CMYK and RGB space.

Each rendering was examined with regard to contain characteristic elements of the landscape of suburban area:

- buildings
- forests
- mid-field greenery
- roads
- water
- hedgerows

In addition the Sleza Mountain was taken into account as a dominant in mostly flat landscape. No atmospheric haze or any different disturbing factors were taken into consideration.

Calculated output data was percentage participation of each element that is seen from each of the centroids. It was determined by analysing the percentage participation of selected color in the picture. Thanks to the digital connector it was possible to link a mathematical result of the corresponding field of TEMKART network and drawing the appropriate value on the map. Classification of values on the map was presented using a logarithmic scale. This was due to the specificity of the results obtained, in which the largest number of observations recorded in the range from 0.0% to 1.0%.

Due to mass amount of data authors used an effective and fast Quicksilver hardware rendering algorithm. It fits the task perfectly forasmuch used model was geometrically simple, despite it contains huge number of polygons and there was no need for images to look realistic. Rendering were performed with the use of GPU Nvidia Quadro 5000, image operations, like binarization, histograms creation, colour segmentation were made in Matlab Image Processing Toolbox and Parallel Processing Toolbox.

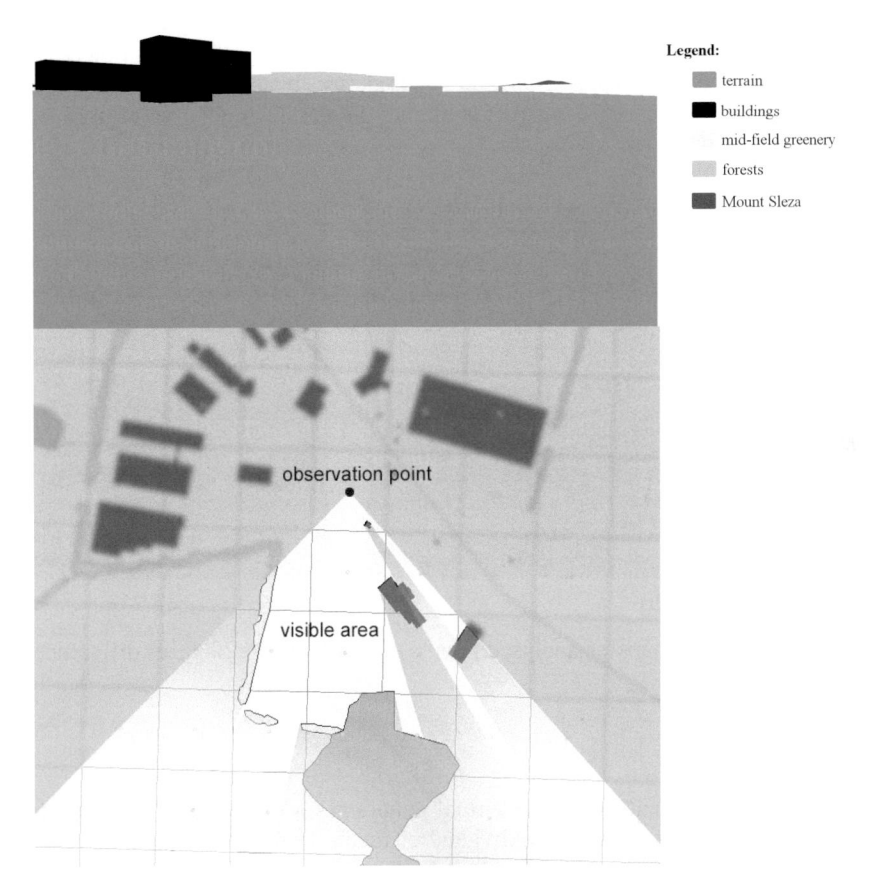

Fig. 2: An example of a virtual panorama (above) and the same view in orthogonal projection (below)

3 Results

3.1 Visibility Map

During the first stage of analysis almost 3500 visibility graphs were achieved. Afterwards they were averaged to create one visibility map, which histogram were stretched to improve its quality. To monitor landscape changes during time, the same procedure was applied to data obtained in three different years – 1982, 2004 and 2009. Resulted maps for 1982 and 2009 are presented in Fig. 3 a and b. They revealed a progressive process of urbanization of analysed area. It is visible especially near the highway turn in the middle of the picture. There is a wide dark area in that region in 2009 (Fig 3.b), which didn't exists in 1982

(Fig 3.a). Dark area means that there is slight number of points of which that area is seen from and, on the other hand, from that area meagre part of analysed terrain could be observed. There exists more similar regions, however they are not so obvious to perceive. After calculating difference between images representing years 2009 and 1982 (Fig 3.c) more congenial territories are disclosed. Enclosed example demonstrated usefulness of presented method in GeoDesign. One could perform such study to perform an objective estimation of changes in landscape character caused by planned investments or modifications that would affect it. Constructed digital model is highly adjustable to different types of GeoDesign tasks.

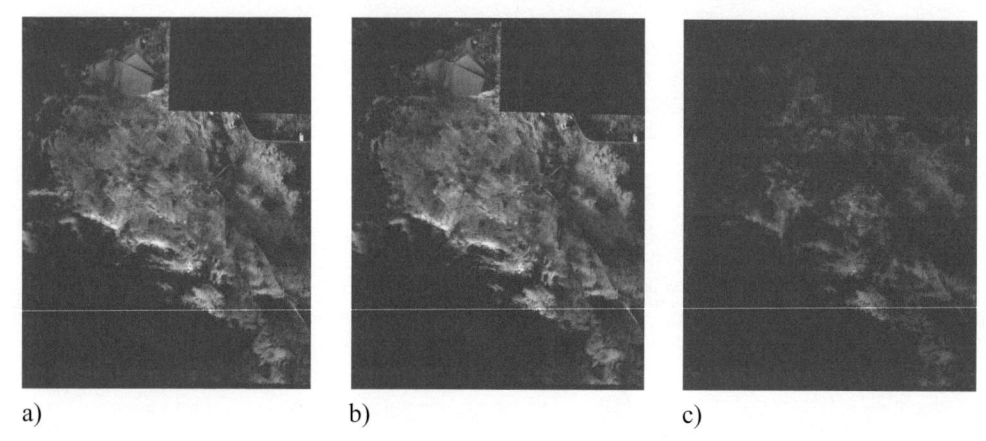

Fig. 3: Achieved visibility maps of analysed area: a) in 1982; b) in 2009; c) difference between *b* and *a*.

3.2 Virtual Panoramas

Owing to the visibility range analysis of the virtual panoramas a set of maps working in GIS was obtained, showing the degree of individual elements visibility of the landscape. (Fig. 5) All maps are stored in a geodatabase and can be enriched by any layer. The resulting map database consists of a set which shows the states of the landscape in the years 1982, 2004 and 2009 for each of the eight characteristic elements including the view of the four sites of the world for each of them. Presented phenomenon shows how an observer standing on the ground at some particular point can see the landscape individual resources (if an item is visible and its share in the panorama). Visual assessment of the landscape can be considered in qualitative terms, however, in this case it was achieved by means of quantitative analysis which gave tangible and objective results. Statistical analysis showed a trend of distribution of observations in the study panoramas. The visible surface of the analysed elements was located predominantly in the range from 0.0 to 0.05 (0% to 5%). The distribution of the data was concentrated closer to zero. An exception is the surface of the ground and the sky which is mainly from 0.4 to 0.6 (40% to 60%) (Fig. 4) The legends in the classification contain 10 intervals in a logarithmic scale with which it is easier to performance the nature of the phenomenon.

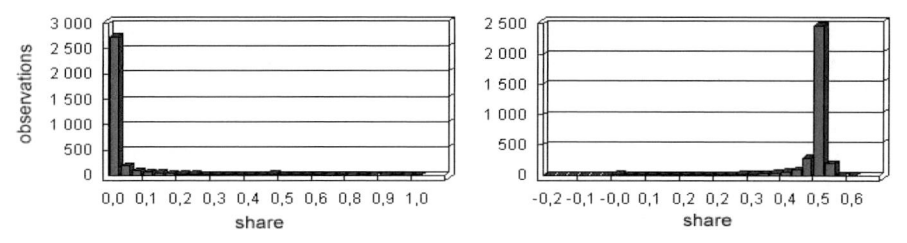

Fig. 4: A histogram showing the distribution of observations in a virtual panorama of the building layer (left) and terrain layer (right)

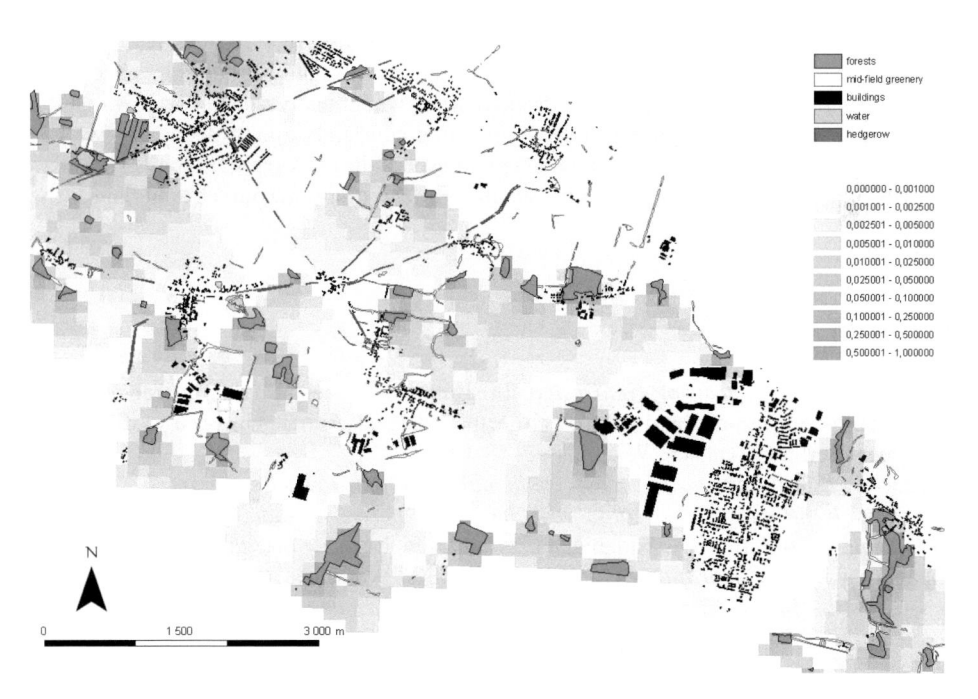

Fig. 5: The landscape resource visibility maps showing the gradient of forests in the north direction

4 Discussion and Conclusions

The virtual landscape model allows performing mass analysis of the visibility in relatively effortless way. It can be modified depending on needs (adding more layers, the analysis in subsequent years). In addition, measurable and equal indicators was used for all the analyzed points on the map, regardless of the test area size. Therefore, the results are objective and give a comparison, which would be impossible for expert analysis in the field, assuming a large area of research. However, this method requires an appropriate choice of parameters to build the model and assumptions in the performance simulation and analysis. This step requires expert work, which conduct the desirable classification of landscape indicators. An important element of the construction of the model is to match the grid TEMKART to specific sites. Grid and related observation points can be compacted depending on terrain, landscapes and adopted to resolutions of the map. The visibility map method which presents passive exposure is based on the binarization viewed area (the object is visible or not, without indicating the intensity of the phenomenon). Passive exposure assigns a higher value for the objects those are seen from larger number of points. This is useful for identifying most characteristic areas and landmarks. This result may be verified by means of the landscape resource visibility maps, which gives outcome in gradient value, therefore, shows the extent to which an element of the landscape is visible. These methods provide objective data to interpret the visual character of the landscape, monitor changes in the landscape by performing systematic analyses, planning and decision making in protection and forming the landscape. However it requires evaluation of an expert who knows the conditions of study area landscapes (natural, cultural, historical). Moreover, visibility maps should be confronted with the subjective perception evaluation and the conclusions can be applied in urban planning and landscape protection as well as providing a guidelines to form aesthetic perceptual aspects in LCA methodology. They also could affecting at the local community awareness about the resources and landscape changes.

References

Ervin, S. (2012), *A System for GeoDesign*. In: Buhmann, E. et al. (Eds.), Peer-reviewed Proceedings of Digital Landscape Architecture 2012 at Anhalt University of Applied Sciences. Berlin/Offenbach, Wichmann, 158-167.

Klimczak H. (2001), *Studia rozmieszczania obiektów punktowych, liniowych i powierzchniowych na przykładzie obszarów leśnych i terenów zadrzewionych*. In: Modelowanie kartograficzne w badaniach przydatności obszarów pod zalesienie, Wrocław.

Kondracki J. (2002), *Geografia regionalna Polski PWN*. Warszawa.

Ode, A., Tveit, M. & Fry G. (2008), *Capturing Landscape Visual Character Using Indicators: Touching Base with Landscape Aesthetic Theory*. In: Landscape Research, 33, 89-117.

Ozimek, P., Ozimek, A. & Łabędź, P (2009), *Digital Analyses of Wind Farms Visual Apsects in South-East Poland*. In: Digital Design in Landscape Architecture 2009, Proceedings at Anhalt University of Applied Sciences.

Swanwick C. (2002), *Landscape Character Assessment. Guidance for England and Scotland*. The Countryside Agency, Scottish Natural Heritage.

Zwoliński Z. (2009), *The routine of landform geodiversity map design for the Polish Carpathian Mts*. In: Geoecology of the eurasiatic Alpides (Eds. Łajczak, A. & E. Rojan), Landform Analysies, 11, 77-85.

Applying 3D Landscape Modeling in Geodesign

Joerg SCHALLER

Keynote: 31 May 2012

Abstract

Sustainable Landscape Planning and Landscape Architecture are using spatial data and are applying GIS methods and technologies to depict spatial and time-related interrelationships.

At one hand GIS data, evaluation and assessment models are needed in the planning process as well as sophisticated tools to display the results of planning efforts. Secondly all developed designs have to be approved for quality using digital 2D and 3D GIS visualization and modeling technologies.

The assessment of the social, environmental and sustainable quality of a landscape development – or project implementation design plan can be carried out by using Geodesign tools in several steps of the Geodesign framework proposed by Carl Steinitz (STEINITZ et al. 2003).

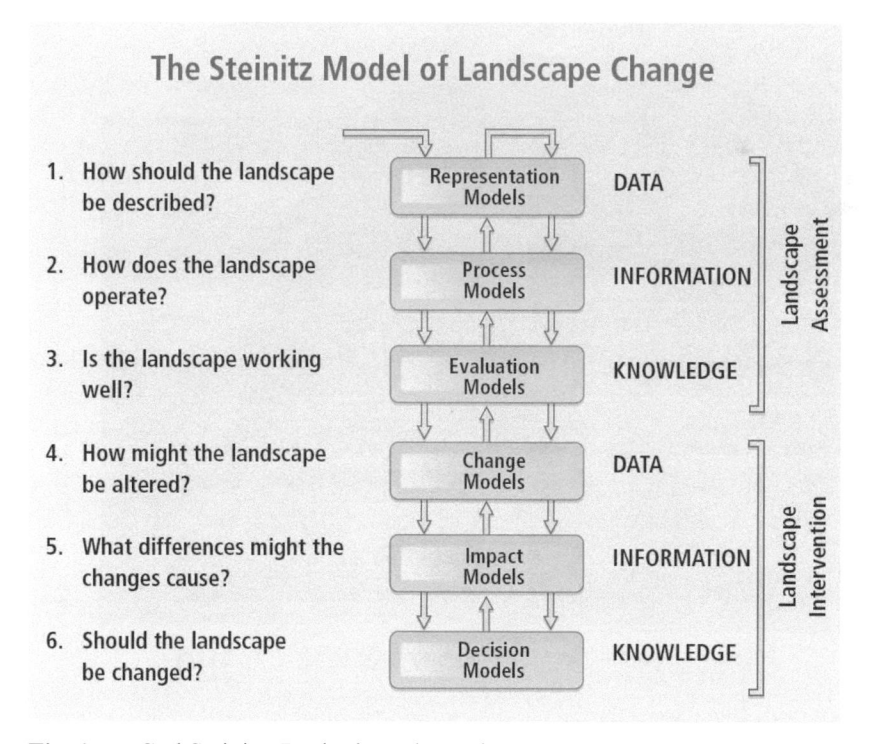

Fig. 1: Carl Steinitz: Designing Alternative Futures

Every environmental assessment of the planning output in form of scenarios or project alternatives requires the application of model coupling in different dimensions to provide the necessary balancing information to decide whether a scenario or a project alternative is convincing related to the selected social and environmental assessment criteria.

The complex requirements of Godesign applications are nowadays supported by high end GIS and remote sensing data gathering methods such as RADAR, LiDAR DTM and DSM data, high resolution satellite data, stereo camera images etc., as well as automated draping technologies for 3D objects.

These technologies allow the creation of synthetic and photorealistic 3D landscape visualizations and simulations to display planning alternatives, scenarios and depict clearly their impact on landscape scenery.

The assessment of impacts on landscape structure and scenery is an important requirement for landscape architects as partners of architectural designers, rural land consolidation planners or engineering planners designing new infrastructure or settlement developments.
3D Landscape modeling is therefore an important tool to speed up Geodesign processes and to support ad hoc decisions of the design quality related to landscape scenery.

Nowadays all stakeholders of a planning process are involved in the decision finding and it is a necessary requirement to react fast on inputs and to depict the consequences of the

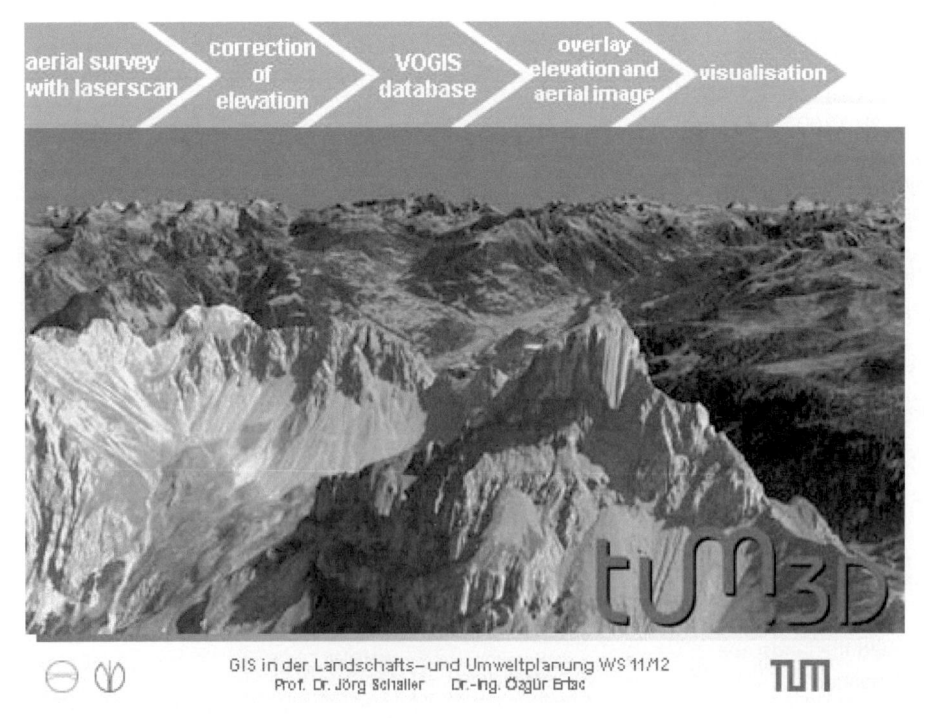

Fig. 2: 3D Visualization of Vorarlberg

stakeholders proposals via ad hoc visualization. If we are able to provide all these tools in the near future (some of them have been already realized) we will create transparency in the public and in consequence the necessary acceptance of the stakeholders to implement the plan.

The presentation will include some actual application and assessment examples of 3D landscape modeling for environmental planning purposes ().

References

Heißenhuber, A., Kantelhardt, J., Schaller, J. & Magel, H. (2004), *Visualisierung und Bewertung ausgewählter Landnutzungsentwicklungen.* In: Natur und Landschaft, 4, 159-166.

Schaller, J. (2002), *Ressourcenschutz und Ressourcenmanagement gelingen nur in Kooperation – Zur veränderten Rolle der ‚Landschaftsexperten'* In: Lehrstuhl für Bodenordnung und Landentwicklung, Materialiensammlung 27/2002. 155-158.

Schaller, J. (2003), *3D GIS-Anwendungen für Kommunale Planungen.* Powerpointpräsentation und Kurzfassung. 3D-Forum Lindau „Die Stadt am Computer". Lindau 25.-26.03.2003.

Schaller, J. (2003), *Bäuerliche Kulturlandschaft im Verschwinden –Fiktion oder bald Realität?* Powerpointpräsentation und Kurzfassung. Bayer. Akademie Ländlicher Raum e.V. in Zusammenarbeit mit dem Bayer. Gemeindetag und der Arge Landwirtschaftliches Bauwesen. Herbsttagung „Partner Gemeinde und Landwirtschaft – Plädoyer für eine bewusstere kommunale Agrarpolitik, im Pfannerhaus in Roßhaupten, 07./08.10.2003.

Schaller, J. (2003), *Ermittlung von Überschwemmungsgebieten mit Laserscanning-Verfahren.* In: Tagungsband Schutz des Wassers – Schutz vor dem Wasser, 246-255. ATV-DVWK Landesverbandstagung am 22. und 23. Okt. 2003 in Fürth.

Schaller, J. (2003), *Geographische Informationssysteme – Stand der Technik und Anwendung für Landes- und Regionalplanung.* Powerpointpräsentation und Kurzfassung. Regionalplanertagung des Landes Sachsen-Anhalt, 19. März 2003, Bernburg.

Schaller, J. (2003), *GIS Application of geoscientific databases for regional sustainable resource management and planning.* 4th European Congress on Regional Geoscientific Cartography and Information Systems in Bologna, Italy, June 17th – 20th 2003.

Schaller, J. (2004), *GIS for Resource Management and Environmental Planning in the Region of Munich.* Colloqium Series. University of Redlands/ESRI, March 11, 2004. CD.

Schaller, J. (2005), *Die Zukunft von GIS in Landnutzung, Umwelt und Naturschutz.* Tagungsband 10. Münchner Fortbildungsseminar Geoinformationssysteme, 2. bis 4. März 2005, TU München.

Schaller, J. (2005), *Regionale Auswirkungen des Landnutzungswandels im Voralpenraum.* In: Landnutzung im Wandel – Chance oder Risiko für den Naturschutz. E. Schmidt Verlag, 33-42.

Schaller, J. (2007), *Sustainable Development of the Region of Munich Using Advanced GIS Geodatabase and Environmental Modelling Technologies.* In: Conference Proceeding. 7th International Workshop on Geographical Information System (IWGIS'07, 14-15 Sept., Peking, China).

Schaller, J. (2007), *The Results of environmental Impact Studies of the Regional planning Procedure Danube River Construction Straubing-Vilshofen; The Revitalisation of the Isar river in Munich.* In: Tagungsband Ministry of Environment, 27.04.2007, Seoul, Korea, 89-141.

Schaller, J. (2010), *Solare Bauleitplanung mittels GIS.* In: Fachforum Photovoltaik-Freiflächenanlagen, BVS, 2. März 2010, Landshut.

Schaller, J. & Baltzer, U. (2006), *Moscow Urban Cadastre Information System on ESRI Platform.* ESRI International User Conference, San Diego, USA, Aug. 2006.

Schaller, J., Bonazountas, M., Kanellopoulos, S., Kallidromitou, D., Caballeros, D., Xanthopoulos, G., Martirano, G. & Tsarouxi, T. (2009), *ArcFIRE: ArcGIS Forest Fire Application in Greece.* 19th Annual MDS/ESRI ARCGIS Conference, Athens, Greece, 19-21 November. www.marathondata.gr.

Schaller, J., Bonazountas, M., Mattos, C., Kallidromitou, D., Martirano, G. & Roumeliotis, L. (2009), *MedIsolae-3D. Mediterranean Islands SDI and 3D Aerial Web-Navigation.* In: Digital Landscape Architecture 2009, 73-85. 21-23 May, 2009, Malta.

Schaller, J., Friot, D., Jolliet, O., Gehrke, T., Aigner, C., Blanc, I., Bonazountas, M. & Sabanegh, R. (2005), *Benchmarking European Regions with GIS Generated and Visualised Sustainability Indicators.* In: Book of Abstracts. Dubrovnik Conference on sustainable Development of Energy, Water and Environment Systems. Dubrovnik, Croatia, June 2005, 41.

Schaller, J., Gehrke , T. & Strout, N. (2009), *ArcGIS Processing Models for Regional Environmental Planning in Bavaria.* In: Planning Support Systems. Best Practice and New Methods. The GeoJournal Library 95, Springer Verlag, 243-264.

Schaller, J. & Mattos, C. (2009), *GIS Model Applications for Sustainable Development and Environmental Planning at the Regional Level.* In: GeoSpatial Visual Analytics, 45-57.

Schaller, J., Meister, D., Wopperer, C.-H., Bauer, M. & Fahnberg, C. (2004), *Günzburg oder Legoburg. Chancen und Risiken eines Identitätsprozesses.* Abschlussbericht im Rahmen des Forschungsprogramms Stadt 2030. meister.architekten ulm + Stadt Günzburg (Hrsg.).

Schaller, J. & Schober, H. M. (2004): *Das Landschaftsentwicklungskonzept (LEK) für die Region München.* In: Landschaftsökologie in Forschung, Planung und Anwendung. Friedrich Duhme zum Gedenken. Hrsg. v. Zehlius-Eckert, W., Gnädinger, J. & Tobias, S. 181.

Internet References

Dangermond, J. (2009), *GIS – Designing Our Future.* ArcNews, Summer 2009. http://www.esri.com/news/arcnews/summer09articles/gis-designing-our-future.html.

Dangermond, J. (2009), *The Vision of a Purposefully Designed Future.* ArcUser, Fall 2009. http://www.esri.com/news/arcuser/1009/geodesignuc.html.

Dangermond, J. (2010), *Harmonising Geography and Design.* Geospatial Today, June 2010. http://emag.geospatialtoday.com/index.aspx?issue=issue19&page=19.

ESRI (October 2010), *Changing Geography by Design – Selected Readings in GeoDesign.* ESRI Press. http://www.esri.com/library/ebooks/geodesign.pdf.

Steinitz, C. (2010), *Ways of Designing*. Redlands, California: GeoDesign Summit, January 6–8, 2010. http://www.geodesignsummit.com/videos/day-two.html.

Steinitz, C., Arias, H., Bassett, S., Flaxman, M., Goode, T., Maddock, T., Mouat, D., Peiser, R. & Shearer, A. (2003), *Alternative Futures for Changing Landscapes: The Upper San Pedro River Basin in Arizona and Sonora*. Washington, D.C., Island Press.

Szukalski, B. (2010), *ArcGIS Online as a Substrate for GeoDesign (and More)*. ArcGIS Online Blog, January 11, 2010. http://blogs.esri.com/esri/arcgis/2010/01/11/arcgis-online-as-a-substrate-for-geodesign/.

Wheeler, C. (2010), *Designing GeoDesign: Summit on new field that couples GIS and design*. ArcUser, Spring 2010. http://www.esri.com/news/arcuser/0410/files/geodesign.pdf.

Concepts for Automatic Generalization of Virtual 3D Landscape Models

Tassilo GLANDER, Matthias TRAPP and Jürgen DÖLLNER

This contribution to the DLA Conference 2011 appeared in gis.SCIENCE – Zeitschrift für Geoinformation, 25. Jahrgang, 1/2012, pages 18-23, published by Wichmann, VDE Verlag Berlin/Offenbach.

Please find the full online contribution at
http://www.kolleg.loel.hs-anhalt.de/landschaftsinformatik/435.html.

Abstract

This paper discusses concepts for the automatic generalization of virtual 3D landscape models. As complexity, heterogeneity, and diversity of geodata that constitute landscape models are constantly growing, the need for landscape models that generalize their contents to a consistent, coherent level-of-abstraction and information density becomes an essential requirement for applications such as in conceptual landscape design, simulation and analysis, and mobile mapping. We discuss concepts of generalization and working principles as well as the concept of level-of-abstraction. We furthermore present three exemplary automated techniques for generalizing 3D landscape models, including a geometric generalization technique that generates discrete iso-surfaces of 3D terrain models in real-time, a geometric generalization technique for site and building models, and a real-time generalization lens technique.

GeoDesign and Participation

Public Participation in Geodesign – A Prognosis for the Future

Carl STEINITZ

Keynote: 2 June 2012

1 Introduction

Landscape planning, in common with all physical design activities, requires answers to six questions:
1. How should the landscape context be described in content, space and time?
2. How does the landscape context function?
3. Is the current context working well?
4. How might the landscape context be changed in the future?
5. What difference might the changes cause?
6. How should the landscape context be changed?

There is a commonly held expectation that the answers to these questions and their coordination into a landscape plan for an area will be undertaken by a team of professionally and scientifically trained persons, guided by a client committee which is frequently a branch of government. My talk will show examples of how each of these questions has been answered directly through public participation. It will raise an awkward question: what will the future professional roles be when today's developing technologies are ubiquitous and enable direct public participation in all aspects of planning and design?

Professional practice and education are changing rapidly and this tendency will continue due largely to changes in political attitudes and in information technologies. For example, consider the implications of what is happening in Europe with regards to landscape planning, something which may become a worldwide model. The European Landscape Convention of the Council of Europe offers a very useful model. In Florence in October 2000, the 47 member countries of The Council of Europe adopted The European Landscape Convention. The Action Plan of the European Landscape Convention was adopted by the Heads of States and Governments of the member states in Warsaw, on May 17, 2005. This treaty has been ratified and is law in most but not all of the member states. As an international treaty, it supersedes national law in its field.

2 The Principal Provisions of the European Landscape Convention's Action Plan Are in Article 5

In Article 5 General Measures, each Party undertakes to
a. recognize landscapes in law as an essential component of people's surroundings, an expression of the diversity of their shared cultural and natural heritage, and a foundation of their identity;

b. establish and implement landscape policies aimed at landscape protection, management and planning through the adoption of the specific measures set out in Article 6;
c. establish procedures for *the participation of the general public*, local and regional authorities, and other parties with an interest *in the definition* and implementation *of the landscape policies* mentioned in paragraph b above (italics mine);
d. integrate landscape into its regional land town planning policies and in its cultural, environmental, agricultural, social and economic policies, as well as in any other policies with possible direct or indirect impact on landscape.

Because The European Landscape Convention codifies the need for stakeholder input into the beginning of defining any future policy or design the design team is then required to organize its work to produce such materials for public review and decision making. The people of the place are not just considered as clients, but are active members of the design team. I believe that this will ultimately and substantially transform professional practice, and it will force us to rethink some of our educational processes.

I have recently completed writing a book, "A Framework for Geodesign", to be published by Esri Press in July. In it, I have argued throughout that design at larger sizes of designing for landscape change – geodesign – is necessarily a collaborative enterprise.

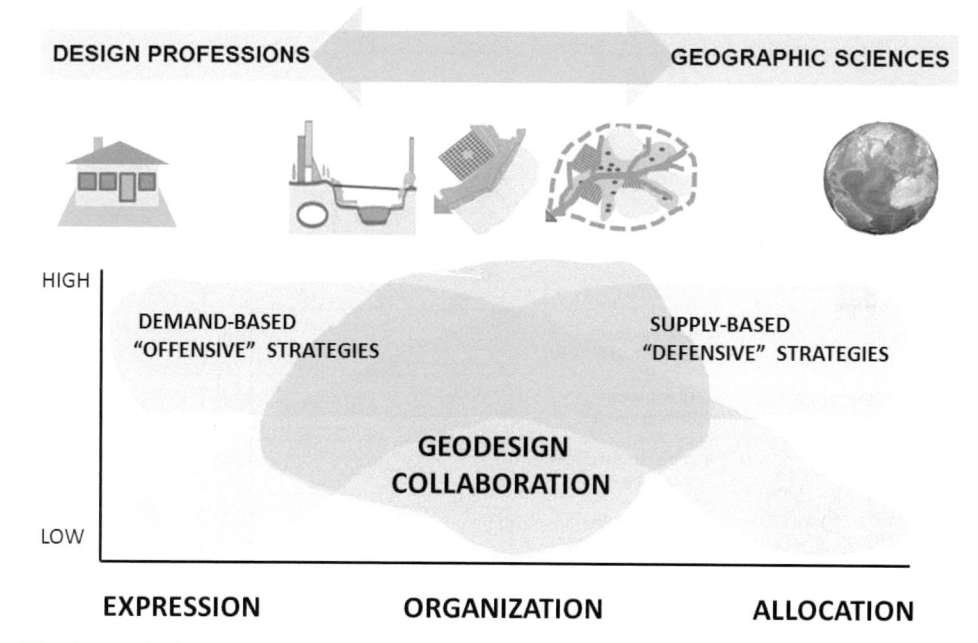

Fig. 1: The basis and need for collaboration, and the emphases as a function of size and scale

This is different from the individualistic assumptions underpinning the majority (but not all) of education in the design professions. The model of practice for traditional design practice is that there is a client (at the head of the table) and a single designer (albeit often supported by "staff") who will help make the design (Fig. 2).

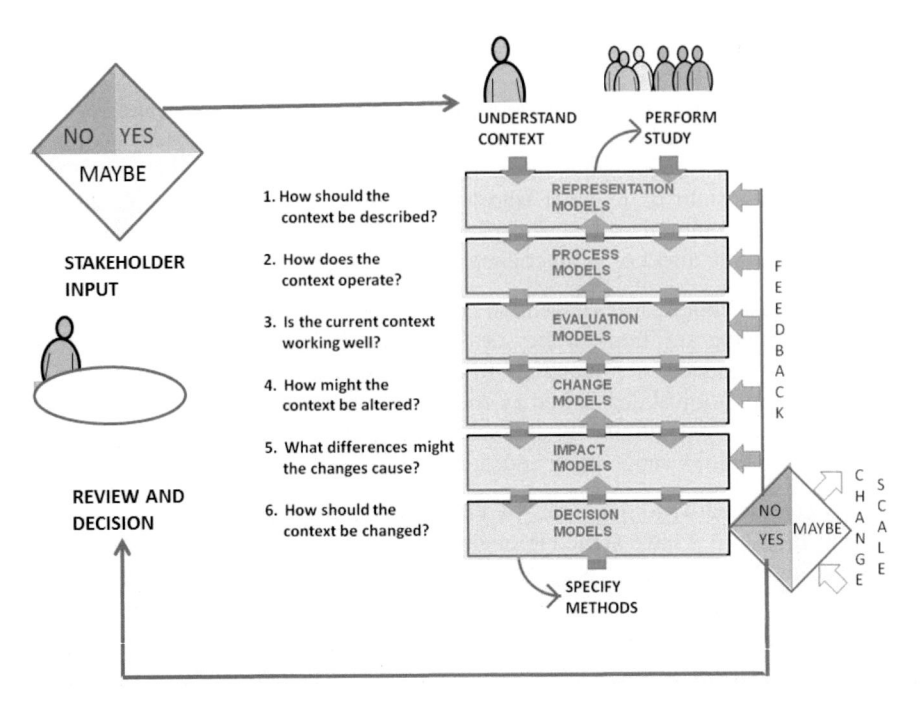

Fig. 2: The client and the designer

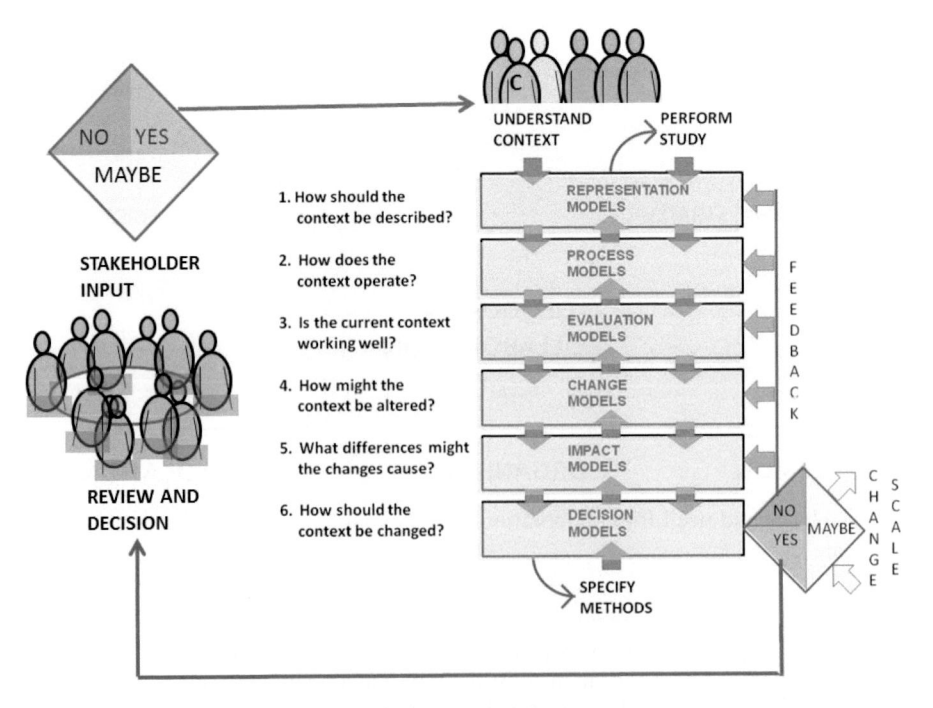

Fig. 3: The people of the place and the geodesign team

Recognizing the collaborative nature required to deal with the obvious complexities of geodesign contributed to this framework being structured as it is, as I have described it in detail in the book and in previous DLA Conferences. The stakeholder group has its necessary roles in input and decision making, while the technical team of designers and scientists and information technologists has the responsibility of carrying out the study (Fig. 3).

The European Landscape Convention broadens the responsibility of the people of the place and stakeholders and legitimizes their direct and deeper involvement with the design team. In the case studies presented in my book, there is at least one example of direct stakeholder involvement in each of the six fundamental questions of the framework. I will summarize and illustrate them in my talk.

The question must be asked, "Why shouldn't the people of the place take over the whole process of geodesign?" After all, it's their place and among them they surely know more than the other members of the geodesign team. Why, as shown in Fig. 4, shouldn't they take *total* responsibility for changing their own geography, and as they see fit?

There are several obvious reasons why this is not likely to happen completely. The stakeholders may be in serious disagreement about what should be done. They may not have any of the relevant experiences, or any interest beyond their own (if that), or the time and energy to devote to what is frequently a long and difficult set of integrated tasks. Especially in a large region, self-guided geodesign would undoubtedly be a very inefficient and unwieldy process. Both the professional literature and the cases I know show direct involvement in the parts, but not in the whole linked process of geodesign.

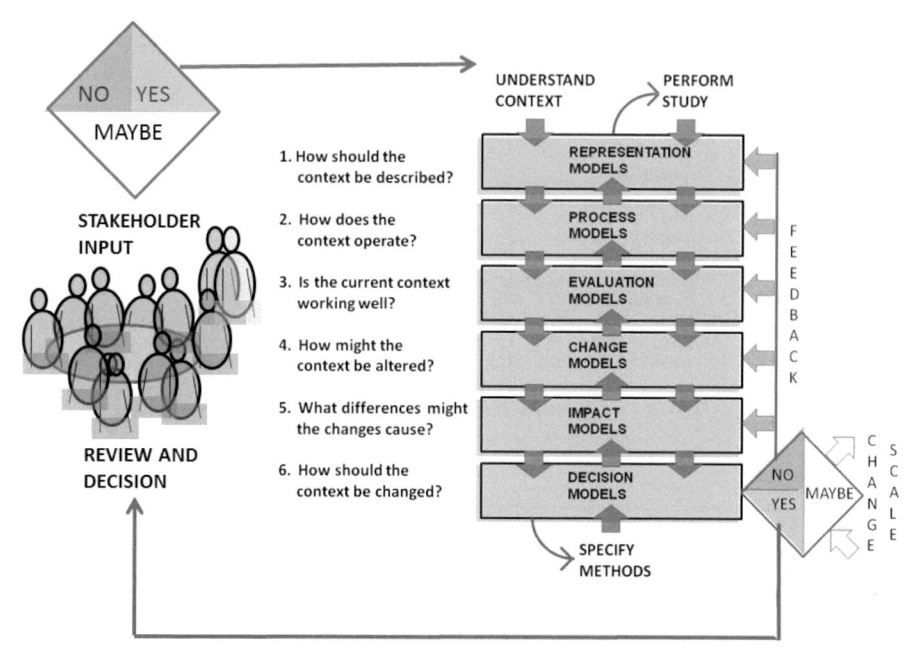

Fig. 4: Complete and direct involvement by the people of the place

However, I have no doubt that the next generation of people in geodesign will see increasing public participation in all aspects, including direct management of the process by the people of the place. I expect a reversal of an important social relationship. Typically, the geodesign team of design professionals, scientists, and information technology specialists work as a team separately from the people of the place. While we meet regularly with stakeholder representatives and communicate on a question-by-question basis during the course of conducting the study, ultimately the process is not a wholly democratic and entirely participatory one. I expect future geodesign studies to involve much more frequent and real-time participation and communication. The extent of that direct participation will vary, principally as a function of size and scale. Smaller projects and simpler methods enable more direct participation (Fig. 5), while larger studies with more complex methods require more significant roles by the design professionals and scientists (Fig. 6). The roles of the "conductor" will become even more central to geodesign, as will the needs for wider and more efficient communication. The balance of activity among design professionals and geographic scientists in applying the framework for geodesign will shift as a function of study area size and scale. The collaborative activities will also see shifting influences among the four essential participating groups as a function of where they are in the framework (Fig. 7). However, one thing should not change, and that is the responsibility of the people of the place to make the final decision to change their geography.

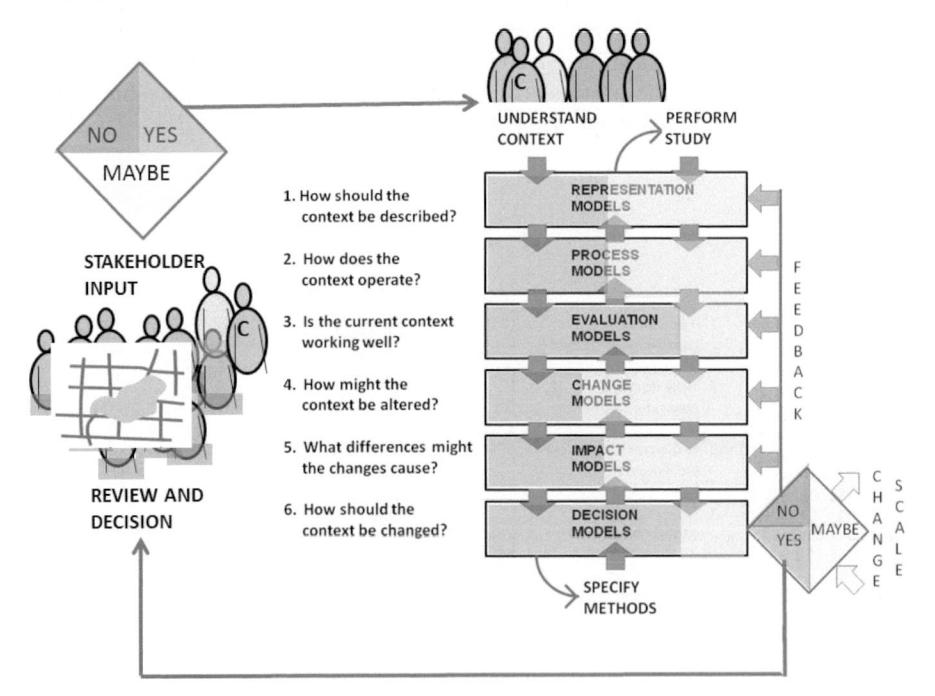

Fig. 5: The people of the place will have a greater role in smaller geodesign studies, whose geographic scale and scope of the project is more limited and manageable. Projects of smaller size and scale may also require less technical expertise, something that is more likely to absent among the people of the place. "Conductors" are indicated with the letter C on their shirts and will be needed both on the geodesign team and also from among the stakeholders.

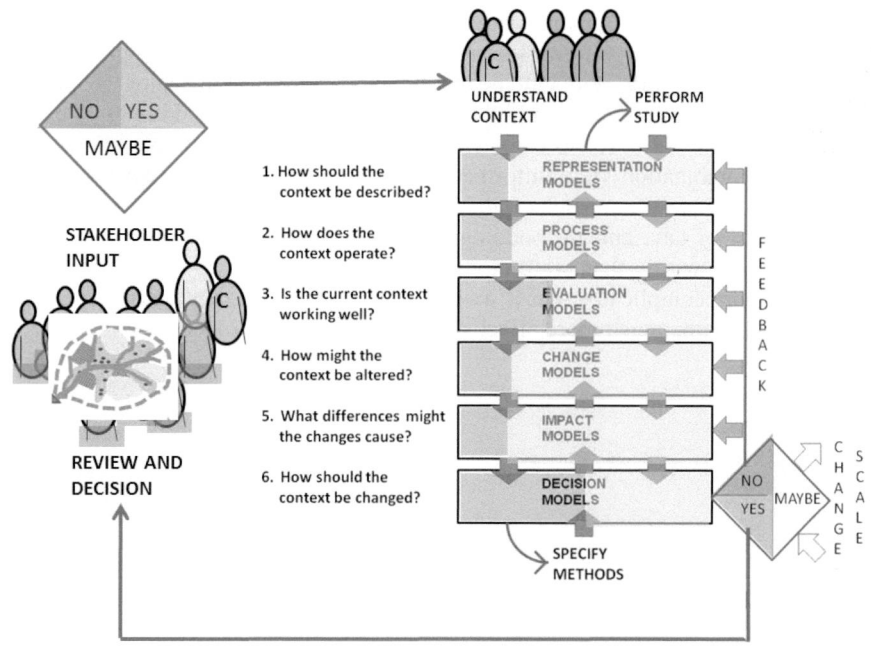

Fig. 6: In the future, larger scale geodesign projects will also have greater involvement from the people of the place, but there will be a need for more technically competent people and they will have to take a more active role.

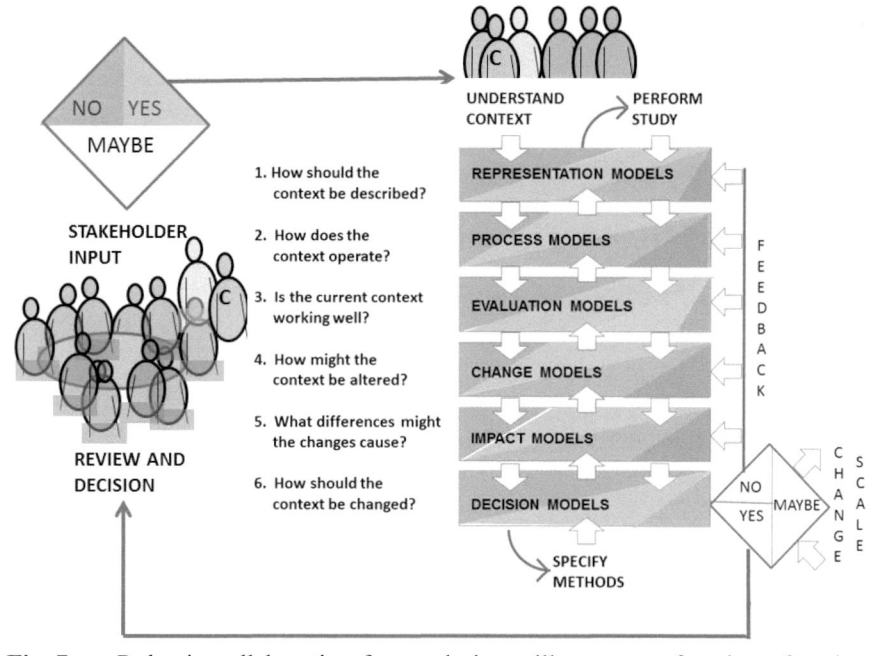

Fig. 7: Roles in collaboration for geodesign will vary as a function of project size and scale

When we consider the research our more advanced students are conducting, and today's technology-driven development, we can see an emerging pattern for geodesign that will develop rapidly in this century. We will be living in a world where major geodesign decisions will be made simultaneously and interactively at several sizes and scales. We will be managing the process (as best we can). Or, in worst case scenarios, we all will be managed by some combination of uninformed decisions and anarchy. There is every likelihood that the students we are teaching today will be practicing in a world where they will have an overload of data and methodological options, and where they will have to choose even much more wisely than we do today. Professional practice – and education – will become much more complicated, but if we can understand and accept that complexity, we are also likely to become much more effective.

Ems3D – Communicating Landscape Change

Jochen MÜLDER and Marina STRICKMANN

1 Introduction

Within complex planning processes, visualisations are a beneficial instrument to communicate content, goals and measures (ORLAND et al 2001, SCHMID 2001). Based on this fact, the subproject "Visualisation" (Ems3D), which is part of the Project "Perspective of a Living Ems Estuary", will provide a visual basis to support the dialogue between all participants within the overall planning process.

The main objective of the project is to develop ecological restoration measures for improving the ecological situation of the Ems estuary. Located in Lower Saxony in the north-west of Germany, the research area includes the last 50km of the river Ems. The surrounding landscape is mainly dominated by meadows, pastures, and farmland. The estuary is influenced by a daily tidal range. During the past three decades, large-scale man-made changes have caused serious environmental problems in the Ems river. Hydrodynamic transformations were conducted to enable the transfer of large cruise ships from a dock in Papenburg to the North Sea. These structural measures had an enormous impact on the local ecological situation. Because of this strong conflict between economy and ecology, the three NGOs World Wide Fund For Nature (WWF), German Association for Nature Conservation (NABU), Friends of the Earth Germany (BUND), and the Technical University (TU) Berlin initiated the project „Perspective of a Living Ems Estuary". In cooperation with Lenné3D GmbH, the TU Berlin develops detailed 3D-visualisations for different planning alternatives. By using an innovative 3D design, the communication processes between all stakeholders should get more efficient and transparent as well as easier to handle for all stakeholders.

Due to low funding and limited time, visualisations are usually prepared at the end of a project as a convincing image of the final planning. Within Ems3D, visualisations accompany the whole planning process, provide an instrument for the communication of work in progress and encourage discussions. These visualisations shall be comprehensible to all participants, independent of their varying knowledge of the actual situation at the river Ems (cp. HAAREN 2002 and APPLETON & LOVETT 2005).

2 Methods and Tools

2.1 Workshops

The interactive process of the project includes workshops over a period of almost two years. Up to now three workshops have been organized to inform experts and laymen of the actual state of planning and to discuss options for future development. The feedback provided is used as basis for subsequent internal expert discussions and workshops.

The process of planning and communication is documented by visualisations of the preliminary work status (e. g. images of the flawed actual state and the subsequent first approaches). Finally the 3D-modelling of measures will be continued until the final state of planning has been visualised.

2.2 Biosphere3D – geodata-based 3D-visualisation

Visualisations with a high level of detail require a substantial workload and computing power. This workload may be reduced significantly by choosing convenient tools and methods. The main challenge for the Ems3D project, beside the richness of detail, is the process-oriented planning process and the large-scale investigation area.

The open-source software Biosphere3D has been selected to render the visualisations. Biosphere 3D can realize projects on all levels of scale, from one's backyard to the whole globe. The program offers direct interfaces to established formats of geo-data like shapefiles, tif-raster and kmz. To model the estuary of the river Ems the required database needs to be prepared with programs possessing a spatial reference system. The data was preprocessed with the software ArcMap by ESRI and Sketchup by Google, and only the final arrangement of the 3D-scenes was carried out with Biosphere3D (cp. Fig. 1).

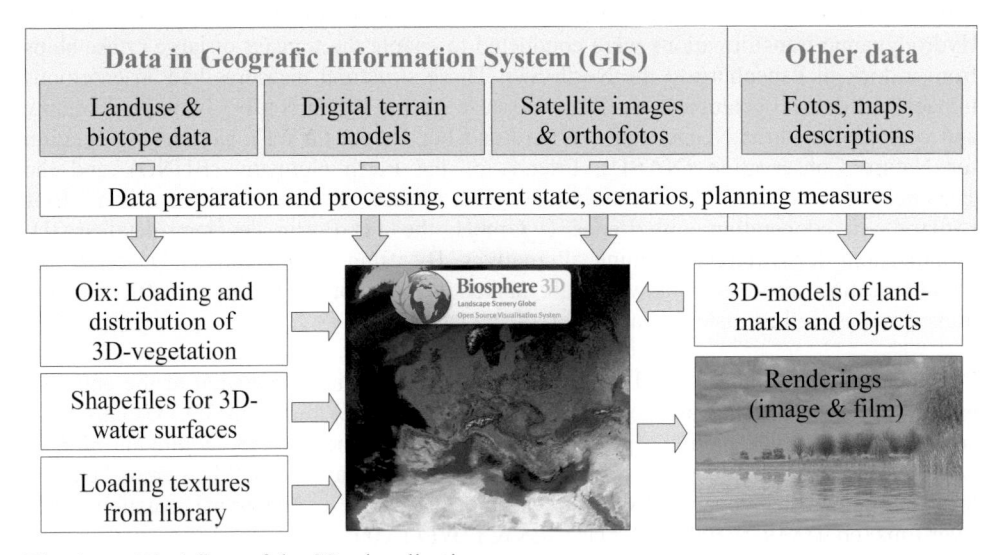

Fig. 1: Workflow of the 3D-visualisation process

2.3 Methods – modelling of the Ems estuary

The main advantage of Biosphere 3D is the ability to combine different kinds of geo-data with specific project-related information. The main elements of the landscape need to be analysed, specified and modelled only once. These elements can be used repeatedly within the planning process and modified if necessary. If a change is done, the model is updated automatically by refreshing or loading the database to the 3D-scene. The 3D-modelling is thus integrated into the planning process and no longer a subsequent, independent element.

Biosphere3D is an open-source software which allows adjusting and adding program features at all times. As the project area covers 50 km of the Ems river and water is the predominant element in the visualised scenes, a new function had to be developed which can display the stream slope. The water surface still has all the regular light reflections.

Being able to update the 3D-model easily is important due to the interactive character of the project, when restoration measures may be newly developed, changed or dropped. In fact, the raw construction of the main landscape elements allows picturing designated perspectives within an acceptably short timeframe.

Only the above mentioned aspects enabled the Ems3D team to constantly produce highly detailed 3D visualisations and support communication processes of the project "Perspective of a Living Ems Estuary".

3 Communicating Landscape Change

Before rendering visualisations it is mandatory to specify the content that shall be communicated. This approach is essential to structure the process and thus select representations and perspectives supporting the dialogue. The manner of introducing and explaining the subject of 3D-visualisations to participants highly influences the effectiveness of the following discussion (MÜLDER ET AL., 2007). Up to now the subproject Ems3D presents visualisations at three workshops. After the presentation of the 3D-planning alternatives the participants start group discussions. The main function of the visualisations is to inform about the planning process and to stimulate the audience before their group work phase.

3.1 Communication concept

For using visualisations as a beneficial instrument within the planning process, some basic aspects need to be considered. SHEPPARD (2005) points to the fact that visualisation should mainly be created in representative and true-to-life perspectives. A pedestrian's or canoeist's eye view provides the best visualisations of the Ems estuary, because they easily allow identifying changes in place. Drawings in 2D do not offer such impressions (ORLAND et al. 2001). However, some pictures and videos are rendered in bird's eye view to provide an overview, especially for large-scale landscape changes. To compare different states of landscape (actual state with future planning, different water levels, various planning alternatives), several images of one point of view are created (WARREN-KRETSCHMAR et al. 2005a).

3.2 Implementing visualisations at the workshops

Laymen generally have little understanding of the methods of scientific landscape modelling. Therefore the presentations at the workshops start off with an introduction to the methods, including technical and methodological possibilities and limitations. This methodological part also includes the "calibration" of people's perception by showing identical perspectives of their well-known landscape as photographs and as visualisations (Fig. 2). It illustrates the level of detail and abstraction applied to the 3D-model. To support the audience's orientation and recognition of the landscape, a map of the area which includes the position and line of sight is presented prior to the visualisations (WARREN-KRETZSCHMAR & TIEDTKE 2005).

Fig. 2: Demonstration of the level of abstraction by comparing a photo (left) with the visualisation of the actual state (right)

4 Visualisation, Perception and Dialogue

At the first workshop visualisations of the actual state were presented for five selected sites. Images and films of these sites illustrate typical characteristics, deficits and values of the investigation area (cp. Fig 3). Participants of this Workshop missed the opportunity to compare different planning measures.

As the planning process advanced, the second workshop was used to demonstrate drafts of planning measures. Compared to the actual state, visualisations of these measures illustrate the consequences of major landscape changes. For some areas different planning alternatives and their impact on the landscape were shown. Figure 4 illustrates, in addition to the actual state, an ecological and a hydrological optimized planning alternative of a planned polder.

Fig. 3: Visualisations of the current state at high tide (left) and low tide (right)

The participants' perception of the visualisations and reactions differed significantly at the two workshops. At the first workshop the participants' reaction could be described as constructive criticism. Certain aspects of the visualisations did not meet the expectations, which seem to rise permanently due to technical advances in computer graphics, especially with regard to the degree of realism. At that time only visualisations of the current state could be presented, and participants missed the opportunity to compare different objectives. This fact appeared to be the main difficulty.

Fig. 4: Comparison of two planning alternatives for a polder: hydrological optimized (left), ecological optimized (right)

The project team assisted the Emd3D team between the workshops to improve the appearance of the 3D-model and to raise the level of detail for certain landscape elements to better represent the Ems river landscape. These improvements helped to spend less time discussing the 'look' of the visualisations and focus on the content and the communication goals. At the second workshop visualisations of potential measures could be presented. This helped participants comprehend the character and objective of 3D-visualisation. The feedback provided was constantly positive. Especially visualisations of large-scale polders (cp. Fig. 4) allowed participants to get an impression of how drastically the landscape might possibly change.

Some participants questioned the visual appearance of certain ecosystems e.g. the floodplain forest. Although the ecologically based plant distribution is close to reality, the forest appears different from far away. This problem was partly solved by adding close-up views of the most important ecosystems (cp Fig. 5).

Fig. 5: The same floodplain forest appears different from close range (left) and from far away (right)

5 Conclusions and Outlook

The activating character of the visualisation caused critical comments, which mostly did not refer to the type of illustration as such. Observers used the opportunity to communicate their personal opinion on the project or to criticise planning aspects. This supports one of the desired tasks of visualisation: to overcome inhibitions, start discussing and thus support the process of communication.

Because of the process-oriented character of the project, participants can exert an influence on planning and design. Although this influence is limited by time and technical conditions it is an important aspect. Permanent communication between project team, experts and laymen leads to permanent changes within the planning concept. To be able to react to these changes as quickly as possible, further developments are necessary to tap the full potential of highly detailed 3D-visualisations in a complex, interactive project that depends on public cooperation.

The decision to use a geo-data based 3D-modelling tool like Biosphere3D turned out to be right. Biosphere3D has a great capability to create 3D-scenes due to the approach of using geo-data to generate the 3D-scenes. However, technical improvements are still necessary, especially to improve large-scale projects with extremely detailed vegetation.

Although a solid introduction was given, the need to discuss the possibilities and limitations of the 3D-model proved to be a key to improving the communication. Limitations include not only time and technical constraints, but also the intention of statement. Picturing many details implies the risk that observers do not discuss the content of the project, but rather the quality of the visualisation. The solution for this dilemma lies in SHEPPARDs (2005) call for accuracy in visualisations based on the available data. The problem of producing and presenting visualisations at a very early phase of the project was omnipresent and was partly solved by a careful moderation.

References

Appleton, K. & Lovett, A. (2005), *GIS-based visualisation of development proposals: reactions from planning and related professionals*. In: Computers, Environment and Urban Systems, 29, 321-339.

Haaren, C. V. (2002), *Landscape planning facing the challenge of the development of cultural landscapes*. In: Landscape and Urban Planning, 60: 73-80.

Mülder, J., Säck-da-Silva, S. & D. Bruns (2007), *Understanding the role of 3D visualisation – The example of Calden airport expansion, Kassel, Germany*. In: Van den Brink, A., van Lammeren, R., van de Velde, R. & Däne, S. (Eds.), Imaging the future. Geovisualisation for participatory spatial planning in Europe. Wageningen Academic Publishers, Mansholt publication series, 3, 75-88.

Orland, B., Budthimedhee, K. & Uusitalo, J. (2001), *Considering virtual worlds as representations of landscape realities and as tools for landscape planning*. In: Landscape and Urban Planning, 54, 139-148.

Schmid, W. A. (2001), *The emerging role of visual resource assessment and visualisation in landscape planning in Switzerland*. In: Landscape and Urban Planning, 54, 213-221.

Sheppard, S. R. J. (2005), *Validity, reliability and ethics in visualisation.* In: Bishop, I. & Lange, E. (Eds.), Visualisation in Landscape and Environmental Planning – Technology and Applications. London, Taylor and Francis, 79-97.

Warren-Kretzschmar, B. & Tiedtke, S. (2005), *What Role Does Visualisation Play in Communication with Citizens?.* In: Buhmann, E., Paar, P., Bishop, I. & Lange, E. (Eds.), Trends in Real-Time Landscape Visualisation and Participation. Proc. at Anhalt University of Applied Science 2005. Heidelberg, Wichmann.

Warren-Kretzschmar, B., Neumann A. & Meiforth, J. (2005a), *Interactive landscape planning – Results of a pilot study in Koenigslutter am Elm, Germany.* In: Schrenk (Eds.), CORP 2005 Geo-Multimedia 05 – Proceedings. Vienna University of Technology.

Extending Virtual Globes to Help Enhance Public Landscape Awareness

Amii HARWOOD, Andrew LOVETT and Jenni TURNER

1 Introduction

The use of internet-based virtual globes has expanded enormously in the past decade and has great potential as a means of representing and communicating landscape characteristics (Tuttle et al., 2008). Virtual globes can help people interpret their present environment and plan for the future (e.g. SHEPPARD & CIZEK, 2009, PETTIT et al., 2011, SCHROTH et al. 2011); they can also provide a window into the past, e.g. through geological modelling (DE PAOR & WHITMEYER 2011, PARASKEVAS 2011). Increasingly it is possible to customise the content displayed to the viewer and this, in turn, facilitates opportunities to support community participation in activities such as GeoDesign (ESRI 2010), particularly by informing people about characteristics of their local landscapes and helping them engage in discussions about how these could alter in the future. This paper describes how virtual globe displays were customised to create tours of an urban fringe landscape in Norfolk, UK and some initial experiences in using these tools as part of public engagement activities.

2 Study Area

The Sustainable Urban Fringes (SURF) project (http://www.sustainablefringes.eu) is part of the Interreg IVB North Sea Region Programme and is partly funded by the European Regional Development Fund. There are 13 project partners, one of which is Norfolk County Council, a local authority in eastern England. In Norfolk the SURF project is focusing on the River Gaywood valley to the east of the town of King's Lynn (see Fig. 1).

The River Gaywood rises from chalk springs and flows some 13 km through a lowland landscape and the middle of King's Lynn to join the River Great Ouse, then via the Wash estuary into the North Sea. Some of the housing estates close to the river on the eastern side of King's Lynn are amongst the most socially deprived parts of Norfolk and there is generally little connection with, or use of, the nearby countryside by residents of the town. The SURF project is seeking to unlock the potential of the river and its catchment as a natural environmental amenity by increasing awareness of local features, enhancing access to green open spaces and improving flood-risk management and opportunities for wildlife through habitat creation and restoration.

Following discussions with members of the SURF project team at Norfolk County Council a set of virtual globe visualisations for the valley were commissioned to help raise awareness and use of the local landscape. Through subsequent technical work and consultations three main themed outputs were developed which provided i) an introduction to the valley, ii) information on geology and past climates and iii) details of green infrastructure.

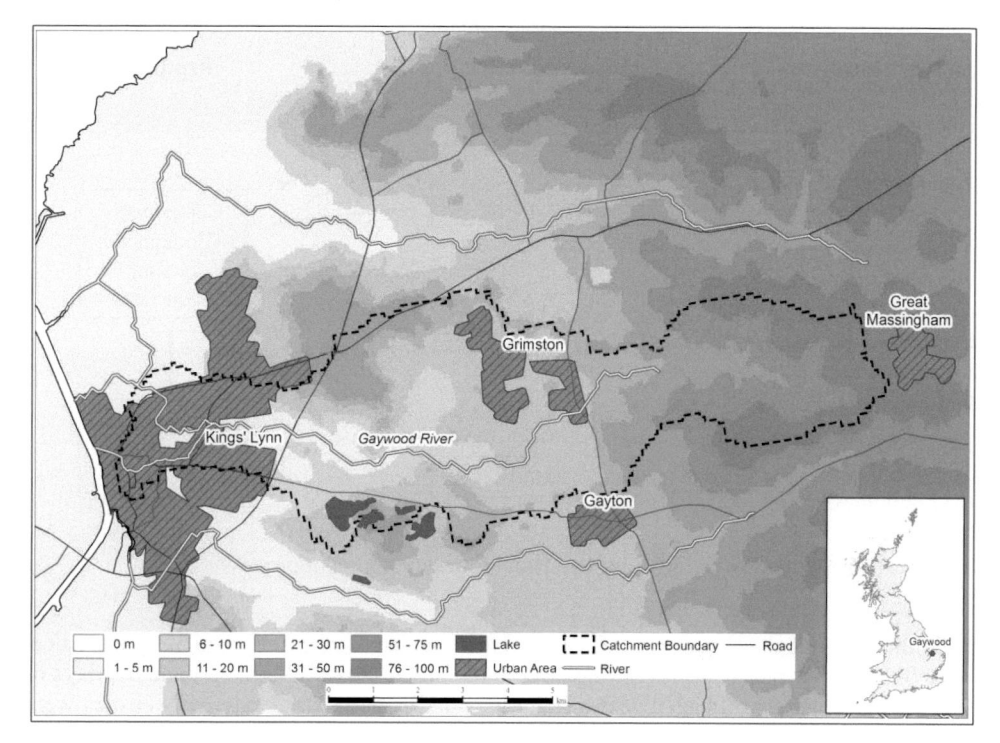

Fig. 1: The location of the Gaywood Valley study area in Norfolk, England

3 Data and Methods

Google Earth was selected as the platform for tool development from a number of available virtual globes (e.g. NASA World Wind, Microsoft Bing Maps, ESRI Arc Explorer, see reviews in TUTTLE et al. 2008 and SCHROTH et al. 2011). The attractions of this platform were that the basic version is free to download (http://www.google.com/earth/index.html), has a huge repository of associated online information, an established support system with user forums, and many additional tools such as SketchUp (http://sketchup.google.com/) to help with customisation. In addition, as arguably the market leader in this type of software it was thought more likely that the general public (especially younger generations) would have some experience of using it at home or school.

Three 'virtual tours' of the Gaywood Valley were produced by customisation of Google Earth version 6 using the KML scripting language (WERNECKE 2009). Fig. 2 illustrates the overall workflow, which involved integrating a number of different types of data.

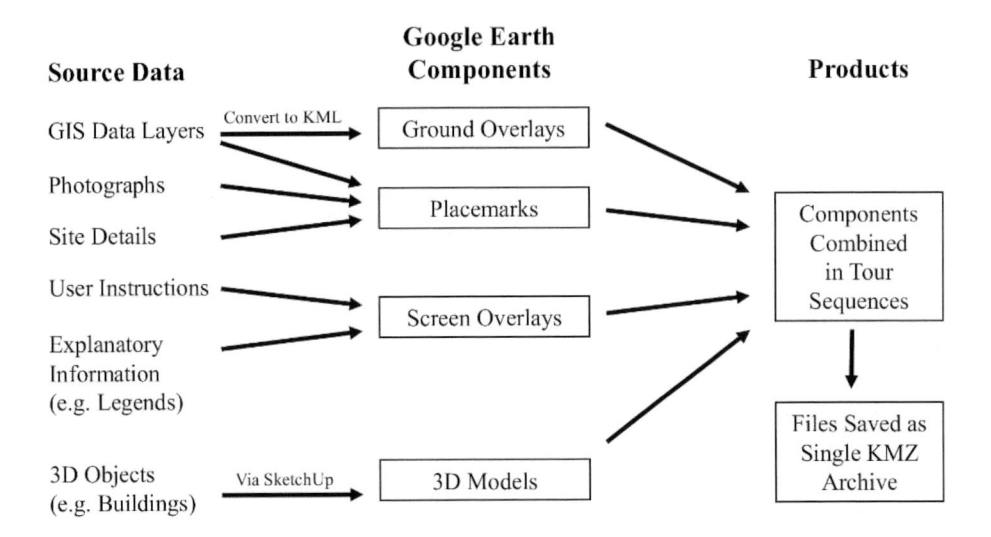

Fig. 2: The workflow adopted in the virtual tour development process

Additional spatial data, such as the catchment boundary and freely available geological maps (http://www.bgs.ac.uk), were imported into Google Earth KML as ground overlays or placemarks. Illustrative images, text descriptions and hyperlinks to external websites were included within placemarks. Screen overlays, images that have a fixed position in the Google Earth viewing window, were used to provide introductory information, instructions for the user or supplementary details on particular features (e.g. a map legend for accessible greenspace). 3D models were used to incorporate a geological cross-section and some key local landmarks. Fig. 3 shows an example of a placemark for a local nature reserve and Fig. 4 uses a number of different overlays and models to illustrate aspects of the local geology.

4   Producing Virtual Tours

The different types of content were integrated via KML code to generate the virtual tours in the form of an on-screen narrative for the user. The first few lines of code set the viewing position and angle for the user so that the study area was framed in the Google Earth window. Some welcoming information was also displayed, followed by instructions for the user supplemented by a series of screen overlays to provide explanations and link to additional sources of information. Fig. 5 illustrates the general style of presentation, with a placemark highlighting the source of the river, a screen overlay in the top-right corner containing a map indicating the viewpoint and other screen overlays proving instructions.

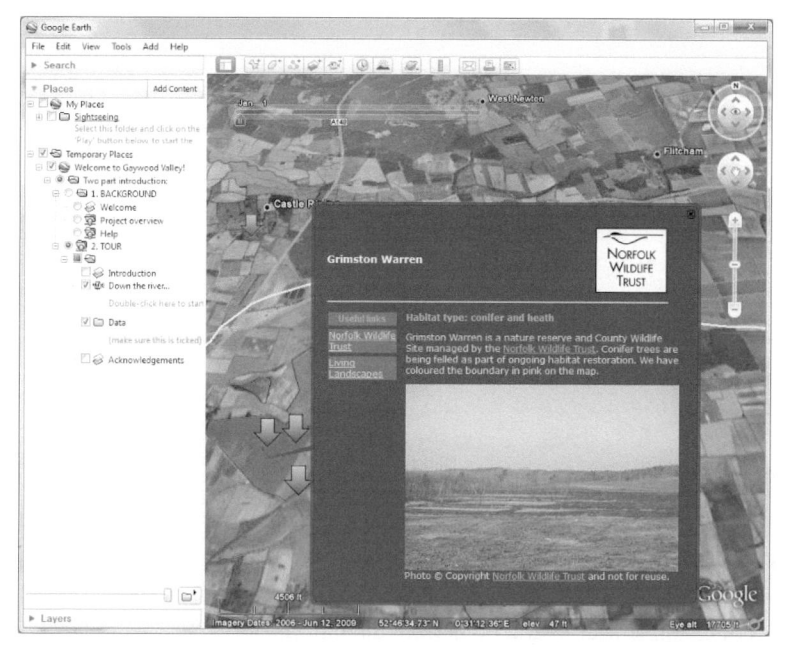

Extending Virtual Globes to Help Enhance Public Landscape Awareness 259

Fig. 3: Example of placemark content for Grimston Warren nature reserve

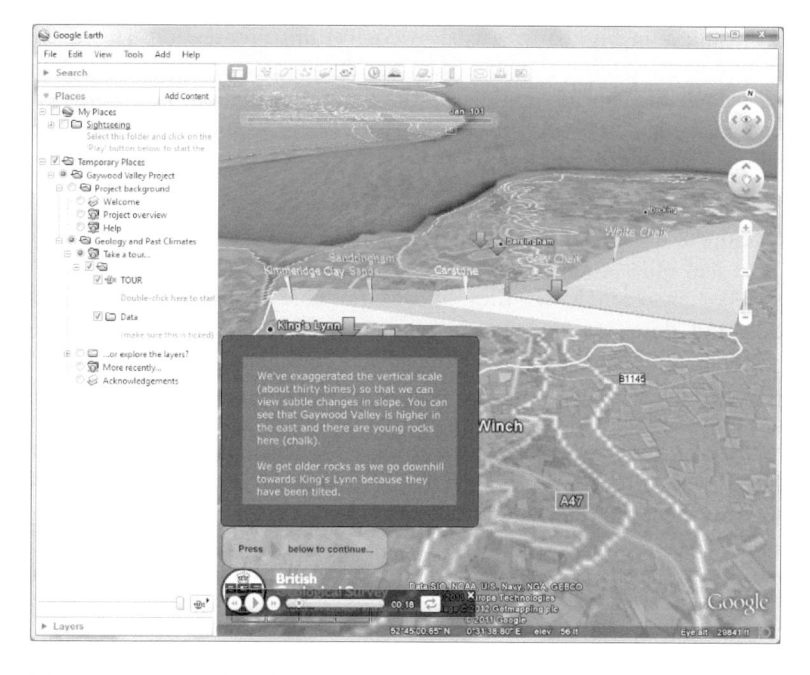

Fig. 4: Screenshot from the geology and past climates tour using a combination of ground overlays, screen overlays, placemarks and 3D models

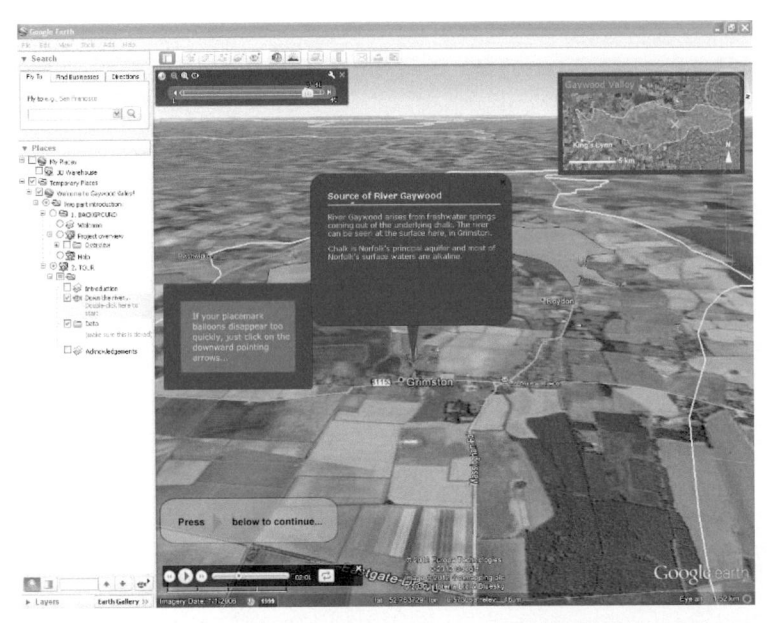

Fig. 5: Screenshot of example user instructions from the introductory tour

In each tour the users were taken on a virtual flight around the catchment which paused at placemarks in a developer-defined sequence. The timeslider feature in Google Earth (in the top-left of the display window in Fig. 5) was used to control this progression. However, flexibility was also built into the interface so that users could pause a tour, explore the data layers for themselves at their own speed, and then resume the presentation. For the green infrastructure theme more emphasis was placed on such self exploration, while the introduction and geology themes involved more of a structured guide accompanied by an onscreen narrative.

As the tours were developed, the content became more complex involving multiple layers of different spatial data and other components. Through the use of 'radioFolders' in the KML and 'network links' between KML files, it was possible to prevent overlap of incompatible elements and thus combine all the content within a single file system (WERNECKE 2009). Simple components were developed first and then revised and nested within a wider structure. To aid dissemination it was possible to zip together a hierarchy of folders and KML files into a single archive in KMZ format. The user could then download and display the content for each theme directly from a single file (each had a file size under 3.5MB). The KMZ archives were developed and tested to run on a Windows OS. At present, compatibility with Linux or Mac computers has not been evaluated.

5 Public Engagement

The public launch of the Gaywood Valley Project in May 2011 allowed testing of draft tools. At this event early versions of the introductory and geology tours were shown in a continuously looping form on a portable virtual reality immersion device. UEA researchers were also available alongside the display to answer questions and receive feedback (Fig. 6).

Fig. 6: Demonstration of the Geology and Past Climates tour at a project launch day in King's Lynn, May 2011

Subsequently, the tools were revised and demonstrated to the project advisory group, interested school teachers and at a meeting with local councillors in King's Lynn (September 2011). The response to the content and the style of presentation was positive so in October 2011 the files were made available to download from a University of East Anglia website (http://www.uea.ac.uk/env/research/reshigh/gaywood). By the end of February 2012 this site had received over 300 unique views, with particular peaks on days following publicity via social media and a short report in a local newspaper. Correspondence with people who have downloaded the files has also helped to improve the instructions provided.

The next phase of engagement will involve day events with several local primary schools and for this purpose a number of worksheets and other teaching materials have been developed. A meeting for local teachers to provide feedback on this experience and further promote the availability of the tours is scheduled for April 2012.

6 Conclusions

The initial stages of creating the Google Earth tours were quite technically demanding, but once some key techniques were mastered and a structure for organising files established progress was relatively rapid. Particular benefits of the approach are that the products can be easily revised, that they rely almost entirely on publicly available data and software, are easy to disseminate and that the existing templates could be readily adapted for other locations. There is also scope for suitably experienced users to add their own content. Evaluation of public use is still in progress, but it is clear that the products have certainly attracted interest and, anecdotally at least, generated positive feedback about both the style of presentation and increasing awareness of the local environment. Customising a virtual globe such as Google Earth in the manner described is also very cost-effective compared to more sophisticated techniques of landscape visualization (LOVETT et al. 2009) and undoubtedly has great potential for supporting initiatives concerned with public participation in landscape planning, including aspects of GeoDesign.

7 Acknowledgements

The Gaywood Valley Sustainable Urban Fringes Project is supported by the Interreg IVB North Sea Region Programme and is partly funded by the European Regional Development Fund. Data were obtained from Norfolk County Council (NCC), the British Geological Survey and the Norfolk Geodiversity Partnership. Special thanks to Gemma Clark at NCC, Tim Holt-Wilson at the Norfolk Geodiversity Partnership for geological advice and Lauren Parkin for creating initial drafts of the teaching materials.

References

De Paor, D. G. & Whitmeyer, S. J. (2011), *Geological and geophysical modelling on virtual globes using KML, COLLADA and Javascript*. In: Computers & Geosciences, 37 (1), 100-110.

ESRI (2010), *Changing Geography by Design: Selected Readings in GeoDesign*. ESRI, Redlands, CA, USA.

Lovett, A. A., Appleton, K. J. & Jones, A. P. (2009), *GIS-based landscape visualization – The state of the art*. In: Mount, M., Harvey, G., Aplin, P. & Priestnall, G. (Eds.), Representing, Modeling and Visualizing the Natural Environment. Boca Raton, FL, USA, CRC Press, 287-309.

Paraskevas, T. (2011), *Virtual globes and geological modeling*. In: International Journal of Geosciences, 2, 648-656.

Pettit, C. J., Raymond, C. M., Bryan, B. A. & Lewis, H. (2011), *Indentifying strengths and weaknesses of landscape visualisation for effective communication of future alternatives*. In: Landscape and Urban Planning, 100 (3), 231-241.

Schroth, O., Pond, E., Campbell, C., Cizek, P., Bohus, S. & Sheppard, S. R. J. (2011), *Tool or toy? Virtual globes in landscape planning*. In: Future Internet, 3, 204-227.

Sheppard, S. R. J. & Cizek, P. (2009), *The ethics of Google Earth: crossing thresholds from spatial data to landscape visualisation*. In: Journal of Environmental Management, 90, 2102-2117.

Tuttle, B. T., Anderson, S. & Huff, R. (2008), *Virtual globes: an overview of their history, uses, and future challenges*. In: Geography Compass, 2/5, 1478-1505.

Wernecke, J. (2009), *The KML Handbook: Geographic Visualisation for the Web*. Upper Saddle River, NJ, USA, Addison-Wesley.

Things Have Changed – A Visual Assessment of a Virtual Landscape from 1900 and 2006

Henk KRAMER, Joske HOUTKAMP and Matthijs DANES

1 Introduction

The landscape of The Netherlands is continually changing under the influence of both natural forces and human activities. There are many possible ways to explore the present landscape, in reality by visiting the actual location or virtually by using tools like Google Earth, Google Street view or bird's eye view in Bing Maps. Both methods offer the user complete freedom of movement and the user can choose his own viewpoint to explore the landscape according to his own desires (HONJO et al 2009).

Exploring the bygone landscape is also possible but the options are more limited. At various moments in time, an artist interpretation of the actual appearance of the historical landscape has been captured in paintings. For the Netherlands, landscape paintings and drawings from many periods and locations are available. Examples of these, that can be viewed online as well as in a book, are the collection off geographical school maps (DONKERS 2006). This collection provides portrayals of historical landscapes with recent pictures of the same location. The accompanying website provides the links to view the locations in Google Earth. But these illustrations only show a small part of the landscape, offering the user only a limited view of the historical landscape. The user is also confined to the specific location and viewpoint chosen by the painter, and does not have the freedom to move around. Also the number of views (paintings) is limited and they do not cover all landscapes in The Netherlands at different moments in time.

Information on historical landscapes is also recorded in historical topographic maps. For the Netherlands, detailed historical topographic maps are available from 1850 and 1900 and from 1950 on with an update for every ten years (KNOL et al 2004, KRAMER & DORLAND 2010). From 2000 on, a four-yearly update is available. These topographic maps cover the whole of The Netherlands, offering a wealth of information on bygone landscapes. However, the only offer 2D information and show the landscape through a cartographic interpretation. The existence of landscape elements is captured in drawn area's using colours (on analogue maps) or thematic value (on digital maps) representing a selection of land cover classes. The maps do not contain the appearance of these landscape elements. When exploring historical maps, the user has to create a mental representation of the actual appearance of the landscape.

The information offered by the paintings and the historical topographic maps can be combined to create a 3D visualisation of the historical landscape (DE BOER 2010, GRIFFON et. al. 2010). These same techniques can be used to make a 3D visualisation of the present landscape. A visual assessment of the changes in the landscape over time can then be made by comparing both visualisations. It is not difficult to allow users to navigate a 3D virtual landscape, but how can the changes in the landscape over time be presented and highlighted to the user?

In this paper we describe the design of an application to facilitate comparison of two landscape models, and discuss the evaluation by a group of expert users from different domains (historians, landscape historians, and archaeologists). The questions we address are: how can two virtual landscapes be presented to reveal specific changes over time? Which problems do we encounter in the construction of the virtual landscapes to make a comparison possible? How can visualisations like this be used to inform historical research? (BISHOP et. al. 2001, KNOWLES, & HILLIER 2003).

For a study area two 3D virtual landscapes were constructed from available topographic maps from 1900 and 2006. Landscape paintings from around 1900 were used to get an impression of the appearance of the historical landscape. The 3D models of the whole landscape allow users to experience the spatial layout of the environment. The user can see the location and dimensions of important elements in a landscape, such as tree lanes, roads, buildings and vegetation, and experience sightlines and vistas. Landscape elements that are outside the modelled area play an important role in the visual perception of the landscape because they are visible on the horizon when viewing from a first person perspective inside the landscape. The topographic maps provided the necessary information on where to construct these elements on the horizon. By comparing the two models the user can detect changes over time that are related to historical events.

In this application, the current situation as well as the historical situation have the same level of abstraction, and show the landscape elements in the same rendering and level of detail. The influence of representational features on the assessment (e.g. high quality photographs versus 3D models), atmospheric circumstances or different seasons, is thus prevented.

The chosen visualisation technique shows the 3D landscapes from both periods side by side, with a map indicating the route as a reference in the bottom of the screen. This should facilitate easy detection of differences between the two visualizations. The viewer can control the pace of the animation to reduce an information overload and to allow closer inspection of interesting locations.

2 The Study Area

An area near Chaam, which is located in the south of The Netherlands, was selected as study area. For this area a historical topographic map from 1900 and a topographic map from 2006 were available in digital format with geo-referencing and thematic values for land cover types (Figure 1).

The maps show the land cover classes that were used as the basis to create the 3D visualizations. In 1900, this area was a mix of natural land cover classes like heath land, swamp, small patches of forest, many tree lanes and small scale agricultural land use (arable land and meadows). In 2006, mainly large scale agricultural land use remains. Heath land, the small patches of forest and tree lanes have disappeared and are replaced by agricultural land use. By 2006, new roads and many new buildings are constructed in the area.

Chaam 1900   Chaam 2006

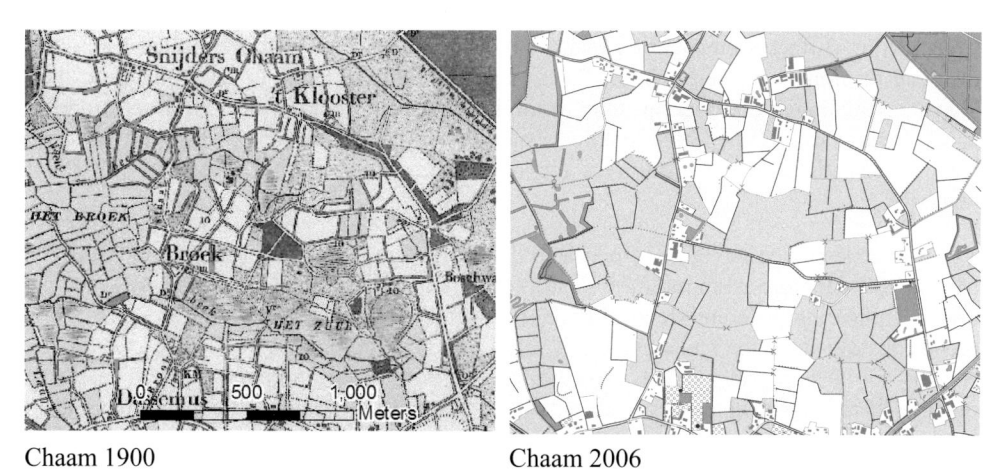

Fig. 1:   Study area Chaam

3 Modelling Approach

For the construction of the 3D model of the landscape we searched for an application that could handle geo-referenced GIS data and in which biotopes can be designed. This makes it easy to create models for large areas, the most time consuming part is the design of these so-called biotopes and connect them to the land-cover types that are available in the GIS data. In practice both Visual Nature Studio (3DNature, 2009) and Landsim3D (BIONATICS 2010) are found to meet these requirements. Landsim3D also allows free navigation through the 3D-scene whereas VNS needs to render the 3D-scene into a movie along a predefined camera path. Eventually this free navigation was decisive to continue with Landsim3D.

In Landsim3D the procedure to design a 3D-scene is as follows: As a start a surface model is imported to form the ground floor. Once the ground floor is determined, additional surface information can be included about different biotopes. For the specific test area in Chaam we recognised the following biotopes: Pine forest, Pastures, Heath land and Arable land. The biotopes already show the rough contours of a landscape, but more point and line features are included for a realistic impression. As line features we included roads, ditches, cycling/walking paths and tree lanes. Buildings were created from the outlines that are available in the GIS as polygons with parameters for height and type. Individual trees were included as point feature with parameters for height and type.

Once the scene is constructed, the final step is to design the 3D-representations, which replace the individual features in the 3D-scene. Such a 3D-representation can include anything, from simple textures to complete 3D-models. In theory each feature could have its own 3D-representation, however, its design would be very time consuming. Instead, also due to a lack of detailed information, we used one representation for buildings, one for heather, two for deciduous trees and also two for pine trees. For a more natural look, we re-used and re-scaled the different vegetation models to visualize a mixture of ages.

Constructing surface elements like roads and ditches from lines that are digitised in a GIS can lead to problems when original features are located too close to each other and their 3D-representations may overlap. One example is when a ditch is located close to and in parallel with a road. When both features are located too close to each other, part of the road may be visualised at the bottom of the ditch. Correcting these issues may be labour intensive.

The 3D-scene can be navigated freely and the movements recorded in a camera path. The camera path can be used in multiple scenes (each representing a different time) to guide the user in comparing different views in time of the landscape. Even though the structure in the scene corresponds to the real situation, it still looks a bit simplistic and artificial. One of the reasons for this is that current visualised biotopes are very homogeneous, whereas in practice there is more heterogeneous with much more local variation. It is possible to adapt the composition of the biotopes but this is time consuming.

Comparing different 3D-scenes in order to experience the changes in the evolving landscape reveals a final issue. The accuracy and the mapping techniques of historic data deviate from data of the current situation. If both maps are digitised and overlaid, small deviations occur. This becomes especially clear whenever the camera path is set to follow a road in the 3D-scene of the historical situation. Once the same path is activated in the 3D-scene of the current situation, the camera keeps the position on the historical path and can show a small offset to the current road features. In a comparison the user might experience such an offset as a geometric error (the road is not digitised correctly), while in reality this is proper change (the road has been rebuilt but not exactly at the same location. On the other hand, it is also possible that visible changes in the landscape models are errors in the source material. This issue can only be solved with detailed historical knowledge of the study area.

One topic that needs to be addressed is the validity of the 3D representation of the historical green landscape elements. We had access to historical landscape paintings from locations nearby the study area that were made at the end of the nineteenth- or start of the twentieth century. We tried to transfer the visible landscape characteristics of these paintings into the 3D visualization of the historical landscape. The maps gave us an indication of the location solitary trees and also the ground dimensions of tree lanes but did not provide information on the size of the trees or the density of the tree lanes. For this, a best guess was made from the information that was available from the paintings and the topographic maps.

We have captured several stills from the two 3D visualizations to give an impression of the two virtual landscapes (Figure 2). Each set of pictures shows exactly the same location, the left picture showing the landscape from 1900 and the right picture showing the landscape from 2006.

The road from 1900 has partially been replaced by agricultural land. In the middle of the right image, just left of the building in 2006, it can be observed that the continuation of the road is still in the same location.

The small patch of coniferous forest from 1900 has disappeared and a new tree lane along the road has emerged. This changes the vista from this point considerably.

A different experience, a closed landscape in 1900 versus an open landscape in 2006

A view on heath land and small patches of coniferous forest in 1900 and a much more open landscape with solitaire trees and tree lanes

Fig. 2: Several stills captured from the virtual landscapes, showing examples of changes in the landscape

4 Evaluation

For the expert consultation we decided to create a movie, captured along a predefined camera path, showing the 3D landscapes from both periods side by side. This did not give the viewers the desired freedom of movement when exploring the virtual landscapes but did result in comparable results for our evaluation. If necessary, the expert could get access to the actual 3D models or be given another movie with different viewpoints. The movies were rendered with Landsim3D for both points of time. Avisynth software was used to synchronize both movies and combine them into one movie. This made the synchronized playback of both movies on a webpage possible. For navigation purposes, a 2D view of the area in Google Maps was included in the website. A java script was created to link the actual location in the movie with a camera-icon shown in the Google Maps view. The user can also use this camera-icon to navigate through the movie. The webpage is shown in figure 3 and is available at http://alterra0125s.wur.nl/gpsvid/tndt.htm.

Fig. 3: Web-based application used for the visual assessment of the virtual landscape from 1900 and 2006

To evaluate the application, we sent the webpage with the animation and a questionnaire to ca 20 prospective users, mainly historians, landscape historians and professionals working in museums. Six reactions were received in time to be included in this paper, from representatives of all three domains.

The questions concerned four topics: the usability of the visualisation, especially the presentation of the two models next to each other; the functionality of the interface, so the control over the navigation (in this case the animation); the quality of the representation; and how this and similar applications could be employed in research or education.

Although the number of respondents is small, their comments are highly informative and consistent with reactions from viewers at earlier stages of development.

Most of them found it difficult to perceive and compare the 3D models and map at the same time; the three frames change continuously and viewers have to find fixed points for comparison. Research on animation has shown that observers often have difficulties with understanding even single animations, because of their transience: it is hard to perceive movements, changes and their timing (TVERSKY & MORRISON 2002). Our visualisation thus makes a high demand on visual attention (KRIZ & HEGARTY 2007).

The respondents suggested several solutions for this problem: to present only the models and show the map on demand; to add a playback option, or lower the walkthrough speed. In this animation, only a play and pause button was offered. It has been proposed by several researchers that interactivity can help viewers in perceiving and comprehending animations by allowing them to view animations at a pace that is congruent with the speed of their comprehension processes (HEGARTY 2004, TVERSKY et al. 2002).

One other solution to alleviate this problem might be to help viewers to focus their attention to the important elements in the three visualisations. This occurs naturally when a user is looking for a specific viewpoint or element in the landscape. Otherwise they might be assisted by so-called visual cueing, for instance highlighting areas that are interesting or important (MAYER & MORENO 2003). However, KRIZ & HEGARTY (2007) found that interactivity and attentional cueing are not always effective in improving comprehension, and that domain knowledge of the viewer plays an important role in the processing of information. One of the respondents suggested including (clickable) points-of-interest, as an instrument to guide viewers through an area of which they have no knowledge. Other options to facilitate comparison, such as toggling between the two periods, mentioned by SCHROTH et al. (2005), may be considered as an alternative.

In this application, the rendering of the environments is sober. The application is still a prototype, and realistic rendering is time consuming and costly. The quality of the 3D environments was also reduced by the conversion into movies that could be played on a website. However, the information required to add details is often not available in the old maps, and adding details that are not contained in the historic maps diminishes the validity of the visualisations. Another reason to preserve an abstract style is that simple and similar, rendering for both periods may facilitate comparison of the landscapes.

The respondents made many remarks about the plainness and lack of visual realism of the represented environments. Some of the textures did not convey the intended types of land use or vegetation clearly. The illumination and shadows are basic, and did not create a convincing 3D experience. Nowadays viewers are used to high quality virtual environments, and in future versions of the application the textures and illumination will be improved. The 3D visualisation of 1900 did not contain indications of use or habitation, like carts, or houses on the horizon. Some viewers mentioned this would make the historical environment more believable or convincing.

Respondents also said they expected a 'historic feel' in the old situation, and suggested not only including more historical artefacts and buildings, but also a different style of rendering to achieve a more engaging representation of the past. As DE BOER et al (2009) explain, post-processing may be used "to evoke a certain period, selecting an artistic-style filter that corresponds to techniques that artists used in that specific time". This would contribute to a realistic experience and evoke the feeling of being present in the past (i.e. *historical sensation*).

The required style however depends on the purpose of the visualisation. For assessment of changes per se, visualisations in an abstract style and without embellishment are preferable. They show the basic landscape elements, their locations and dimensions, without visual distractions that may increase the cognitive overload mentioned earlier. Visualisations used for educational purposes and museums have different requirements; the 'ambience' of the environment should meet with expectations of the viewers, and create a sense of engagement. The developer has more artistic freedom to adapt the environment for this purpose.

Finally, the respondents were all enthusiastic about the possibilities of the application. They enumerated a range of uses involving comparing scenarios, assessing and understanding changes in landscape over time, and showing elements in the landscape that are now lost. They expected that applications like this could be used by many different users, such as policy makers, stakeholders in landscape planning projects, professionals working in landscaping and architecture, researchers in landscape history or cultural history, museum visitors, and tourists.

5 Conclusions and Outlook

In this study we showed that 3D virtual landscapes for two moments in time can be constructed from accompanying topographic maps. We created two movies from these virtual landscapes and presented them side by side to a group of experts through a web application. All of the respondents were enthusiastic about the possibilities of the application. They enumerated a range of uses involving comparing scenarios, assessing and understanding changes in landscape over time, and showing elements in the landscape that are now lost. The application can already be improved by adding the option to control the play back speed, making it easier for the user to slow down at interesting parts and speed up whenever desired. However, the amount of information offered by the application is overwhelming. When comparing virtual landscapes of different moments in time, the viewer needs guidance to know where to look. Offering a fixed path is not enough. For example, the viewer might be assisted by so-called visual cueing, by highlighting areas that are interesting or important. Another useful type of guidance is the placement of markers in the virtual environment that the viewer can select to navigate to specific locations of interest. This user guidance is even more important when the user is given more freedom of movement through the virtual landscape.

The graphic quality of the 3D environment matters although a high quality representation is not strictly necessary for the assessment of change in the two virtual landscapes. Nowadays viewers are used to photorealistic virtual environments presented in games and movies and apparently take this for granted. However, the information required to add a high level of detail is often not available in the old maps, and adding details that are not contained in the historic maps diminishes the validity of the visualisations. Another reason to preserve an abstract style is that simple and similar rendering for both periods may facilitate a better comparison of the landscapes.

Future research will focus on how to guide the user through the virtual landscapes so they recognize and appreciate the way the landscape has evolved over the years. Improvement of the graphic quality of the 3D environment can be expected from the builders of the software. Better graphics cards and faster computers will make this available. The creator of the virtual landscape should focus on the details of the location specific 3D models of buildings and vegetation.

6 Acknowledgements

This research is part of the strategic research program "Sustainable spatial development of ecosystems, landscapes, seas and regions" which is funded by the Dutch Ministry of Agriculture, Nature Conservation and Food Quality, and carried out by Wageningen University Research centre.

Thanks are due to all respondents who took the time to evaluate the application and complete the survey.

References

3DNature (2010), *Visual Nature Studio V3 software*.
Bionatics Europe. (2010), *Landsim3D v2 software*.
Bishop, I. D., Wherrett, J. R. & Miller, D. R. (2001), *Assessment of path choices on a country walk using a virtual environment*. In: Landscape and Urban Planning, 52.
De Boer, A. (2009), *Creating realistic 3D GeoVEs of cultural landscapes using historical maps and drawings*. In: De Maeyer, P., Neutens, T. & De Ryck, M. (Eds.), Proceedings of the 4th International Workshop on 3D Geo-Information, 63-68. Ghent, Ghent University.
De Boer, A. (2010), *Processing old maps and drawings to create virtual historic landscapes*. e- Perimetron, 5 (2).
Donkers, H. (2006), *Verdwenen Nederland*. Noordhoff Uitgevers B.V.
  (online: http://verdwenennederland.bosatlas.nl/index.htm)
Griffon, S., Nespoulous, A., Cheylan, J-P, Marty, P. & Auclair, D. (2010), *Virtual reality for cultural landscape visualization*. In: Virtual Reality, 14.
Hegarty, M. (2004), *Dynamic visualizations and learning: getting to the difficult questions*. In: Learning and Instruction, 14.
Honjo, T., Umeki, K., Lim, E., Wang, D.-H., Yang, P.-A. & Hsieh, H.-C. (2009), *Landscape Visualization on Google Earth*. pma, Third International Symposium on Plant Growth Modeling, Simulation, Visualization and Applications, 445-448.
Knol, W. C., Kramer, H. & Gijsbertse, H. A. (2004), *Historisch grondgebruik Nederland: een landelijke reconstructie van het grondgebruik rond 1900*. Alterra (Alterra-rapport, 573)
Knowles, A. K. & Hillier, A. (2008), *Placing history: how maps, spatial data, and GIS are changing historical scholarship*. ESRI Inc.

Kramer, H. & Dorland, G. J. van (2009), *Historisch grondgebruik Nederland 1990 een landelijke reconstructie van het grondgebruik rond 1990.* Alterra (Alterra-rapport, 1327)

Kriz, S., & Hegarty, M. (2007), *Top-down and bottom-up influences on learning from animations.* In: International Journal of Human-Computer Studies, 65.

Mayer, R. E., & Moreno, R. (2003), *Nine ways to reduce cognitive load in multimedia learning.* In: Educational Psychologist, 38.

Schroth, O., Lange, E. & Schmid, W. A. (2005), *From Information to Participation – Applying Interactive Features in Landscape Visualizations.* In Buhmann, E. et al. (Eds.), Trends in Real-time Visualization and Participation. Heidelberg, Wichmann.

Tversky, B., Morrison, J. B. & Betrancourt, M. (2002), *Animation: can it facilitate?.* In: International Journal of Human-Computer Studies, 57.

Using 3D Virtual GeoDesigns for Exploring the Economic Value of Alternative Green Infrastructure Options

Sigrid HEHL-LANGE, Lewis GILL, John HENNEBERRY, Berna KESKIN, Eckart LANGE, Ian Caleb MELL and Ed MORGAN

1 Introduction

Green spaces and green infrastructure are perceived as important factors promoting quality of life in cities and towns. The provision of green infrastructure does however have attached costs for planning, realisation and maintenance. The VALUE research project ('Valuing Attractive Landscapes in the Urban Economy'), funded by the EU Interreg IVB programme, aims to establish an economic value for investments in green infrastructure. The Sheffield case study undertaken by VALUE focussed on a number of green investments along the River Don in The Wicker on Blonk Street.

The investments are including a new footbridge over the River Don and walkways along the river to facilitate access and movement for pedestrians and cyclists, further sluice gates to react to flooding events caused by extreme weather and the planting of new trees along the river. These investment projects were developed and realized by the South Yorkshire Forest Partnership (SYFP) utilising VALUE funding. The VALUE investments in Blonk Street were completed in 2011 (see Fig. 2). Further actions to prevent flooding through river channel modifications were also delivered. These resources were not funded through VALUE and are not considered elements of the VALUE investment.

2 Methodology

Stated preference techniques are a common method of undertaking economic valuations of non-market resources, such as environmental goods (BATEMAN et al. 2002, LAING et al. 2005). A stated preference technique was developed to underpin a quantitative survey used to estimate the Willingness-to-Pay (WTP) of residents, commuters, business owners and employees for alternative green infrastructure development scenarios in a case study site at the Wicker in Sheffield, UK.

2.1 Quantitative Survey

The economic evaluation undertaken for the Sheffield VALUE investment utilised a contingent valuation (CV) experiment methodology. In the CV survey visual cue cards showing 3D computer-generated images conveying the nature of the proposed investment on Blonk Street were used. Based on computer-generated images the WTP of different user and interest groups for the different options were assessed in a survey. It is considered within the research literature that cue cards showing multiple-choice responses, Likert

Scales and 3D visualisations offer broader scope and choice to the researcher than maps and provide more robust data for analysis (CROMPTON 2001). This is compared to the use of structured written statements outlining the investments proposed (LINDSEY 1994).

The WTP elicitation question was framed as a regular monthly increased payment in rent or mortgage. In the questionnaire the average cost per month of rent/mortgage for a two-bedroom apartment in the case study area was stated. The respondents were then asked: If you were living in the area and it had the characteristics shown in the image, how much extra would you be willing to pay in rent/mortgage each month to have this view? Respondents were therefore provided with the opportunity to state a specific payment. Alternatively they were also able to express the following options: nothing, I would not pay, refused, don't know. 'Nothing' was classified as genuine zero. The three other options were identified as protest zero (BERNATH & ROSCHEWITZ 2008).

An on-site survey was conducted by Ipsos-Mori, a social research company, over a six-week period in August and September 2011 using a face-to-face interview technique. In total 1939 people were asked to participate in the study and 510 responses were achieved, representing a response rate of 26 %.

2.2 Qualitative Survey

As a follow-up to the quantitative survey, qualitative workshops were organized and held in November 2011. The sessions each lasted 1-1.5 hours. In these sessions we used PowerPoint presentations with static and dynamic visualisations on a large screen and additional real time visualisations. The venue was a seminar room in a University building. Participants were recruited through the quantitative survey conducted in summer 2011 by asking respondents at the end of the questionnaire if they would have an interest in attending follow up workshops investigating green infrastructure. 186 people gave us their name and telephone number of which 25 people declared that they would have the time and interest to participate. One of the workshops was held on a Saturday morning, two others were held in the evening on weekdays.

Several people are required to facilitate the visualisation workshop (HEHL-LANGE & LANGE 2005, SCHROTH et al. 2011). In this case one person acted as moderator and also represented the hosting institution, another person acted as the computer facilitator presenting all the visualisations, two additional people were used to take minutes and document the workshop by taking photos and finally there were the invited participants. For the preparation of a visualisation workshop considerable more time than for a conventional workshop was required (SCHROTH 2009). This reflects the time needed to prepare the visual inputs of the workshop including taking views from each computer model for static representations, preparing pre-recorded animations as walkthroughs and identifying how to present the real time interactive visualisation. Sufficient time was also needed to set up the hardware and to verify that all the visualisations ran properly. Seating was U-shaped around a table oriented towards the screen. Lighting had to be dark enough to see the projected landscape visualisations but bright enough for participants, while watching them, to fill out a questionnaire with the same core questions as were used in the quantitative survey. Brighter lighting also supports the discussions among the participants (SCHROTH 2009). Two Info sheets, one with information about the VALUE project and one reflecting the URSULA project, which also contained a consent form were distributed at the beginning of the workshop. As a warm-up exercise, participants were shown firstly all images in random order, an animation (pre-recorded flythrough) and a demonstration of what could be expected in the real time navigation. We used in each workshop in addition to the status quo a different alternative scenario, once 'Past', once 'Sheffield City Council' and once 'Streets'. It would have been too time consuming to show each group all the scenarios. Because of time constraints, only the facilitator could navigate through the real time landscape models. Instead of the 8 to 9 expected participants per workshop only a total of 10 participants took part in the three workshop events.

2.3 3D Visualisation

The 3D visualisations used in the surveys were part of the URSULA (Urban Rivers and Sustainable Living Agendas) project, a joint research project utilising the same case study site in Sheffield. Its goal is to provide innovations, tools and supporting evidence for the redevelopment of urban river corridors to create places where people want to live and work, now and in the future. A virtual landscape model of the case study site was modelled for the status quo (Fig. 2) alongside three different future scenarios. The 'Sheffield City Council' model (Fig. 3) represents the planning proposal of the Sheffield City Council. There are two further scenarios 'Streets' (Fig. 4) and 'Floods'. The scenario 'Floods' was not used in the surveys of Blonk Street but instead a model representing the past (Fig. 1), before the VALUE investments – the new footbridge and the walkways – were built. Visualisations or future scenarios are helpful, as in this way people can share the same vision (LANGE et al. 2008, LANGE & HEHL-LANGE 2010).

Within the URSULA case study site the virtual landscape model was built using different software. Simmetry3d, a real-time visualisation software, which runs on a PC was used for the interactive landscape visualisation. Within Simmetry3d interactive walkthroughs at eye-level are possible based on the technologies used in the computer game industry (MORGAN et al. 2009). Simmetry3d has excellent compatibility for importing GIS data, image data, and for importing data from the other software used in the URSULA project: SketchUp and LENNÉ3D. From vector GIS data building footprints were used as the basis for manually constructing building models in SketchUp. Also, in SketchUp the perspective corrected photographs were draped as textures onto the building volumes. Building models were also built from photographs using the "PhotoMatch" feature in SketchUp (MORGAN et al. 2009). The site plans of alternative scenarios were drawn by the URSULA design team conventionally by pen on paper. In SketchUp these two-dimensional site layout plans were digitized by hand, assigned with land usage and the buildings were extruded up into three dimensions by creating massing models, with floor areas modelled. Simmetry3d has its own modelling tool and library for vegetation; the focus is on trees but not on shrubs and perennials. As the software LENNÉ3D has very realistic species specific tree and plant models, Simmetry3d has built-in compatibility for importing LENNÉ3D vegetation. In the URSULA virtual landscape model vegetation of both software packages are used. The virtual model of the Sheffield-Wicker site consists of a digital terrain model (DTM) provided by the Environment Agency with a horizontal resolution of 0.5 m and a height accuracy of 0.25 m on which an aerial photo of 0.2 m resolution from Cities Revealed was draped within Simmetry3d. All built form like building footprint, roads, paths and river channels and land usage date were imported as GIS vector data from Ordnance Survey MasterMap. All these data were available through licensing agreements through the University.

Fig. 1: Past, before the VALUE investments

Fig. 2: Status quo, with the VALUE investments

Fig. 3: Scenario, 'Sheffield City Council' with the VALUE investment and additional greener investment

Fig. 4: Scenario 'Streets'

Fig. 5: On request of a participant the computer facilitator navigates over the new footbridge

3 Results

The analysis of the quantitative survey shows that preference and WTP are clearly linked with greener investments options (see Fig. 6). To find out respondents' preference and opinion about green infrastructure they had to give answers to the statements 'I like the look of this image' and 'I think there is enough greenery in this image'. Therefore respondents had to check a box on a five-point rating scale from 'I strongly agree' to 'I strongly disagree' in the questionnaire. When the scenarios are compared the level of vegetation with which they agree appears to have the biggest impact. Interestingly respondents were satisfied the most with the 'Past', before the VALUE investments and were willing to pay £ 10.81, e.g. the most.

In the qualitative survey the questions about preference and greenery had to be adapted, that they were equally valid for the static as well for the dynamic visualisations. Instead of 'I like the look of this image' we used 'I like the urban river landscape'. And for the vegetation we wrote simply 'I think there is enough greenery'.

As there were only a limited number of participants in the workshops a quantitative analysis was not possible. But there is the same tendency, that the 'Past', indicates a higher preference against the two other options. For the Status quo, which was the only model evaluated by all 10 respondents, we can suggest that preference is rated higher when they watched the pre-recorded animations, and event better when they experienced the real-time navigation. One of the respondents was willing to pay £ 25 for the 'Past', equally in case of an image, an animation or the real time navigation. In the qualitative survey there are more protest zero than genuine zero in the WTP answers.

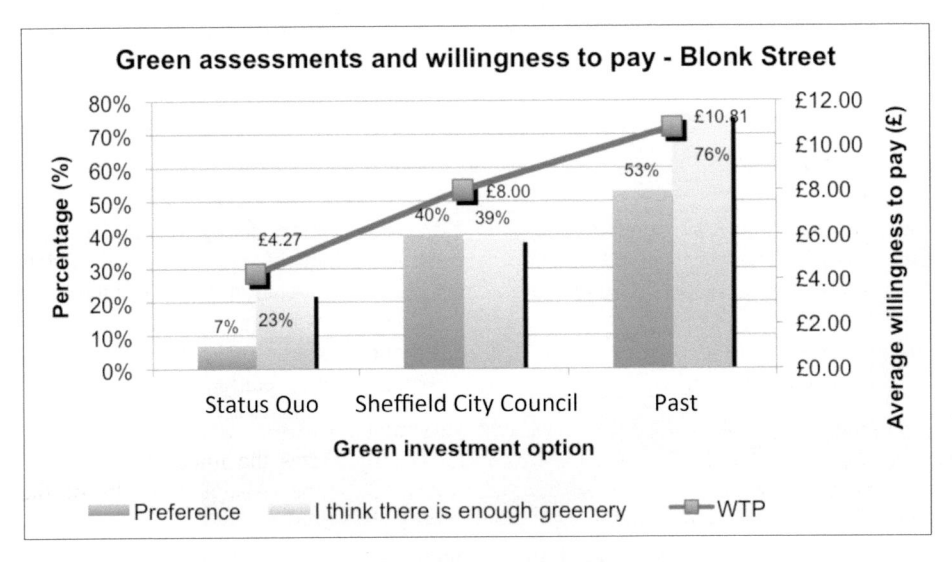

Fig. 6: Preference, assessment of greenness and WTP

In the discussions of the three workshops it was clear that the majority of participants preferred more natural and wild vegetation. In general their opinion was, the more green the better. This can help to explain the preferences for the 'Past' alternative, where the river channel is fairly unmanaged and shows a high proportion of spontaneous vegetation with mostly willows combined with invasive species growing at the river banks. Consequently the Status quo, showing the VALUE investments – the new footbridge and the walkways – was the least preferred option. Most of the spontaneous vegetation and invasive plant species such as Japanese Knotweed (Fallopia japonica) and Himalayan Balsam (Impatiens glandulifera) in the riverbed was recently taken out by the River Stewardship Company (WILD et al. 2008). Another reason that was explored in the discussions was that although the new footbridge was completed it was still closed for the public. This meant that the new footbridge was at the time of our surveys visually available but not physically accessible. Walkers and cyclists therefore gained no additional benefits from the investment. The new walkways along the river are presently not in an attractive state as on the other side of the

River Don new buildings are under construction. One participant also argued that the new footbridge blocks the view.

Two participants only evaluated the Scenario 'Sheffield City Council'. In addition to the three models used for the quantitative research we also integrated for the qualitative survey a fourth model, the Scenario 'Streets', to enquire whether the VALUE investment would profit more from natural wild vegetation. The scenario 'Streets' has like the 'Past' spontaneous vegetation in the riverbed. The explanations provided by five respondents who evaluated the Scenario 'Streets' were similar to the result of the scenario 'Past'. The one respondent who was willing to pay £ 25 for the 'Past' was equally willing to pay the same amount for the scenario 'Streets'. This respondent reported that she used to live in a house with no view at all. Now their house has a gorgeous view. When a council officer came to allocate the tax band for her house, he also commented on the view. And our participant said: You will not charge me for this view. For her it was clear that there is a direct relationship between the quality of the view and WTP.

4 Conclusions

The greener the landscape models are, the higher they are rated by respondents for preference and WTP values. The workshop data suggests, that those visualisation models that contain a greater proportion of natural unmanaged vegetation are considered by respondents to have higher ecological benefits. What was not tested was a scenario with equally the same amount of green as in the 'Past' and 'Streets' but with a lot of new formally planted and managed trees instead of the spontaneous vegetation.

The dynamic visualisations were appreciated very much by the participants and were deemed helpful to the discussions. DANAHY (2001) emphasizes the importance of peripheral vision for person's visual experience in the landscape. Also the results of the evaluation were in some cases higher with the dynamic visualisations.

The selection of the people for the workshops seems to be random, however it is a very selective process. We only had people who are very interested in the site; some came from far away on their own cost. Because of the limited number of participants and the bias in the sample more general conclusions are not supported by this study.

5 Acknowledgement

The VALUE project is funded by the European Regional Development Fund (ERDF) under its INTERREG IVB North West Europe programme. The URSULA project is funded by EPSRC (Engineering and Physical Sciences Research Council). Laurence Pattacini, part of the design team of URSULA, developed the scenario 'Streets'. A big thank to all participants of the survey, especially to those who participated in both surveys.

References

Bateman, I. J., Carson, R. T., Day, B., Hanemann, M., Hanleys, N., Hett, T., Jones-Lee, M., Loomes, G., Mourato, S., Ozdemiroglu, E., Pearce, D., Sugden, R. & Swanson, J. (2002), *Economic valuation with stated preference techniques: a manual*. Cheltenham, UK, E. Elgar.

Bernath, K. & Roschewitz, A. (2008), *Recreational benefits of urban forests: Explaining visitors' willingness to pay in the context of the theory of planned behavior*. In: Journal Environmental Management, 89 (3), 155-166.

Crompton, J. L. (2001), *The effects of different types of information messages on perceptions of price and stated willingness-to-pay. In:* Journal of Leisure Research, 33 (1), 1-31.

Danahy, J. W. (2001), *Technology for dynamic viewing and peripheral vision in landscape visualization*. In: Landscape and Urban Planning, 54, 125-137

Deliverance Software, Simmetry 3d, version 3.1. Available at: http://www.simmetry3d.com.

Google, SketchUp, version 1.6. Available at: http://sketchup.google.com.

Hehl-Lange, S. & Lange, E. (2005), *Ein partizipativer, computergestützter Planungsansatz für ein Windenergieprojekt mit Hilfe eines virtuellen Landschaftsmodells*. In: Natur & Landschaft, 80 (4), 148-153.

Laing, R., Davies, A. M. & Scott, S. (2005), *Combining visualization with choice experimentation in the built environment*. In: Bishop, I. & Lange, E. (Eds.), Visualization in Landscape and Environmental Planning. Technology and Applications. London, Spon Press, Taylor & Francis, 212-219.

Lange, E., Hehl-Lange, S. & Brewer, M. J. (2008), *Scenario-visualization for the assessment of perceived green space qualities at the urban-rural fringe*. In: Journal of Environmental Management, 89 (3), 245-256.

Lange, E. & Hehl-Lange, S. (2010), *Making visions visible for long-term landscape management*. In: Futures, 42 (7), 693-699.

Lenné3D, Flora3D. Available at: http://www.lenne3d.com.

Lindsey, G. (1994), *Market Models, Protest Bids, and Outliers in Contingent Valuation*. In: Journal of Water Resources Planning and Management, 120 (1), 121-129.

Morgan, E., Gill, L., Lange, E. & Romano, D. (2009), *Rapid Prototyping of Urban River Corridors Using 3D Interactive, Real-time Graphics*. In: Buhmann, E., Kieferle, J., Pietsch, M., Paar, P. & Kretzler, E. (Eds.), Proceedings Digital Landscape Architecture 2009, 21 – 23 May, 2009, Anhalt University of Applied Sciences, Malta, 198-205.

Schroth, O. (2009), *From Information to Participation – Interactive Landscape Visualization as a Tool for Collaborative Planning*. Dissertation, ETH Zürich, vdf Hochschulverlag, 234 pp.

Schroth, O., Wissen Hayek, U., Lange, E., Sheppard, S. R. J. & Schmid, W. A. (2011), *Multiple-case study of landscape visualizations as a tool in transdisciplinary planning workshops*. In: Landscape Journal, 30, 53-71.

URSULA Urban River Corridors and Sustainable Living Agendas http://www.ursula.ac.uk/.

Wild, T. C., Ogden, S. & Lerner, D. N. (2008), *An innovative partnership response to the management of urban river corridors – Sheffield's River Stewardship Company*. 11th International Conference on Urban Drainage, Edinburgh, Scotland, UK.

Transdisciplinary Collaboration Platform Based on GeoDesign for Securing Urban Quality

Ulrike WISSEN HAYEK, Noemi NEUENSCHWANDER,
Antje KUNZE, Jan HALATSCH, Timo VON WIRTH,
Adrienne GRÊT-REGAMEY and Gerhard SCHMITT

1 Introduction

Guiding urban agglomerations towards more sustainable development is a great challenge (UN-HABITAT 2009). Existing concepts and rules for generating urban patterns of high urban quality do not work in today's agglomerations and societies, and concepts for new (sub)urban qualities are needed (CARMONA et al. 2006, MODARRES & KIRBY 2010). New approaches need to facilitate collaborations between science and a variety of public and private stakeholders (DE JONG & SPAANS 2009, SCHOLZ 2011, KUNZE et al. 2011), enabling them to make decisions on urban development taking into account multiple dimensions and values, alternative development possibilities, performance indicators and uncertainties. Furthermore, developments in complex urban systems take place at multiple planning scales and time frames and therefore demand for a solid mutual understanding of spatially explicit information (DE JONG & SPAANS 2009, ERVIN 2011, NEUENSCHWANDER et al., 2011).

Simulation and modelling in a functional way might support such interactive and iterative collaboration processes. This process should be characterized by cooperation of non-academic stakeholders (society) and science, mutual learning and knowledge integration (SCHOLZ 2000). Such processes are defined as transdisciplinary, i.e., they integrate "disciplinary scientific and non-academic knowledge into an inter- and transdisciplinary research framework for complex problem solving" (ENENGEL et al. 2012, 106).

This paper presents the concept of a collaborative urban planning platform supported by a suitable combination of simulation tools for identifying sustainable urban patterns. Preliminary results of setting up this platform are shown exemplarily with a focus on urban green space patterns. A central element for the collaboration process is a library of quality criteria and rules for urban patterns from coupled design, social, ecological and economic perspectives. These specifications are encoded and made directly applicable as digital modelling rules. Combining disciplinary rules to interdisciplinary rule sets according to the situation of a specific urban case, GIS-based 3D visualizations of the urban patterns are generated and can be interactively explored. In this paper, details on gathering and operationalizing specific rules are given on the example of suburban green space patterns for the Swiss case region Limmattal, an urban agglomeration in the greater Zurich area.

2 Material and Methods

2.1 Urban quality

As urban quality depends on the respective disciplinary perspective, it has to be defined according to all dimensions. How an urban area presents itself can be described with the dimensions "Structure", "Gestalt" and "Form" of the natural and built environment (ALEXANDER et al. 1977, SCHAEFER, 2011). "Structure" is determined by the topography, the distribution of land uses and land use densities, and the infrastructure. "Gestalt" is defined as the character and identity of each quarter and settlement. Urban design, architecture and the design of open spaces give the environment a culturally shaped "Form" (SCHAEFER 2011).

Quality of life in urban areas has an objective and subjective component. Within objective criteria for quality of life, spatial and structural characteristics but also functional, social aspects in the urban realm play a crucial role. From an ecological or landscape perspective, the services provided by the urban ecosystems can significantly support human wellbeing (BOLUND & HUNHAMMAR 1999). Major problems in urban areas caused by traffic and soil sealing, e.g., air pollution, wastewater disposal, urban heat islands, and noise, have to be solved locally. The urban ecosystems can contribute to reducing these problems if on one hand structure, age and composition of green spaces is adequate (quality). On the other hand, size, amount and spatial distribution on the regional and local level (quantity) is important, which can be increased, e.g., by enhancing ecological networks (OPDAM et al. 2006). In addition to services rather depending on the functioning of the environmental systems, cultural services such as possibilities for people's identification are relevant ("Gestalt" and "Form"). For securing and sustaining these urban qualities, an intense societal examination of landscape aesthetics is required (RODEWALD 2011).

Ecological objectives and solutions, however, cannot be viewed independent of social, economic and design requirements and solutions. The obvious lack in quality of current cities and agglomerations is the result of insufficient implementation of interdisciplinary concepts for urban development in practice (KURATH 2011).

2.2 Traditional city planning methods

The complexity of urban and regional systems, the speed of urbanisation, and the uncertainty of developments with growing problems such as transport bottlenecks, housing shortage and shortage of space for recreation, exceed the abilities of traditional city planning methods (RATCLIFFE & KRAWCZYK 2011). EISINGER (2009) and KURATH (2011) argue that securing urban quality requires both, abolishing sectoral thinking and robust processes of "producing the city". New planning methods are needed, which help in designing novel urban patterns, securing urban quality. Such transdisciplinary research settings should organize a future oriented mutual sustainability learning and capacity building process among regional stakeholders, urban planning experts and scientists (see SCHOLZ 2011, 379).

2.3 Current simulation and modelling tools for urban development

Urban typologies are commonly used by urban planners for defining flexible frameworks in urban development (SCHIRMER et al. 2011). 3D visualization of the urban landscape is a straightforward method to provide these rule sets in a comprehensive form for stakeholders from science and practice. Particularly for fast interactive urban visualization of large areas, ESRI's CityEngine system has shown to be a valuable tool for encoding typological rules to shape grammars, which generate 3D geometries of the urban layout at various scales (HALATSCH et al. 2008, SCHIRMER et al. 2011).

In addition, integrated land use, transportation and environmental planning requires evaluation of different scenarios and policies, testing land use regulations, development subsidies or costs in relation to the transport system and the land use patterns. The complexity of spatial interactions has motivated the development of a number of urban simulation systems that forecast the behaviour of each component. For example, relations between land use patterns and traffic congestion can be analysed implementing agent-based transport simulation models (www.matsim.org) or computable general equilibrium models based on economic causality (RUTHERFORD 1995). UrbanSim is one example for a sophisticated integrated urban simulation model providing reasonable and intuitive results to policy scenario inputs (WADDELL 2011). Current research also develops models based on satisfaction of human needs in the urban living environment. This social dimension in urban simulation has only recently gained further attention (FEITOSA et al. 2011).

Combining these simulation and visualization tools can facilitate a transdisciplinary urban collaboration platform (SCHROTH et al. 2011) by continuously accompanying the development of urban systems in a design-simulation-collaborative cycle.

3 Region Limmattal – An Agglomeration of Zurich (CH)

The Limmattal is the valley of the river Limmat bordered with rolling hills and extending over 24 km from east to west from Zurich city centre to the city of Baden (Fig. 2). It comprises an area of about 18'600 hectares, a population of about 165'000 inhabitants and 118'000 employees in the year 2010.

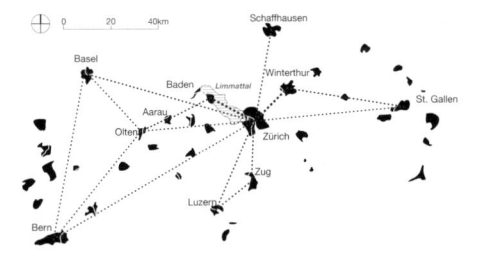

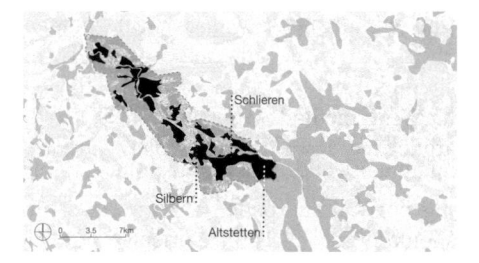

Fig. 1: Left: Position of the "Region Limmattal" in the Swiss urban system (nodes: settlement areas) and strength of its networking (lines: technical infrastructure). Right: Schlieren, Silbern and Altstetten are focal areas allowing further analyses on local level (black/dark grey = settlement areas).

The Limmattal is a typical suburban region challenged by the negative effects of urban sprawl, directly perceivable by aesthetic degradation, loss of environmental quality, cultural heritage and reduced quality of life (SCHUMACHER et al., 2004). The region, however, experiences high migration rates, with an estimated overall increase of about 30'000 people and 20'000 jobs by 2030. A comprehensive development concept for the whole region, guiding sustainable urban development paths, does not yet exist. Thus, the region provides a suitable case study for elaborating a regional development vision implementing the urban collaboration platform.

4 Results

4.1 Concept of a transdisciplinary urban collaboration platform

The overall goal is the development of a collaborative planning platform, where stakeholders of all planning levels and disciplines collaboratively analyse the current urban situation, develop scenarios and assess those alternative urban development patterns. This is supported by interactive 3D visualizations and indicators for urban quality. Through an iterative assessment and adaptation process of the virtual urban patterns on local and regional scales, sustainable urban patterns can be identified in a transdisciplinary dialogue (Fig. 1).

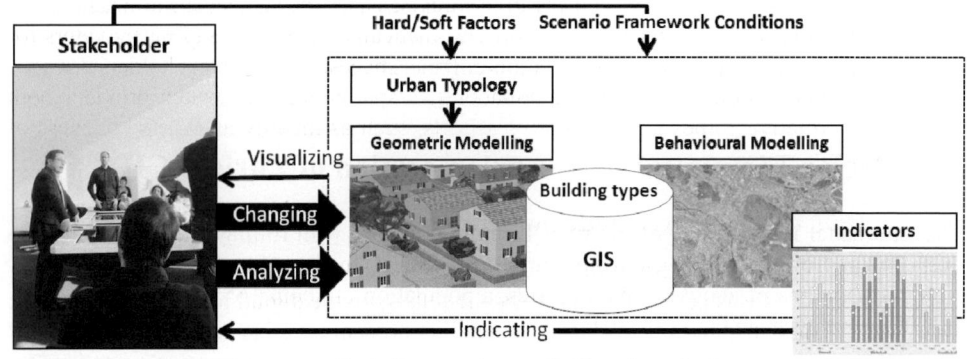

Fig. 2: Concept of the transdisciplinary urban collaboration platform

First, the diverse criteria for urban quality are made interoperable. Hard factors of building regulations (e.g. street widths, building heights, or target values for recreational space per inhabitant) and soft factors, such as elements for identification (e.g. cultural sites), are formatted systematically. In order to support clear imaginations of the future urban patterns, it is important to visualize and assess these patterns from the regional (conceptual) up to the building level (experiential). To this end we adapt and implement an urban typology (CATS 2009) and define 10 building types. An urban area is classified into building types and patterns. In addition, significant aspects of the existing buildings and respective green space types are mapped in the field. With this typology hard and soft factors are generalized into design principles that are transferable to any urban area.

In the next step, the general quantitative and qualitative design pattern rules are encoded to Computer Graphics Architecture (CGA) shape grammar rules. In ESRI's CityEngine system (GIS-based procedural urban model) these are then implemented for geometric 3D modelling of urban patterns with roads, buildings and green spaces. These digital modelling rules provide a library of urban patterns, which initially include design and ecological specifications. Implementing the procedural urban model and changing interactively these CGA rules leads to new urban patterns.

The building types are the least common denominator in the GIS database allowing for linking the results of the behavioural modelling to the geometric, procedural urban model. We employ a land use transportation microsimulation (combination of UrbanSim and MATSim) as well as a Computable General Equilibrium Model (see section 2.3) to simulate and test possible consequences of policies, alternative transport systems and land use scenarios under complex system conditions. The modelling results deliver plausible future 2D urban patterns on a quarter to parcel level for a case study area, as well as indicator values, e.g., employment, population, traffic and congestion times. The 2D pattern maps are used as input to the geometric model.

When running the collaboration platform, the stakeholders define scenario framework conditions, which are operationalized for the modelling systems. 3D visualizations of the urban development scenarios as well as corresponding indicator values are used to conduct detailed analyses of urban design, economic, ecological and social impacts. The latter serve as a "reality check" for the scenarios, discovering how good different social needs of target groups are fulfilled. Furthermore, a trade-off model based on the stakeholders' weighting of urban qualities based on indicator values can automatically change the input parameters for the size and distribution of buildings and urban green spaces. Various stakeholders can thus analyse possible options available in a collaborative spatial analysis, which provides both debatable normative values and qualitative aspects, such as identity and distinctiveness of new urban patterns.

4.2   Setting up the digital modelling library

In this paper, preliminary results of integrating ecological criteria into the urban typology are presented. This was part of the first steps for setting up the digital modelling library, i.e. the digital CGA shape grammar rules for generating procedural urban patterns, which allow for integrating specifications of spatial attributes from different disciplinary perspectives.

Table 1 shows an example of how specifications of building regulations and information on the green space's potential for providing relevant ecosystem services (see Section 2.1) can be linked to the building type "Multi-family Houses". Furthermore, an example of a typical, generalized pattern of multi-family houses and their respective green spaces in the Limmattal is given. Exemplars of several alternative patterns for this building type in the case study area are mapped and specified. In addition, the typical vegetation of the green space type is mapped. The information on relevant urban ecosystem services that can be provided by the green space type is the basis for defining the required qualitative and quantitative characteristics of the green spaces (= rules). According to these characteristics the green space type can be optimized in quality.

Building regulations (selection) and a generalized pattern	Typical vegetation species (selection)	Relevant ES potentially provided	Requirements for increasing ES provision (selected aspects)
Building – Street: 6m Trees – Street: 2m Hedge – Street: 0.5m Flat roofs have to be vegetated Maximal amount of paved area (without building area): 33%	Lawn • Lolium perenne • Poa pratensis Trees • Prunus serrulata • Abies spec. • Fraxinus excelsior • Acer platanoides Hedges • Prunus laurocerasus • Carpinus betulus • Thuja spec. Shrubs • Pinus mugo • Corylus avellana • Sorbus aucuparia	Microclimate regulation Aesthetics/ Recreation Identity Habitat Infiltration	High amount of vegetation, particularly large trees;… Diversity, specific character and scenic beauty; accessibility… High vegetation age; existence of impressive single trees;… High amount of native species; structural diversity;… High amount of un-paved area;…

Tab. 1: Illustration of the urban typology on the example of the building type "Multi-family Houses" (ES = Ecosystem Services)

The patterns and their specifications were encoded to CGA shape grammars following the approach of WISSEN HAYEK et al. (2010). Executing the shape grammar rules on given parcels in the virtual model of the case study area, ESRI's CityEngine system generates 3D urban patterns (Figure 4). In a next step, relevant quantitative indicators for assessing the quality of the green spaces will be linked to these visualizations allowing for an optimization of the patterns in participatory processes (NEUENSCHWANDER et al. 2011). Furthermore, the grammar will be augmented with rules of further disciplines.

Fig. 3: 3D visualization of current and alternative future urban patterns suitable for interactive exploration in the procedural urban model

5   Conclusions and Outlook

For sustainable regional and urban development, norms and values regarding objectives of all disciplines on the local up to the regional scale should be taken into account by various actor groups such as local authorities, planners, urban designers and architects, real-estate developers, nature protection, social or cultural organisations. The presented urban collaboration platform is thought to facilitate developing objectives for transformation of suburban areas towards a more sustainable development. It can provide the information on essential interdisciplinary norms and values – as shown on the example of recreational and ecological objectives – in form of digital modelling rules and 3D urban visualizations illustrating the spatially explicit implementation of the rules. Through allowing the actors to define desired values for specific needs, such as amount of building floor space or recreational space, the consequence of individual demands on all other needs can be demonstrated with the 3D visualizations, which will be linked with quantitative indicators for the local and regional scale. In this way, mutual learning processes of all actors can be initiated. The model cannot design sustainable urban patterns. Using such simulations in collaborative planning processes, however, can support taking into account a rather high degree of urban complexity required for securing urban quality.

References

Alexander, C., Ishikawa, S. & M. Silverstein (1977), *A Pattern Language: Towns, Buildings, Construction.* New York, Oxford University Press.

Bolund, P. & Hunhammar, S. (1999), *Ecosystem Services in Urban Areas.* In: Ecological Economics, 29, 293-301.

Carmona, M., Marshall, S. & Stevens, Q. (2006), *Design codes: their use and potential.* In: Progress in Planning, 65, 209-289.

CATS – Center for Applied Transect Studies (2009), *SmartCode Version 9.2.* The Town Paper, Gaithersburg, USA. http://transect.org/codes.html (accessed 17.01.2012).

De Jong, J. & Spaans, M. (2009), *Trade-offs at a regional level in spatial planning: Two case studies as a source for inspiration.* In: Land Use Policy, 26, 368-379.

Eisinger, A. (2009), *Urban Reports: Urban strategies and visions in mid-sized cities in a local and global context.* gta Verlag, Zurich.

Enengel, B., Muhar, A., Penker, M., Freyer, B., Drlik, S. & Ritter, R. (2012), *Co-production of knowledge in transdisciplinary doctoral theses on landscape development – An analysis of actor roles and knowledge types in different research phases.* In: Landscape and Urban Planning, 105, 106-117.

Ervin, S. (2011), *A System for GeoDesign.* In: Buhmann et al. (Eds.), Peer-reviewed Proceedings Digital Landscape Architecture 2011, Anhalt University of Applied Sciences. Berlin/Offenbach, Wichmann, 145-154.

Feitosa, F. F., Le, Q. B. & Vlek, P. L. G. (2011), *Multi-agent simulator for urban segregation (MASUS), A tool to explore alternatives for promoting inclusive cities.* In: Computers, Environment and Urban Systems, 35 (2), 104-115.

Halatsch, J., Kunze, A. & Schmitt, G. (2008), *Using Shape Grammars for Master Planning.* In: Gero, J. S. (Ed.), Design Computing and Cognition DCC'08. Springer, 655-673.

Kunze, A., Halatsch, J., Vanegas, C. & Jacobi, M. M. (2011), *A Conceptual Participatory Design Framework for Urban Planning: The case study workshop 'World Cup 2014 Urban Scenarios', Porto Alegre, Brazil.* eCAADe, University of Ljubljana (SI), 895-903.

Kurath, S. (2011), *Stadtlandschaften Entwerfen? Grenzen und Chancen der Planung im Spiegel der städtebaulichen Praxis.* Bielefeld transcript Verlag.

Modarres, A. & Kirby, A. (2010), *The suburban question: Notes for a research program.* In: Cities, 27, 114-121.

Neuenschwander, N., Wissen Hayek, U. & Grêt-Regamey, A. (2011), *GIS-based 3D Urban Modeling Framework Integrating Constraints and Benefits of Ecosystems for Participatory Optimization of Urban Green Space Patterns.* In: Schrenk, M. (Ed.), REAL CORP 2011, Essen, Germany, 365-374.

Opdam, P., Steingröver, E. & van Rooij, S. (2006), *Ecological networks: A spatial concept for multi-actor planning of sustainable landscapes.* In: Landscape and Urban Planning, 75, 322-332.

Ratcliffe, J. & Krawczyk, E. (2011), *Imagineering city futures: The use of prospective through scenarios in urban planning.* In: Futures, 43 (7), 642-653.

Rodewald, R. (2011), *Kulturlandschaft zwischen Ästhetik, Biodiversität und Geschichte. Was ist eine schöne Landschaft?* In: Max Himmelheber-Stiftung (Hrsg.), Scheidewege, Jahresschrift für skeptisches Denken. Stuttgart, Hirzel, 231-240.

Rutherford, T. F. (1995), *Extensions of GAMS for complementarity problems arising in applied economics.* In: Journal of Economic Dynamics and Control, 19 (8), 1299-1324.

Schaefer, M. (2011), *"Standortmosaik Zürich" or the ecology of access.* In: anthos, 2, 50-53.

Schirmer, P. & Kawagishi, N. (2011), *Using shape grammars as a rule based approach in urban planning – a report on practice.* eCAADe, University of Ljubljana, Architecture (SI), 116-124.

Scholz, R. W. (2000), *Mutual learning as a basic principle of transdisciplinarity.* In: Scholz, R. W., Häberli, R., Bill, A. & Welti, M. (Eds.), Transdisciplinarity: Joint Problem-Solving among Science, Technology and Society. Zurich, Haffmans Sachbuch Verlag, Vol. Workbook II, 13-17.

Scholz, R. W. (2011), *Environmental literacy in science and society: from knowledge to decisions.* Cambridge, Cambridge University Press.

Schroth, O., Wissen Hayek, U., Lange, E., Sheppard, S. R. J. & Schmid, W. A. (2011), *Multiple-Case Study of Landscape Visualizations as a Tool in Transdisciplinary Planning Workshops.* In. Landscape Journal, 30 (1), 53-71.

Schumacher, M., Koch. M. & Ruegg, J. (2004), *The Zurich Limmattal. Steps of a servant valley towards emancipation.* In: Dubois-Taine, G. (Ed.), From Helsinki to Nicosia, PUCA and COST C10, Paris, 215-236.

United Nations Human Settlement Programme (UN-HABITAT) (2009), *Planning Sustainable Cities – Global Report on Human Settlements 2009.* London, Earthscan.

Waddell, P. (2011), *Integrated Land Use and Transportation Planning and Modeling: Addressing Challenges in Research and Practice.* In: Transport Reviews, 31 (2), 209-229.

Wissen Hayek, U., Neuenschwander, N., Halatsch, J. & Grêt-Regamey, A. (2010), *Procedural modeling of urban green space pattern designs taking into account ecological parameters.* 28th eCAADe Conference, ETH Zurich (CH), 339-347.

Adaptive e-Learning DLA Course – A Framework

Athanasios STYLIADIS and Nikolaos HASANAGAS

1 Introduction

Adaptive, intelligence or personalized e-learning is recognized as, probably, the most interesting research areas in distance learning on-line Web-based Education (WBE) (KIM 2007, MARLOW et al. 2009). In particular, for the WBE LA education, so far, there are not well-defined and commonly accepted rules on how the learning material should be designed (metadata-based content development), organized in reusable Learning Objects (LO), selected and sequenced to make *"instructional on-line sense"* in a Web-based Digital Landscape Architecture course (WBE DLA). Hence, the goal of this paper is to shorten the gap between the established traditional e-learning management systems and the modern adaptive & intelligent WBE tutoring for the benefit of the LA education (WATSON 2009).

In particular, the proposed WBE DLA framework and methodology incorporates a number of **reusable LO** (related to topics which are common in LA courses, e.g. *digital design*) with learners' profiles and techniques for personalized content **selection** and finally for adaptive course **sequencing**. Actually, the proposed methodology benefits from a LO's novel metadata-based design and XML implementation with embedded LA functionality, according to the cognitive style of learning needs and preferences in LA education. The proposed learning rules are *generic* (i.e. domain-, view- and user-independent), hosting, accordingly, lecturing functionality for relative courses (e.g. WBE Digital Architecture or Geomatics Engineering courses).

In paper's Second Section (WBE DLA: The Architecture) the proposed architecture for an adaptive & intelligent WBE DLA is presented. In Third Section (WBE DLA: Content Development, Selection & Sequencing) the novel metadata-based structure of the proposed reusable LO is presented and the techniques for personalized content selection and adaptive course sequencing are discussed. Finally, in Conclusions and Outlook possible future extensions are presented.

2 WBE DLA: The Architecture

In the proposed WBE DLA personalized e-learning framework, seven key WBE components (modules) have been identified, namely: (1) Data acquisition; (2) Content development (data processing based on reusable LO with metadata tags); (3) local or *Internet databases* (reusable LO repositories); (4) *Personalized content selection* (LO retrieval & analysis according to learners' profiles); (5) *Adaptive on-demand sequencing of the selected LO*; (6) *The intelligent WBE DLA course* (GUI format publication); and (7) *The e-Learning Management Control Software* (XML or PHP coding implementation). The interactions between these seven WBE DLA modules are shown in Figure 1.

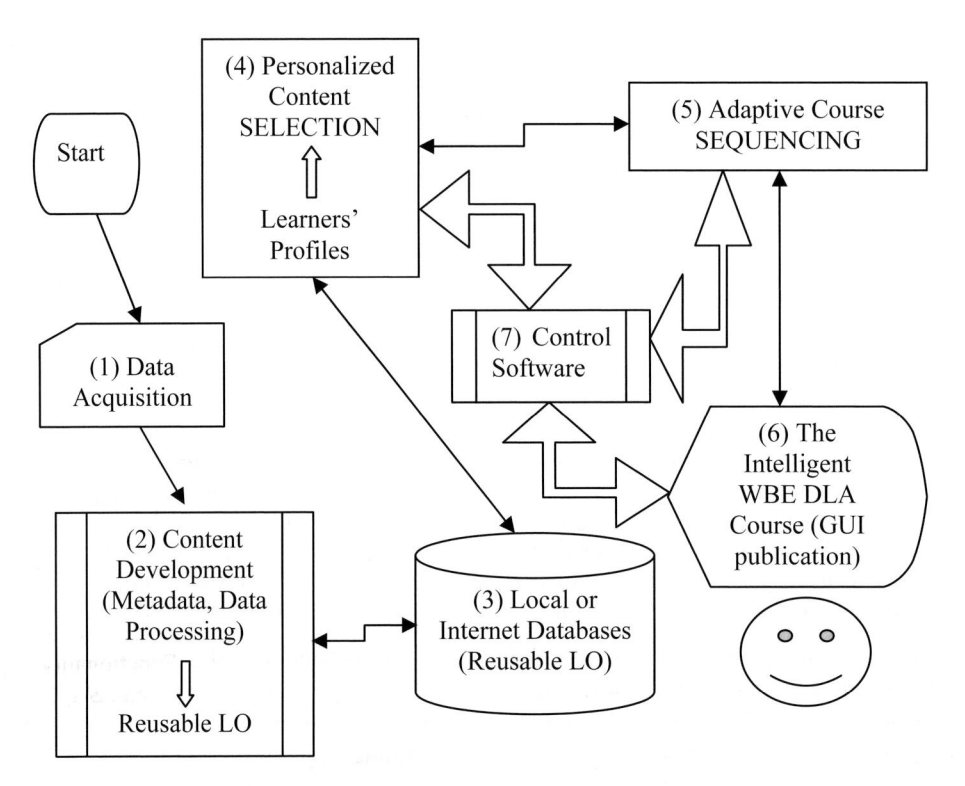

Fig. 1: Framework: The 7 main components of the proposed WBE DLA methodology

3 WBE DLA: Content Development, Selection and Sequencing

Data Acquisition

For data acquisition a number of resources could be used, like: books, Internet sources, lecturing notes, etc. (CARTER 2002). Also, for this purpose some simple task-defined user-friendly GUIs are needed for the data-entry procedure (STYLIADIS 2002).

Didactic requirements for e-learning form the size of the reusable learning units (NOTHHELFER 2009). So, the LO must be as small and concise as needed, in order to support content-authors to focused on the didactic features it contains, and then to help them to identify easily the pointers to the specific content descriptions for usability support (metadata-based pedagogical functionality).

3.1 The Reusable Learning Object

In the proposed WBE DLA the LO is the basic reusable unit and in order to be adaptive and searchable must have a metadata description. The IEEE Learning Object Metadata (LOM) standard (IEEE LTSC 2002) does not support LA topics, so a novel metadata schema is needed for the LO representation. Also, for the LO implementation, the XML markup language or the PHP scripting one, seems to be a reliable answer (Internet functionality, object-oriented programming). In particular, the XML or PHP tags make the data meaningful, so the LO items can be searched, extracted, printed in PDF, published and reused in a number of ways on demand (personalization functionality).

The proposed new metadata schema (Fig. 2) is related to the DLA LO tags and it is consisted from 5 fields with 10 ASCII characters (6 digits & 4 alphanumeric): The first 3 digits are for the LO title's ID (**100** for the *Digital Design I* LO; **200** for the *Digital Design II* LO, etc.). Following, the next 2 digits are for the specific concept's ID (**00** for a generic topic in digital design; **10** for AutoCAD; **20** for MicroStation; **30** for 3ds Max; **40** for Google SketchUp, etc.). Following, the next 1 digit is for the course level (**0** for an ease e-learning course; **1** for a moderate one; and **3** for a difficult demand course). Following, the next 2 characters are denoted to e-learning course specialization (**00** for a generic course; **LA** for a course in Landscape Architecture; **AR** for a course in Architecture; **GE** for a course in Geomatics Engineering, **EP** for an Environmental Policy one, etc.). Finally, the last 2 characters are denoted to required LA applications functionality (**00** for generic LA applications; **DD** for digital design LA applications; **SA** for spatial analysis LA applications; **EP** for environmental policy LA applications, etc.).

LO title's ID (e.g. 100)	Specific Concept's ID (e.g. 10)	Course Level (e.g. 1)	Course Specialization (e.g. LA)	LA Applications Functionality (e.g. SA)

Fig. 2: The metadata schema for the reusable Learning Object

The structure of the proposed reusable LO:
- **Title**: The LO title (e.g. Digital Design I, Spatial Analysis II, Environmental Policy).
- **Generic Core Content**: The definition of the core educational concept (content) for which the LO is designed accompanied by some generic examples and applications. Also, a number of pointers -embedded intelligently into the text- should point to the specific content descriptions according to the introduced metadata schema.
- **A number of Specific Content Descriptions**: To support reusability – The detailed descriptions of specific educational concepts into the Title domain (e.g. in case of a *Digital Design I* course: Theory of design, AutoCAD tutoring, MicroStation tutoring, R/T Landscaping Architect tutoring), accompanied by some examples and applications for each description. The tags of these specific descriptions will follow the introduced metadata schema.
- **Test:** The test element used for e-learning evaluation. It includes several (e.g. 10) test items relevant to learning difficulty level and to the concept of the LO. Also, for a fair judge examination policy, the actual test item is assigned to the learner randomly by the i=random(1..10) function. So, the assessment function TEST (i, text-script); where: i=1..10 and text-script={*AutoCAD 2D; Digital Design for LA generic applications; Moderate CAD knowledge*, etc.} is needed.

Figure 3 presents the XML implementation of the WBE DLA *digital design I* LO. It includes one title and one generic core content item (moderate level for LA e-learning with Spatial Analysis applications functionality), three specific description items to support LO reusability on demand with 2 options (2D AutoCAD, 3ds Max), and two test elements, one with a resource reference to an external XML examination-file and the second one representing the way to form an examination from the available LO's test items' in this case the examination-test includes the test items from test03 to test06, which is the combination of all available LO's test items related to LA spatial analysis applications functionality (i.e. the SA parameter).

```
<LearningObject Digital_Design_I="http://www...."
<Title id="100">Digital Design I</title>

<Generic_Core_Content id="100001LA00">
        <paragraph> The digital design in general ...
        AutoCAD 2D modeling ... [ptr→Tag:100101LASA] ...
        Visualization 3ds Max ... [ptr→Tag:100301LASA] ...
        ......... </generic_core_content>
<Specific_Description id="100101LASA">
        <paragraph> AutoCAD...2D modeling....
        ......... </specific_description>
.......................................................
<Specific_Description id="100301LASA">
        <paragraph> 3d Studio Max ... visualization ...
        ......... </specific_description>
.......................................................
 <TestItem identifier="testSA" identifierref="http://www.xxx.xx/exam-SA.xml"/>
.......................................................
<TestItem identifier="testSA">
        <beginItem>test03</beginItem>
        <endItem>test06</endItem>
</testItem>
```

Fig. 3: The XML-implementation of the Digital Design I reusable Learning Object

Similar LO descriptions could be defined for Digital Design II, Digital Design III, Spatial Analysis, Environmental Policy, etc. covering the LA curriculum.

Content Development

For reusable content development a number of GUI open-source environments are available, like: (SCORM 2003), IEEE Learning Object Metadata (LOM) standard (IEEE 2002; http://ltsc.ieee.org), Metavist 2005 s/w which creates FGDC-compliant metadata (Federal Geographic Data Committee; http://www.fgdc.gov), OMEKA project (http://www.omeka.org), AICC (http://www.aicc.org), ADL (http://www.adlnet.org), etc. The above content development tools are generic with limited functionality for e-learning LA content development. So, for a particular WBE DLA framework, there are two options: manual data-entry or the development of some new GUI data-entry forms implemented in XML, PHP or MS-Access (STYLIADIS 2002).

Reusable LO Repositories

After the content development procedure, the reusable LO should be added to local or Internet databases or learning object repositories for collection, sharing and reusing of distributed LOs (SAMPSON & KARAMPIPERIS 2004). The proposed metadata schema, because of its simple ASCII format, could easily be adopted by these databases for an effective LO access and management (ARAPI et al. 2007).

3.2 The Personalized Content Selection Procedure

The Learners' Profiles (LP) include education, background, cognitive style, learning preferences, needs, teaching & evaluation rules, and details on progress & performance in related LOs. These LPs are mapped to specific WBE selection criteria (pedagogical module), which are used as an input to e-learning management control s/w for forming the appropriate metadata pointers to the specific content descriptions. Then, the on-demand LOs are composed and, finally, a content menu for the requested WBE DLA course is dynamically synthesized (SAMPSON & KARAMPIPERIS 2004).

Pedagogical Module

In most WBE systems that incorporate course sequencing techniques, a pedagogical module is responsible for setting the principles and rules of content selection and instructional planning. In the proposed framework this pedagogical module is a part of the management control s/w, and the selection of content (i.e. the DLA learning objects) is based on a set of selection criteria according to LPs. Most of these selection criteria are *generic rules* (KRAMER et al. 2010), and there are no well defined and commonly accepted criteria on how the DLA learning objects should be selected and how they should be sequenced to make *"instructional value"* with LA functionality. So, in order to design highly adaptive WBE DLA systems a set of selection criteria rules is required, since the involved dependencies between the educational characteristics of the DLA learning objects and the potential learners' profiles are very complex (ARAPI et al. 2007).

In the proposed framework, the metadata schema has the potential to a semantically more accurate retrieval of content data, and hence the LO selection problem could be addressed by proposing a selection criteria module as a challenge for design pedagogy (theory & knowledge), that instead of "forcing" an educational material designer to define the set of the selection rules in a traditional way; supports an on-demand metadata-based decision that actually simulates the decision process of the educator (GIRVAN & SAVAGE 2010).

3.3 The Adaptive Course Sequencing

The next step is the on-demand **sequencing** according to the learning LA communities. So, WBE DLA adaptive course sequencing is defined as the process that selects LA learning objects from a local database, an Intranet or a global Internet-based digital repository (i.e. huge knowledge databases with design and LA functionality) and sequence them on-demand (i.e. in such a way which is appropriate) for the targeted LA learning community or individuals interested in WBE LA. Personalized learning trends to present the LA learning objects associated with a WBE on-line course in an optimized order for sequencing (BRUSILOVSKY & VASSILEVA 2003, KARAMPIPERIS & SAMPSON 2005, CHEN 2008).

WBE-Testing & Self-Assessment

For e-learning course testing including self-assessment quizzes the following software routines are proposed (KOSTONS et al. 2010; LAZARIDIS et al. 2010):

Name	Type	Description
test.html	HTML page	HTML form that contains the quiz questions
score.php	Application	Script to assess learner's answers
rtf.php	Application	Script to generate an RTF certificate from the template
pdf.php	Application	Script to generate an PDF certificate from the template

The e-Learning Management Control Software

The following application example is used for the e-learning management control s/w (the 7th framework component) demonstration and presentation. So, supposed that:

- Learner's profile (good knowledge of ICT; moderate knowledge of CAD).
- Learner's needs (digital design for Landscape Architecture, AutoCAD I, SketchUp, GIS/Spatial Analysis applications functionality).
- Examination rules: University/tutor/course regulations or learner's willing for self re-examination (moderate, e.g. N=5).

Hence, after applying a simple mapping procedure, the information about the data (i.e. the LO metadata contents) are found; and they are the values of the pointers (pointing to specific descriptions): **100101LASA, 100401LASA**. In Figure 4, the control software for the adaptive course sequencing structure in XML batch coding is displayed. Also, in the case of a PHP scripting implementation (Linux, Apache, MySQL5, PHP5), the course sequencing structure will be modular (routine-based) instead of the batch/goto programming logic.

Step	XML Implementation
1:	<Title id="**100**">Digital Design I</title>; counter=0; N=5;
2:	TEST (random(1..10), moderate CAD knowledge); if (FAILED) goto Step-13
3:	<Generic_Core_Content id="100001LA00">
4:	TEST (random(1..10), digital design for LA generic applications); if (FAILED AND counter++<N) goto Step-3 else if (counter=N) goto Step-13
5:	... [ptr→Tag:100**101**LASA] ... (AutoCAD 2D) goto Step-7
6:	... [ptr→Tag:100**401**LASA] ... (SketchUp) goto Step-10
7:	<Specific_Description id="100**101**LASA">
8:	TEST (random(1..10), AutoCAD 2D); if (FAILED AND counter++<N) goto Step-7 else if (counter=N) goto Step-13
9:	goto Step-6
10:	<Specific_Description id="100**401**LASA">
11:	TEST (random(1..10), SketchUp); if (FAILED AND counter++<N) goto Step-10 else if (counter=N) goto Step-13
12:	END (pass); Stop
13:	END (failed); Stop

Fig. 4: Control Software: adaptive course sequencing (XML batch programming)

The GUI intelligent WBE DLA course (GUI formatted publication component)
This delivery module (the 6th framework component) includes XSL files, for transforming the XML files into a number of publication documents, like: PDF documents, RTF documents, HTML files, Wiki forms, etc. So, in the WBE DLA client/server environment, every time when a request from a client for a specific content to be presented to the learner is received, this publication component invokes the right XSL file for the appropriate transformation. Hence, the on-demand content will appear on client's browser in a user-friendly GUI format.

4 Conclusions and Outlook

The presented work, which is just a framework and not an e-learning ready-to-use system, is an attempt to address the content development and the LO selection and sequencing problem in intelligent learning systems with LA functionality. For the content development a novel metadata schema was introduced and incorporated in LO structure as a challenge for design pedagogy, so reusability is supported and the LO are designed in a highly de-contextualized manner. The proposed methodology provides the framework for designing intelligent e-learning systems in DLA, in digital architecture, or in geomatics engineering.

In future extensions, learning characteristics like content difficulty or semantic functionality, which affects both selection and sequencing of reusable LO must be defined (KARAMPIPERIS & SAMPSON 2005). Also, future research would related to LO intelligent selection and decomposition from existing WBE courses in similar disciplines, allowing reuse of these disaggregated LO in different disciplines (while preserving e-learning functionality and the educational characteristics they were initially designed for).

References

Arapi, P., Moumoutzis, N., Mylonakis, M., Theodorakis, G. & Christodoulakis, S. (2007), *A Pedagogy-Driven Personalization Framework to Support Automatic Construction of Adaptive Learning Experiences*. In: Leung, H., Li, F., Lau, R. & Li, Q. (Eds.), Advances in Web Based Learning – ICWL 2007: 6th International Conference, Edinburgh 2007, LNCS 4823 Berlin/Heidelberg, Springer, 55-65.

Brusilovsky, P. & Vassileva, J. (2003), *Course Sequencing Techniques for Large-Scale Web-based Education*. In: International Journal of Continuing Engineering Education and Life-long Learning, 13.

Carter, J. (2002), *A framework for the development of multimedia systems for use in engineering education*. In: Computers & Education, 39 (2).

Chen, C.-M. (2008), *Intelligent Web-based learning system with personalized learning path guidance*. In: Computers & Education, 51 (2).

Girvan, C. & Savage, T. (2010), *Identifying an appropriate pedagogy for Virtual Worlds: A Communal Constructivism case study*. In: Computers & Education, 55 (1).

IEEE LTSC (2002), *Draft Standard for Learning Object Metadata*. Learning Technology Standards Committee (LTSC), IEEE 1484.12.1-2002.
http://ltsc.ieee.org/wg12/files/LOM_1484_12_1_v1_Final_Draft.pdf

Kim, W. (2007), *Starting Directions for Personalized E-Learning*. In: Leung, H., Li, F., Lau, R. & Li, Q. (Eds.), Advances in Web Based Learning – ICWL 2007: 6th International Conference, Edinburgh 2007, LNCS 4823, 2008. Berlin/Heidelberg, Springer, 13-19.

Karampiperis, P. & Sampson, D. (2005), *Adaptive Learning Resources Sequencing in Educational Hypermedia Systems*. In: Educational Technology & Society, 8 (4), 128-147.

Kostons, D., van Gog, T. & Paas, F. (2010), *Self-assessment and task selection in learner-controlled instruction: Differences between effective and ineffective learners*. In: Computers & Education, 54 (4).

Kramer, H., van Lammeren, R. & Ruyten, F. (2010), *Favouring four dimensions in landscape design*. In: Buhmann, E. et al. (Eds.), Peer Reviewed Proceedings of Digital Landscape Architecture 2010, Anhalt University of Applied Sciences. Berlin/Offenbach, Wichmann.

Lazaridis, F., Green, S. & Pearson, E. (2010), *Creating personalized assessments based on learner knowledge and objectives in a hypermedia Web testing application*. In: Computers & Education, 55 (4).

Marlow, C., Motloch, J., Calkins, M. & Hunt, M. (2009), *Design Education in Transition*. In: Buhmann, E. et al. (Eds.), Digital Design in Landscape Architecture 2009: Proceedings at Anhalt University of Applied Sciences, 2009. Heidelberg, Wichmann.

Nothhelfer, U. (2009), *Interactive Learning Beyond Knowledge-based Education*. In: Buhmann, E. et al. (Eds.), Digital Design in Landscape Architecture 2009: Proceedings at Anhalt University of Applied Sciences, 2009. Heidelberg, Wichmann.

Sampson, D. & Karampiperis, P. (2004), *Reusable Learning Objects: Designing Metadata Management Systems supporting Interoperable Learning Object Repositories*. In: McGreal, R. (Ed.), Online Education Using Learning Objects. Taylor & Francis Books Ltd.

SCORM (2003), *Best Practices Guide for Content Developers*. The Sharable Content Object Reference Model, Cernegie Mellon University, Learning Systems Architecture Lab. http://www.dokeos.com/doc/thirdparty/ScormBestPracticesContentDev.pdf.

Styliadis, A. D. (2002), *Programming the User Interface in Human-Computer Interaction: A Computing GIS Perspective*. Ziti editions, Thessaloniki, Greece.

Watson, D. (2009), *Are landscape programmes meeting the challenge of educating the second generation of digital landscape architects?*. In: Buhmann, E. et. al. (Eds.), Digital Design in Landscape Architecture 2009: Proceedings at Anhalt University of Applied Sciences, 2009. Heidelberg, Wichmann.

Concept for Collaborative Design of Wind Farms Facilitated by an Interactive GIS-based Visual-acoustic 3D Simulation

Madeleine MANYOKY, Ulrike WISSEN HAYEK, Thomas M. KLEIN, Reto PIEREN, Kurt HEUTSCHI and Adrienne GRÊT-REGAMEY

1 Introduction

Planning of wind farms appears to be a complicated matter in Switzerland and all over Europe, causing growing government and business frustration. The implementation often fails on the local level when it comes to choosing a suitable location for a wind farm, although the public generally supports wind power (DEVINE-WRIGHT 2005, WOLSINK, 2005). Social acceptance is a key issue for successful wind energy market development. The choice of the location is most crucial for public stakeholders and there is growing awareness amongst policy makers that not only the physical characteristics of a wind farm but also the process of planning a wind farm is an important factor influencing public acceptability (DEVINE-WRIGHT 2005). However, 75% of all stakeholders in Switzerland state that there is a lack of planning instruments to improve or support social acceptance (BFE 2009). Thus, participatory wind power planning needs adequate new instruments.

This paper presents the concept and preliminary results of the development of a visual-acoustic simulation integrating realistic acoustic soundscape modeling into GIS-based 3D landscape visualizations. Movie and sound recordings were generated for a reference site of an existing wind farm at the Mont Crosin (Canton Berne, CH) as basis for validating the visual and the acoustic simulation. A first interactive GIS-based 3D visualization with high level of detail was generated using a game engine, offering sophisticated tools for animating 3D objects or realistic representation of lights and shades. This 3D visualization will be further developed to a prototype of an audio-visual reproduction system. Then, visual-acoustic simulation models for specific areas with different landscape characteristics will be established and used in virtual reality choice experiments for valuation of alternative wind farm scenarios. The final simulation tool will allow for an improved impact assessment in strong collaboration with the public, which provides a better, more comprehensible decision basis for designating suitable locations for wind farms.

2 Theoretical Background

In Europe, there is a strong demand for renewable energy. Wind is an important energy source, however, the development proceeds very slowly in Switzerland. One of the most significant factors explaining acceptance or rejection of wind farms is the impact of the new infrastructures on a specific type of landscape characterized by aesthetic quality and a sense of place (DEVINE-WRIGHT 2005, WOLSNIK 2005). Besides these factors, noise made by rotating turbine blades is the most prominent environmental annoyance factor, which is strongly correlated with the visual impact on the landscape (PEDERSEN et al. 2008). Research results show that the soundscape can have a substantial impact on aesthetic and affective assessments of visual landscapes, and particularly anthropogenic, technical sound can impact scenic enjoyment and thus the recreational quality of locations (BENFIELD et al. 2010). But it is less known how to adequately represent a multi-sensory environment, how such representations might influence landscape assessments, or how they could influence decision-making in planning the environment (LANGE 2011).

There is common agreement in the point that technical information about noise in form of dB values alone is inaccessible to the public (MCDONALD 2009). Noise is an inherently psychological perception wherefore reliance on just physical measures of sound is not sufficient (MACE et al. 2004). The sound level can be rated very objectively, however there are also other factors such as expectation, source attribution, prior experience, motives, and difference thresholds, which have an impact on the subjective response and evaluation of the environment (MACE et al. 2004). Thus, in environmental assessment of wind turbine noise, including the human perspective regarding the acceptability of noise in certain landscapes is mandatory. With regard to visual aesthetic quality assessment of landscapes studying from a psychological perspective is required too, including objective judgments as well as human perception (DANIEL 2001). Hence, approaches are needed to assess both, the response to the visual and the acoustical impact of a wind farm scenario taking into account human perceptions and preferences. In this context, visual-acoustic simulations have high potential to facilitate a more comprehensive appreciation of values and for detecting acceptable places for wind power technologies.

2.1 Current simulation and modelling tools

Current planning tools including 2D maps, tools and techniques available for determining visibility and simulating wind farm projects (e.g. WindPRO, www.emd.dk) and data for noise levels fail to adequately integrate visual and acoustic factors into site planning for collaboratively identifying suitable places for wind power technologies. However, GIS-based 3D visualizations have proved to facilitate the communication between various stakeholders, professionals and the public in the context of participatory wind power development (LANGE & HEHL-LANGE 2005).

With today's visualization software programs virtual landscapes of tremendous visual realism as well as interactivity can be generated (PAAR & REKITTKE 2005). The rapid development of hard- and software for interactive visualization is mainly pushed by the growing market for computer games (HERWIG et al. 2005). 3D game engines are optimized for real-time navigation in a virtual environment (HERWIG & PAAR 2005). Benefits are the possibility of interactively experiencing the virtual environment and the high level of detail. Particularly a high level of detail in the foreground is required in order to allow for reliable visual landscape assessment (APPLETON & LOVETT 2003, LANGE 2001). Therefore 3D game engines are interesting alternatives to professional GIS-based 3D landscape visualization software such as, for example, Visual Nature Studio (www.3dnature.com).

Crytek's CryENGINE (http://mycryengine.com), for example, is a sophisticated 3D game engine, which offers suitable functions for developing a visual-acoustic simulation of wind farms. It is possible to import a digital terrain model and orthophoto (both raster data) and to animate landscape objects such as wind turbines. The wind speed parameter can be increased in the virtual landscape and can affect the speed of the turbine's blade rotation, shaking of leaves etc. accordingly. Even sound can be integrated into the virtual reality.

Auralization is the technique of creating audible sound files from numerical (simulated, measured, or synthesized) data (VORLÄNDER 2008). As an engineering tool it has recently been discovered for environmental noise applications. The main factors influencing sound propagation of wind turbines are well understood today. There are engineering models such as ISO 9613-2 (1996) or HARMONOISE (2005) available that take into account ground reflection and in a simplified manner the effects of inhomogeneous atmosphere due to varying meteorological conditions. For research purposes reference models exist which solve the wave equation in the propagation region considering the situation specific boundary conditions (SALOMONS 2001, HEUTSCHI et al 2005). Such instruments are the basic tools to acquire adequate auralizations of wind turbine noise.

In conclusion, there are sophisticated software tools and simulation models available for either landscape visualization or auralization. 3D landscape visualizations providing realistic, accurate and evaluable representations of the real-world environments with integrated spatially explicit noise emissions of wind turbines are not yet available.

3 Material & Methods

3.1 Development and implementation of a visual-acoustic simulation

The overall goal is the development of a combined virtual 3D visual and acoustic simulation tool for wind power plants to assess and adequately discuss choices of locations with experts and public stakeholders. The workflow presented in Figure 1 is divided into 2 phases: (1) development of the integrated visual-acoustic simulation model (VisAsim model) and (2) implementation of the VisAsim model for landscape assessment.

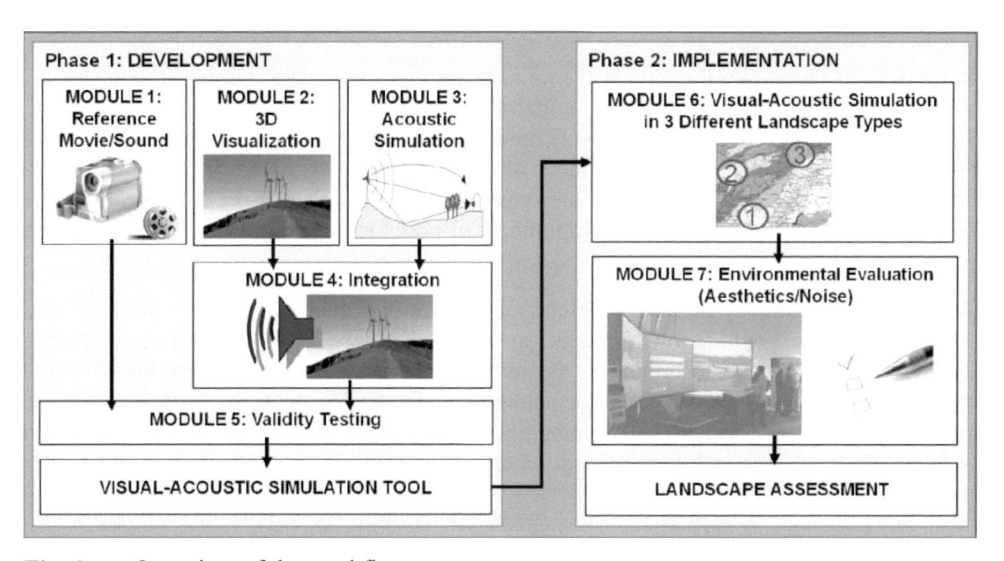

Fig. 1: Overview of the workflow

In Phase 1, we deal with the question how to adequately link and display the 3D landscape model with acoustic sound of wind power plants providing an adequate level of detail as well as ensuring the correlation of noise variations and spatial movement (e.g. synchronization of moving objects and noise modulations). In order to reduce complexity of this task, we focus on the validation of the integrated simulation and its implementation at selected viewpoints. Based on the spatially referenced parameters in the GIS-based landscape model, the spatially explicit sound signals are linked with the visual landscape representation. In a later phase, an interface will be developed that links the information about the three dimensional position and orientation in the 3D landscape model to the soundscape model. This will allow stakeholders to walk through the virtual landscape and to have a more personal role in the planning and evaluation process of wind turbine placement (BISHOP & MILLER 2007).

In order to test the validity of the VisAsim model it has to be compared to environment conditions close to reality. For this purpose we are simulating an existing wind farm in the first phase. As reference area, the existing wind farm at Mont Crosin (Canton Berne, Switzerland) was selected because it serves as a model example for a successful wind farm and is highly accepted in the public.

"Validity refers to the degree that something is as it purports to be." (PALMER & HOFFMAN 2001, 154). Thereby it can be distinguished between response validity and accuracy in the representation (SHEPPARD 2005). With regard to the acoustic simulation the same sound has to be generated as in the real environment and thus, the stimuli can be analyzed through a direct listening comparison. However, since the real world comprises much more visual detail than a computer generated representation can provide, 3D visualizations are always an abstraction from the real environment. For this reason, it is not possible to generate precisely the same stimulus as the real environment, but one can test if the response is the same for both stimuli (WILLIAMS et al. 2007). Therefore, we will check if the test persons' assessments (response equivalence) based on the VisAsim model are the same as the ones based on the video of the reference site (accuracy in representing sound, physical and visual qualities).

In the second phase, we implement the validated tool for assessing public preferences for wind farm options, accounting for the impacts on landscape aesthetics and noise in specific landscape contexts. The methods, workflows and tools elaborated in Phase 1 are then applied for generating VisAsim models of three focus areas with different landscape characteristics and of different landscape sensibility. Three scenarios per focus area will be simulated with low, medium and high numbers of possible wind turbines in the respective perimeter.

We want to find out, what are the effects of the simultaneous presentation of visual and acoustic effects of wind power plants on the preferences for a wind farm scenario. Furthermore, is there a difference in the influence of the noise of wind turbines on the assessment of landscape aesthetics depending on the landscape type? To answer these questions, the (mutual) effects of the integrated visual-acoustic simulation on the evaluation of landscape aesthetics and noise perception are investigated in experiments.

Developed in the early 1980s, discrete choice experiments are increasingly being used by economists to elicit preferences for different non-market goods and services, such as landscape aesthetical or recreational quality (CHAMP et al. 2003). Drawing upon socio-psychological concepts such as motives or expectation, choice experiments can be used to find out about segments of respondents showing different preference levels. Segmentation analysis such as latent class analysis is especially useful to evaluate the impact of different management scenarios and to conclude on optimizing actual wind turbines planning projects (BIROL et al 2006). Therefore, we will conduct a virtual reality choice experiment (BATEMAN et al. 2009) with selected experts and lay people in a laboratory experiment to determine the impact (aesthetics/noise) of the scenarios on the landscape.

4 Results

4.1 Reference Data

Movie and sound recordings were generated for the site of the existing wind farm at the Mont Crosin as reference for the visual and the acoustic simulation (see Figure 1, Module 1). Suitable locations for these recordings were chosen based on the following visual and acoustic criteria: From a visual point of view the choice of location is based on the behavior of pedestrians, ideally along roadsides (BRAUN & ZIEGLER 2006). In the recordings should be no disturbing objects, e.g. cows, buildings and other complex objects that may be difficult to visualize. Different contents of views were defined: a frontal wind turbine, a wind turbine in the background, several wind turbines and no wind turbine. From an acoustical point of view, locations fulfilling the following criteria are needed: an emission recording (in close vicinity of a wind turbine), an ambiance recording (close to a forest edge), a propagation recording (up to 500m distance to a wind turbine) and a multi-source recording (two wind turbines from different directions). The acoustic criteria are based on the principle of auralization comprising the basic elements of sound generation, propagation and reproduction (VORLÄNDER 2008). The accomplished reference recordings meet these predefined visual and acoustic criteria ensuring an optimal modeling and validation of the visual-acoustic simulation tool.

The videos and the sound recordings were produced when the vegetation was fully foliated (Figure 2). A Soundfield microphone and a single-lens reflex camera with movie recording capability were used as an acoustic and visual recording system. For recording the movie, a 10-20 mm lens was used and fixed to 10 mm to capture a field of view of 100° which comes close to a human's field of view. Simultaneously, wind speed measurements at 10 m above ground were gathered. In order to assess the visual and acoustic impact of the wind turbines on the landscape under different wind conditions, we took recordings at a wind speed of 3 m/s and over 9 m/s. The recordings serve as basic information for developing and testing the VisAsim model (see Figure 1, Module 5).

In addition, aerial images from the reference site were captured using an unmanned aerial vehicle (UAV) (Figure 3). UAVs are usually equipped with different sensors for navigation, positioning, and mapping such as still-video cameras. The position of the UAV as well as the camera shutter can be controlled remotely from the ground (MANYOKY et al 2011). The aerial images were photogrammetrically processed and a digital terrain model of high resolution in decimetre range was calculated. This data allows for highly detailed and up-to-date landscape visualizations (see Figure 1, Module 2).

Fig. 2: Simultaneous video and sound recordings at the wind farm Mont Crosin (BE)

Fig. 3: Acquisition of aerial images using an UAV

4.2 First interactive GIS-based 3D visualization integrating sound

A first interactive GIS-based 3D visualization with high level of detail was generated using the Sandbox-Editor of Crytek's CryENGINE Version 3.2.1 (Figure 4). The developed prototype is a 3D landscape simulation based on a digital elevation model and an orthophoto. Objects of landscape elements such as vegetation and 3D models of wind turbines were added according to the corresponding coordinates of their actual locations. The user is able to go to any location in the visualization in order to explore the visual and acoustic situation. Moreover the lighting and the daytime can be changed interactively so that the scenery can be experienced at night time as well.

Audio recordings of actual wind turbine immissions served as base data for sound implementation. We calculated sound levels using a simplified physical acoustic noise propagation model. The audio files were then modified according to these sound levels with FMOD Designer (www.fmod.org), a software to create interactive audio. These audio files were integrated into the visualization model in CryENGINE where the sound can be reproduced on a multichannel surround system. At any location in the visualization the estimated ambient circumstances can now be experienced (Figure 1, Module 2).

Fig. 4: Prototype of the visual-acoustic simulation tool, wind farm at Mont Crosin (BE)

However, based on the result it was decided not to integrate audio recordings directly into CryENGINE. The applicable sound models are too simplified and do not represent realistic soundscapes yet. For example, the sound propagation model does not take into account reflections at landscape objects such as buildings or vegetation (for further details see MANYOKY 2011). In the next step, thus, a sophisticated auralization will be developed and linked to an enhanced version of the 3D landscape model. Major challenge is thereby to synchronize the movement of the animated landscape objects with the sound depending on the wind speed and wind direction. Particularly the specific sound frequencies of blades passing the turbine pole ("swish-swish" sound) have to be synchronized. Failures in this synchronization were immediately perceived and criticized by test persons. Furthermore, the georeferenced locations as well as width and height of all 3D landscape objects (wind turbines, vegetation) in the scene have to be reported as input for the auralization model.

5 Conclusions and Outlook

The presented concept for the development of an interactive GIS-based visual-acoustic 3D simulation provides a feasible approach of integrating spatial noise emission into virtual landscapes to assess possible locations for wind farms. Thereby the chosen game engine CryENGINE turned out as suitable software for visual landscape representation offering high level of detail and interactivity required for comprehensive visual landscape assessment (objective and subjective). However, with regard to sound representation the CryENGINE does not provide enough functionality for adequate auralization yet. Thus, a separate auralization has to be established and linked to the 3D visualization.

Implementing this tool for evaluating the impact of different wind farm scenarios in virtual reality choice experiments (VRCE) will result in stated preferences of the public. The assessment results will reveal the impact of different wind farm scenarios in different landscape types. These will allow for recommendations for wind farm planning, informing decision-makers on the various planning levels about the general potentials of the wind farms' societal acceptability in specific landscapes.

Beyond that it may be suggested that the VisAsim tool can facilitate participatory wind farm planning. The final visual-acoustic simulation tool will allow for an improved impact assessment on the perceived landscape quality in participatory processes, which provides a better, more comprehensible decision basis for designating suitable locations for wind farms (Figure 5). Combining the results of both, the choice experiment and local participatory approaches, realistic visions for landscape development with wind farms can be elaborated taking into account the public needs with regard to aesthetical and acoustic landscape quality. The methods developed for VisAsim will be transferable to assess also other landscape changes with visual and acoustic impacts such as new streets or high voltage power lines.

Fig. 5: Visual-Acoustic simulation of wind turbines for a more comprehensible decision basis for choosing suitable locations for wind farms

6 Acknowledgements

The presented concept is the basis of the interdisciplinary project "VisAsim – Visual-Acoustic Simulation for landscape impact assessment of wind farms" (2011-2014) funded by the Swiss National Science Foundation.

References

Appleton, K. & Lovett, A. (2003), *GIS-based visualisation of rural landscapes: defining 'sufficient' realism for environmental decision-making.* In: Landscape and Urban Planning, 65, 117-131.

BFE (2009), *Energieforschung*, Swiss Federal Office for Energy, Bern. http://www.bfe.admin.ch/dokumentation/energieforschung/index.html?lang=de&publication=10017.

Bateman, I. J., Day, B. H., Jones, A. P. & Jude, S. (2009), *Reducing gain-loss asymmetry: A virtual reality choice experiment valuing land use change.* In: Journal of Environmental Economics and Management, 58, 106-118.

Benfield, J. A., Bell, P. A., Troup, L. J. & Soderstorm, N. C. (2010), *Aesthetic and affective effects of vocal and traffic noise on natural landscape assessment.* In: Journal of Environmental Psychology, 30, 103-111.

Birol, E., Karousakis, K. & Koundouri, P. (2006), *Using a choice experiment to account for preference heterogeneity in wetland attributes: The case of Cheimaditida wetland in Greece.* In: Ecological Economics, 60, 145-156.

Bishop, I. D. & Miller, D. R. (2007), *Visual assessment of off-shore wind turbines: The influence of distance, contrast, movement and social variables.* In: Renewable Energy, 32, 814-831.

Braun, S. & Ziegler, A. (2006), w*indLANDSCHAFT – Neue Landschaften mit Windenergieanlagen.* LAREG – Schriftenreihe des Fachgebiets für Landschaftsarchitektur regionaler Freiräume an der TU München, Wissenschaftlicher Verlag Berlin.

Champ, P. A., Boyle, K. J. & Brown, T. C. (Eds.) (2003), *A Primer on Nonmarket Valuation. The Economics of Non-Market Goods and Resources.* Dordrecht, The Netherlands, Kluwer Academic Publishers.

Daniel, T. C. (2001), *Whither scenic beauty? Visual landscape quality assessment in the 21st century.* In: Landscape and Urban Planning, 54, 267-281.

Devine-Wright, P. (2005), *Beyond NIMBYism: towards an Integrated Framework for Understanding Public Perceptions on Wind Energy.* In: Wind Energy, 8, 125-139.

Harmonoise (2005), *Final Technical Report.* AEA Technology Rail BV, Utrecht, The Netherlands.

Herwig, A. & Paar, P. (2002), *Game Engines: Tools for Landscape Visualization and Planning?* In: Buhmann, E., Nothelfer, U. & Pietsch, M. (Eds.), Proceedings at Anhalt University of Applied Sciences, Trends in GIS and virtualization in environmental planning and design. Heidelberg, Wichmann, 162-171.

Herwig, A., Kretzler, E. & Paar, P. (2005), *Using games software for interactive landscape visualization.* In: Bishop, I. & Lange E. (Eds.), Visualization in Landscape and Environmental Planning – Technology and Applications. London/New York, Taylor & Francis, 62-67.

Heutschi, K., Horvath, M. & Hofmann, J. (2005), *Simulation of Ground Impedance in Finite Difference Time Domain Calculations of Outdoor Sound Propagation.* In: Acta Acustica united with Acustica, 91, 35-40.

ISO 9613-2 (1996), *Acoustics – Attenuation of sound during propagation outdoors – Part 2: General method of calculation.*

Lange, E. (2001), *The limits of realism: perceptions of virtual landscapes.* In: Landscape and Urban Planning, 54, 163-182.

Lange, E. (2011), *99 volumes later: We can visualise. Now what?* In: Landscape and Urban Planning, 100 (4), 403-406.

Lange, E. & Hehl-Lange, S. (2005), *Combining a Participatory Planning Approach with a Virtual Landscape Model for the Siting of Wind Turbines.* In: Journal of Environmental Planning and Management, 48, 833-852.

Mace, B. L., Bell, P. A. & Loomis, R. J. (2004), *Visibility and natural quiet in national parks and wilderness areas – Psychological Considerations*. In: Environment and Behavior, 36/1, 5-31.

Manyoky, M. (2011), *SoundLandScape – Linking GIS-based 3D landscape visualization and spatial explicit ambient noise made by wind turbines*. Master thesis, ETH Zurich.

Manyoky, M., Theiler, P., Steudler, D. & Eisenbeiss, H., (2011), *Unmanned aerial vehicle in cadastral applications*, In: The International Archives of the Photogrammetry, Remote Sensing and Spatial Information Sciences, UAV-g congress 2011, accepted.

McDonald, P. (2009), *Auralisation and Dissemination of Noise Map Data Using Virtual Audio*. Euronoise Conference, Edinburgh.

Paar, P. & Rekittke, J. (2005), *Lenné3D – Walk-through Visualization of Planned Landscapes*. In: Bishop, I. & Lange, E. (Eds), Visualization in Landscape and Environmental Planning. Technology and Applications. London/New York, Taylor & Francis, 152-162.

Palmer, J. F. & Hoffman, R. E. (2001), *Rating reliability and representation validity in scenic landscape assessment*. In: Landscape and Urban Planning, 77/1-4, 149-161.

Pedersen, E & Larsman, P. (2008), *The impact of visual factors on noise annoyance among people living in the vicinity of wind turbines*. In: Journal of Environmental Psychology, 28, 379-389.

Salomons, E. M. (2001), *Computational Atmospheric Acoustics*. Kluwer Academic Publishers.

Sheppard, S. R. J. (2005), *Validity, Reliability and Ethics in Visualization*. In: Bishop, I. & Lange, E. (Eds), Visualization in Landscape and Environmental Planning. Technology and Applications. London/New York, Taylor & Francis, 79-97.

Vorländer, M. (2008), *Auralization: Fundamentals of Acoustics, Modelling, Simulation, Algorithms and Acoustic Virtual Reality*, Springer.

Williams, K. J. H., Ford, R. M., Bishop, I. D., Loiterton, D. & Hickey, J. (2007), *Realism and selectivity in data-driven visualizations: A process for developing viewer-oriented landscape surrogates*. In: Landscape and Urban Planning, 81, 213-224.

Wolsink, M. (2005), *Wind power implementation: The nature of public attitudes: Equity and fairness instead of 'backyard motives'*. In: Renewable and Sustainable Energy Reviews, 11, 1188-1207.

Collaborative Landscape Assessment and GeoDesign

Boris STEMMER

1 Introduction

Collaborative Landscape Assessment aims to help integrate the public's views into landscape design and planning, particularly at large-scale, such as in collaborative GeoDesign. This paper presents an approach using the Web-GIS 'KuLaDig' to solicit landscape related information from people who represent different interest groups. Members of the public are invited to identify and assess landscapes that they are familiar with and that they like. The Web-GIS 'KuLaDig' offers a variety of features that support users in identifying an area and marking its location on a map and then performing landscape assessments. Personal commentaries may also be supplied to the area identified and assessed. Results from public assessments are compared, first with one another, by employing standard GIS analysis. The product is a map depicting a synopsis of all public assessments made. This map indicates where members of the public agree on areas and landscape elements that they like, and where they share common landscape values. In a next step this evaluation is compared with landscape assessments made by experts, such as in landscape character assessments and the inventorying of historic landscape elements. Again, a map is created that indicates where, in this case the public and some experts agree on important areas, and what particular values they attach to them.

The concept of collaborative landscape assessment is discussed where both experts and members of the public contribute to assessment and evaluation in a complementary way. The hypothesis is that differences exist, between public and expert assessments, and also within the groups that represent the public. It is important, with respect to principles of good governance and democratic decision making, to discuss how to best accommodate such differences in planning. Some proposals for solutions are made. A preliminary conclusion on the role of collaborative landscape assessment in the context of GeoDesign is also proposed.

The collaborative assessement concept is tested first with students of landscape architecture and planning. A second testing is conducted with members of the civil society representing NGOs, Churches, sport clubs, etc. This second testing is done in the real life environment of Cologne (Köln-Chorweiler), Germany. The City of Cologne and the LVR (Landschaftsverband-Rheinland) are local partners supporting this project by identifying key-persons to be included in the testing and by providng access to the Web-GIS-Platform KuLaDig.

This paper includes the preparations done to conduct the students and the public test applications of the proposed method. Preliminary survey results will be presented during the DLA 2012 conference.

2 The 'KuLaDig' System (KuLaDig – Kultur. Landschaft. Digital)

KuLaDig is a digital landscape information system originally designed for the inventorying of historic cultural landscape and landscape heritage elements. This system has been developed further, by the Rhineland Regional Council (LVR Landschaftsverband Rheinland) and the Hesse Department for Monument Preservation (Landesamt für Denkmalpflege Hessen), to allow wider applications. While KuLaDig still functions as a WebGIS based system for the documentation of historic cultural landscapes it now also offers the opportunity for everybody to participate in the further development of the database. A WebGIS interface is used for this purpose. In this way the new and innovative approach, presented here, is both participatory and expert based at the same time.

Thus, the idea is to use the KuLaDig system for the collecting of information pertaining not only to historic landscape elements but to all kinds of landscape phenomena. If the extended applications could also include information on what appears to be important to members of the public then this would be an important step towards doing landscape assessments in accordance with the European Landscape Convention (JONES 2011).

3 Landscape Perception

Current landscape concepts are based on multifunctional models (HABER 1971, MANDER, WIGGERING & HELMING 2007) while landscape planning and management are using ecosystem and landscape services concepts (DAILY 1997, TERMORSHUIZEN & OPDAM 2009). Important functions and ecosystem and landscape services are pertaining to the identification of people with a region and to their attachment to specific areas. These, as well as recreational aspects, are closely related to human health and wellbeing. Especially in densely populated areas these functions and services are considered of high importance to making human environments liveable. The assumption is that if a landscape affords a sufficient variety of identification objects – for example historic landscape elements (MOORE & WHELAN 2007, 2008 printing) – it is also able to fulfil central function such as local attachment and recreation.

Expert approaches to assessing landscape functions and services are based on the idea that landscape perceptions and preferences are more or less predictable. The assumption is that all people who live in a specific region collectively share what they feel is important in their environment. Expert assessments are thus interpretations of such collective preferences. In addition to cultural landscape analysis scenic landscape evaluations are done to provide evidence of which landscape objects are probably important for people's identification with their environment. However, for very simple reasons, expert methods of landscape inventorying and assessment alone will never be able to explain the process of identification and feeling of attachment to an area that becomes home. When dealing with a common good like landscape no single person or a small elite groups can possibly know what everybody else thinks and feels. Planners are such a small elite group (KÜHNE 2011: 174) and, based on their education and knowledge, will make decisions on based on their special interest. They do not necessarily share the public's emotions an aesthetic preferences that cannot be explained by any modeling (DEMUTH 2000, 100; KÜHNE 2006, 150; TESSIN 2008, 8).

Adopting Kühne's constructivist landscape concept a new approach to integrating landscape perception into planning must be taken. Very few aspects of landscape may be evaluated on the basis of 'objective' criteria applied to the 'physical landscape' alone. Examples are soil properties and their ecosystem functions. Even some landscape properties might be analyzed 'objectively', for example in order to measure landscapes for their diversity. However, to penetrate the realm of people's landscape perception, one has to depart from the physical landscape and start to understand its social dimensions. According to Kühne's concept these include three dimensions of the 'adopted physical landscape', the 'social landscape', and the 'individually adopted social landscape'. First, the 'adopted physical landscape' includes all of the material objects and elements that people perceive in their surroundings. The 'social landscape' is the product of the evaluation of the 'adopted physical landscape' by the public, while the 'individually adopted social landscape' is a single person's perceived landscape that is generated from the social landscape as well as from the individual's landscape experiences. Most landscape evaluations are not made with reference to the 'physical landscape' but to the 'social' and to the 'individually adopted social landscape'. Consequently, decisions on landscape issues should be made during a participatory process that help to find out which values people see in the landscape.

4 Online Participatory Visual Landscape Assessment

It is common practise to conduct visual landscape assessments based on expert lead methods. Participatory approaches are rarely applied. Research on participatory landscape assessments are more advanced than their practical application. A variety of options exist. One of the most common method is based on the assessment of landscape photographs. Such methods have been established in research as early as the 1970s (KAPLAN 1975, 93; DANIEL & BOSTER 1976). Recent research using photographs was conducted by ROTH (2005) and STEINITZ (20 Nov. 2009). While methods employing photographs are easy to use one of their disadvantages is that the images are chosen by experts before the actual assessment occurs. Through this preselecting of landscapes, perspectives and of images, perceptions and assessments are limited to what researchers provide. Results do not provide a complete spectrum of the potential variety of the public's landscape conception. Another participatory and collaborative approach to landscape assessment in planning is to use GIS (webGIS, PPGIS). A well know German example of online participation in planning is the so called 'Interactive Landscape Plan' for the municipality of Königslutter (Interaktiver Landschaftsplan Königslutter) (VON HAAREN et al. 2005). Visual assessment is one aspect among several topics related to landscape planning. At the time when the policy making process in Königslutter started the accessibility of the Internet was not as advance as it is today. Only a small portion of all of the relevant participatory activities where done via the Internet. But for those statements received via new IT media it can be stated that these were more precise than most others, and they also had a concrete spatial reference (VON HAAREN et al. 2005, 229). More recently, KAHILA & KYTTÄ (2009) have introduced what they call a 'softGIS Approach'. It has a special focus on urban planning and employs a webGIS method to collect 'soft' data. The term 'soft' is used in contrast to the 'hard' physical data that experts use traditionally. Similarly to them, also NORDIN & BERGLUND (2010) and BERGLUND & NORDIN (2007) have reported on experiences using GIS for generating and processing 'soft' data, in their case in the context of children's participation in planning. It seems that, even for young children starting with the age of 10, they have no trouble to locate sites of their everyday landscape experiences on maps and aerial photographs. Brown and Weber (in preparation) did a survey on national park experiences, also using a webGIS. As it turned out in this case, the PPGIS (public participatory GIS) method could successfully be used to measure visitors' experience and evaluation of different spots within the Greater Alpine region of Victoria, Australia.

These examples suggest that most people – including young children - are able to work with GIS and that they are also able to express their landscape experiences using maps and aerial photographs. Therefore it seems quite reasonable to use a WebGIS (in this case KuLaDig) to generally include people into the assessment of landscapes. These methods and technique are related to traditional mental map approaches (GOULD & WHITE 1974, LYNCH 1960). The aim of combining WebGIS and mental mapping is for planners to get to know the landscapes that are in people's minds, and to make them visible on maps.

5 Study Area: 'Cologne Chorweiler'

For the purposes of this study an area within the District No. 6 of the German city of Cologne was chosen. This district is called 'Chorweiler' and it is part of the greater Cologne agglomeration. It is located at its western edge. This edge situation includes all of the many different spatial phenomena that are known to be typical for similar agglomerations in Germany, and many other countries. It might be described as a transition area, a sort of interface between rural and urban zones. The term 'Zwischenstadt' has been coined for such 'in-between' situations (SIEVERTS 2001). In this particular case the dynamics of turning open space into urban development are very high.

While the Cologne agglomeration edges appear to typically represent the 'Zwischenstadt' phenomena, at the same time, and in many ways, there is also a number of special issues to be considered. The City of Cologne is located on the north-south transportation axis that follows the Rhine River. All modes of transportation, such as motorways and train lines, have extremely high traffic volumes. The development of new, and the improvement of existing, transportation infrastructure is thus a mayor political issue. Currently, Cologne already has two major 'Autobahn' rings, and a third ring road system is now in the planning stages. The changed traffic flow will dramatically affect the agglomeration's edge at Chorweiler. The third ring will be build by using existing roads and streets. The Trans-European traffic will be channelled towards the new highway ring and it will, at the same time have substantial influence on all other forms of transportation infrastructure. Regional and local traffic that has been using the roads of the third ring so far will now try and find alternative routes. This diffusion of huge amounts of vehicular traffic into the finer networks of local streets will cause a landscapes change of hitherto unknown proportions, and its effects will be felt all over the agglomeration's edge.

A second important issue is pertaining to the development of renewable energy. Recent governmental decisions include energy production through agricultural uses; this decision will have an even more dramatic change on landscape as has ever been known to date. In fact, these changes are equal to introducing a fundamentally new phase in the era of industrialization of agriculture. The changing of energy production systems is accompanied by changes in the energy supply infrastructure. Not all of the energy production occurs near to where the major consumers are located. For Germany, for example, energy that is produced in the north might be needed by consumers who live and work further south.[1] In the particular case of Cologne it again is the western agglomeration edge that is affected the most: it has been designated to, somehow, accommodate a number of new high voltage power lines.

Resulting from these and other developments, planning in this area does not have a local dimension any longer. The driving forces for development have assumed a regional, national and, in the case of trans-border transportation, an European dimension. Municipalities have no choice but to collaborate and create regional concepts. Therefore, the new challenge is that public participation also has to be done across municipal borders.

Köln-Chorweiler has been identified as the study area. It is considered to typically represent the agglomeration edge, and it is affected by a multitude of different developments. The spectrum of spatial and land use types is large and includes high density residential and commercial areas, such as those near the former village Chorweiler, mosaic of rural areas such as those of the 'Worringer Bruch', and also the highly industrialised complexes of the globally active Bayer corporation at Dormagen. At the same time there is a long history of human settlement and traditional agriculture. A large variety of relicts have been documented, witnesses of former land use, including old country estates, chapels, shrines and crucifixes, but also historic vista-lines and archaeological monuments, the most important of them being the Battle of Worringen (MOLITOR & UTZERATH 2009). All of the historic landscape elements and monuments are endangered to succumb to the enormous pressure exercised by the development described above. In fact; there is a high risk for many places of altogether losing their identity. When a place is turned into a faceless settlement or infrastructure the landscape does no longer tell the story of its multi-faceted genesis.

6 Aim of the Project

The aim of this project is to try and learn to understand if and how it might be possible to undertake public landscape assessment using a WebGIS. In doing so one specific question is how and to what extent public landscape assessment results differ from results gained from expert landscape assessments. Detailed investigations are made to analyse the matching and mismatching of the different forms of assessment. Finally, the ways of integrating (informally done) public landscape assessment results into (formal) planning and also into (statutory) landscape policy.

---

[1] For detailed information please refer to the Energy concept of the Federal Republic Government Bundesministerium für Wirtschaft und Technologie and Bundesministerium für Umwelt, Naturschutz und Reaktorsicherheit (October 2011).

7 Methods

The survey will be conducted using a web interface (www.landschaftsbild.org) that introduces to the project and is offering a lot of information on how to work with the KuLaDig System. Besides this it was designed not to be too complicated and is lacking all that is not really necessary to participate in the project.

For purposes of conducting public landscape assessments individual people are invited to use the tools provided by the information system KuLaDiG. In particular, individuals are asked, for every single landscape or landscape element they wish to introduce into the system, to draw a polygon on a map and to provide a written description. The narrative should include an explanation of what it is that they like about this special place.

The output of this survey includes a number of polygons and texts. To analyse theses outputs, a 'content analysis' will be done for the texts, using CAQDAS (SILVERMAN 2010, 251). Attributes and qualities of these places will thus be identified, while the polygons will be used for a geographical overlay analysis. If place qualities and attributes are attached to the polygons an evaluation map can easily be produced using GIS-Software. Recurring words and items will be used to create thematic maps. At this point, mainly for technical reasons, it is not possible to integrate empirical tools provided by social sciences. For example, it would be good to make use of a Semantic Differential, a tool which would make it much easier to gain data that are immediately comparable with one other.

After a map containing results from the public landscape assessment has been made available, and after this map has been checked and verified, it is ready to be compared with results generated by other landscape assessment procedures. In this particular case all results from assessments will be included into the comparative analysis that can be found in plans and policies of Cologne planning administrations. These included statutory policy documents and plans such as the official landscape plans and the zoning plans, including environmental reports. In addition, informal documents and plans will be includes, such as the 'Interkommunale Integrierte Raum-Analyse' (IIRA).

8 Conclusions and Outlook

The outcome of the survey was not available when this paper was written, but it will be presented on the DLA2012 conference. Meanwhile only preliminary conclusions can be made. Public participation via the internet seems to have a great perspective. Most people use this media and as shown before are able to use WebGIS to express their landscape experiences. It still is questionable if a survey will produce outcomes that are usable for a landscape assessment. Still not solved is the problem of combining experts and public evaluation, especially if both are contrary. Who then is right? Which decision making mechanism will be used in this case?

References

Berglund, U. & Nordin, K (2007), *Using GIS to Make Young People's Voices Heard in Urban Planning*. In: Built Environment, 33,469-81.

Brown, G. & Weber, D., *Public Participation GIS: A new method for national park planning*. In: Landscape and Urban Planning (in preparation).

Bundesministerium für Wirtschaft und Technologie & Bundesministerium für Umwelt, Naturschutz und Reaktorsicherheit (Oktober 2011), *Das Energiekonzept der Bundesregierung 2010 und die Energiewende 2011. Energiekonzept für eine umweltschonende, zuverlässige und bezahlbare Energieversorgung.*

Daily, G. C. (1997), *Introduction: What are ecosystem services?*. In: Daily, G. C. (Ed.), Nature's services. Societal dependence on natural ecosystems. Washington, DC: Island Press, 1-10.

Daniel, T. C. & Boster, R. S. (1976), *Measuring landscape esthetics: the scenic beauty estimation method.* Fort Collins, Colo.

Demuth, B. (2000), *Das Schutzgut Landschaftsbild in der Landschaftsplanung. Methodenüberprüfung anhand ausgewählter Beispiele der Landschaftsrahmenplanung.* Berlin, Mensch-und-Buch-Verlag.

Gould, P. & White, R. (1974), *Mental maps.* Harmondsworth [et al.], Penguin Books.

Haaren, C, v., Oppermann, B., Friese, K.-I., Hachmann, R., Meiforth, J., Neumann, A., Tiedtke, S., Warren-Kretzschmar, B. & Wolter, F.-E. (2005), *Interaktiver Landschaftsplan Königslutter am Elm. Ergebnisse aus dem E+E-Vorhaben "Interaktiver Landschaftsplan Königslutter am Elm" des Bundesamtes für Naturschutz.* Bonn-Bad Godesberg, Bundesamt für Naturschutz.

Haber, W. (1971), *Landschaftspflege durch differenzierte Bodennutzung.* Von Haber, W., Bayerisches Landwirtschaftliches Jahrbuch 48, Sonderheft 1, 19-35.

Jones, M. (2011), *The European Landscape Convention – Challenges of Participation.* Edited by Jones, M. & Stenseke, M. Dordrecht, Springer Science+Business Media B.V.

Kahila, M. & Kyttä, M. (2009), *SoftGIS as a Bridge-Builder in Collaborative Urban Planning*. In: Geertman, S. & Stillwell, J. (Eds), Planning support systems. Best practice and new methods. Dordrecht, Springer, 389-412.

Kaplan, S. (1975), *An Informal Model for the Prediction of Preference*. In: Ervin, S. Zube, H., Brush, R. O. & Fabos, J. G. (Eds.), Landscape assessment. Values, perceptions, and resources. Stroudsburg, Pa., Dowden Hutchinson & Ross [et al.], 92-101.

Kühne, O. (2006), *Landschaft und ihre Konstruktion. Theoretische Überlegungen und empirische Befunde*. In: Naturschutz und Landschaftsplanung, 38, 146-52.

Kühne, O. (2011), *Die Konstruktion von Landschaft aus Perspektive des politischen Liberalismus. Zusammenhänge zwischen politischen Theorien und Umgang mit Landschaft.* In: Naturschutz und Landschaftsplanung, 43, 171-76.

Lynch, K. (1960), *The image of the city. Kevin Lynch.* Cambridge, Mass, The Technology Pr & Harvard Univ. Pr.

Mander, Ü., Wiggering, H. & Helming, K. (Eds.) (2007), *Multifunctional Land Use. Meeting Future Demands for landscape Goods and Services.* Berlin/Heidelberg, Springer.

Molitor, R. & Utzerath, M. (2009), *Zukunftsinitiative StadtRegion Köln-Rhein-Erft. Interkommunale Integrierte RaumAnalyse IIRA – Zwischenstand.*

Moore, N. & Whelan, Y (Eds.) (2007, 2008 printing), *Heritage, memory and the politics of identity. New perspectives on the cultural landscape*. Aldershot, England, Burlington, VT, Ashgate.

Nordin, K. & Berglund, U. (2010), *Children's Maps in GIS: A Tool for Communicating Outdoor Experiences in Urban Planning*. In: International Journal of Information Communication Technologies and Human Development (IJICTHD), 2. 1-16.

Roth, M. (2005), *Online Visual Landscape Assessment Using Internet Survey Techniques*. In: Buhmann, E. et al. (Eds.), Trends in online landscape architecture. Proceedings at Anhalt University of Applied Sciences 2004; [compromised of contributions from speakers at the International Conference on "Landscape Architecture Online", held in Dessau, Germany, May 13th and 14th 2004, fifth in a series of conferences on "New Trends in Landscape Architecture"; contributions of the Seminar "Landscape Architecture Online . New Communication Techniques for Landscape Architecture and Architecture"]. Heidelberg, Wichmann.

Sieverts, T. (2001), *Zwischenstadt. Zwischen Ort und Welt, Raum und Zeit, Stadt und Land.* Gütersloh/Berlin/Basel, Bertelsmann Fachzeitschriften; Boston/Berlin, Birkhäuser.

Silverman, D. (2010), *Doing qualitative research. A practical handbook*. Los Angeles, CA [et al.], Sage.

Steinitz, C. (20.11.2009), *An Assessment of the Visual Landscape of the Autonomous Region of Valencia, Spain:. a case study in linking research, teaching, and landscape planning*. Ljubljana.

Termorshuizen, J. W. & Opdam, P. (2009), *Landscape services as a bridge between landscape ecology and sustainable development*. In: Landscape ecology, 24, 1037-1052.

Tessin, W. (2008), *Ästhetik des Angenehmen. Städtische Freiräume zwischen professioneller Ästhetik und Laiengeschmack*. Wiesbaden, VS Verlag für Sozialwissenschaften/GWV Fachverlage Wiesbaden.

"Beep-Scape" – Using Applications for Mobile Devices to Communicate the Landscape

Fernando BUJAIDAR, Diana SANTA CRUZ and Gabriel SEAH

1 Introduction

Modern technologies and innovations have propelled us into unprecedented ways of travelling. We are able to reach our destination at speed that we could never have imagined some 50 years ago. The breakthrough undisputedly can be considered as one of the greatest human interventions; however, reaching our goal at such a speed also has its pitfall. One of the most significant areas of concern shall be mentioned as the missing experience in between the destinations, the loss of landscape perception, if we may call it. Our natural ability to perceive the landscape has simply failed to catch up with the speed of technology advancement. In most cases, elements in between destinations are only blurry images that one can hardly remember.

Fig. 1:
beep-SCAPE

To quote an example, we can place train travelling as the focus. As a mass rapid transport, rail lines transect many landscapes kinds. Many of which are of amazing cultural, historical and natural landscapes, but if asked, how many people do actually remembered and are able to appreciate these beauty? In respond to the issue, another form of technology might be able to reduce the impact of such phenomena. GPS features in smart phones have given us the ability to locate our position, providing us with critical navigation information, combined with the application (Apps) feature; they are of multiple possibilities, hence, opportunities.

The new technology has opened up the possibility of a new realm, a chance to use technology to aid and enhance the experience of perceiving the landscape. Looking at the current situation, most people today own and operate a smart phone; on the other hand, landscape communication strategies is a rising area with new challenges for the landscape architecture requiring interdisciplinary cooperation and fresh ideas to broaden the field of action. Putting these elements together a landscape tool in the form of an app can therefore reach and provide mass communication services to inform people of a prominent landscape while travelling in trains or automobiles

This paper goes into the first steps of a normal landscape architecture task that came out with a design approach which results in the need of outputting a draft design of such tool.

2 Background and Method

The framework of this idea in first instance came from a previous project carried out in the Rems valley region of Stuttgart, as a 3rd trimester main project of the IMLA program of the HfWU Nürtingen. The task given by the Supervisors (Prof. Dipl.-Ing., Sigurd Henne and Prof. Dr. Christian Küpfer) came as an approach for "Enhancing Linear Landscapes along Mayor Traffic Routes in the Urban Region of Stuttgart", where the aim of the project was to ponder the point of view of landscape architecture towards an area, communicate it to the general public and at the same time add a value to the region. From there, the answer for the task as group took shape as an analysis of the visual perception from the train line.

The method developed for the analysis through train window was approached the same a picture or panting can be perceived and composed; where the elements are located in the perspective are arranged according to a hierarchy. This approach basically divides the composition in three different grounds where main objects tend to be in the mid distance, while some detail elements are found in the foreground and the complement for the composition is in the background while the whole is always contained in a frame.

From this perspective the situation is not different when we look at the window of a train, as it is framing the landscape outside we are looking at. The difference stands in the fact that we are looking into a dynamic picture and the speed factor changes the rules for the observer: meanwhile in a static picture the foreground shows the details, in the train we don't have time for distinguish anything but colours and changes in rhythms, the middle ground is easier to understand because the changes are happening in a slower pace so some details can be perceived; and finally the background is more steady and helps to get a fix point of reference.

On that ground, the analysis went on in two parts: the first phase was a study made from the point of view of the landscape architect, going to the area and studding the landscape from the train line applying the method of the moving landscape for identifying the elements composing the picture. The output was an analysis video, a catalogue of elements and a summary of the trip in a travel storyboard that shows an overview of the complete trip and how the vistas are working together whit a list of opportunities and constraints. The second phase of the analysis was a survey conducted among the train passenger to see what views and what elements in its different combinations are the most liked or recalled.

Afterwards, an improvement proposal was carried out based on a visual value map created with the help of GIS, output of the analysis phase as base for the further design criteria and for a master plan to enhance the visibility from the train. All together with a scenario simulation showing the variations in the landscape and a 3d visualization of the trip after the application of the master plan.

Finally, the communication concept to transmit the character of the region and the implementation of the ideas was developed. Improve the perception of the users towards a better experience of the train travel in the Remstal is the base of the proposal. The idea took shape after a presentation and discussion of ideas with Dipl.-Ing. (FH) Werner Rolf; who supervised the GIS analysis for the project. The result was a theoretical design of an application for mobile devices that could work in real time highlighting the main features of the landscape as people travels on the train.

3 Design and Operation

The essence of the application is to make people aware of the landscape outside the train, so they can get to know about it and learn from what they see. The program should be able to tell about environmental features, historical places, famous or important characteristics of the area, relevant sites, etc. and give further explanations and directions on how to get to the points of interest of the user if necessary. It can be also used as a planning tool to underline certain changes that have been possible thanks to planning and design of the landscape.

This app will be programmed with data and information of a train journey taking mainly the transected landscapes into account. The data will highlight important and prominent landscapes as critical information. In order to set up the base data, the analysis method survey method explained before, will be used as base to obtain information of the perception of the landscape. The main goal is to determine the aesthetic quality of a landscape and to communicate it. Analysis and evaluation will be carried out leading to the highlighted landscapes in the app.

The combination of the GPS function from the mobile device will work in real time recognizing the location of the user in relation to the significant landscapes and informing when a particular interesting feature is about to come along the rail lines. The app will act like a alerting and information device, releasing a beep as well as providing information and further details of the landscape feature. A second function will be to ask the program what is special about certain feature that is visible at the moment and the system will give the information.

In summary, the app will aim not only in the enhancement of train journey experience but also at raising appreciation for cultural, historical and natural landscapes, through a fast moving perspective.

The target group for this application is any person that is making a new route in the country for a holyday and has several hours of train ahead, so instead of watching a movie in the mobile device looks out the window.

Another possible approach for the use of the app would be that some billboards are built in the train stations and in some cases, in the actual features, so the person can also gather this information outside the train and be able to see it at any time.

Fig. 2:
The touch screen

The *functioning of the application* is sketched in the following steps:

1. A person with a smart phone, computer o tablet gets inside the train.

2. once the route is logged in or recognised, his/her mobile device will beep if; a) There is a code system already installed in that line from which the user can retrieve all the information about interesting features in that specific route. b) There are some features that might be interesting for the subject according to his pre-established user preferences, for example, if he sets his phone to beep when there are historical castles nearby, etc.

3. During the journey, the phone will beep to alert the user that he is near his element of interest, as well as provide general and specific information about its features and information on how to get there, starting with the right stop to leave the train.

4 Outlook

The present paper outlines the base idea that can become a bigger scale project because combines up to date tools and topics from both, global information systems and landscape architecture, and combines them into an common use feature for the general public, bringing closer the planners and the stake holders giving a media to communicate the landscape in a very simple to understand and fun way.

It is an opportunity to develop technology, research, planning projects and even participatory processes and tourism. It can also help to the development of small regions from the fact that when the traveller learns something about a place he is going through it might as well want to get closer or stay.

Another potential of this idea are the possible parties interested in investing in this development are, like for example any governmental planning offices, due to the possibility of use this application for communicating people about their work in the regions and therefore attracting people and also the railway companies are can draw the attention in the idea since they have many leisure travel offers so this can be a plus for them.

The strength of the idea is that is virtually applicable in every place where a railway can be found. The starting process will need collection and processing of many sorts of data and along with the assistance of GIS systems. Nevertheless, once a prototype is running in a small scale the repetition of the method will make the implementation in other places easier. Mainly it can change the leisure train travelling, from the current situation of using the train merely to get to the place, to an ideal state where a trip is about the enjoyment of the journey and not only about the destination point.

References

Augé, M. (1995), Non-Places: Introduction to an Anthropology of Supermodernity.
Grava, S. (2003), Urban transportation Systems.
International Master in Landscape Architecture: The Valley of the River Rems; project outline (2011), Main Project II Trimester 3.
Lynch, K. (1960), The image of the city Cambridge.
Sieverts, T. (2000), Cities without cities.
Strolollogy, http://www.spaziergangswissenschaft.de/.
Virilio, P. (2005), Negative Horizon: An Essay in Dromoscopy.

Landscape Modeling

Interdisciplinary Research and Education in a Virtual Cultural Landscape Laboratory

Ralf BILL

Keynote: 26 May 2011

Summary

In recent years the transfer of old documents (books, paintings, maps etc.) from analogue to digital form has gained enormous importance. Numerous interventions are concentrated in the digitalisation of library collections, but also commercial companies like Microsoft or Google try to convert large analogue stocks such as books, paintings, etc. in digital form. Data in digital form can be made accessible more easily to a large user community, especially to the interested scientific community.

The aim of the described research project is to set up a virtual research environment for interdisciplinary research focusing on the landscape of historical Mecklenburg. Old maps from 1786 covering the entire town of Mecklenburg were georeferenced and should be combined with current geo-information, satellite and aerial imagery to support spatial-temporal research aspects on different scales in space (regional 1:200,000 to local 1:25,000) and time (nearly 250 years in three time steps, the last 30 years also in three time slices).

The Virtual Laboratory for Cultural Landscape Research (VKLandLab) is designed and developed by the Chair of Geodesy and Geoinformatics, hosted at the Computing Centre (ITMZ) and linked to the Digital Library (UB) at Rostock University. VKLandLab includes new developments such as wikis, blogs, data tagging, etc. and proven components already integrated in various data-related infrastructures such as InternetGIS, data repositories and authentication structures. The focus is to build a data-related infrastructure and a work platform that supports students as well as professional researchers from different disciplines in their research in space and time.

1 Motivation

The modern knowledge and information society with its various possibilities of efficient communication and easy access to very large amounts of information and powerful computing technology is a new challenge to science. The chances to achieve better scientific results, both qualitatively and quantitatively, with the new methods have increased considerably in parallel, but also the difficulty to control such distributed, dynamic system components. Under the heading "e-science" a network-based science, "grid-based science" or "digitally enhanced science" is becoming increasingly established. This requires the systems development and the organizational structures of a network and middleware infrastructure, with which computational resources, information resources, application programs can be offered, requested and allocated for community-specific applications (BMBF Science Management 1/2005).

With the promotion of virtual research environments the German Research Foundation (DFG) aims to further develop integrated information infrastructures for grid-based research. The new communication technologies and publication procedures permanently change the existing information infrastructure and the traditional publication process. Digital information and communication networks provide the technical requirements for time-and location-independent collaboration. Essential for the development and effective use of new communication and publication networks are powerful tools and infrastructure to support the scientific work processes. These include the virtual research environments, platforms for network-based collaborative work processes that support new forms of cooperation and result in an easier access to scientific data and information. They provide both a central access to each subject-specific resource, data and document as well as the necessary conditions for a substantive link between the various information units.

2 Cultural Landscape Research

The cultural landscape analysis – as an example of historical-geographical, land use change and land improvement research – requires the modelling of landscape and socio-economic processes over time in their history. In assessing the sustainability of landscape developments the spatial distribution patterns and the mosaic of key elements in the landscape need to be considered. For this purpose, the cultural landscape research uses historical data sets. Large amounts of data collected in electronic form over the last decades and centuries in statistical offices, museums, archives and numerous historical and geographical research projects can be set in value using modern IT methods. For questions of landscape monitoring (NEUBERT & WALZ 2002, WALZ et al. 2004) or for inventarisation of the cultural landscape (PLÖGER 2003) since many years historical maps are investigated with modern IT methods.

2.1 Historical maps and geo-information data sets in Mecklenburg

For over a decade the professorship of Geodesy and Geoinformatics is dealing with the exploitation of historical maps (sometimes also called old maps) in digital form as a source of interdisciplinary landscape research. Around the year 2000 for the first time an attempt started to offer maps on the area of the historic Mecklenburg in a homogeneous digital form. Scanning the old maps of Wiebeking (1786-1788, 48 sheets, 1:24,000) and Schmettau (1788, 16 sheets, 1:50,000) and the topographic base maps (1877-1889, 168 sheets, 1:25,000) made available the oldest topographic maps of large parts of the federal state Mecklenburg-Vorpommern (more precisely, of the duchies of Mecklenburg-Strelitz and Mecklenburg-Schwerin) in an excellent quality (GROSSE & ZINNDORF 2001, GROSSE 2003).

KRESSNER (2009) evaluated these scanned maps in terms of their quality and their geometric origin. He investigated the thematic suitability for scientific analysis, especially in relation to landscape research issues. These old maps are georeferenced and their geometric quality is described. They can be deployed and integrated into GIS in the sense of historical digital primary research data holdings of the science. This is the main data set for the designed virtual research environment for the modern cultural landscape research. With a territorial extension of 13,000 square kilometres and a time horizon of around 225 years, there is nothing comparable in the world. Combined with the spatial data of today (digital topographical maps in raster (DTK 10) and vector form (DLM) from the Authoritative Topographic-Cartographic Information System (ATKIS)), and embedded in modern information and communication technologies (ICT), new research avenues and forms of access for cultural landscape research are becoming available. The potential of such a virtual platform for cultural landscape research is illustrated by KRESSNER (2009) with small case studies and student projects on spatial-temporal changes in the region of Ribnitz, a small city in Mecklenburg.

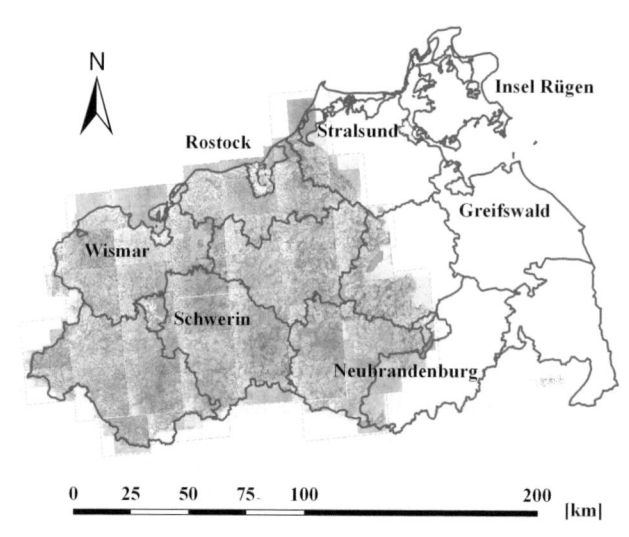

Fig. 1: Area coverage of Mecklenburg from 1786 in the today's boundaries of Mecklenburg-Vorpommern

3 The Project VKLandLab

3.1 Technology partners in the project

The research project VKLandLab is processed by several partners in common. On the one hand, the underlying technology is designed and developed from the chair of Geodesy and Geoinformatics (GG). As infrastructure units the University Computing and Media Centre (ITMZ) and the University Library (UB) are participating. ITMZ provides network-based central and distributed information and communication services for university research, teaching and administration. Here, various database applications are operated and maintained, including a central image database and the digital library of the University. This ensures a full provision of online documents such as digital research data and results literature of the library, and online databases.

At the University Library (UB) the Digital Library is developed as a central, important strategic infrastructure component of the university. The document server RosDok (http://rosdok.uni-rostock.de) as part of the digital library – a joint project and cooperative service of the University Library, the ITMZ and the department for databases and information systems of the university – includes all forms of electronic publications to be researched, presented, and permanently archived. RosDok's available documents are accessible by metadata and are free (open-access) on the Internet. Digitised sets of historical documents are also stored and structured according to the METS format, allowing them to be integrated and visualised in the developed viewer. For the exchange of metadata with OAI search engines, the OAI-PMH interface can be used. A conversion to common metadata formats (e.g. Dublin Core, XMetaDiss) can be done. To increase the visibility of the documents they are registered in the catalogue of the library and in the catalogue of the Common Library Network (GBV).

3.2 Research partners in the project

The technological platform is evaluated and used by an interdisciplinary team of students and researchers for their research lines. Different scientific disciplines of the University of Rostock are involved:

Historical sciences interrelate a database for Mecklenburg census of 1819 in the Grand Duchy of Schwerin to Schmettau`s maps and a census of 1867 in the Grand Duchy of Schwerin on the first topographic maps of the Prussian surveying and mapping.

Settlement planners and experts for preservation of monuments dedicate themselves to the investigation of village forms, e.g. parish forms and farmyard types. For the region a canon of typical, representative and satisfactory phenomena (local forms and their stages of development) should be derived and prepared for a systematic and appropriate generalization.

Landscape ecologists, dealing with the analysis of areal distribution, severity and location of the woodland and marshland in Mecklenburg over time, try to edit and create a basis for large-scale analysis of the structure (biodiversity) and function (humus storage) of forests and moor.

Landscape planners study the implication of the historical-genetic cultural landscape development in the tools of today's spatial planning and management of historic landscape features. The dynamics of land changes is an important basis for defining sustainable models and general principles for spatial planning and regional policy.

Hydrologists generate hydro meteorological and hydrological relevant basic parameters for further analysis in the cultural landscape, such as for grass reference evapotranspiration, the climatic water balance, the expansion of drainage systems and water development. This will create the basis for the coupling with model-applications (e.g. water and nitrogen balance models).

Geodesists and geomorphologists investigate the kettle holes, hollow forms (Sölle), caused by melting out of an enclosed sediment block and distributed especially in the younger Pleistocene areas. In Mecklenburg-Pomerania about 90,000 hollow forms occur, nevertheless their form of origin and development is under discussion.

The spatial-temporal analysis of the development of habitat fragments, the analysis of the spatial distribution pattern of these kettle holes in their historical development, the floristic inventory of selected hollow forms particularly is of interest to *biologists*. The comparison with existing historical data and the analysis of the relationship between landscape dynamics (e.g. reduction in the number of kettle holes) and floristic composition of plant communities in kettle holes may allow to derive long-term preservation strategies of species-rich communities in kettle holes.

4 Concept and implementation

4.1 Components of the platform

Following the basic principles of e-science the technical objectives of the project are to create a central portal application that support researchers in their collaborative work with essential tools and resources for dealing with spatial data. A central component in addition to components of communication and workflow management is a spatial data infrastructure (SDI), which includes a basic collection of technologies, policies and comprehensive agreements in order to make spatial data consistently available and accessible (BILL 2010). The Open Geospatial Portal Reference Architecture defines a SDI as a central access point to geographic information resources. One of the intended project objectives is to provide the opportunity to investigate, view, generate and administer content and spatial information for different user groups at one (virtually) central location. The OGC reference architecture is thus a good basis to formulate technical demands on the development and coordination of the portal application. The components of the total information system (see Figure 2) can be divided into the following categories:

Portal services provide a central entrance point and access to all relevant functions and tools of the portal application. In addition to the aggregation and abstraction of all available resources, they also allow access to management and administration of the portal itself. To implement this functionality, combined with the design of fixed and variable content of the web site of the project, a free Content Management Framework (TYPO3) is used. In addition to predesigned layouts in the corporate design of the University of Rostock, the flexible PHP framework supports the embedding of dynamic visualization and processing components of the portal application.

Data services and view services allow the provision of spatial data for processing by the users and their visual map-based treatment as a foundation for a detailed assessment of the content. Therefore the open source WebGIS framework kvwmap is used (BILL, KORDUAN & RAHN 2008). kvwmap is built on top of the UMN MapServer development for collecting, processing, analyzing and presenting geospatial information. The user interface of a Web-based client offers similar functionality as desktop GIS for viewing and processing of spatial objects and specialized describing data.

Catalogue services permit the discovery of spatial data based on searchable parameters for the data theme, origin, and appearance. The search in a central meta information directory put users in the position to identify and allocate necessary data and make them accessible independent on type and location of their storage. For this purpose the free web-based application catalogue GeoNetwork Open Source is used. Portal users can query metadata to existing research records and create own metadata for new records. Metadata records can be presented in various standardized forms. For geospatial information this means the use of ISO 19115/19139 (with any possible INSPIRE compliance).

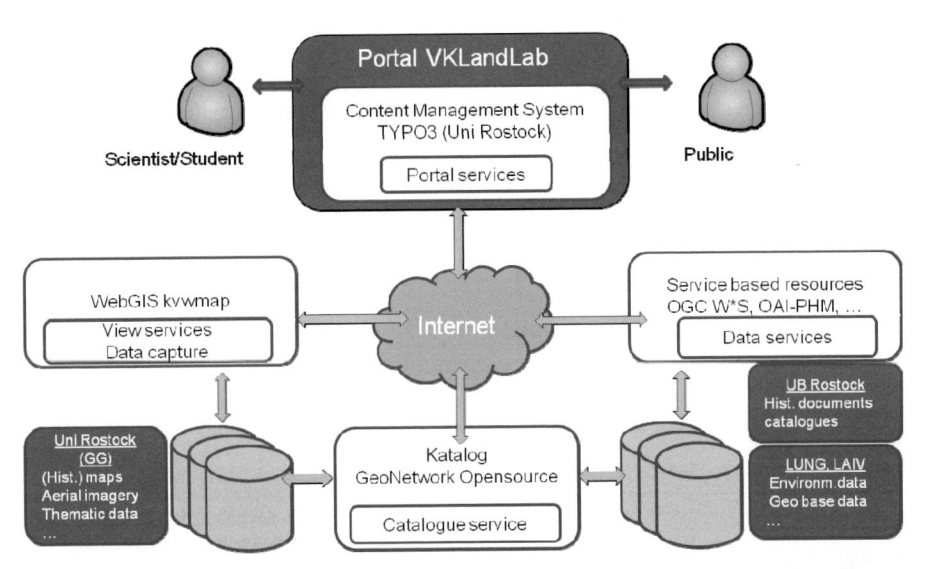

Fig. 2: Portal components with related services and data sources (following the OGC Geospatial Portal Reference Architecture)

Furthermore, in the database of the catalogue references are kept to historically-documents from the holdings of the University Library. This is achieved through a continuous alignment between GeoNetwork and the catalogue of the library system through the OAI-PHM protocol.

All the above mentioned components are available as standalone applications, but they can also be integrated strongly intertwined with each other because of their consistent and standardized approach to service-based OGC/WWW interfaces. Thus, a higher-level abstraction is achievable for user applications. The modularity of the individual application also enables future flexibility and scalability of the entire portal for sharing and modification of individual components.

At present, another component of the category data services, but here treated separately as a special feature of the historic nature of the portal application listed, is in the design and development phase. This is a *historical place names directory*. Local references do exist in metadata records of many cultural and historical projects or reference to catalogues of libraries. Nevertheless their existing query mechanisms for place names, both geographically and chronologically, are less suitable for visualization and automated analysis. Within the project a Web-based local name service (a *gazetteer service*) is implemented in compliance with relevant interface standards (Web Feature Service). This service should provide a high resolution search in time and space in Mecklenburg for local historical terms and synonyms over the last 250 years. Ambiguities are resolved here in as much detail as possible and links to current/recent official administrative units are produced. The underlying data set is combined from a series of freely available contemporary (such as genealogienetz.de), administrative (State Office of Internal Administration (LAIV) MV, Federal Agency for Cartography and Geodesy (BKG)) as well as historical name inventories (Wossidlo Archive, Historical Census composed 1819/1867).

Due to the resulting spatial referencing it is possible to link records to spatial research services giving users expanded opportunities for access. Furthermore, the automated integration into higher level data infrastructures of the geoinformation and library community is possible.

In addition, ICT-related resources for project coordination, internal communication and workflow management, such as Microsoft Sharepoint and mailing lists are available.

4.2 Data sets and data modeling

The InternetGIS kvwmap retrieves data, both from a local PostGIS spatial database system as well as from external data services (according to the OGC-based standards such as Web Feature and Web Map Service (WFS/WMS)). At this time data offered include more than 30 layers (Fig. 3) of basic data sets as well as special thematic data, historical maps, and

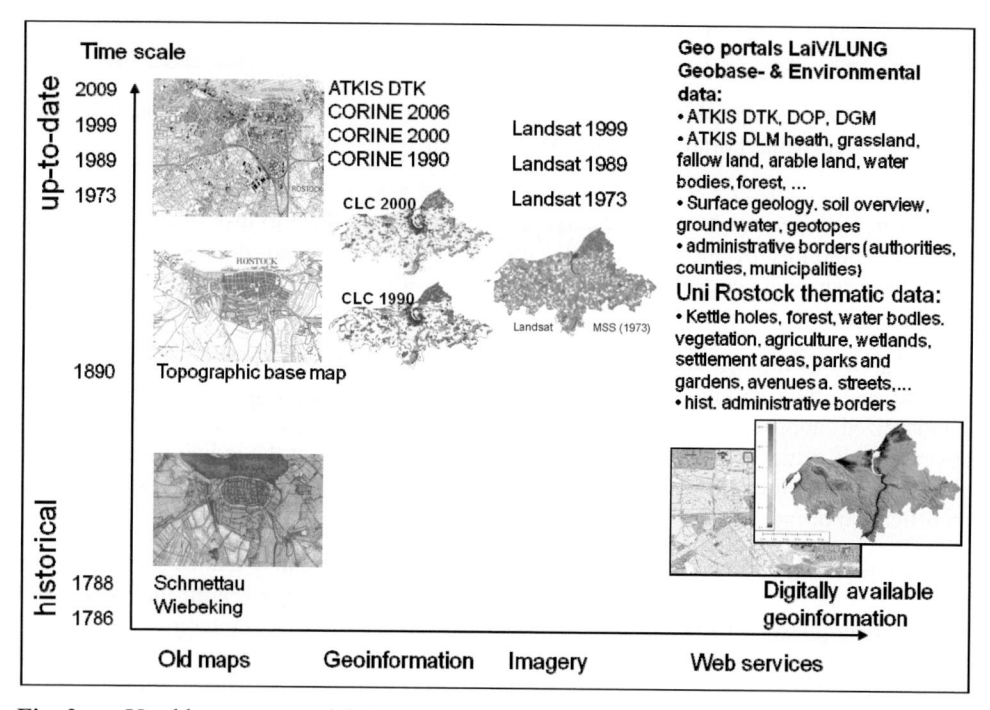

Fig. 3: Used base maps and data sources

12 environmental thematic layers embedded via WMS/WFS of the State Office for the Environment, Nature Conservation and Geology (LUNG) and LAIV Mecklenburg-Vorpommern. Further resources are various historical data sources, such as Digital Atlas of the Historic Mecklenburg on land use and settlement pattern in the 18th Century and census data for the Census in Mecklenburg-Schwerin in 1819 and 1867 (based on original cards) in the treatment.

The individual layers of the subject-specific research topics are based on 13 technical data models (see Figure 4 on the example Forest areas), which include for example attribute values on time section of the captured cards to the scientific information. To date, in research and student projects over 32000 categories of spatial objects are digitized.

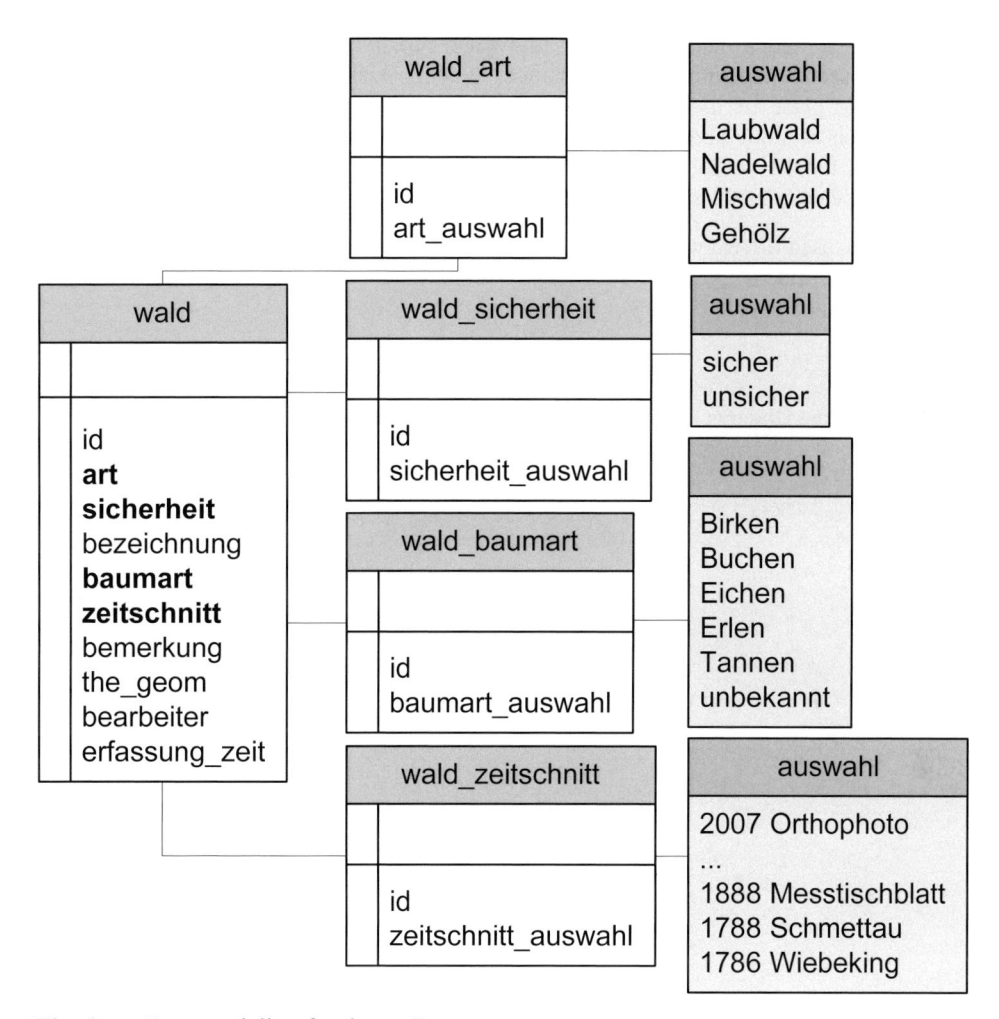

Fig. 4: Data modeling for theme Forest areas

5 Students case studies

The virtual research platform was used in master student teaching and student training activities in this winter term in the module "Geoinformatics" in the specialisiation "Integrated local planning" (Figure 5 and 6). Each student had to digitize a tile size of 10*10 kilometres in Mecklenburg. Different objects, such as the settlement area, the wetlands, vegetation and agricultural use for the 3 time slices were captured within the virtual research laboratory and the InternetGIS kvwmap. Following this, a summary evaluation was done for each object class, and the results are displayed by using visualizations in form of maps and diagrams. Therefore the students used ArcGIS and Excel, in most cases by terminal server at their student apartment.

In addition each student had to solve an individual research question given by the above mentioned research disciplines. For instance students dealing with kettle holes had to investigate the place of the kettle hole in relation to the digital terrain model (DTM). Attributes such as slope and aspect had to be generated from DTM and appended to the attribute table for the kettle holes. Thus the student could analyse whether the kettle hole lies in the slope or in the plane, in the sink or on a summit.

Other students used different landscape metrics measures (such as nearest neighbour, shape index, proximity index) to derive information on the functioning and the interaction of the individual digitized objects.

Beside a lot of problems caused by the parallelism of developing the platform while students digitized and different versions of ArcGIS in the computer laboratory and via terminal server the students evaluated the use of the Internet platform in general as positive.

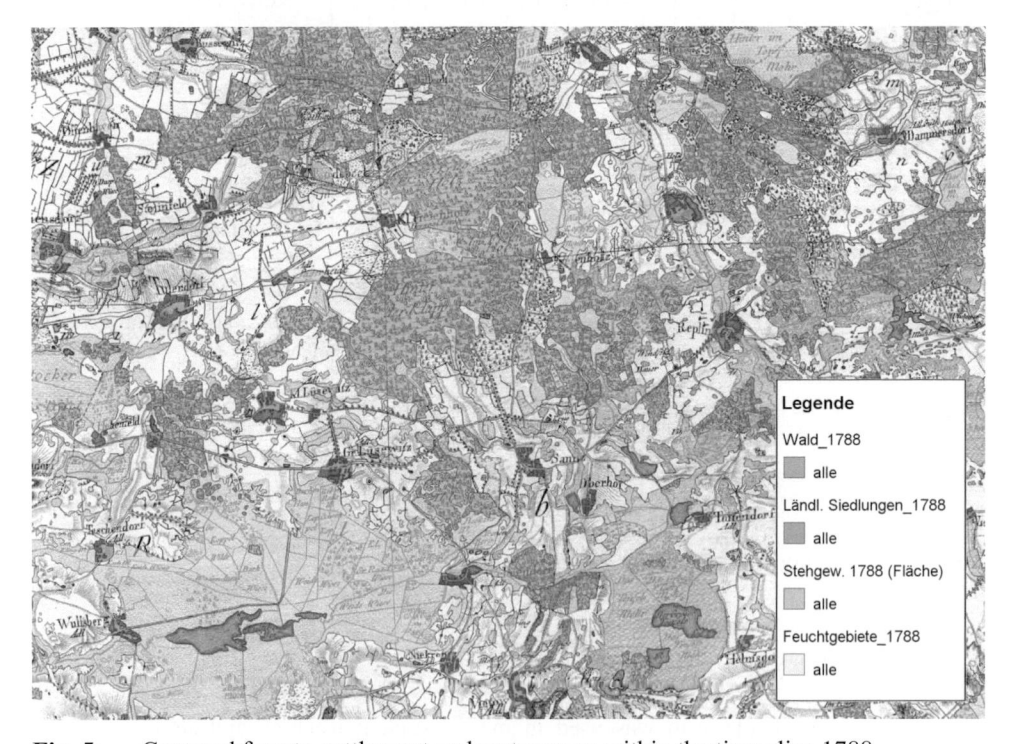

Fig. 5: Captured forests, settlement and water areas within the time slice 1788

Fig. 6: Captured features within the feature class "Kettle holes"

References

Bill, R. (2010), *Grundlagen der Geo-Informationssysteme*. 5. Auflage. Berlin/Offenbach, Wichmann, 814 p.

Bill, R., Korduan, P. & Rahn, S. (2008), *kvwmap – GIS-Entwicklung für Kommunen und Landkreise*. In: Transfer, Das Steinbeis Magazin, 02, 10-11.

Große, B. & Zinndorf, S. (2001), *Möglichkeiten und Grenzen der Nutzung von Altkarten, mobiler Scan-Technik und GIS-Anwendungen in der Landschaftsforschung*. In: Scharfe, W. (Hrsg.): 10. Kartographiehistorisches Colloquium, Bonn 2000.

Große, B. (2003), *Bedeutung digitaler Altkarten für GIS-Anwendungen in der Landschaftsforschung*. Jahrestagung der Kartenkuratoren (D, CH), 26.05.2003, Freiburg.

Neubert, M. & Walz, U. (2002), *Auswertung historischer Kartenwerke für ein Landschaftsmonitoring*. In: Strobl, J., Blaschke, T. & Griesebner, G. (Hrsg.), Angewandte Geographische Informationsverarbeitung. XIV Beträge zum AGIT-Symposium Salzburg. Heidelberg, Wichmann, 396-402.

Kreßner, L. (2009), *Digitale Analyse der Genauigkeit sowie der Erfassungs- und Darstellungsqualität von Altkarten aus Mecklenburg-Vorpommern – dargestellt an den Kartenwerken von Wiebeking (ca. 1786) und Schmettau (ca. 1788)*. Dissertation, Universität Rostock.

Plöger, R. (2003), *Inventarisierung der Kulturlandschaft mit Hilfe von Geographischen Informationssystemen (GIS). Methodische Untersuchung für historisch-geographische Forschungsaufgaben.* Dissertation, Rheinische Friedrich-Wilhelms-Universität Bonn.

Walz, U., Lutze, G., Schultz A. & Syrbe, R. U. (Hrsg.) (2004), *Landschaftsstruktur im Kontext von naturräumlicher Vorprägung und Nutzung – Datengrundlagen, Methoden und Anwendungen.* IÖR-Schriften, 43. Dresden.

Acknowledgements

The author thanks the Deutsche Forschungsgemeinschaft (DFG) for funding the project within the funding virtual research laboratories programme (support code Bi 467/21-1).

Visualizing Wetland and Meadow Landscapes

Howard HAHN

1 Introduction

In highly energy and water subsidized countries, "green" site planning and engineering is moving toward greater water and energy conservation. One outcome of this movement is the transition of once manicured landscapes to a looser appearance where reconstructed wetlands and meadows are managed in their naturalistic form for ecosystem function. Unfortunately, the public often regards these landscapes as messy or weedy without cultural cues of ordered care (NASSAUER 1995). Despite design efforts to elevate the aesthetics of these unkempt landscapes, developers are sometimes resistive to implement conservation developments that potential buyers cannot easily visualize prior to construction. Engineers meeting with municipalities and other reviewing agencies face similar visualization needs when trying to explain the layout and anticipated appearance of naturalistic bio-filtration areas aimed at improving water quality. In my experience, engineers and reviewers are moving beyond the generalities of "eyewash" renderings toward a more exacting scrutiny of scene specifics: plant species selection, compositional mixes, growth heights, densities, and other complexities like those described by DEUSSEN et al. (1995). Furthermore, as a working tool, these renderings may need to undergo several iterations of design refinement.

Several visualization methods can be used: precedent photographs of similar built projects, hand-generated renderings, photo-montages, computer modeling/rendering (typically architecturally oriented), and synthetic landscape generation programs. This paper briefly mentions these methods, but focuses on a fourth-generation synthetic landscape generation program, e-on Software's Vue™ (www.e-onsoftware.com), that truly mimics real-world scenes in visual richness and detail. Originally developed for artists and the entertainment industry, this software is now being applied to more precise landscape design and engineering environments. Vue is particularly well-suited for wetland and meadow landscapes where thousands or hundreds-of-thousands, of unique 3D plants can be distributed and rendered in a virtual world of digital landform, water, understory shrubs, trees, clouds, and atmospheric lighting. This paper will summarize how this new digital technology (Vue) has the potential to transform the rendering of wetland and meadow landscapes to enable visualization of looser plant forms characteristic of "green" design.

2 Purpose and Need

In recent years, there has been an increased emphasis on creating or restoring wetland and meadow landscapes to increase stormwater runoff infiltration, reduce overall landscape water use by new land development projects, or improve surface water quality through bio-filtration. In the United States, sustainable development strategies promulgated by both the Leadership in Energy and Environmental Design–Neighborhood Design (LEED-ND) and the Sustainable Sites Initiative (SSI) have been increasingly applied (voluntarily) to achieve water use and quality objectives. When I informally advocate meadow landscapes, one common complaint is the perception of a "weedy" appearance which may be hindering wide-spread adoption. With proper site selection and analysis, thoughtful species selection, deliberate design, proper planting, and informed maintenance practices, the aesthetics can be striking. One of the best recent books to provide detailed guidance is *Urban and Suburban Meadows* (ZIMMERMAN 2010).

Apart from the water conservation and ecosystem functions of these landscapes, there is also a financial incentive in both life-cycle maintenance costs and higher development returns for these landscapes if implemented correctly. The Urban Land Institute (ULI) has documented many case studies of green buildings and sustainable landscapes as being highly desired and commanding premium prices (ULI 2007, 13).

As more attention is paid to designing, constructing, and marketing sustainable design incorporating wetlands and meadows, there is a greater need for enhanced visualization techniques. Traditional watercolor illustrations, or even computerized photo-simulation style renderings, may not be detailed enough or accurately depict grass or wetland plant species in regards to composition, density, scale, color, or form. Many illustrators also cater to architectural clientele and may be unfamiliar with the intricacies of landscape elements other than context entourage for framing buildings or development.

Another shortcoming of manual and computerized visualization methods to date is arduous iterations of design refinement. Once the perspective view is blocked out and the illustrator receives verbal direction – and perhaps reference photographs – wetland or meadow plants are generally introduced in mass using suggestive artistic strokes. If layers were used, some amount of control is retained, but major changes require a total rework of layers at prohibitive time and cost.

The new digital tool, e-on Vue, shows promise of being a transformative tool for landscape depiction of wetland and meadow landscapes. Four case studies will be used to highlight early explorations with Vue. Although engineering approvals and the public process do not yet require detailed visualization, two case studies in a practitioner setting will show how Vue allowed greater scrutiny of planting and engineering details earlier in the process.

3   Emergence of a New Tool: e-on Software's Vue™

Synthetic landscape generation software has steadily evolved over several decades. Early programs like Perspective Plot by Devon Nickerson were capable of rendering un-shaded 3D views of wireframe mesh terrains (NICKERSON 1980). Vegetation was limited to stick trees used to evaluate forestry cut patterns. Software capable of rendering realistic terrain, atmospherics, and architectural forms were early milestones. However, mass planting and highly realistic 3D vegetation required the convergence of high performance hardware and sophisticated algorithms. A major step forward was automated vegetation placement according to local terrain conditions (slope, elevation, and aspect) introduced by 3D Nature with the release of the World Construction Set program in 1994 and a GIS-oriented product, Visual Nature Studio, in 2001 (3D NATURE 2010). Although these programs were revolutionary in landscape depiction, plants were not true 3D geometry. Rather, the approach taken was alpha mapping plant images on 2D billboards viewed in 3D, even though some variation was introduced through randomized scaling, rotation, and mirroring.

The approach by e-on Software was to use true 3D plants, which in the opinion of the author looked rather cartoonish in early software versions. Continued advances in hardware, algorithm development, and introduction of automated planting increased the power and realism of Vue (E-ON SOFTWARE 2010). Today, the standalone program Vue Infinite, or the Vue xStream version which integrates with Autodesk 3ds Max, Maya, Softimage, Newtek Lightwave, or Cinema 4D is emerging as the de facto standard for artistic landscape depiction as evidenced by its widespread use by artists and film-makers.

As a former landscape architect practitioner (now academic) who has used numerous terrain and vegetation generation programs over several decades, I first began applying Vue to constructed wetlands in an urban context to illustrate water quality treatment chains being proposed by project engineers. Since Vue primarily appealed to landscape artists, significant early time was spent determining how to integrate Vue into the workflow of more technically inclined projects. Issues included the proper scaling of imported engineering data, determining the adequacy of available plant material for the intended wetland environment, and testing combinations of input/output formats.

Vue is a complex program capable of producing imagery of any real or imagined natural scene. A brief summary of the Vue program editors include:

Terrain Editor: Enables freeform sculpting of terrain with a variety of tools applied through an airbrush and global effects (erosion, fluvial, etc.). Terrains can be automatically generated from within the program as procedurals, or imported as grid/TIN objects, USGS DEMS, or height field maps.

Atmosphere Editor: Multiple atmosphere models are offered including standard, volumetric, spectral, and environmental mapped. The editor provides full control over sun properties, clouds, sky/fog/haze, wind, and special effects (stars, rainbows, and ice rings).

Light Editor: Controls supplemental light sources (other than the sun) placed in the landscape. Controls include lens flares, light gels, volumetric effects, shadows, light attenuation/colors, and the influence of light on scene objects.

Plant Editor: Three-dimensional plants can be created by modifying the branching habits and leaf characteristics of prototype Solid-Growth™ forms. When placed, each plant instance will be unique in terms of branching pattern and scale. Alternatively, it is possible to import plant objects from other programs, but the detail of the forms will not be unique.

Material Editor: Using this editor, sophisticated procedural materials can be created. As a variation, a material can be defined as a layered ecosystem where placement of objects is automatically controlled according to environmental conditions of slope, elevation, aspect, and other influences. Additionally, object locations can be "painted" according to a compositional mix and density, or manually and individually placed. Ecosystem objects can be plants, fallen leaves, rocks, or any modeled creation.

Function Editor: As a hallmark of sophisticated programs, the function editor allows extended control of scene properties or object placement, appearance, and motion. This is accomplished through constructing a process map of linked functions controlled by mathematical algorithms manipulated using a graph. Although this editor takes experience to master, it offers immense power.

Animation Editor: Any object or global condition (wind, clouds, water, etc.) can be animated.

Python Scripting: An industry standard, cross-platform, and object-oriented scripting language enabling the creation of customized Vue processes and integration with other programs. This feature is oriented to programmers.

Fig. 1: Same Vue scene rendered under four different lighting conditions: pre-dawn, afternoon, sunset, and moonlit night (Rendering: K. Kleinschmidt and H. Hahn)

4 Methods

Application of this program was tested with four case study projects: two wetland design projects in Los Angeles (real), a student project depicting a Louisiana bayou (academic), and a meadow-based residential project (academic). These projects required importing terrain data and crafting structures in 3D modeling programs. Considerable time was spent preparing the terrain for planting and testing plant composition, scaling, and density. Summary details of all case studies are shown in Table 1.

Case Study 1 (CS1) – South Los Angeles Wetlands Park: Conceived in the early stages of engineering, this first personal application of Vue was directed at illustrating how storm water influent would enter a forebay, and then progress through three treatment cells (ponds) intended to provide bio-filtration for improving water quality. Although terrain sculpting is possible within Vue, a precise triangulation was imported to achieve a pond edge profile consisting of a deep pool, shallow littoral bench (2-3 feet deep), and upslope areas. The site was divided into five zones (open water, emergent marsh, riparian scrub, riparian woodland, and upland) and a plant list for each zone guided plant selections. To precisely control automatic plant placement, the triangulated terrain was sub-divided into discrete areas. Large trees were manually placed to control exact positioning. All other 3D objects such as buildings, fencing, bridges, benches, and other site accouterments were modeled using Google SketchUp and imported via the .obj format to preserve textures.

Case Study 2 (CS2) – Los Angeles River Natural Park: This project was another naturalistic park (former urban golf/tennis facility) used for treating urban stormwater. Construction techniques replicated the workflow used to create the first case study. However, a grading plan had not yet been developed by project engineers, so much latitude existed regarding the overall wetland grading, trail system/bridge layout, fencing, amenities, plant selection, and planting design details. Workflow consisted of developing a rough concept model to review several viewpoint options before undertaking more detailed modeling and plant mix compositional tests.

Case Study 3 (CS3) – Bayou Bienvenue (KLEINSCHMIDT 2010): The focus of this academic project was to visually depict a series of bayou water quality treatment cells under various operational and dredging conditions. A site area of 700 acres (not all planted) was a good test of Vue's data capacity and rendering speed. Like the previous case studies, a triangulated terrain model was constructed and various plant selections were tested under different water depth conditions. Planting design was entirely done using Vue with no paper-based intermediate plans. The observation tower, pavilions and boardwalks were constructed in SketchUp and imported into Vue. Unlike the other wetland case studies, procedural textures were tested and applied to cell bottoms to simulate how the exposed soils might appear during dredging operations.

Case Study 4 (CS4)– Stagg Hill Residential Development: For this project, terrain was imported into Vue by converting a 2m LIDAR digital elevation model into a height field map. Since the map's dynamic range is limited to 256 gray levels, vertical accuracy of the LIDAR data was reduced from seven inches to two feet. Vegetation ranged from woodland slopes to open meadows. Most of the digital plant models were selected from the default Vue plant library. Using the Vue plant editor, a few additional meadow species were custom created from scans of actual local prairie grasses collected during a field trip.

Case Study	Site Area (acres)	Model Size (polygons)	Simulations (variations)	Production Time (hrs)*
CS1 – LA Wetlands Park	9	215 million	5 stills	140
CS2 – LA River Natural Park	17	625 million	1 still	75
CS3 – Bayou Bienvenue	700	625 billion	3x 2var 2x 6var	40
CS4 – Stagg Hill Meadows	31	1.1 billion	20 stills 1 animation	8 8

* As inexperienced users, much extra donated time (in the case of commercial projects) was required for technique testing and detailed modeling to meet engineering requirements. Production times will dramatically decrease with experience and the compilation of plant libraries. Production times do not include unattended machine rendering.

Table 1: Case Study Summary

5 Results

5.1 Vue Compared to Other Visualization Methods

After using the case study projects to gain experience, the following section is a general assessment of how Vue compares to other visualization methods performed by myself, students, or hired illustrators.

Traditional Media: Methods like watercolor have long been used to illustrate wetland- and meadow-type landscapes and can be very visually appealing. For general public communication, these traditional methods remain a viable – and sometimes preferred – option. Lack of overall detail can be an asset for design still in the "visionary stage" where detail is implied, but not explicitly shown. Vue offers the advantage of producing much higher levels of detail sometimes necessary when communicating with a technical audience of engineers and limnologists. Vue images can also appear "painterly" if correct lighting, materials, and forms are used. Post rendering filters can also be applied to achieve a more artistic appearance. Lastly, once a terrain/base model is developed in Vue, it is easier to rapidly produce several planting scheme iterations that would be difficult in traditional media without major rework or starting over.

Photo Montage Techniques: Within the last decade, photo montages have become very popular because of production speed and suggestive visual character. Strange juxtapositions and scale distortions might be part of the communication appeal to imply unresolved visual or user activity possibilities. It has been my experience that audiences sometimes view montages too literally or get confused. Vue represents the opposite end of the accuracy/realism continuum.

Photo/Model Simulation: Today, most development illustrations are based on computer modeling and digital photo compositing/graphic rendering. Computer modeling establishes accurate scale and perspective, and digital rendering (by computer or manually done) provides visual richness. As a shortcoming, depicting detailed plant forms and mixes across expansive areas is difficult using architecturally-oriented software. Non-Vue programs can be overwhelmed by detailed or vast planting. For simulations skewed towards landscape depiction like wetland and meadow scenes, Vue excels and brings realism to levels matching architecturally-oriented software. Vue also enables rendering with more light and weather variation, which is not easily achieved through supplemental photo compositing. Since Vue scenes are three-dimensional down to individual plants and landscape forms, lighting and weather conditions apply to every minute scene element in an integrated world.

Other Landscape Creation/Rendering Software: Prior generation landscape creation programs using 2D plant billboards yield similar results to Vue, but with less visual richness in terms of detail and sophistication. Some of these programs retain the advantage of close integration with Geographic Information Systems (GIS) which enables visualization of GIS analyses and better quantification of plant materials. Vue data interchange with GIS software may be possible via python scripting, but it is untested to the author's knowledge.

5.2 Critique of Case Studies

Case Study 1 (CS1) – South Los Angeles Wetlands Park: As my first use of Vue, the level of detail and realism far exceeded other visualization methods I had used and accelerated engineering review. Unprecedented in my experience, visualization at this complexity was enabled after the conceptual design process in which some amount of engineering had been done, but before the initiation of detailed engineering design. Much of the project design aesthetics including wetland pond shapes, bridge locations and appearance, overall layout, and planting appearance were first proposed through Vue renderings. Although plant selection was guided by a recommended plant list, no detailed planting plan was prepared so Vue became the planting design tool. Vue also enabled several fairly rapid review iterations with project engineers to fine tune wetland and upland plant selections, density, and growth heights which would have been difficult using other visualization methods. Some structural elements like bridge materials were scaled back for public review to avoid excessive financial commitments before detailed design and cost estimates were undertaken. The project was unanimously approved by the City Council and received a write-up in the Los Angeles Times in which a Vue rendering was the feature graphic (LEOVY 2008).

Fig. 2: Vue rendering (1 of 5) of South Los Angeles Wetland Park delineating water flowing from treatment Cell#1 to Cell#2 (Source: Psomas 2008; rendering by H. Hahn)

Negatively, production time for this project was high (Table 1). Much initial modeling time was spent in external modelers in preparation for import into Vue. Additional time can also be attributed to first time use of Vue, terrain subdivision to precisely control planting, plant preparation and testing, and test renders for materials and lighting. Once the model was satisfactorily developed and populated with plants, multiple viewpoints (5 in this case) offset the cost per rendering.

Case Study 2 (CS2) – Los Angeles River Natural Park: No new techniques were developed through this project, but the site was nearly twice as large as CS1. Even though the polygon count for the CS2 model was three times larger than CS1, rendering only required one-fifth the time due to next generation hardware processors and improvements in Vue software algorithms. Terrain modeling followed the same workflow as CS1. Similar to CS1, rendered results went through several iterations enabled by Vue. The detailed and accurate depictions of both plants and engineering structures (spillways, bridge piers, etc.) allowed engineers to "think a bit more about the engineering aspects early on in the planning process" (D. Beck, Psomas project engineer, personal communication, January 25, 2011).

Fig. 3: Vue rendering of Los Angeles River Natural Park showing water treatment chain (Source: Psomas 2010; rendering by H. Hahn)

Case Study 3 (CS3) – Bayou Bienvenue: Workflow followed the techniques of CS1 and CS2, so a major focus of this case study was to see how well Vue handled huge amounts of data (625 billion polygons). If planting design is well organized by zone and file structure, and plant groups are not all turned on at once to avoid exceeding graphic card memory, the program has no problem rendering large areas. Rendering times ranged from two to eleven hours per image for five different scenes each having multiple variations of planting complexity. For the amount of terrain modeled and planting design time invested, production was extremely efficient considering 18 renderings were produced in extreme detail covering six bayou successional stages. Based on my career experience, few, if any, visualization methods or software could replicate these results.

Fig. 4: Vue scenes from observation tower showing 4 of 6 successional stages of Bayou Bienvenue (Rendering: K. Kleinschmidt and H. Hahn)

Case Study 4 (CS4) – Stagg Hill Residential Development: This academic case study demonstrated how Vue can be similarly applied to meadow landscapes with impressive results. Other techniques can produce visually appealing results, but no other technique or software tool can produce the realism, complexity, or precision that Vue offers. Grass ecosystems are high-polygon count models, but again, Vue exhibited no problems. Vue enables many possibilities of rendering meadows under varied lighting or weather conditions which could expand design review considerations. Although not yet tried, grass and various flowering herbaceous mixes could be virtually tested in Vue to evaluate how the aesthetic design and appearance over time and seasons might be improved to foster greater public acceptance (also not yet surveyed).

Fig. 5: Vue scene of existing Stagg Hill meadow prior to development (Rendering: D. Cross)

6 Conclusions

Case study results using Vue have been impressive and summary conclusions are:
- Use as Visualization Tool: For wetland, meadow, and other complex landscapes, Vue is an excellent visualization tool and in some instances might be considered the exclusive tool compared to other visualization methods.
- Data/Program Integration: Vue is robust in its import/export formats and integrates well with other 3D modeling, CAD, and imaging software. It has potential to integrate with GIS through python scripting, but this has not yet been attempted.
- Project Scales: As evidenced in the case studies, Vue was stable and can be used productively at multiple project scales and model sizes.
- Iterative Review/Design: Automatic ecosystem planting enables multiple design iterations with minimal effort. Since detailed visualization can be transitioned to earlier stages of the design process, visual outcomes can be more readily influenced.

- Use of Vue as a Design Tool: As a design tool, Vue can be used to interactively sculpt terrain. Although not attempted in the case studies, sculpting is probably best suited at the conceptual design stage before terrain needs to conform to precise engineering standards. All detailed planting design can be solely performed using Vue.
- Expanding Rendering Possibilities: As a fully integrated world of lighting, atmospherics, and weather, Vue greatly expands the realm of visualization possibilities in both still and animated formats.
- Learning Curve: Impressive results can be achieved by just using environmental presets and commercially available plant libraries. Vue is also a complex program, and fully utilizing its immense power (like the function editor) requires a high level of commitment. Simpler, artistic versions are available, but professional designers and illustrators will prefer the full feature set of Vue Infinite or Vue xStream.

References

3D Nature (2010), *Brief history of computer 3D landscapes and 3D Nature.* Retrieved 19 December, http://3dnature.com/history.html.

Deussen, O., Colditz, C., Coconu, L. & Hege, H. (2005), *Efficient modeling and rendering of landscape.* In: Visualization in landscape and environmental planning: Technology and applications, edited by Bishop & Lange. New York, Taylor and Francis Group.

e-on Software (2010), *e-on Software brief history.* Retrieved 19 December, http://www.e-onsoftware.com/about/?page=1.

Kleinschmidt, K. (2010), *Discovering the bayou: Successional restoration of Bayou Bienvenue.* Masters report, Kansas State University.

Leovy, J. (2008), *Urban wetlands park to be developed in South L.A.: City council approves project at former MTA yard.* Los Angeles Times, Apr. 24.

Nassauer, J. (1995), *Messy ecosystems, orderly frames.* In: Landscape Journal, 14 (2), 161-169.

Nickerson, D. (1980), *Perspective Plot, An Interactive Analytical Technique of the Visual Modeling of Land Management Activities.* U.S. Department of Agriculture, Forest Service, Pacific Northwest Forest and Range Experiment Station, Portland, Oregon.

Zimmerman, C. (2010), *Urban and Suburban Meadows: Bringing Meadowscaping to Big and Small Spaces.* Silver Spring, Maryland, Matrix Media Press.

Urban Land Institute (ULI) (Gause, J. A., Ed.) (2007), *Developing Sustainable Planned Communities.* Washington, D.C., ULI Press.

Landscape Architecture Design Simulation Using CNC Tools as Hands-On Tools

Pia FRICKER, Christophe GIROT,
Alexandre KAPELLOS and James MELSOM

1 Introduction

The innovative and integrative use of digital CNC (Computer Numerically Controlled) technologies in the field of landscape architecture is, for the most part, quite new when compared with the field of architecture. This is due to the fact that the focus of the work of landscape designers has recently shifted to large-scale urban spatial developments and their associated dynamic behaviour in complex urban spatial situations.

The following paper focuses on new techniques for visualizing work processes and developments for large-scale landscape designs. The integration of these processes within a teaching environment stands at the forefront. In this context, the use of programmed tools and the immediate translation of preliminary design ideas to models using design tools such as CAD/CAM (Computer-Aided Design/Computer-Aided Manufacturing) technologies, i.e. the Mini Mill, allow students to investigate and test new approaches.

Taking the MAS LA (Master of Advanced Studies in Landscape Architecture) program of the Chair for Landscape Architecture of Professor Christophe Girot (ILA) at the Department of Architecture at the ETH Zurich (CH) as a case study, the paper illustrates the potential of the introduced technologies. Through intensive work with the latest software in the area of modelling and visualization, MAS graduates are capable completing complex design tasks as well as developing new forms of design method. The chosen CAD programs are particularly appropriate for the visualization of large-scale landscape designs and offer the possibility for export to computer-steered milling machines.

Especially with the 2010 established visualization and modelling laboratory (LVML), the Chair offers an outstanding centre of expertise within the fields of 'Landscape Visualization' and 'Landscape Modelling'. Under the patronage of the Chairs for Landscape Architecture (GIROT, ILA) and Planning of Landscape and Urban Systems (GRÊT-REGAMEY, IRL), a lab could be established that researches new methods for the depiction, modelling and visualization of large-scale landscapes. Here, various software and hardware solutions are combined experimentally: for example a 3D landscape scanner with 1 km range is being used in order to investigate new boundaries of perception and illustration of the built environment.

Professional partnerships to the developers of software and hardware solutions as well as experts in the areas of landscape and urban planning allow for hands-on examination and implementation in the various research areas.

Fig. 1: True-color 3D point cloud model with a geo-referenced site mesh (Pascal Werner)

Successful research projects in collaboration with city authorities clearly show the interest and the necessity for the implementation of these technologies. We are therefore in a position to critically reflect on the work done the past few years and define new concepts for teaching and research through acquired experience. The goal of this paper is illustrate the new orientation of application areas for CAD/CAM technologies and their associated potentials within teaching and research projects.

2 From Representation to Integration

2.1 Overview

Although CNC (computer numerical controlled) technologies were originally used primarily as representational tools, our current activities focus on how to use them as integrative ones. Especially in large-scale landscape architecture projects, there is a strong need to develop technologies during the design process, which already integrate digital machines as supportive tools at an early stage (HAMPE & KONSORSKI-LANG 2010). Our experiences have shown that the practical handling of CNC milling machines often requires considerable preparatory work, and the actual making of the model requires a lot of experience as well as time. As a result, we are testing out a 'mini-mill', which is both portable and requires less experience to operate. These portable CNC milling machines can be easily used in studio, at workshops, or at meetings with clients to explore new readings of landscape architectonic parameters and spatial concepts, as well as to sensitize perception (GERSHENFELD 2008).

2.2 Intention of the MAS Program

The key aim within the realm of landscape visualisation and modelling is to consider the impact of combining traditional methods with various digital tools and methodologies. These traditional methods, such as sketching, photography, and diagrammatic analysis, share a level of intuition and inspiration, which can be combined with digital tools and methodologies to combine these characteristics with heightened accuracy, feedback, and real-world calculations.

The digital tools in use are numerous, but can be combined into key methodologies: site-scanning, modelling, visualisation, CNC-milling, and projection. Each can be emphasised or favoured depending on the design project and context, or as the design itself develops. Key to the workflow of the students is that none of the methodologies is tied to a stage of the project – all are equally weighted in terms of their applicability from the onset until the conclusion of the design (MITCHELL 1990).

3D landscape models are generated from early in the process, yet relate more closely to the architectural concept of the 'sketch model'. Such models are characterised by their speed of creation, simplicity and lack of detail – they should not be precious, but be cut, broken, and cannibalised by new design ideas. Essential in this process are fluidity of design concept and the relationship to scale and the existing site is not lost.

These initial sketch design proposals which test individual tools and processes can later be revisited with new, highly accurate data, captured on-site by the students themselves using the Terrestrial Laser Scanner. This also enables the designer to add more detailed topographical data exactly where it is needed as the design develops. These iterations in both the design process and site data density allow a continual deepening and questioning of previous design steps and assumptions, and result in a comparative approach through which the application an success of individual tools, methodologies and the design itself can be gauged.

The software is chosen based on flexibility and ease of use, and ability to communicate easily with other packages and formats. In the case of topographically variable terrain, a modelling application should ideally support both strong polygon and nurbs (mathematically defined surface) capabilities. In our work, Rhinoceros is an excellent choice as it combines these characteristics with robust 2D CAD functionality and customisation through scripting and nodes (grasshopper). The interaction with further digital means of output, such as CNC milling, point cloud interoperability and rendering flexibility (MERTENS 2010).

The new methodologies are taught in modular format of between 4 and 6 weeks. This timeframe allows the transition from experiment to implementation of each of the new methodologies, and the modules are arranged in an order that allows the learning of terminology, technique, and application to be gradual, continually referring and integrating aspects of the previous modules.

2.3 Landscape Modelling and Visualisation Methods

The MAS students were gradually introduced to the tools and methodologies using a local site, most recently the widening and further excavation of the Gubrist Tunnel site in Zurich, Switzerland. The site is an ideal choice as it features many of the problematics core to the contemporary practice of Landscape Architecture, such as manipulation of topography, integration with transport infrastructure, and direct interface with urban and agricultural networks. The excavation material also provides directly the material with which to transform the site and its context. The final in-depth thesis work should be applied to a site with which the students are already familiar, and which already has an existing research and design proposal. Within the course structure, the goal is therefore to re-visit and re-examine the project using new techniques and experimenting with various scale levels of intervention.

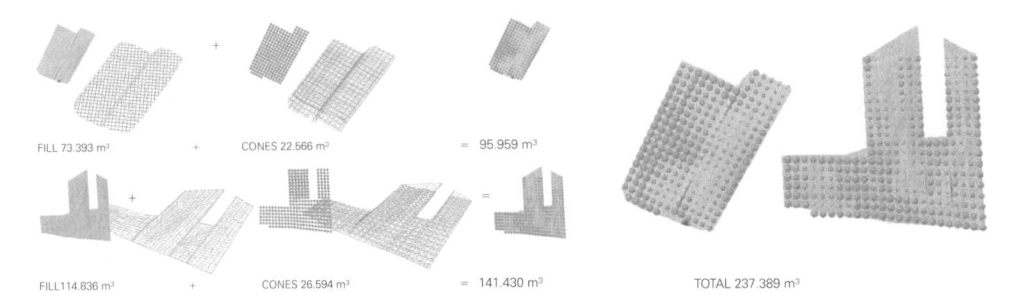

Fig. 2: Analysis showing the various volume calculations generated from site measurements (student Christine Baumgartner)

The application of these new methodologies should extend from large-scale context and influence to the resolution and representation of small-scale intervention and detail resolution. In addition, the final thesis work allows the students to experiment with the combination of various methodologies to provide new insights and design conclusions.

The format of the course, aimed to include landscape and architectural professionals concurrently in offices, offers further potential – that of the further integration of academic research and the practice of Landscape Architecture. An alternative project path is offered – to take a similarly explored site from previous or ongoing professional project. It is important that such projects address appropriate landscape architecture issues and thematics, and that a design or research exists with which to contrast the new methodologies and outcomes.

The major challenges and potentials in the current arena of evolving software and hardware possibilities are in the areas of flexibility and speed. Designers can choose to resist becoming reliant on particular tools and methods, but allow for a flexible approach that can gradually integrate new data types and scales of data sets. Rather than using fixed methods and having to reinvent an approach once old methods are considered obsolete, tools can be gradually exchanged, and issues of compatibility and efficiency avoided.

In addition the general transition not only towards open source software, but also open source data opens new possibilities for the generation and transformation of landscapes and urban spaces. Detailed site data is becoming increasingly easy to access and manipulate, which enables both those at the professional and student to have access to high quality data and resulting landscape design outcomes which are both refined and locally applicable. The impact of the transition to a methodical process of teaching has already changed the manner in which the students present their projects. Central to the description of the project is a full understanding of both the stages of design, and the process which guided its evolution.

Fig. 3: Sketch visualisation generated directly from a stage within the design process (student Christine Baumgartner)

Fig. 4: Critique presentation of method, process and design outcomes

The process of landscape modelling is integrated into the overall 'digital chain' allowing a fluid exchange of data between the different phases of the project, and the generation of analytical drawings, such as plans, sections, and views.

It is possible to output any stage of the project at any given time, either in studio via the Mini Mill, or the large-scale CNC router. In this scenario, the physical model becomes the verification tool, a frozen moment in the design process.

This ongoing body of work has allowed us to explore other facets of digital and manual design production, both data output and data acquisition. The sculptural potential of sand models can supplement the data set through 3D scanning. In the communication and potential for both site generation and presentation, integration with Google Earth allows students to integrate and contextualize their designs at any stage of the design process.

2.4 Large Scale Landscape Modelling and Milling

Within the MAS program the students are introduced to 3D modelling and milling in the very first stage of the year-long course. Module 1 is a 5-week workshop where the students are introduced to two new aspects of large-scale project design: NURBS modelling and CNC milling (BISHOP & LANGE 2005).

Part one involves learning and becoming familiar with the software, Rhinoceros 4.0 and RhinoCam 2.0. Although most postgraduate students have little or no knowledge of advanced modelling packages beyond AutoCAD, Vectorworks or ArchiCAD, the learning curve of Rhino is fast: within a one-day introduction course, students are able to create and manipulate basic NURB geometries and explore the different tools available to them within the software.

The students are asked to model a project they have already worked on previously: a design from their master studies diploma or a project they are working on, or have worked on, in their office. The pedagogical intention is to provoke a new attitude towards landscape design, in the light of the new tools available.

Part two of the module involves getting the students familiar with the 3-axis CNC router and preparing the G-code files. This step is for the students slightly more daunting as it involves creating geometries and textures that do not exist as such in their files, but are "built" through the creative use of "step control" and "stock to leave" parameters and varying milling bit diameters.

To overcome the initial intimidation, students are taken through the whole milling process with a tutor, setting up the machine, etc, and a simple model is outputted (GERSHENFELD 2008, RAMGE 2008, SENNETT 2008).

This first introductory part lasts 2 weeks, giving the students the necessary knowledge and confidence to model and mill on their own. The second half of the module is dedicated to refining modelling and milling methodologies, experimenting and understanding the extraordinary potential of CNC technology. Whereas models are usually understood to be static representations of the final and finished stage of a design, students are highly encouraged to cxpcriment with other materials and milling techniques. Project versioning allows for project representation based on parameters such as time, water level or sediment deposit for example or even various stages of the project design. This "digital chain" allows for multiple versions to be easily and quickly outputted. Our didactic intention is to push students to see large scale design more as dynamic process instead of just an end result, where multiple, on-going parameters are involved.

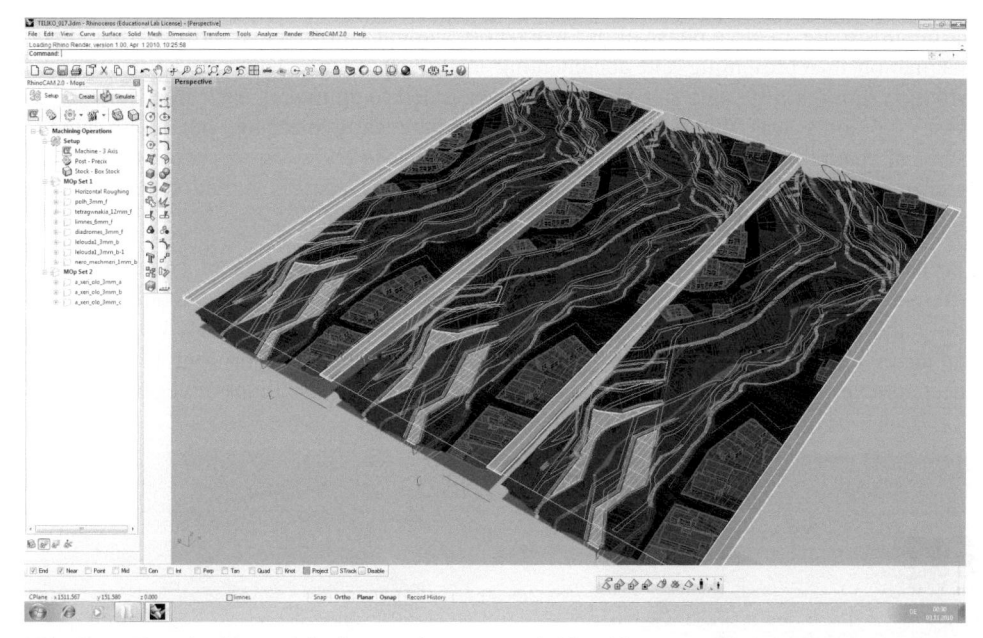

Fig. 5: Despite its complexity, students are quickly able to model complex topographies in Rhino 4 (student Dimitris Manolis)

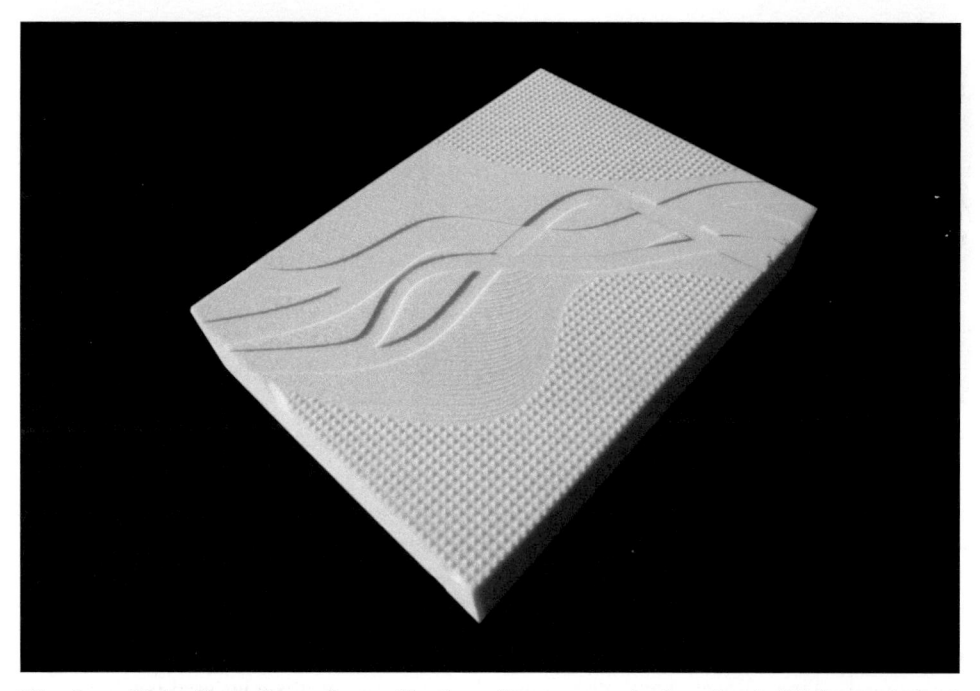

Fig. 6: RhinoCam allows the application of textures and other simple 3D forms through the creative use of milling bit diameters, step size and stock to leave (student Karin Aemmer)

During this more creative phase of our workshop, students showed extraordinary talent and imagination in representing and documenting landscape design. To represent his projects evolution through time, one student milled a four-sided foam model, each side showing on phase. Not only was this method clever and aesthetically pleasing, but represented a technical challenge in terms of milling (illustr. 6). So as to show various project iterations, another student milled a base or project context, and by using a milled cast, presented multiple plaster inserts.

The use of 2 materials clearly distinguished the two parts. The use of colour to show different water levels or qualities of water was used in another model. The foam block was first painted and the "pockets" of water were then carved out. Finally, in a very beautiful but again technically challenging example, using two-sided milling and playing with different levels of opacity one student was able to show various qualities of vegetation.

Fig. 7: Molded plaster inserts to show various design strategies (student Salome Kuratli)

Within the context of the "digital chain" students are shown that files generated in Rhino are also used further down the design process, in other applications to generate 2D visualisations in other modules of the course or integrated with Google Earth so as to allow the students to see their project in context.

It is important to state that the intention of the course is not to create CAD-CAM professionals. Time allocated and other issues do not allow for this. Our goal is to show the potentialities of the tools and show the large palette of possible applications available to professionals within the field of landscape design. Despite the complexities of software packages such as Rhino, our experience shows that even uninitiated students can in no time use the tools made available to them to their best advantage and produce work of high quality.

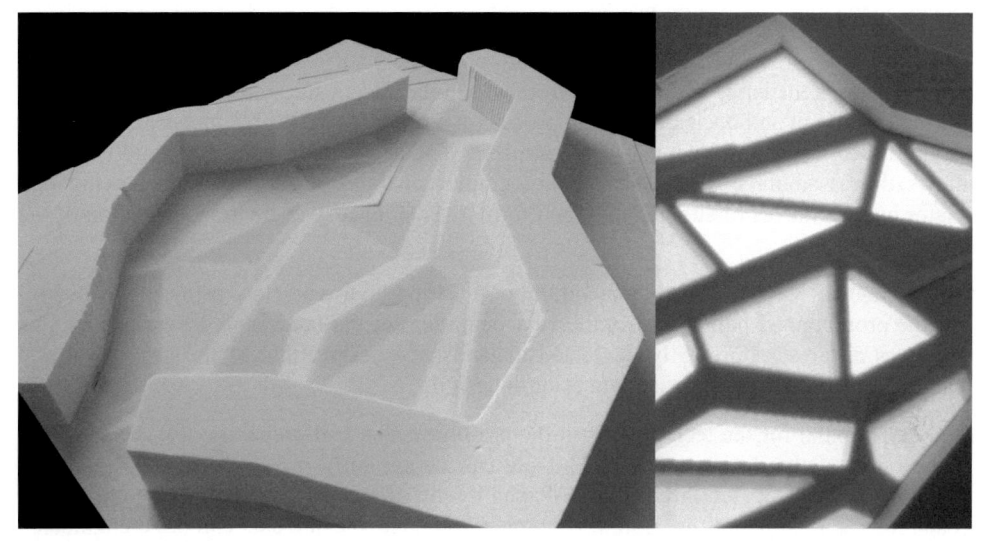

Fig. 8: Double-sided milling and use of light to indicate soil qualities (student Lorenzo Figna)

3 Conclusion and the next Steps

The applications of the described processes have varied implications for landscape design education and the design practice. One aspect within a common design studio, the result potential is the focus and heightening of crucial factors to be addressed and design problems to be solved. This mode of working, which begins as directive and focused, brings the possibility to push the design project far further than is traditionally possible within the academic semester. The ability to represent and contrast landscape systems, both existing and potential, directly influences the design process, and tighten the iterations of decision-making.

The impacts for the profession of landscape architecture begin with multiple representations of site, to integrate the interests and focus of specialists and other involved stakeholders. Rather than oversimplifying landscape systems, these systems should be focused and magnified, drawing direct attention to design challenges and facilitating the comparison of landscape systems. As such, the systems should not only rely on a common base and reference, as GIS systems allow, but also effect and compare to one another in a non-destructive manner.

Advanced modelling and visualizing techniques are used at every stage of the design process and combined with on-site preparatory tests and recordings of the environmental impact of local seasonal variations. The adaptation of an artificial topology within its surroundings is where the extreme precision generated by point cloud scans becomes essential. The density of technical and visual information inside the point clouds allows for highly informed design decisions. Alternatively, the development of filters to deal with overly saturated datasets maintains an efficient workflow, and allows for efficient processes of data acquisition and design-use.

Through these processes, the material and physical reality of large-scale projects are rendered comprehensible and operable from within the design studio itself.
During the design and decision-making process, a multitude of possible physical, visual and natural aspects of the project can be scrutinized. The geo-referenced point cloud base also allows the assemblage of landscape photographs, enabling a form of site viewing that relates back to the art of site panning, and form a visual history of the transformation of the site.

At this moment we have been concentrating on setting up a workflow using the Mini Mill and the programmed height tool in the very beginning of the landscape design phase. This addresses an area traditionally neglected in landscape design, that of the generation and manipulation of the existing site data prior to design.

A great potential for the technique is in the pre-processing and preparation of the site data, at the onset of the design process. The site can be similarly processed in the manner in which design projects are directed and shaped by the design brief: a document that sets the goals, focus and limits of the design outcome. In practice, the site data can be re-engineered to display existing and potential activators, historic and topographical potential, as well as react directly to an applicable design brief.

4 Acknowledgements

The research and teaching work described in this paper are the result of a very committed teaching team at the "Chair of Christophe Girot", special thanks to all of them (Susanne Hofer, Ilmar Hurkxkens, Pascal Werner).

The authors would like to thank the MAS LA students of 2010-2011: Aemmer Karin, Baumgartner Christine, Figna Lorenzo, Goula Ioulia, Kuratli Salome, Manolis Dimitrios, Stavrotheodorou Vasiliki, Ankita Thaker, for their dedicated participation in the program.

References

Bishop, I. D. & Lange, E. (2005), *Visualization in Landscape and Environmental Planning, Technology and Application.* New York, Taylor & Francis, London.
Burmeister, K. (2006), *Vom Personal Computer zum Personal Fabricator.* Hamburg, Murmann Verlag.
Gershenfeld, N. (2008), *Fab, Basic Books.* Jackson.
Girot, C., Kapellos, A. & Melsom, J. (2010), *Iterative Landscapes.* In: Michael, H. & Konsorski-Lang, S. (Ed.), The Design of Material, Organism, and Minds: Different Understandings of Design. Berlin. Springer-Verlag.

Hagan, S. (2008), *digitalia*, N.Y., Routledge.
Kolarevic, B. (2003), *Architecture in the Digital Age. Design and Manufacturing.* N.Y., Spon Press.
Mertens, E. (2010), *Landschaftsarchitektur visualisieren.* Basel/Boston/Berlin, Birkhäuser Verlag.
Mitchell, W. (1990), *The Logic of Architecture.* MIT, USA
Ramge, T. (2008), *Marke Eigenbau.* Frankfurt, Campus Verlag.
Sennett, R. (2008), *The Craftsman.* London, Yale University Press.

Detecting Greenery in Near Infrared Images of Ground-level Scenes

Piotr ŁABĘDŹ and Agnieszka OZIMEK

1 Introduction

An increasing demand for landscape digital analyses entails a rapid progress in the numerous fields of computer science, including image processing. A wide spectrum of problems was solved using remote sensing techniques, albeit applied algorithms operate on satellite or aerial views (ADAMS & GILLESPIE 2006). However, our environment is usually perceived by observers from the eye-level; therefore, ground-level views and panoramas seem to constitute landscape representation, which corresponds with the human visual sensations in the best way.

Scenic photographs, adequately transformed, provide important information about land cover. These images consist of components that can be divided into three groups: the natural or cultural substance and the background. An automatic identification of the sky in digital photographs can usually be accomplished on the basis of its colour. After reduction of this element, the view contains only natural (prevailingly vegetal) and cultural objects. Effective image segmentation, aimed at their distinction is often hampered by similarities in the colours of vegetation and man-made objects, like house walls and roofs, café parasols or park benches. The further difficulties are caused by variations in the colour of vegetation with changes in season (particularly in the autumn) and lighting conditions (due to solar filtering, intensity, direction and orientation, in relation to the camera). Therefore, resulting images have to be controlled and corrected manually (OZIMEK & OZIMEK 2009).

In bi-level (black-and-white) images, with the distinction of cultural substance (Fig.1), quantitative and qualitative parameters of the view can be calculated. Basing on these data, research focused on landscape evaluation can be conducted (UNWIN 1975).

Application of vegetation indices (KRIEGLER et al. 1969, HUETE et al 2002) combined with thresholding operation (converting the image into a binary mode) potentially allows for automatic image pre-processing, with the object of greenery detection. The negative image would present only anthropogenic elements. It should be noticed, that this method does not take into account the other natural elements of view, such as water, rocks or bare soil.

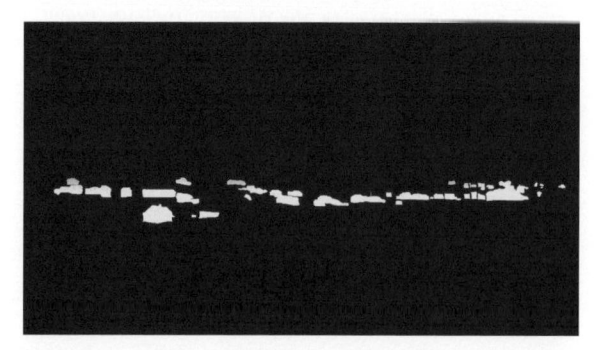

Fig. 1: An example of bi-level image with buildings marked with white colour

This paper explores the effectiveness of vegetation indices (used in remote sensing) to automatic detection of plants using near-infrared photographs taken at ground-level. In particular, it examines the differences in various lighting conditions (sunny and cloudy days) and diverse leaves colours (green, yellow, red, brown).

2 Material and Methods

2.1 Vegetation indices

In remote sensing, NDVI (Normalized Difference Vegetation Index) is calculated basing on the equation:

$$NDVI = \frac{NIR-R}{NIR+R}$$

where: NIR – means the near infrared channel,
R – stands for the red channel (KRIEGLER et al. 1969).

As it can be seen from the graph of spectral characteristics of plants (Fig. 2), the algorithm can bring entirely correct results, as far as the distinction of verdure from the other "green" objects is concerned. Nevertheless, the problems related to the atmospheric effects (water vapour content) appear, especially in distant views.

In order to improve the results, EVI (Enhanced Vegetation Index) was put into practice, which is not only "chlorophyll sensitive", but also takes into account the blue channel, responsible for atmospheric clutter (HUETE et al. 2002).

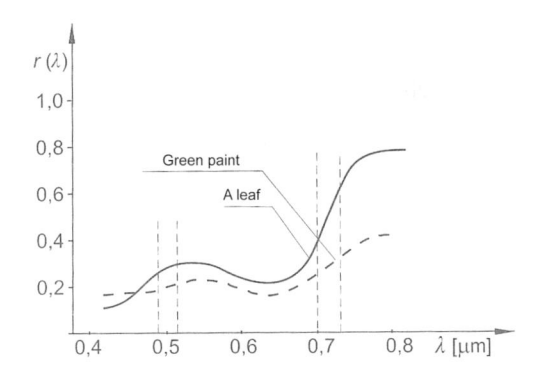

Fig. 2: Spectral characteristics of plants versus other green objects (SANECKI et al. 2006)

It is computed from the equation:

$$EVI = G \frac{NIR-R}{NIR + C_1 R + C_2 B + L}$$

where: NIR/R/B – colour channels: near infrared, red and blue, respectively, L – the canopy background, C_1 and C_2 – coefficients, considering aerosol resistance in the atmosphere, G – gain factor, L – soil adjusted factor. In the most popular MODIS EVI (Moderate Resolution Imaging Spectroradiometer) implementation they are: $L = 1$, $C_1 = 6$, $C_2 = 7.5$, and $G = 2.5$. The results are calibrated for the specialist equipment (ADAMS AND GILLESPIE 2006).

2.2 Input data

The equations show that both algorithms operate on non-spectral range, including near infrared. In this case, process of the image acquisition requires usage of the professional devices. Nonetheless, the standard CCD matrix in a photo camera is sensitive for the electromagnetic wavelength between 350 and 1200 nm. This range is limited to the visible spectrum (380 – 760 nm) by means of filters. After their removing and using IR filter instead (> 850 nm), it is possible to register near infrared, crucial for the computations. These data have to be supplemented with colour channels (red: 600 ~ 760 nm, green: 500 ~ 600 nm, blue: 380 ~ 490 nm) from the second photograph of the same scene, made in the visible range. In the initial phase of the study, various techniques of image formation (JPG with "white balance" setting, or data from RAW format) have been checked, in the context of the results correctness.

Beyond question, the process of photography registration plays a crucial role in the final effects. It embraces, above all, sufficient image resolution (at least matching sight resolution) (BISHOP 2003), its correct sharpness, white balance and exposition. The calculations have been made for images 4302 × 2860 pixels with the equivalent focal length equal to 120 mm, redundant for human perception (the sufficient image size for this focus is 1600 × 1200 pixels). The photographs have been registered in the RAW format and converted into TIFF standard.

2.3 Modelling approach

The first step of the applied algorithms was a colour channel separation, which enables calculations on 2D matrices. NDVI and EVI indices have been computed, following the equations, cited in the section 2.1.

Afterwards, image binarization should be conducted, in order to obtain an image with the black background and plants marked with white colour. The value of threshold can be understood as the level, with which intensity of every pixel is compared, and points that are darker than the threshold take minimum value (black). The remaining pixels are converted into the maximum value (white). The main difficulty lies in the accurate choice of this threshold value. In other case, the result can be wrong.

The Otsu method has been chosen, since it guarantees comparatively correct results. In the initial phase of this technique, the histogram of image brightness is calculated. For every threshold, which divides image into two classes (objects and background), the between-class variance and the inner-class variance is calculated. As a result, the value of threshold is chosen, for which the inner-class variance is the smallest and, at the same time, the between-class variance is the biggest (RUSS 2002, PETROU & PETROU 2010, MALINA & SMIATACZ 2005). In the case, when the effect of this operation is negative, the manual choice of the threshold is possible, basing on the histogram features. Frequently, selection of the local histogram minimum brings about effective class separation (JAYARAMAN et al. 2009, MALINA & SMIATACZ 2005).

In the resulting images numerous errors appear, in particular, tiny spots that do not correspond to objects, but occur as the effect of insignificant differences in image local brightness. In order to remove them, morphological operations can be applied. Two basic operations, dilation and erosion, conducted successively, constitute the more complex transformation, known as closing (JAYARAMAN et al. 2009, MALINA & SMIATACZ 2005, NIENIEWSKI 1998). The outcome is more generalized; a picture loses in details, small holes are filled and the objects contours are smoother (Fig. 3).

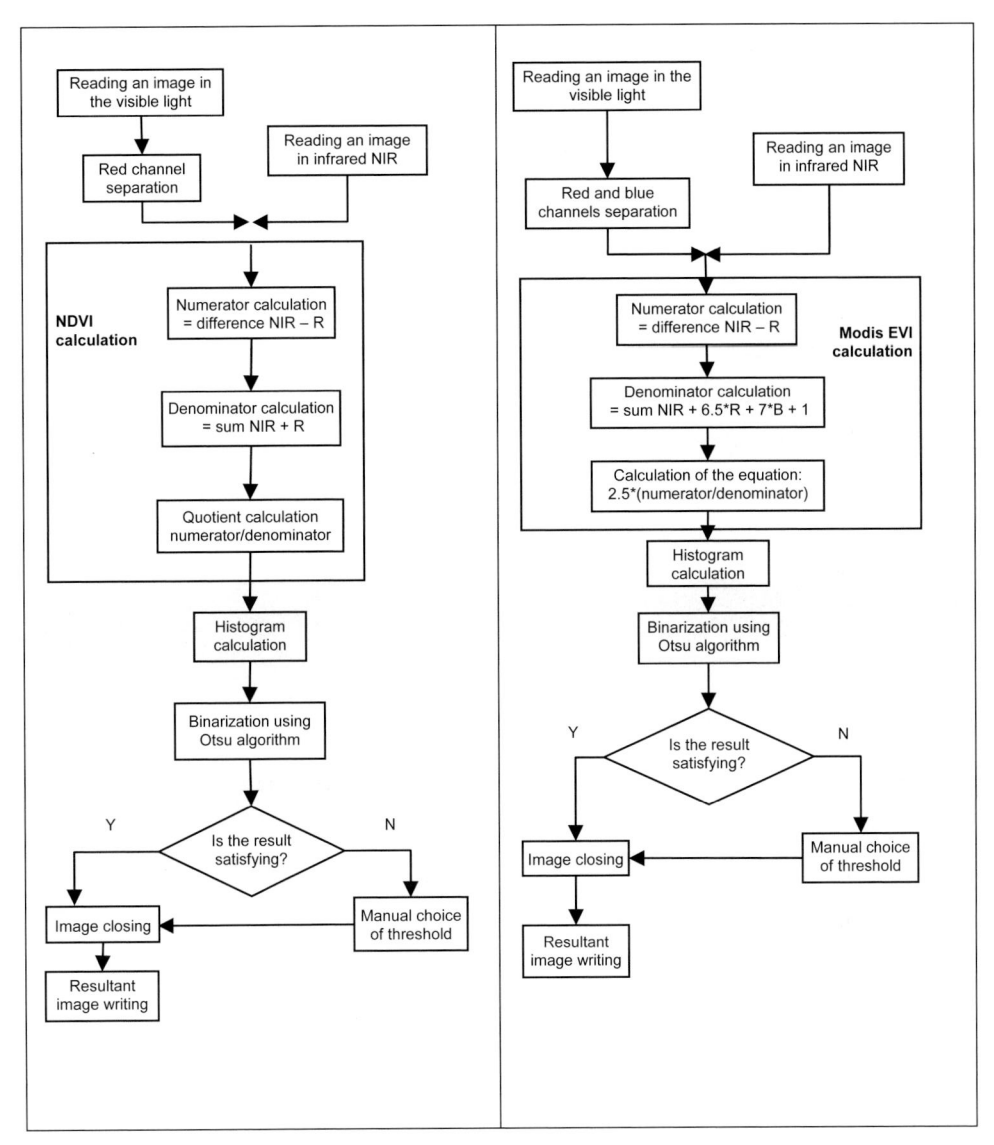

Fig. 3: Algorithms of greenery distinction

3 Results

3.1 NDVI versus EVI

Fig. 4: An image in the visible spectrum

Fig. 5: The same scene in the infrared (NIR)

Fig. 6: Normalized Difference Vegetation Index (NDVI)

Fig. 7: Enhanced Vegetation Index – MODIS implementation (MODIS EVI)

In the first experiment two egetation indices (NDVI) have been compared. Fig. 4 shows the image in the visible spectrum and the next one (Fig. 5) – the same scene in near infrared (NIR channel). The greenery is characterized by the high intensity, as well as the bright building seen in distance.

Fig. 6 and 7 present automatically binarized images obtained as the products of NDVI and MODIS EVI calculation. They seem to suggest that the NDVI (Fig. 6) promises better outcome; however, it should be taken into consideration that MODIS EVI (Fig. 7) is scaled for specific equipment (spectroradiometers). After elimination of calibrating coefficients (Fig. 8) both indices provide similar results. Fig. 9 confirms this fact, showing insignificant difference between the resultant images. The final effect (after closing operation – Fig. 10) presented in the Fig. 11 has been obtained by the operation of the logical negation and the background elimination.

Fig. 8: EVI without calibration coefficients

Fig. 9: A difference between NDVI and EVI

Fig. 10: An image after closing operation

Fig. 11: Elements of greenery filtered out

3.2 Different lighting conditions

Fig. 12: The photograph taken during the sunny day

Fig. 13: The same scene photographed during the cloudy day

In order to examine the effects of automatic greenery detection in different lighting conditions, photographs were taken during the sunny and cloudy days (Fig.12 and 13). In addition, the same scene was photographed in a moment, when the sun was temporarily obscured by clouds. The autumn season was chosen, because in this case, plants distinction was the most problematic. There are numerous obstacles, when the photographs are taken against the light and the branches are only partially covered with leaves. Like in the previous example, the calculations of NDVI and EVI (without calibrating coefficients) for a specific image gave almost identical results (Fig. 14 and 15). The difference in automatic greenery distinction in various lighting condition is noticeable (Fig.16).

Fig. 14: Vegetation indices for a sunny day

Fig. 15: Vegetation indices for a cloudy day

Fig. 16: A difference between photographs taken in the sunny and cloudy day

Fig. 17: Cultural elements distinguished using the proposed method

3.3 Various leaves colours

When leaves change the colour in the autumn, their reflection in near infrared remains at the high level, thanks to the fact they still contain some chlorophyll. Simultaneously, the content of carotenoids is increasing. These substances are responsible for red, orange and yellow hues appearing in this season (Fig. 18). They reach high values in the red channel, which has the negative impact on vegetation indices calculations (Fig. 19). Like in the former experiments, in the last image (Fig. 20) an attempt was made to distinguish man-made objects in the analysed view, but results are burden with errors. In this photograph, beside the problems concerning greenery, some faults occur as a result of reflections in windows and other glossy surfaces (Fig. 20).

Fig. 18: Plants with green and yellow leaves

Fig. 19: Vegetation indices

Fig. 20: An attempt of man-made object distinction

4 Conclusions and Outlook

In this paper an effort was made to evaluate the usefulness of NDVI and EVI for automatic greenery detection in the eye-level views. This determines the initial step for cultural substance distinction, which is crucial for landscape parameters calculation and would help in its evaluation. As it was proved by the experiments, both indices (NDVI and EVI without calibrating values) provide similar results.

In the resultant images some errors occurred, in particular:
- The absence of chlorophyll in trees trunks and branches, as well as in dry grass or fruits cause that they are not indicated as parts of plants. In particular, it is visible in the foreground (Fig. 4, 11, 12, 13, 17).
- As a result of seasonal colour changes (yellow leaves), the level of reflectance in the red channel rises; thus, the values of vegetation indices become low (Fig.20).
- Some inaccuracy occurs in the objects contours, prevailingly, as the consequence of pixels values interpolation (antialiasing) (Fig. 11).
- Glossy surfaces (glass, water) reflect waves, not only in the visible spectrum, entailing their resemblance to the source objects (Fig. 20).

- The applied algorithms are dedicated to the "green parts" of plants, therefore the other natural elements (soil, rocks, water) are not detected (Fig. 12, 17).
- The problems with plants indication appear, while photographs are taken against the light; particularly, when the branches are not fully covered with leaves.

On the other hand, the algorithm proposed in the part 2.3 brings several advantages:
- Distinction between the verdure and green man-made objects is effective (Fig. 3, 5, 7, 9).
- Fragments of plants located in the shade are characterized with the high level of reflectance in near infrared, and low in the red channel; therefore, they are properly distinguished by the vegetation indices (Fig. 11, 13, 17, 18).
- The algorithm brings correct results in distant parts of the view (Fig. 6, 8).
- The Otsu method provides positive outcomes. Seldom, was the manual correction of the threshold level necessary (Fig. 9, 11, 14, 16).

The aim of the research has been realised to some extend. While the distinction between anthropogenic objects and distant areas covered with trees or grass is effective and can automate the phase of image processing, a considerable number of faults appear in the foreground.

What was to be expected, only fragments of plants that contain chlorophyll (leaves) can be indicated using the algorithm proposed. Therefore, the method does not guarantee correct results in the separation of the other natural landscape components (trees branches and trunks, water, soil, rocks).

The further difficulties are due the reflections in glossy objects. Spectral similarity between the mirror images and real objects has the negative effect on the outcomes.

Some errors are integrally connected with the raster image structure. Antialiasing, accomplished in order to smooth objects contours, causes inaccuracies that appear at the edges, in particular, when the local gradient is high.

References

Adams, J. B. & Gillespie A. R. (2006), *Remote sensing of landscapes with spectral images: a physical modelling approach.* Cambridge University Press,

Bishop I. D. (2003), *Assessment of visual qualities, impacts, and behaviours, in the landscape, by using measures of visibility.* In: Environment and Planning B: Planning and Design 2003, 30, 677-688.

Huete, A., Didan, K., Miura, T., Rodriguez, E. P., Gao, X. & Ferreira, L. G. (2002), *Overview of the radiometric and biophysical performance of the MODIS vegetation indices.* In: Remote Sensing of Environment, 83 (1-2).

Jayaraman, S., Esakkirajan, S. & Veerakumar T. (2009), *Digital Image Processing.* Tata McGraw Hill.

Kriegler, F. J., Malila, W. A., Nalepka, R. F. & Richardson W. (1969), *Preprocessing transformations and their effects on multispectral recognition.* Proceedings of the Sixth International Symposium on Remote Sensing of Environment.

Malina, W. & Smiatacz, M. (2005), *Metody cyfrowego przetwarzania obrazów (Methods of digital image processing)*. Oficyna Wydawnicza EXIT.

Nieniewski, M. (1998), *Morfologia matematyczna w przetwarzaniu obrazów (Mathematical morphology in image processing)*. Akademicka Oficyna Wydawnicza PLJ.

Ozimek, A. & Ozimek, P. (2009), *Algorytmy przetwarzanie obrazu w wyróżnianiu tworzywa kulturowego i przyrodniczego na fotografiach krajobrazowych (Image processing algorithms in cultural and natural substance distinction in landscape photographs)*. Nauka, przyroda, technologie, Tom 3, zeszyt 1, #12, Wydawnictwo Uniwersytetu Przyrodniczego w Poznaniu.

Petrou, M. & Petrou, C. (2010), *Image Processing: The Fundamentals*. John Wiley & Sons.

Russ, J. (2002), *The image processing handbook*. CRC Press.

Unwin, K. I. (1975), *The relationship of observer and landscape in landscape evaluation*. In: Transactions of the Institue of British Geographers, 66, 130-133.

Replicating Fractal Structures with the Reverse Box Counting Method – An Urban South-east Asian Example

Songlin WEI, Cyril FLEURANT and Jon Bryan BURLEY

1 Introduction

Planners, designers, academics, governmental agencies, and citizens are interested in replicating complex spatial patterns for both landscape preservation/conservation and blending new built environments with existing environments. Studying fractal patterns is one general approach to modelling these spatial patterns; however calculating the fractal pattern has been relatively simple and widely employed (DAUPHINÉ 2011, THOMAS et al. 2010a, THOMAS et al. 2010b, MA et al. 2008, DAUPHINÉ 1995, FRANKHAUSER 1994). Fractals can be relatively easily calculated by employing software such as through Fractalyse 2.4 (TheMA 2012). However, replicating the pattern and applying the fractal number in an applied manner has been more elusive.

A French team consisting of geographers, a physicist, and a hydrologist began employing the box counting method to measure and then replicate fractal patterns with the reverse box counting method (the replicating processes for planning and design applications is the relatively new part of the scholarly progression in fractal studies) (DUCHESNE et al. 2002). Recently this box counting approach has been employed to study forest patterns in Michigan for surface mine reclamation (FLEURANT et al. 2009), specific individual tree species patterns in the Upper Peninsula of Michigan (LEHMANN 2009), and to describe and replicate Chinese gardens in Suzhou, China (YUE & BURLEY 2011). These papers review the literature in detail leading up to our study. In addition, software has been compiled by the French team, written in C# to replicate patters based upon fractal numbers.

2 Study Area and Methodology

In our study we examined the spatial pattern of buildings on Lamma Island (Fig. 1), a part of the Hong Kong island chain, P. R. of China. In Chinese, the island is known as 南丫島), also known as Pok Liu Chau (Chinese 博寮洲) or Pok Liu. We were interested in measureing and establishing a pattern that could be replicated for urban planning and design applications. In China there is interest in making the pattern of new built environments blend with older traditional settings.

To apply the box counting method, one makes a box, and then divides the box into four boxes and continues this reduction method until at least one box is empty. Then the length of the box and the number of filled boxes are recorded. Then the box size is reduced gain, the length recorded and the number of filled boxes recorded. The process is stopped when each point is in its own box, the final length is recorded and the final number of boxes with

Fig. 1: A photograph on Lamma Island overlooking the bay and looking to the west

points is counted. The length of the box is transformed in equation 1 and the number of boxes is transformed by equation 2. These pairs of numbers are regressed with Ln(1/r) being the regressor, resulting in the slope of the equation representing the fractal number, equation 3. In our study we mapped the locations of building sites in 2008 (Fig. 2) and visited the island in the summer of 2009 for field corroboration.

$$V1 = Ln(1/r) \tag{1}$$

Where: r = the length of a box
V1 = the box length variable for regression

$$V2 = Ln(N) \tag{2}$$

Where: N = the number of boxes with points
V2 = the number of boxes variable for regression

$$Ln(N) = slope\ Ln(1/r) + intercept \tag{3}$$

The process is reversed to construct a pattern that is similar to the one measured. Currently there is no mathematical proof demonstrating that the process is reversible. The procedure assumes that the process is reversible. The reverse box-counting method employs the smallest box size used to construct the measured fractal pattern and then a random numbers table is utilized to identify which boxes contain the item of interest and which boxes remain

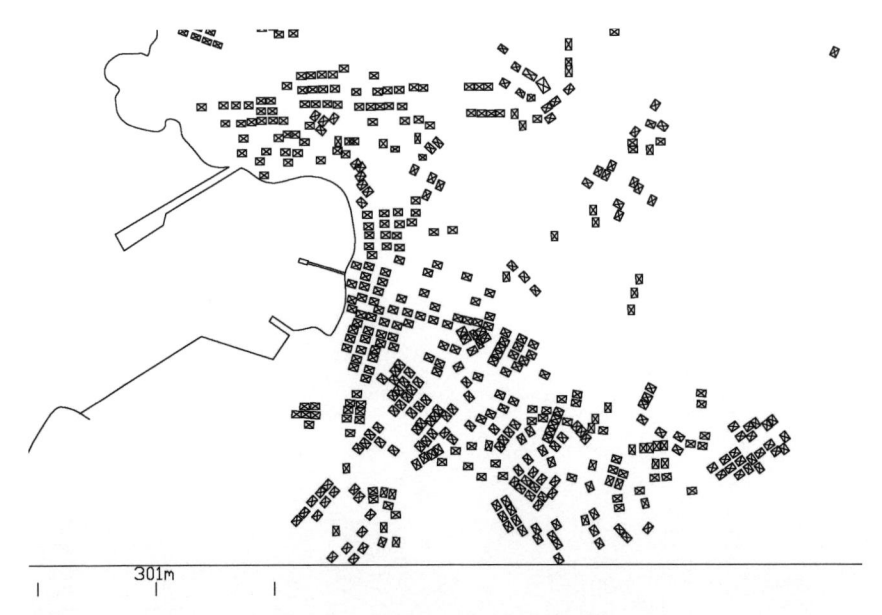

Fig. 2: A map of the building locations on Lamma Island

empty. For example if 95% of the boxes are measured to be empty, then the random numbers tables are set to generate an assignment where approximately 5% of the boxes are filled. FLEURANT et al. (2009) illustrate this approach to produce pattern of trees based upon measurements of tree distributions on xeric environments in the Upper Peninsula of Michigan, USA.

3 Results

Table 1 presents the values employed to predict the spatial pattern of the structures in the study area. The first empty boxes occurred when the box size was 162.5 meters (Fig 3). When the box size was reduced to 20.3125 meters, each structure was in its own box (Fig. 4). In our study the resulting fractal number was 1.115, with 480 boxes filled across 4,096 boxes. Therefore approximately 11.7% of the boxes are filled.

Box Length m	Filled Boxes	V1	V2
162.5	48	-5.090678002	3.871201011
81.25	139	-4.397530821	4.934473933
40.625	323	-3.704383641	5.777652323
20.3125	480	-3.01123646	6.173786104

Table 1: Values measured and transformed to predict the fractal number of structures in Lamma Island. V1 and V2 are regressed to compute the fractal number with V2 as the dependent variable in the regression.

Fig. 3: In this enlargement of a portion of the boxed grid, at least one box is empty and the process of measuring the box length and counting the number of boxes containing at least one point in the box is initiated

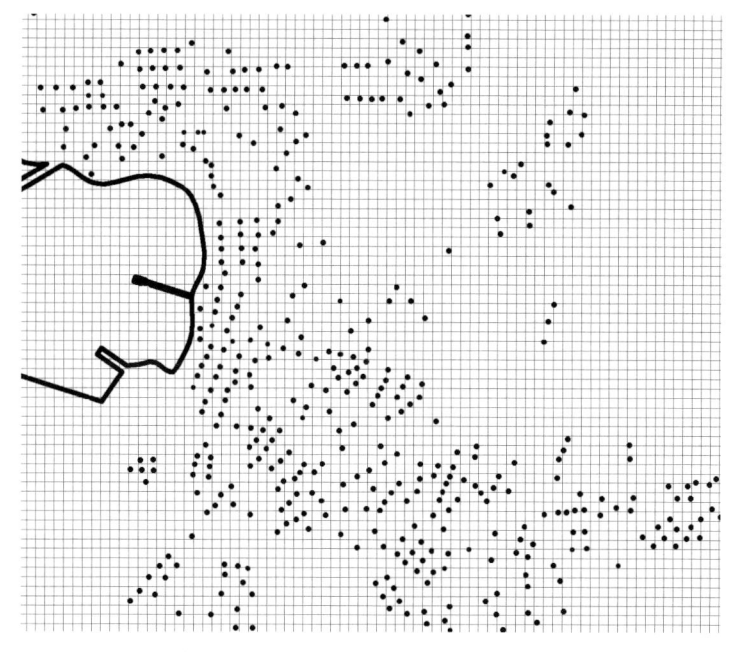

Fig. 4: In this enlargement of a portion of the boxed grid, each point is now in its own box and the process is then stopped

4 Discussion/Conclusion

One can replicate this pattern by randomly filling approximately 480 boxes in a 128 by 128 box grid at a length of 20.3125 meters (see FLEURANT et al. 2009, LEHMANN 2009) for more precise details and examples. The method is simply a tool to assist in emulating a pattern. It does not mean that the pattern must or should be replicated.

23	96	75	5	35	88	87	89	84	63
84	94	67	85	71	66	29	93	70	31
2	77	52	71	92	17	92	93	42	65
89	38	7	39	99	89	97	30	46	23
63	61	3	100	37	63	85	93	89	57
41	1	8	78	38	82	55	73	9	9
6	50	97	15	1	21	50	88	22	28
66	93	16	85	53	25	69	32	32	4
12	73	13	95	30	65	51	55	31	79
98	52	5	75	40	95	98	57	39	46

Table 2: 100 random numbers from a random number table

To replicate the pattern, suppose one had a 100 box grid area with 20.3125 m lengths for each box. Then approximately 12 of the boxes would need to have locations for structures. Table 2 presents 100 random numbers with each number assigned to a box. Since 12% of the boxes need to be filled, the numbers equal to or less than 12 represent locations where the structures are located and numbers greater than 12 are empty. In the first column, the numbers 2, 6 and 12 are equal to or less than 12 and represent the location of structures in rows 3, 7, and 9. Figure 5 presents the location of structures based upon Table 2. This figure is strictly a pattern that represents the fractal approximation of 1.115. Notice that there are 13 locations, not 12. In addition, this pattern is quite a different basis for structure location than typical environmental concerns, such as not being in a floodway (physical) or on a sacred site (cultural). Figure 5 is simply a guide suggesting form. It is not the complete answer or solution to any landscape plan.

The method is supposed to be independent upon the size of box implemented to initiate the measurements. YUE, WEI & BURLEY (unpublished, in review) chose a different size box to initiate the process, including more structures for Lamma Island and calculated a fractal number of 1.158 with the smallest box size of 18.75 m and 556 filled boxes. This replication should give some insight into the variance, reliability, and robust nature of the reverse box-counting process. In other words, the exact box size is not necessarily essential, neither is the exact fractal number, but rather, it is a neighborhood of similar ranges that appear consistent. These results are the first report indication that the reverse box-counting method has some related similarity.

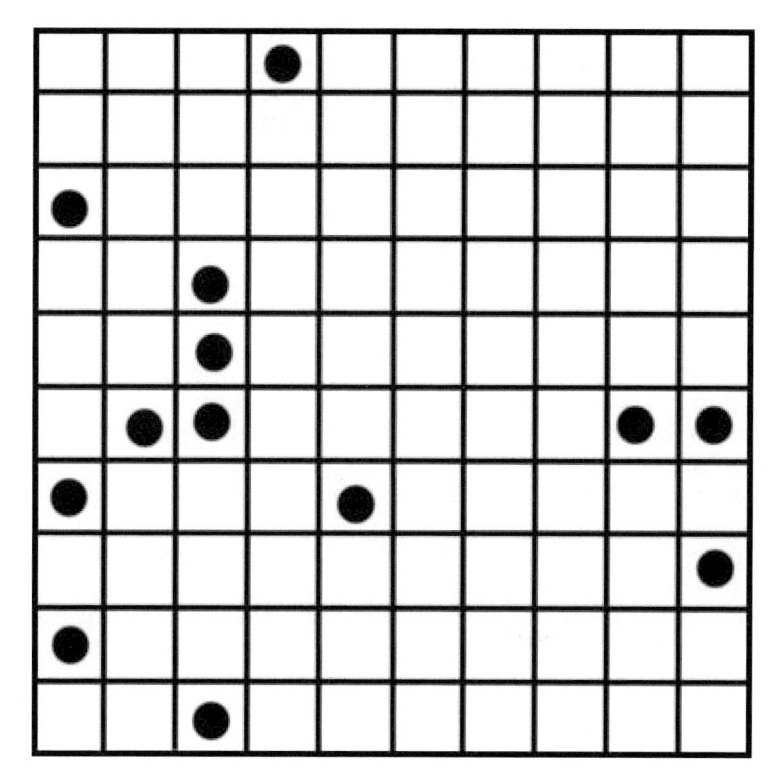

Fig. 5: A replicated fractal pattern from based upon measurements on Lamma Island

YUE, WEI & BURLEY (unpublished, in review) have also examined the reverse box-counting method in conjunction with other spatial measures such as with GIS and logistic regression to explain the physical properties of spatial elements. At Lamma Island, we noticed the development did not occur on hill-tops (Fig. 1). Thus the pattern of housing on Lamma Island could be possibly refined with spatial models that combine fractal methods and other physical spatial variables. We believe there are numerous opportunities to combine spatial methods such as auto-correlation measures, logistic regression, and fractal patterns to explain more of the variance in spatial phenomena.

We would encourage other investigators to employ such methods to study the replication properties of their urban and natural areas. We believe that there is much to explore concerning the properties and creation of spatial patterns.

References

Dauphiné, A. (1995), *Chaos, fractales et dynaminques en géographie*. Montpillier, France, GIP Reclus [in French].
Dauphiné, A. (2001), *Géographie fractale*. Paris, France, Hermes Science Publications [in French].

Duchesne, J., Fleurant C. & Tanguy F. (2002), *Brevet d'un procédé d'élaboration d'un plan d'implantation de végétaux, plan d'implantation de végétaux obtenu et système informatique pour l'élaboration d'un tel plan (Patent of a process of development of a plan of establishment of plants, plan of establishment of plants obtained and computing system for the development of such a plan)*. INPI, European patent N° 02 07836 [in French]

Fleurant, C., Burley, J. B., Loures, L., Lehmann, W. & McHugh, J. (2009), *Inverse box-counting method and application: a fractal-based procedure to reclaim a Michigan surface mine*. In: WSEAS Transactions on Environment and Development, 5 (1), 76-89.

Frankhauser, P. (1994), *Le Fractalité des Structures Urbaines (The Fractals of the Structures Urban)*. Paris, France, Anthropos [in French].

Lehmann, W. J. (2009), *Replicating Natural Tree Stand Patterns in a Northern Michigan Rock Outcrop Landscape: a Fractal Based Method and Application for Reforstesting a Reclamied Michigan Surface Mine*. Master of Arts, Environmental Design, Michigan State University, East Lansing, Michigan.

Ma, R., Gu, C., Pu, Y. & Ma, X. (2008), *Mining the urban sprawl pattern: a case study on Sunan, China*. In: Sensors, 8 (10), 6371-6395.

TheMa (2012), *Fractalyse 2.4. Théoriser et Modéliser pour Aménager*. CNRS – Universités de Franche-Comité et de Bourgagne.

Thomas, I., Frankhauser, P. & Badariotti, D. (2010a), *Comparing the fractality of European urban neighbourhoods: do national contexts matter?* In: Journal of Geographical System, 1-20.

Thomas, I., Frankhauser, P., Frenay, B. & Verleysen, M. (2010b), *Clustering patterns of urban built-up areas with curves of fractal scaling behaviour*. In: Environment and Planning B (Planning and design), 37, 942-954.

Yue, Z. & Burley, J. B. (2011), *Non-euclidian methods to characterize the Masters of the Nets Garden, Suzhou, China*. Recent Researches in Energy, Environment and Landscape Architecture. Proceedings of the 7th IASME/WSEAS International Conference on Energy, Environment, Ecosystems, and Sustainable Development (EEESD '11) and Proceedings of the 4th IASME/WSEAS Intenrational Conference on Landscape Architecture (LA '11), Angers, France, November 17-9, 2011, 2011, 122-128.

Yue, Z., Wei, S. & Burley, J. B. (unpublished, in review), *Non-euclidan methods to replicate urban and garden patterns in P. R. of China*. WSEAS Transactions on Environment and Development.

Landscape Analysis Using GIS for Ecologically Oriented Planning in Costa Rica

Marcela A. GARCIA PADILLA

Abstract

The current investigation was originated by the detection of voids within management plans (MPs) of protected areas in Costa Rica, which, over time, has prevented an effective landscape planning from taking place. On such account, this paper presents an approach to apply useful tools for the landscape evaluation and suggests an analysis using ecological considerations and GIS, which is illustrated in a Costa Rican case study: The National Marine Park of Las Baulas (PNMB), aimed to protect the nesting sites of the Leatherback Turtle, *Dermochelys coriacea*. The method uses strategic criteria for handling available data and applies land-suitability analysis, risk analysis and potential land-use suitability. All the latter shall contribute to support further land use policies such as landscape zoning. The aim is to evidence potential frameworks for the consolidation of landscape planning by using digital tools, as both a practical and academic opportunity, for biodiversity protection.

1 GIS and Ecologically Oriented Planning

Landscape planning lies at the root of planning and decision making processes. This investigation was based on ecologically oriented planning (KAUIE 2000) for a careful analysis of the landscape stressing that an ecological perspective, as a major task, is to be undertaken. Although there are problems that go beyond planning, a lot of other major knotty situations might as well be solved with an identification of priorities and threats of the current status, together with an evaluation and spatial planning. The importance of classifying data for planning purposes is essential to achieve a comprehensive analysis of land-uses and natural resources that respond to contextual peculiarities. Furthermore, there are different ways of aggregation and they all refer to the spatial relationships which help to improve the decision making process (KAUIE 2000).

Additionally, Geographic Information Systems (GIS) analytical tools include procedures for handling and doing such spatial aggregation of the available information. Therefore, the combination of analysis techniques together with the landscape ecology perspective, which in turn reveals a high understanding of the environment, is proposed for this investigation.

2 Data Input for the PNMB Landscape Analysis

There is a significant and sufficient amount of graphic information available, given that several interdisciplinary studies, regarding the management of the PNMB, have been elaborated. Therefore, the task of integrating the available digital data (see Table 1) was from the very beginning assumed. Correspondingly, the jpeg and pdf files were traced and georeferenced using the software ArcGIS 9 for the gathered information to be integrated and analyzed with the already available shape files, in order to define zones within the protected area.

Two main spatial considerations were particularly included for the PNMB regarding the spatial boundaries during the landscape analysis: (1) the watershed influence area and (2) the buffer zones; both relevant for the location in terms of flows and exchange of nutrients. Both considerations refer to the integrity and compatibility with the surroundings (STEPHAN et al. 2002). Thus, the identification of areas beyond the immediate buffer zone, and yet within the watershed areas of influence, was used as the location limits. However, it is to be recognized that the analysis shall be done at different spatial levels and scales according to concepts of land mosaics (FORMAN 1995)

JPG format	PDF format	SHP files	
– Buffer zone with 500m – Orthophotographs illustrating 100-cm sea level rise, estimated for the end of the century. The image recreates the expected 50-m displacement of the beach landward in response to sea level rise. – Local Photographs (2000-2010)	– Roofed areas – Beach access – Discharge into the watershed system – Location of waste water – Location of preferred nesting sites – Fragility areas – Underground water recharge areas – Archaeological sites	– marine area and continental area – roads – buffer zone 500m – forest area – watershed area – political-adm division – Local towns	
		– Land-cover 2000	Mangrove Second degree Forest Forest Lagoon Salty Lagoon Island Pasture Estuary Beach Plantation Open space Scrubland "Charral"
		– Rivers and watersheds	Rivers Matapalo, San Andrés, San Francisco and Pinilla. Tamarindo, San Francisco and Ventanas Estuary.
		– Fragility Areas	High, moderate and low
		– Conflicts of use	Overuse, Subuse and adequate use

Table 1: List of available digital data and base maps used as reference for the investigation. Sources: Tropical Scientific Centre (2004), GeoCAD (2009), Fonseca, A. & C. Drews (2009) and The Leatherback Trust.

3 Land Suitability, Risk Analysis and Potential Land-use Suitability Using Ecological Principles

Among the methods in planning processes suggested by KAUIE (2000) to evaluate existing environmental factors, *Land Suitability, Risk Analysis* and *Potential land use suitability* were selected. For the first, the existing land use, land cover and land use capability were included. For the second one, the identification of potential threats involved considering the conflicts of use (overuse, adequate or sub use), fragility areas (erosion; hydrological and biological, urban influence and extreme vulnerability) together with saturated conditions of groundwater recharge areas. As to the third one, the potentials areas are aimed at restoration and connectivity of the landscape elements by establishing:

Priority area for high ecological value: where high impact construction should not be allowed due to extreme vulnerability and fragility of erosion.
Potential area for protection: identifies biological (inland) fragility areas and the zones with priority for reforestation inside farming areas and outside the PNMB.
Area of urban influence: aggregates the areas near the existing urban developments that might be subject of expansion and thus dominate the local function of the zone.
Area of hydrological fragility: zones where water recharge capacity is threatened.

As a descriptive map interpretation, the existing lands use confirms that, as a result of former land-use practices (e.g., livestock farming), the surrounding area of the PNMB is mostly classified as pasture; whereas as for the conflicts of use the map shows that many of these can be used as potential areas for reforestation. However, using ecological principles provides a further lecture of the analysis when the areas are considered patches, corridors and buffers (DRAMSTADT, OLSON and FORMAN 1996). This is evidenced when the ecological principles are applied in the map of potential land use suitability for the improvement of the local landscape-ecology (TURNER 2001). For example, the urban patch at the south of the PNMB separates the unique zones of mangroves-estuary and causes the urban influence fragility to increase, which in consequence makes the park even more vulnerable to isolation. The possibility of interconnection supports the existing forest patches of primary and secondary level with intermittent scrublands and identifies potential areas for integration with other reforested areas. This could also help repair the evident lack of green coverage near most of the river corridors, increase the area of green buffer and, thus, reduce the hydrological fragility (saturated levels of recharge for groundwater) located in the unique zones. Therefore, the areas for high ecological value connect the mangrove-estuaries and the coast together with the forest areas. These priority areas (even outside the existing PNMB boundaries) enhance connectivity in order to strengthen the integrity of the habitat and for it not to be absorbed by the urban use.

4 Application of Landscape Analysis

4.1 Definition of Landscape planning objective

The landscape analysis, using ecological principles, reinforces the need to increase connectivity, avoids spatial fragmentation and isolation of high ecological value areas and reduces the edge effect of external influence (TURNER 2001). Such situation clearly establishes the landscape planning objective for the case of the PNMB and surrounding areas:

To reinforce the landscape conditions of the PNMB through the suggestion of zones which integrate the existing and potential areas with ecological value.

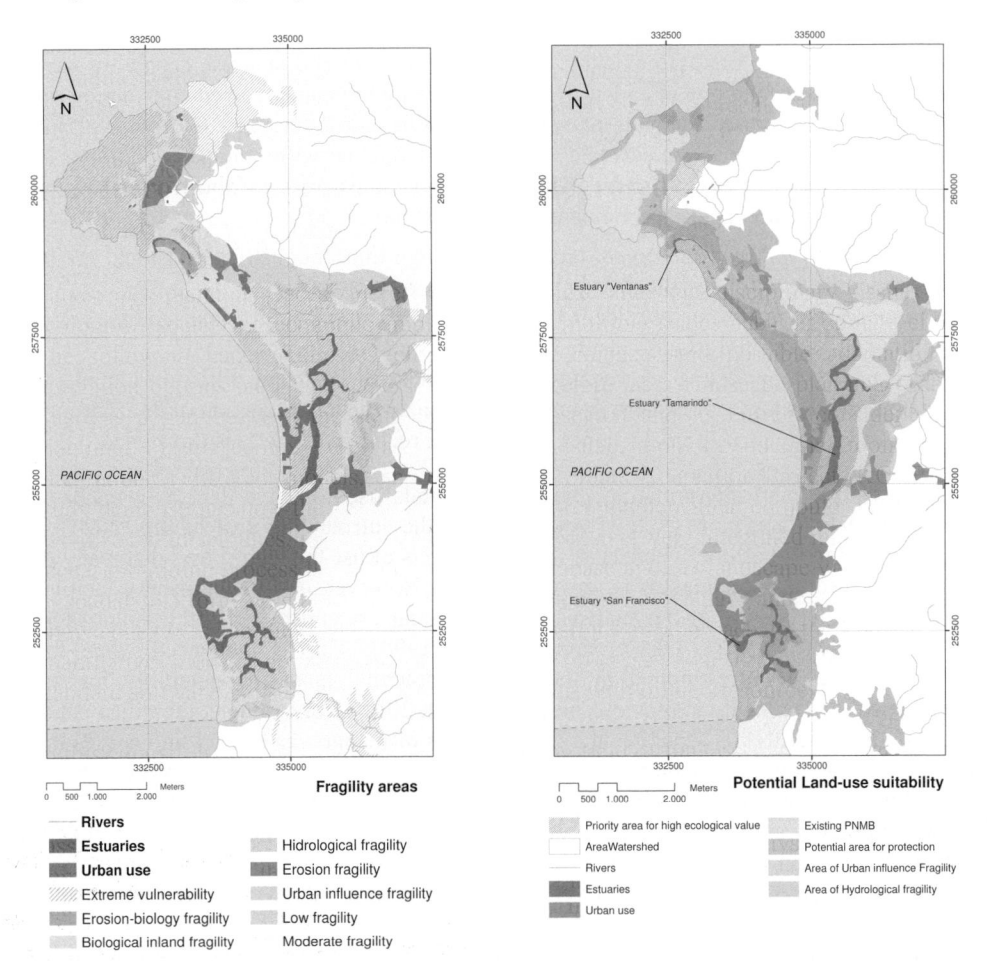

Fig. 1-2: Maps showing the Fragility areas and Potential Land-use suitability

4.2 Landscape zoning proposal

The landscape planning objective can be included in the PNMB Management Plan (MP) and translated into a landscape zoning proposal through a clear spatial framework which describes what can or cannot occur where and how (PADILLA 2010). In other words, the zoning determines what is allowed to take place (e.g. activities, type of land-use, etc.) throughout the different geographical areas of the protected territory. *Zoning plans* can be seen as an important instrument to facilitate the conservation and use of protected areas, since they are recommended by international guidelines from the IUCN (THOMAS, LEE & MIDDLETON 2003).

As a main contribution of this investigation, a proposal for a landscape zoning map for the PNMB is suggested (see Fig. 3) which basically translates the spatial limits through the definition of six zones according to the landscape use which are: protection, restoration, potential for protection, limited development areas, archaeological sites and buffer zones. Each of these zones is defined to have purposes according to landscape ecology criteria.

For instance, a major re-definition presented by the landscape-zoning proposal is the core zone; which is the current legally constituted PNMB, and is surrounded by buffer zones and potential areas for protection. The landscape planning objective is spatially translated as an increase of the PNMB area. This is justified through the criteria of ecological robustness using protective transition membranes that permeate the adjacent urban dominance over the fragility areas. Therefore, the zoning map reflects the relevance of considering the landscape behaviour beyond the existing park's limits. There are also the buffer zones that are specifically intended for the unique areas such as the estuaries and help to consolidate the green corridors along the coastal fringe.

5 Conclusion

Planners can deal with integrative considerations using a holistic vision about the involved factors and, at the same time, apply an analytical process. GIS does, indeed, offer a valuable and practical input when it comes to the aggregation, storage, analysis and display of information (e.g., the overlaying of data of diverse nature). However, the selection of criteria for effective analysis is a key factor for handling the information, particularly when there are issues on the data availability and the planner has to elaborate effective and creative classification criteria. For example, this research demonstrated that buffer areas are much more than an "offset" command. The integration of scientific knowledge is necessary for the technical definition and justification of buffer zones and not simply using the software tool. Possible more substantial achievements can be reached by having clear planning objectives supported by theory concepts.

Nevertheless, there are still many difficulties for establishing reliable data sources in countries such as Costa Rica where there is a lack of institutional framework for regional planning and technological support. Also, the integration of socio-cultural factors and definition of spatial planning scales in protected areas is still a bigger challenge. There is a lack of tools for linking, implementing and monitoring management policies in terms of spatial conditions according to technical analysis. However, it is clear that a definition for the adequate land-use is not possible without an interdisciplinary approach.

All the above represent an enormous potential for landscape professionals in planning to reaffirm their contribution beyond perception approaches. Territorial planning using digital tools can open new opportunities for introducing the evaluation of the landscape near protected areas and carrying out multi-criteria analysis using GIS systems in ecological rich countries to reduce the loss of biodiversity. Costa Rica has a quarter of its territory under protection and has made great efforts for the implementation of laws. Nevertheless, the country still needs suitable applications for the data evaluation before consolidating national policies.

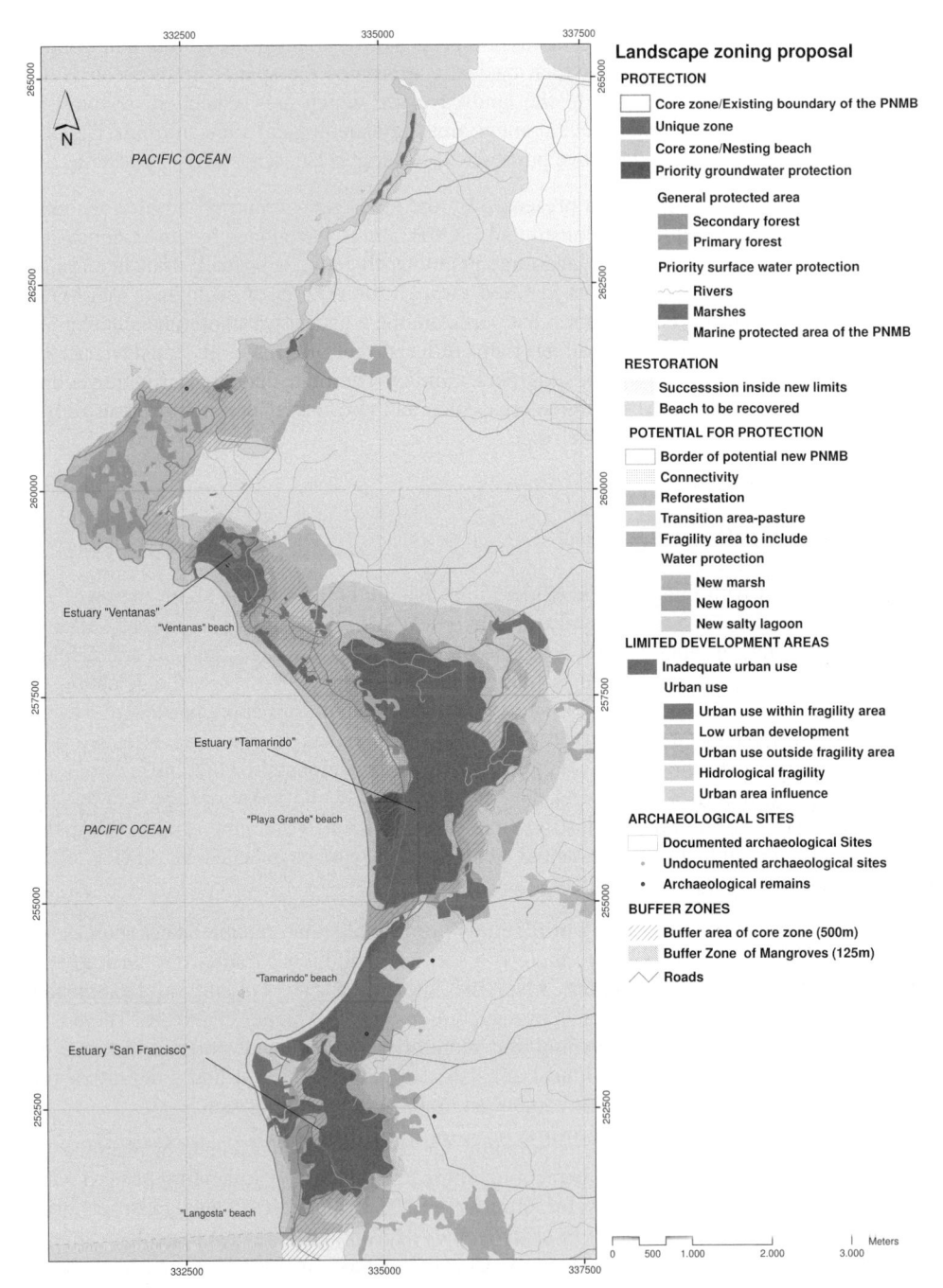

Fig. 3: Map showing a landscape zoning proposal which includes areas beyond the park's limits

References

Amend, S. et al. (2002), *Management plans. Concepts and Proposals.* Page 31. IUCN and GTZ. http://www.conservation-development.com/rsFiles/Datei/MPI_FILE_PDF.pdf (accessed in October 2009).

Dramstad, Olson & Forman (1996), *Landscape Ecology Principles in Landscape Architecture and Land-Use Planning.* Harvard University School of Design, USA.

Fonseca, A. & Drews, C. (2009), *Rising sea level due to climate change at Playa Grande, Las Baulas National Park, Costa Rica: inundation simulation based on a high resolution, digital elevation model and implications for park management.* WWF/Stereocarto Report, San José.

Forman, R. (1995), *Land Mosaics. The ecology of landscapes and regions.* Cambridge University Press, The United Kingdom.

Garcia Padilla, M. A. (2010), *Integrating landscape planning: zoning map proposal for "Las Baulas" Marine National Park, Costa Rica.* Master Thesis, Anhalt University of Applied Sciences, Bernburg, Germany.

GeoCad (2009), *Integral Study about the impact of building constructions and tourism, urban development in the buffer zone of Las Baulas Marine National Park, Guanacaste, Costa Rica.* http://www.setena.go.cr/WEB-BAULAS/documentos/ (accessed in October 2009).

Kauie, G. (2000), *Ecologically Orientated Planning.* Frankfurt/Main [et al.], Peter Lang Publishing Group.

Thomas, L. & Middleton, J. (2003), *Guidelines for Management Planning of Protected Areas.* IUCN Gland, Switzerland and Cambrigde UK.

Tropical Science Center (CCT) (2004), *Action Plan of Las Baulas de Guanacaste Marine National Park´s.* Management Plan, Costa Rica.

Turner, M. G., Gardner, R. H. & O'Neill, R. V. (2001), *Landscape Ecology in Theory and Practice.* New York, NY, USA, Springer-Verlag.

http://www.bosquesnuestros.org
http://www.inbio.ac.cr/en/biod/bio_biodiver.htm
http://www.setena.go.cr
http://www.sinac.go.cr/planesmanejo.php

Soaring Spaces – The Development of an Integrated Terrestrial and Bathymetric Information System for the Maltese Islands

Saviour FORMOSA

1 Introduction

In a rapidly developing world where the introduction of massive online information systems has enabled both the scientist and the general public to interact with remotely-located data from across the globe, the reality of access to data and eventually to information is slowly bringing forth the realisation that decades-old barriers to access to data still need to be overcome. Data availability suffers from a plethora of scourges that have left entire countries with a dearth of reliable baseline information, particularly small states which have limited human capacity to manage the whole datacycle in the physical, social and environmental domains. The main limitations include the fact that there are few homogeneous structures in operation, which governance situation has rendered data gathering agencies as a series of independent hoarding kingdoms, where data 'ownership' is seen as a private not as a corporate or a national affair thus the main users instead of being custodians transform themselves as the private owners of such data.

Other more technical issues include the fact that there are too many standards to follow, data is not dynamic (gathered ad hoc as a one-off and not real-time), data is not quality assured/controlled, queries are not organised and recorded, data is not secured – ('illegal' use of storage on personal storage devices and other digital media) and that versioning is not practiced. In addition, even where the data is available, there is an upsurge in requests for access to such data which has increased drastically since TIM BERNERS-LEE'S (1989) world wide web (WWW) proposal changed society as never before. The WWW changed a medium that was at best techno-centric to one that is now essentially socio-technic. Increasing requirements for bandwidth has resulted in a need for a reanalysis of DAHRENDORFF's (1990) access issue in contemporary worlds, both real and virtual, where not all society has access to the information through on-line services.

The other most important issue relates to the access to standardised processes for information-creation which is being tackled from various legislative loci such as the Data Protection Act (OJ 1995), the Århus Convention (OJ 2003a, OJ 2003b), the Freedom of Information Act (OJ 2003c) and the INSPIRE Directive (OJ 2007), the SEIS initiative, as well as other guidance documents that are targeted to enable the smooth and free flow of effective information.

In this paper, the Malta case study is brought to the fore with an analysis of the approaches taken to ensure that despite the limitations imposed on a small country, access is being made to the diverse users of data ranging from data in the physical and social domains through the fruition of a decades-long drive that culminates in the dissemination of environmental and 3D data to users for free. This following situation as analysed in 2006 where most datasets were not inter-connected, basemaps had a time-tag of 1988, environmental data capture was ad hoc and data dissemination was available through a mapserver with a date tag of 2000, whilst actual data could be requested by the public on a ad hoc basis (CONCHIN et al. 2010).

The need was felt to acquire a comprehensive set of new baseline information, through nation-wide digital terrain and bathymetric scanning, acquisition and analysis of environmental and spatial data within an interactive medium and the free dissemination of all information through an INSPIRE compliant online tool. This scope was targeted to ensure project integrativity and the need to ensure compliance with the legislative requirements (MALTA ENVIRONMENT & PLANNING AUTHORITY 2009).

2 State of Affairs: a 20-Year Approach

2.1 Laying the Structural and Academic Foundations

The need for change has been felt for some time since the mid-1990s when a review was carried out on the GIS legacy that had been acquired since 1964, which experienced rapid changes in visualization and data transfer (GATT & STOTHERS 1996). With major changes experienced in geographic information systems (GIS) from 1985 onwards, two main phases were identified, those related to Digital Mapping/data collection phase and those related to the application of GIS in an operational context (Figure 1). The process entailed the setting up of a national mapping agency (NMA) in 1988, a transition to a fully digital scenario between 1994 and 1998 which also saw the introduction of GIS, and the launch of the mapserver in 2000 (MEPA 2000).

This drive identified several hardware restrictions and high storage costs, which caused the NMA to truncate all the spatial data (Malta falls within 1 zone on the UTM projection) by removing part of the northings and eastings which were the same within all the national datasets. Whilst such a decision was aimed to ease the rolling-out costs, it has obstructed all the EU/EEA data reporting to the various agencies, since such datas need to be reconverted to the original full UTM. In order to reconvert all the resultant 20-year data legacy as required for convergence, costs are estimated to have overtaken any savings made over the decades; a situation that requires immediate action.

A new wave of data creation initiatives brought about by EU membership gave a boost to the drive to give birth to a phase through creation of baseline datasets against which all new information would be gauged against. This process was tested for its potential through the creation of datasets such as Corine Landcover 1990-2000-2006 and the relative change analysis, elevation maps, environmental protection maps (EEA CDDA, Natura 2000, SAC)[1] and in 2011, a Land use/cover area frame survey (LUCAS) field survey[2].

---

[1] European Environment Agency, Common Data Repository. Available online:
http://cdr.eionet.europa.eu/mt (accessed on 29 February 2012).
[2] European Commission, Land use/cover area frame survey (LUCAS) Decision 1445/2000/EC. Available online:
http://epp.eurostat.ec.europa.eu/statistics_explained/index.php/Glossary:Land_use/cover_area_frame_survey_(LUCAS) (accessed on 29 February 2012).

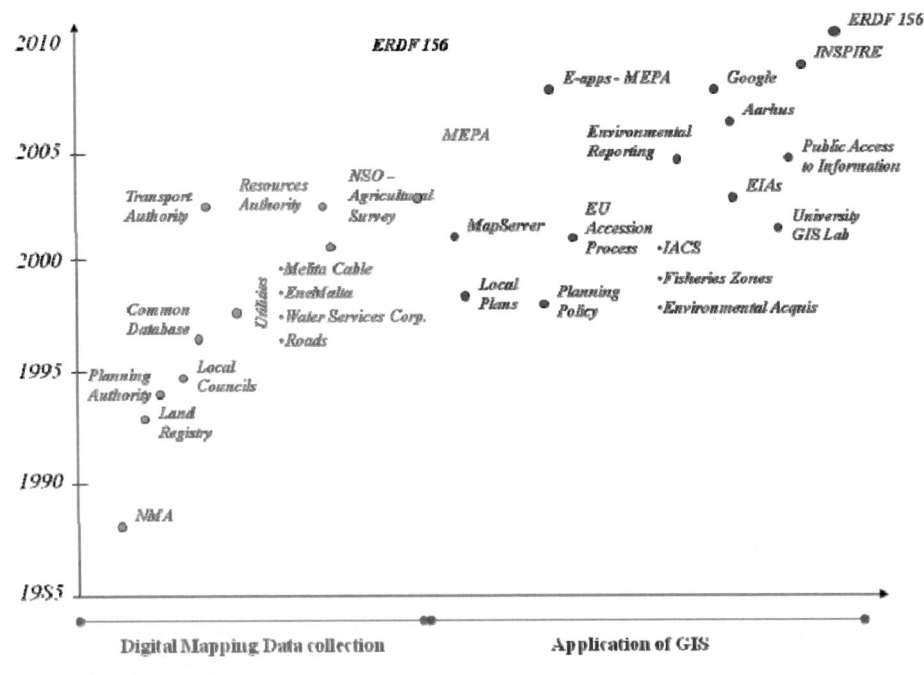

Fig 1: The GIS data process

On the academic side, various research initiatives were initiated. These included studies on organisational change (GATT et al 1996), remote sensing (TABONE ADAMI 1998), census web-mapping (FORMOSA 2000), environmental-landuse (TABONE ADAMI 2001), GML-related (AGUIS 2003), ethics in GIS (VALENTINO 2004), 3D GIS for spatial planning (CONCHIN 2005), environmental (FARRUGIA 2006), quality improvement cycle (RIZZO NAUDI 2007), through to socio-technical approaches to GIS (FORMOSA et al. 2011). The iterative studies have helped to consolidate the GIS-based data and operation structure which eventually shaped the foundation for a national geographic information data structure which culminated in the ERDF project discussed in this paper.

A number of projects in the spatial planning and environmental domains resulted in the launching of a comprehensive strategy based on the data cycle; such projects included Structural Funds 2007-13, EAFRD 2007-13, Transition Facility Programme for Malta, Pre-Accession Funds, and other programmes3. Having laid out the groundwork, the next step identified the need for the integration of the environmental domain with the spatial development domain through the implementation of information resources and information technology infrastructure in line with the Aarhus, INSPIRE and SEIS requirements as well as the outcomes of such projects as Plan4all (BEYER & WASSERBURGER 2009).

---

[3] Internationally Funded Projects at MEPA. Available online: http://www.mepa.org.mt/internationally-funded-projectsatmepa (accessed on 29 February 2012).

2.2 The Engine to Be: ERDF

The project that emanated from this exercise is co-financed by the European Regional Development Fund under Operational Programme 1 – Cohesion Policy 2007-2013 – Investing in Competitiveness for a Better Quality of Life. The authors are implementing the project through the Malta Environment and Planning Authority (MEPA) in collaboration with the Malta Resources Authority (MRA), the Department of Environmental Health, the National Statistics Office (NSO) and the University of Malta.

The project will result in the procurement of equipment, information management systems, environmental baseline surveys, training of staff, and the enhancement of the national monitoring programmes in the environmental themes of air, water, noise, radiation and soil.

The following outputs will be delivered:
- Comprehensive assessment of the environmental monitoring requirements in the areas of air, water, radiation, noise and soil, an environmental monitoring strategy and detailed monitoring programmes will be designed and drawn up by Q2 of 2013 to cover all monitoring requirements;
- Installation of air, noise and radiation equipment, information resources systems and infrastructure procured, installed, tested and commissioned, and relevant staff trained in their operation by the Q4 of 2012;
- Baseline studies with 100% scan coverage of the Maltese Islands conducted in the areas of water, radiation, noise and soil, together with terrestrial spatial surveys and bathymetric surveys of coastal waters within 1 nautical mile by Q2 of 2013;
- A Shared Environmental Information System (SEIS) designed and implemented by Q2 of 2013; The results of the latter will raise public awareness of environmental issues, which participation and enhanced policy decisions.

3 Acquisition and Access

The initial status research was initiated in 2006, which study highlighted the need to enhance the national monitoring programmes in the five environmental themes through the identification of information gaps in monitoring processes and filling data gaps, carrying out environmental baseline surveys and through the procurement of monitoring equipment & information management systems and finally the training of staff.

The project will result in the integration of the requirements for EU environmental reporting through the employment of the INSPIRE Directive for the spatial component, the use of the Aarhus Convention as the conveyor for the dataflow and ultimately the employment of a tool pertaining to the SEIS requirements for the eventual online dissemination. The latter will employ standards comprising; OGC WMS, OGC WMS-T, OGC WFS, OGC WCS, ANSI SQL, INSPIRE, Z39.50 and CSW.

The project outcomes can be structured into 4 sectors: Environmental Acquisition, Spatial Constructs, Dissemination Media and Access as per Table 1.

This review ensured that a new baseline is created from which to launch Malta's new data capturing exercises across the different themes. Terrestrial and bathymetric data would be available at higher resolutions suitable for environmental modelling and EU reporting

Environmental Acquisition	Spatial Constructs
(1) Air Strategy and Baseline Study (2) Water Strategy and Baseline Study (3) Noise and Radiation Strategy (4) Soil Strategy (degradation processes and contamination in diffuse sources)	(1) Full LIDAR Scan: Terrestrial and Bathymetric (2) Ground truthing for sea substrate type (3) Oblique aerial & satellite imagery (4) An address point dataset
Dissemination Media	**Access**
(1) Online information service (2) Online mapservice – SEIS (3) Statistical backing for experts – inc. spatial stats	(1) All Data are to be disseminated for FREE

Table 1: The 4 Project Outcomes

purposes. As the non availability of high quality 3D spatial data hinders comprehensive land use planning, environmental monitoringthe project will deliver a series of scans using different technologies in order to provide a comprehensive seemless dataset. The delivery includes high resolution 3D terrestrial data coverage for the Maltese Islands using a combination of oblique aerial imagery (Figures 2 a-b)and Light Detection and Ranging (LIDAR) data (Figure 2c), as well as through a bathymetric survey of coastal waters within 1 nautical mile (nm) radius off the baseline coastline, using a combination of aerial LIDAR surveys, acoustic scans and a physical grab sampling survey as well as GIS-based noise maps (Figures 2d). These technologies, as well as other fieldwork technologies, has equipped the researchers with a launching pad for the diverse physical, environmental and social studies that are undertaken in relation to social and environmental health.

The outputs from the project include the following services and supply shows Table 2.

Fig. 2a: Oblique Imagery Valletta 2011

Fig. 2b: Oblique Imagery Fort Manoel 2011

Fig. 2c: LIDAR Scanning Exercise 2012

Fig. 2d: Noise Mapping Exercise MEPA 2011

Services	Supply
(1) LIDAR Scan: Terrestrial (Topographic Light Detection and Ranging (LiDAR)) Digital Surface Model (DSM) and Digital Terrain Model (DTM) (316 km.sq) – Figure 3a (2) Bathymetric LIDAR aerial survey – depths of 0m to 15m within 1 nautical mile from the Maltese coastline (38km.sq) – Figure 3b (3) Bathymetric Scan: Acoustic (side scan sonar) Digital Surface Model and an acoustic information map of sea bed (361 km.sq) – Figure 3c (4) High resolution oblique aerial imagery and derived orthophoto mosaic and tiled imagery of the Maltese Islands (316km.sq) (5) Satellite imagery (6) A complete address point database	(7) Remote GPS Cameras (Remote capture GPS receiver) (8) 3D scanner (9) GIS Handhelds (10) Global Navigation Satellite System Station and geodetic receivers Figure 4 outlines the remotely-sensed, aerial and in situ data capture data capture processes employed in the project.

Table 2: Project Outputs

Fig. 3a: Maltese Islands coast inclusive of 1 nautical mile boundary from the baseline coastline

Fig. 3b: Maltese Islands showing coastal water area with depths of 0m to 15m within 1 nautical mile from the Maltese baseline coastline

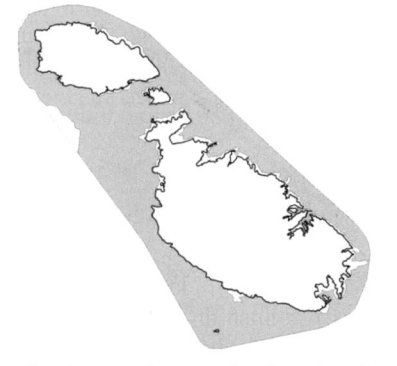

Fig. 3c: Maltese Islands showing coastal water area with depths of 15m to 200m within 1 nautical mile from the Maltese baseline coastline

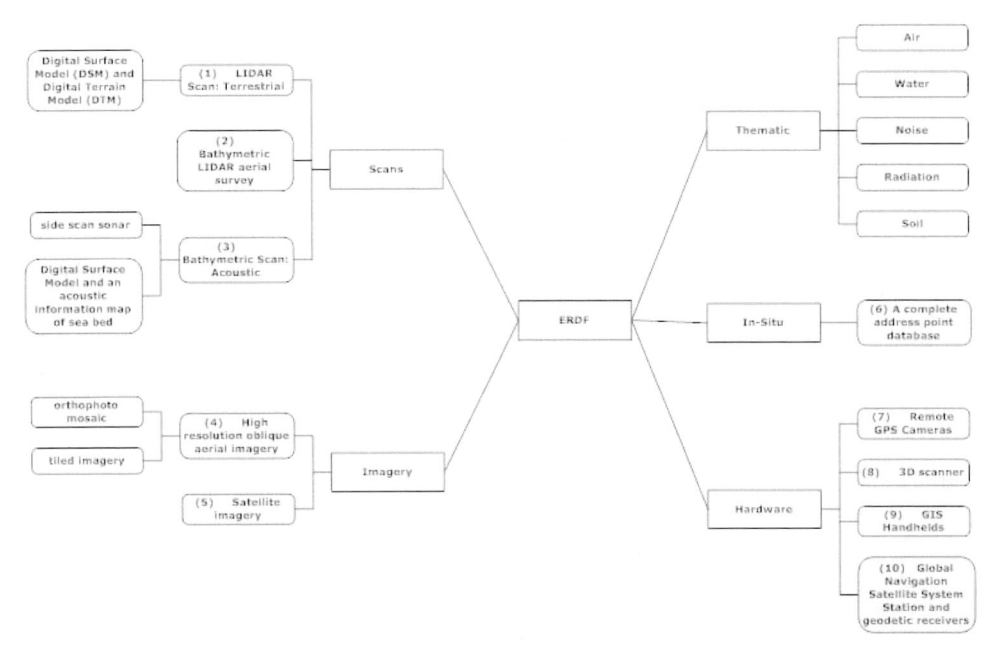

Fig. 4: Remotely-Sensed, Aerial and In-situ Data Capture Components

4 Taking the Project to the Masses

The data dissemination phase will be based on a SEIS-based strategy and implementation through an online tool. The tool was planned to develop and implement the requirements outlined by SEIS, which outputs will include a dedicated geoportal based on ArcGIS geodatabase design as based on ArcGIS server architecture.

The phase is aimed to review the state of play of the current developments with respect to the Shared Environmental Information System (SEIS), including the following:

- EU Directives supporting the EU's SEIS initiative and any proposed recommendations of the EEA, JRC, EUROSTAT;
- Commission's Communication COM (2008) 46 Final "Towards a Shared Environmental Information System";
- SEIS developments by the European Environment Agency (EEA);
- Overview and updates on the SEIS-BASIS (Shared Environmental Information System Baseline and Evolution Study) project which aims to provide guidance on how to improve the comparability and quality of environmental data, as required by SEIS;
- The outputs of the NESIS project and roadmap developments on how to move from the current information systems of EU's environment agencies towards an INPSIRE-SEIS based system. To include relevant results for the NESIS State of Play study on examples of best practice as a source of guidelines for MEPA's proposed SEIS as informed by recent developments;

- Relevance of the INSPIRE Directive (Directive 2007/2/EC) and the Aarhus Convention to the EU's SEIS;
- Linking of an integrated reporting system is required in line with the EEA Reportnet initiative and its CDR (Common Data Repository) structure to SEIS;
- An analysis of the existing Maltese information management systems and platforms, as well as an assessment of the present institutional capacity necessary for the operation of the Maltese component of the SEIS;
- New or emerging reporting standards currently being adapted, such as XML-related standards, to which the SEIS should conform.

The proposal was to design a geodatabase data model that is flexible, caters for potential expansion, easily adaptable by the environmental agency and supports migration from current data structures. In order to improve the development of the geodatabase schema the project entailed; the identification of data sources and key data themes; development of representation specifications and relationships of the geodatabases; data capture definition procedures; geodatabase design documentation and; the employment of ArcSDE in order to to manage the underlying geospatial data that will be stored in Microsoft's SQL Server RDBMS (Figure 5) (BONOZOUNTAS & KARAMPOURNIOTIS 2012).

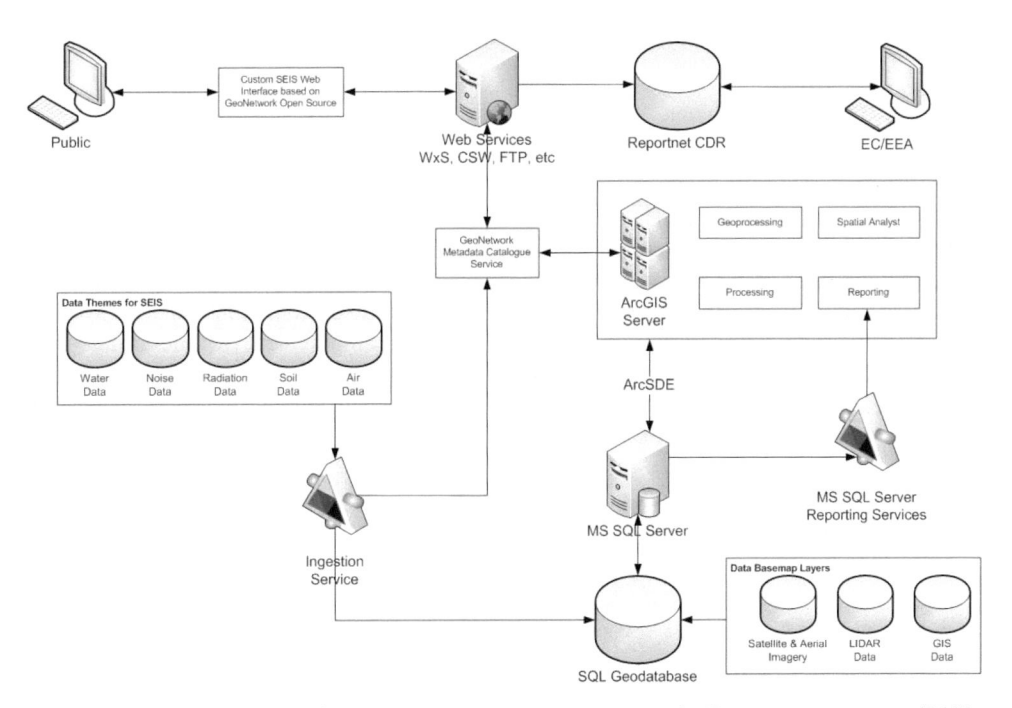

Fig. 5: Malta-SEIS Architecture: Source: BONOZOUNTAS & KARAMPOURNIOTIS (2012, 17)

The project datacycle, from data capture, analysis, reporting and dissemination, outlines a structure that conforms to the Aarhus, INSPIRE and SEIS requirements. The main functions cover in sequence: Web interfacing and security (allowing the viewing, data querying, reporting and analysis of all spatial datasets identified in the five themes and the information resources datasets); Front end flexibility (allowing for increasing levels of geospatial data within a changing and dynamic environment); Discovery services/functions (provision of access to resources via product-neutral visualization and downloading services; Geoportal (linked to the National Spatial Data Infrastructure (SDI) and eventually to the EU SDI under development; Spatial dataset standards (to conform to the INSPIRE Directive (Directive 2007/2/EC) standards and; Spatial data transfer (ensuring available processes are thoroughly understood and depicted into the model with a target to establish a coherent data integration system that will allow the further transmission through the online portal).

The final leap will be to integrate all these systems with others linked to domains not normally taken up by the physical domains, those pertaining to the social disciplines. The system will further strive to understand physical-social dataset integration as identified in the Plan4all Malta case study (FORMOSA et al. 2011). The study investigated the potential for an INSPIRE-based system aimed at integrating information from the physical, social, criminological, psychological and health domains. The studies of the impacts of environmental factors on social and health issues are expected to lead to public participation and knowledge gain in the Maltese Islands as people become more aware of readily-available realtime information irrespective of their location.

5 Conclusion

In conclusion, the Maltese Islands, whilst having a legacy of data creation, have taken the bold step to create baseline data for integrative future analysis. The project took on a challenging aspect in that it covered the whole state, which in the economies of scale was an ambitious one. The processes undertaken ensured that the information gain was an integrated one as it conceptualises a nation-wide strategy targeting access to data employing various Directives and Conventions. The employment of the Arhus and INSPIRE Directives, as well as the SEIS initiative ensured that the various components governing spatial and thematic information come together into one freely-available location.

References

Aguis, C. (2003), *Using GML to represent Spatial Environmental Information.* Unpublished MSc GIS thesis, University of Huddersfield, Huddersfield.
Berners-Lee, T. (1989), *Information Management: A Proposal.* Available online: http://www.w3.org/History/1989/proposal.html (accessed on 29 February 2012).
Beyer, C. & Wasserburger, W. (2009), *Plan4all Deliverable 2.2. Analysis of innovative challenges.*
Bonozountas, M. & Karampourniotis, I. (2012), *MALTA-SEIS: Deliverable D2.1 Report of Analysis and Detailed Proposal for SEIS.* CT3067/2010 – 02, Malta.

Conchin, S., (2005), *Investigating the development of 3D GIS technologies for Spatial Planning: A Malta study.* Unpublished MSc GIS & Environment thesis, Manchester Metropolitan University, Manchester.

Conchin, S., Agius, C., Formosa, S. Rizzo Naudi, A.(2010), *Does visualisation of digital landscapes serve itself? How topographic, planning, environmental and other thematic information is integrated and disseminated via web GIS.* In: Buhmann, E. et al. (Eds.), Peer Reviewed Proceedings of Digital Landscape Architecture 2010, Anhalt University of Applied Sciences. Berlin/Offenbach, Wichmann. ISBN 978-3-87907-491-4.

Dahrendorff, R., (1990), *The Modern Social Conflict: An Essay on the Politics of Liberty.* Berkeley, CA, University of California Press.

European Commission, *Land use/cover area frame survey (LUCAS) Decision 1445/2000/EC.* Available online: http://epp.eurostat.ec.europa.eu/statistics_explained/index.php/Glossary:Land_use/cover_area_frame_survey_(LUCAS) (accessed on 29 February 2012).

European Environment Agency, *Common Data Repository.* Available online: http://cdr.eionet.europa.eu/mt (accessed on 29 February 2012).

Farrugia A. (2006), *Implications of EU Accession on Environmental Spatial Data: a Malta Case Study.* Unpublished MSc GIS Science thesis, Manchester Metropolitan University, Manchester.

Formosa, S. (2000), *Coming of Age: Investigating the Conception of a Census Web-Mapping Service for the Maltese Islands.* Unpublished MSc thesis Geographical Information Systems. University of Huddersfield, United Kingdom. Available online: http://www.tcnseurope.org/census/1995/index.htm (accessed on 29 February 2012).

Formosa, S., Magri, V., Neuschmid, J. & Schrenk, M. (2011), *Sharing integrated spatial and thematic data: the CRISOLA case for Malta and the European project Plan4all process.* In: Future Internet 2011, 3 (4), 344-361, doi:10.3390/fi3040344.

Gatt, M. & Stothers, N. (1996), *The Implementation and Application of GIS in the Planning Authority of Malta.* Geographical Information, Second Joint European Conference and Exhibition on Geographical Information, Barcelona, Spain, ISBN 90-5199-268-8 (IOS Press), ISBN 4-274-90098-3 (OHMSHA).

Internationally Funded Projects at MEPA. Available online: http://www.mepa.org.mt/internationally-funded-projectsatmepa (accessed on 29 February 2012).

MEPA mapserver (2000), Available online: http://www.mepa.org.mt/mepa-mapserver (accessed on 29 February 2012).

Malta Environment & Planning Authority (2009), *Developing National Environmental Monitoring Infrastructure and Capacity.* MEPA, Floriana, Malta.

Malta Environment & Planning Authority (May 2011), *Data Flow 7 Supplementary report Noise Action Plan (Summary).* MEPA, Floriana, Malta.

Official Journal of the European Union (24 October 1995), *Directive 95/46/EC of the European Parliament and of the Council of 24 October 1995 on the protection of individuals with regard to the processing of personal data and on the free movement of such data.* L 281 , 23/11/1995.

Official Journal of the European Union (28 January 2003a), *Directive 2003/4/EC of the European Parliament and of the Council of 28 January 2003 on public access to environmental information and repealing Council Directive 90/313/EEC.* L 041, 14/02/2003.

Official Journal of the European Union (26 May 2003b), *Directive 2003/35/EC of the European Parliament and of the Council of 26 May 2003 providing for public participation in respect of the drawing up of certain plans and programmes relating to the environment and amending with regard to public participation and access to justice Council Directives 85/337/EEC and 96/61/EC – Statement by the Commission.* L 156, 25/06/2003.

Official Journal of the European Union (17 November 2003c), *Directive 2003/98/EC of the European Parliament and of the Council of 17 November 2003 on the re-use of public sector information.* L 345, 31/12/2003.

Official Journal of the European Union (25 April 2007), *Directive 2007/2/EC of the European Parliament and of the Council of 14 March 2007 establishing an Infrastructure for Spatial Information in the European Community (INSPIRE.*, L108, Vol. 50

Rizzo Naudi, A. (2007), *A Continuous Quality Improvement Cycle for Geographic Information Systems within the Malta Environment & Planning Authority* (MEPA). Unpublished M. Sc. dissertation, Manchester Metropolitan University, Manchester.

Shared Environmental Information System. Available online: http://ec.europa.eu/environment/seis/ (accessed on 29 February 2012).

Tabone Adami, E. (1998), *Corrections for the estimation of chlorophyll concentrations in coastal waters from remotely sensed data*, MPhil thesis. University of Cambridge, Cambridge.

Tabone Adami, E. (2001), *Integrated modeling of nutrient transfers from land-based sources for eutrophication assessment of Maltese coastal waters.* PhD thesis, University of Cambridge, Cambridge.

Valentino, C. (2004), *Developing a Coherent Approach to Ethical Use of Geographical Information in Malta.* Unpublished MSc GIS & Management thesis, Manchester Metropolitan University, Manchester.

Hyper-localism and Parametric Mapping for Collaborative Urbanism

Raffaele PÉ

1 Introduction

The paper presents a study on transforming data harvested from social networks into spatial explicit information, which can be relevant for urban redevelopment. In particular, spatial disposals for urban growth and their attractiveness for local users can be identified and visualized. Integrating these social aspects into the urban planning and design process is crucial to inform sustainable paradigms of urban development. Aim of the experiment is to weave sources of knowledge derived from spatial planning practices and the ones produced by the stakeholders, to achieve socially acceptable orientations in urban design (SCHOLZ 2011). The modelling proposed is an example of how these aspects can be integrated into GeoDesign processes in digital form..

An attempt to report and communicate these geo-political contents is influenced by the need of understanding the potential embedded in the lenses through which we observe the city and we represent it (WALDHEIM 2006). The notion of context, as a means through which we acquire knowledge about the city, nowadays is widened by the use of the internet and social media. Organized networks are comparable to self-organized communities in spontaneous relationship, that build their activities on principles like participation and shared action. Their deep diffusion and appeal on a large number of users sets up the premises for the construction of new form of institutions in contemporary society (ROSSITER 2006). The inclusion of the information collected through organized networks in a mapping study for the city shows the intention of the planner to analyse an urban context learning mutually from theory and practice, according to a trans-disciplinary idea of collaboration (GUATTARI 1995). The method then intends to represent some aspects of the complexity related to urban regeneration process, for the deduction of latent conditions and spatial behaviours.

The term Hyper-localism is a word of recent coinage used in journalism to express the news coverage of local events referred to a specific geographical area and its inhabitants. An "hyper-local" event is often fostered through web communities (Yeah Hackney http://www.yeahhackney.com Dalston People http://www.dalstonpeople.co.uk, Neighbourhoodr http://www.neighborhoodr.com) to ensure a more efficient and radical diffusion of the information at a global scale, providing a considerable amount of data available to large audiences as an expression of social interactions within a certain neighbourhood. Hyper-local websites define a clear area of interest in relation to a physical place and help the users in identifying relevant local actors on the basis of communal trust. In social media in fact the uploaded material is offered to the users as an editable content, which is trustable on the basis of its on-line ratings. These networks advertise the aspirations of individuals or groups of individuals translating their sense of belonging and appreciation for their urban environment in measurable spatial parameters. If we look at websites' rating systems (i.e. Google Trends http://www.google.com/trends), phenomena and places are analysed in diagrams as quantifiable variables of appreciation of the contents among their users.

The use of hyper-local sites to gather information on the local for planning purposes implies the acknowledgement of their ability to promote social inclusion and spatial awareness as drivers of urban development. In order to disclose such dynamics, the employment of parametric maps can provide a tool for the detection of self-generated patterns of urban appropriation. Organized networks, exploited as instruments for urban monitoring, structure a correspondence between biographies of places and the agenda of their users, recreating an objective interweave between history and geography. GeoDesign applications represent an efficient device for a proactive and participative management of planning issues identifying emerging social patterns in suburban environments. Referring to Gadamer's position in The Relevance of the Beautiful (1986), we see the role of these digital applications as "objects of art" for landscape framing, through which the processing of data aims to become a playful engine of social imagination and spatial integration.

The presented work is a mapping exercise conducted at the Royal College of Art, London - Architectural Department for the regeneration of the suburb of London Hackney. The case study aims to trace the key-points of a research that perceives the issues related to urban transformation as a matter of systemic interpretation (BERGER 2006) of relevant social phenomena in local colonization, promoting spatial awareness to envision urban re-configuration.

2 Material and Methods

2.1 Input Data

The choice of exploring an urban context from the point of view of an hyper-local network requires first of all the collection of relevant data from the websites and then the geo-referencing of such data to their actual location on a map, to provide a geographic background to local social behaviours and spatial practices.

The selection of an urban quadrant as a case study is important to disclose these conditions. For the following exercise the suburb of London Hackney has been selected for its emerging character of fast developing neighbourhood, due to its proximity to the site for the Olympics 2012, the remarkable concentration of young people and artists among its population, and the possibility to work with an existing hyper-local website (Yeah! Hackney). Despite all these positive features the neighbourhood presents several elements of spatial decay, including scarce connectivity to other parts of London, low quality of dwellings and residential facilities and the absence of major public spaces and services. This work aims to identify available drivers of regeneration studying spatial behaviours as expression of emergent social patterns of local colonization.

A satellite image of the area taken from Google Maps has been used to localize activities and information on the map. The exercise will exploit RGB colours from the quadrant for the following modelling phases in Grasshopper. The quadrant includes the geographic

Fig. 1: Map of London Hackney taken from *Google Maps* as a basis for this exercise

boundaries of Hackney as established by the local council, covering an area of approximately 20 sq km as illustrated in Fig. 1.

Among the data gathered from the website, main *local businesses*, *NGOs* active on the area and listed *local communities* have been chosen as attractors for the construction of a parametric map. Each attractor in the website shows a chart with the number of viewers as a degree of appreciation of the attractors by the network's users. These values have been used as indicators of appreciation of the urban attractors as they express a sense of belonging of the users to the attractors they have promoted on the website.

2.2 Modelling approach

The experiment uses Rhinoceros 4.0 and the latest version of its plug-in Grasshopper as basic tools to manipulate and to model numeric data.

Once imported the selected quadrant in Rhinoceros, the image is processed in Grasshopper using the Image Sampler component, which detects the RGB colours of the satellite image, allowing the recognition of green spaces, dwellings, squares and streets. Open spaces, gardens and fields are presented as available spaces for the development of the attractors, as they can outline areas of future growth and transformation. The dichotomy between colonized areas and open territories has been rendered creating a NURBs surface for which the peak points symbolize clusters of disposable land for future development. The surface embodies an idea of "transfigured landscape" that expresses deep potentialities for urban transformation included in the quadrant. The landscape is analysed dividing the surface in contour curves in order to measure such phenomenon. In parallel, the attractors chosen from the local network are geo-referenced on the map and then included in the scripting process in the shape of red crosses as shown in Fig. 2.

The tessellation of the surface through a Voronoi diagram has the aim of clarifying the neighbouring relationships between spatial opportunities and the attractors. In a Voronoi diagram the distance between two objects in space is their shortest path in the network rather than their Euclidean distance (AA.VV. 2008). The Voronoi cells (dark green network in Fig. 3) are generated from a set of points obtained dividing the contour curves in a constant number of segments. The end points of each segment become the "sites" for the construction of the Voronoi cells. Bisecting with a perpendicular the lines that connects each site we obtain the network of cells. The more the cells are small and concentrated, the greater will be the spatial availability for urban growth.

Starting from the same set of sites, the cells obtained are offset according to their proximity to the attractor points (orange cells in Fig. 3). This parametric operations allows to compare spatial disposal and the physical possibility of growth of each attractor. The diagram in Fig. 3 shows that where the attractors are more concentrated the spatial disposal for growth is actually quite low (the orange cells almost overlap with the green network), therefore the current urban configuration will be unable to receive a greater number of users on the long term. On the contrary, on the area of the Olympics, where the space for urban growth is available, a smaller number of attractors identifies smaller possibilities of regeneration and urban colonization. The use of a Voronoi diagram in the exercise helps in creating an abstract topography of the place where all the described relationships can be displayed more directly and then synthesized.

The possibility of increasing the complexity of a generative map represents a relevant practice in outlining some hidden qualities of the city. For this reason the vertical extent of the map has been modelled exploiting numeric values collected from the social network which display the appreciation of the attractors by their users. The offset cells (orange cells in Fig. 3) are exploited as bases for a set of extrusions whose height is directly affected by the appreciation that users have shown on the social network by clicking on the attractors' page or linking the page to other networks. The extrusions are also proportional to their proximity to the attractors, in the way that the values are apportioned on the whole quadrant of study. The maps obtained are then overlapped and integrated in an axonometric view as in Fig. 4.

To summarize the generative map is influenced by the following parameters:
- local attractors or centralities (local businesses, NGOs, local communities shown on the hyper-local network)
- spatial disposal for urban growth (open spaces, fields, left-over)
- users' appreciation of the local attractors

Fig. 2: Map of the spatial disposal for urban growth. The darker green colour indicates peak areas of spatial disposal in relation to the attractors.

3 Results

The results of this experiment are shown in Figure 4. The outcome of the modelling process is an augmented image of the city which shapes some of its intrinsic characters and envisions its possible future development. The generative map displays the relationships existing in Hackney between what is perceived by its inhabitants as local attractors and the surrounding spatial operators. In particular on the East border of the quadrant, where the site of the Olympic Games 2012 is located, the sense of attraction to this operator in quite low in comparison to the great offer of public space and commercial facilities. This condition is even more relevant if we think that at the time of the experiment the main shopping mall of the Olympics, Westfield, was already open. The inhabitants of Hackney

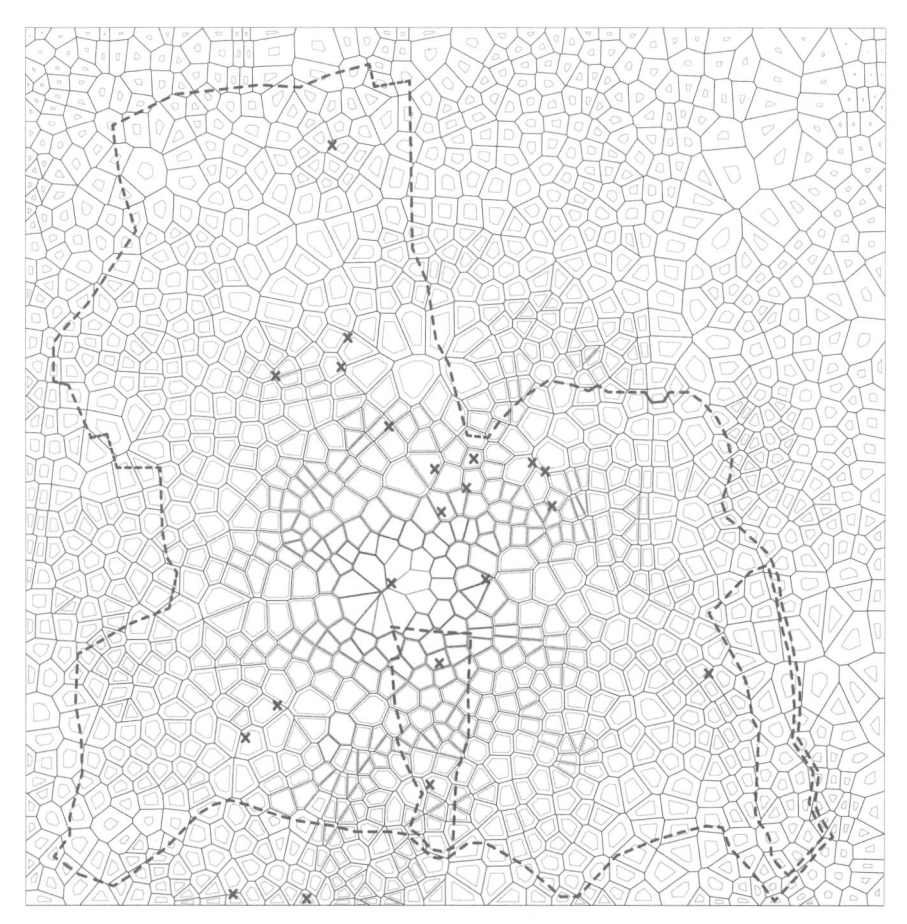

Fig. 3: Generative map of the spatial attractions. The map shows the appreciation of the attractors by the hyper-local users in relation to the spatial availability for future development.

don't seem to be affected by the presence of such a global infrastructure nearby. The intentions of the mayor to land the Olympics in this context following more consolidated practices in urban planning does not appear to be an effective strategy to provide urban regeneration. The extrusions of the local attractors' appreciation clearly indicates the central park of London Fields as a very appealing location for the growth of the neighbourhood, together with some isolated centralities towards Whitechapel (South-West) and the water reservoirs of Walthamstow (North-East). The green fields of Victoria Park also result as fertile places for urban transformation as well as the transversal corridor of locations along the Regent's Canal towards the City (West). Furthermore, the peaks display new emerging centralities as drivers of urban change that can be identified for example in the areas surrounding the food market of Broadway rd. near London Fields, the Wetherell rd. market near Victoria Park, or the MOT International Gallery for contemporary art on the waterfront of the Canal.

4 Discussion and Conclusions

The presented work is an attempt to demonstrate the benefit offered by the exploitation of GeoDesign tools like parametric mapping in urban design and planning. The map shows in fact the necessity of finding capillary drivers for the regeneration of Hackney, like local markets or public spaces for art and culture, regardless the presence of a global scale operator such as the site for the Olympic Games 2012.

This approach allows to detect latent conditions of urban environments, hence to generate alternative planning practices that include the view of the inhabitants on the transformation of their own neighbourhood. The use of data collected through a hyper-local social network aims to inform design methods and policies which are more collaborative and participative in the sense that they should integrate the vision of their users in a stage that anticipates the act of transformation. In this respect the use of such applications and the role of participants

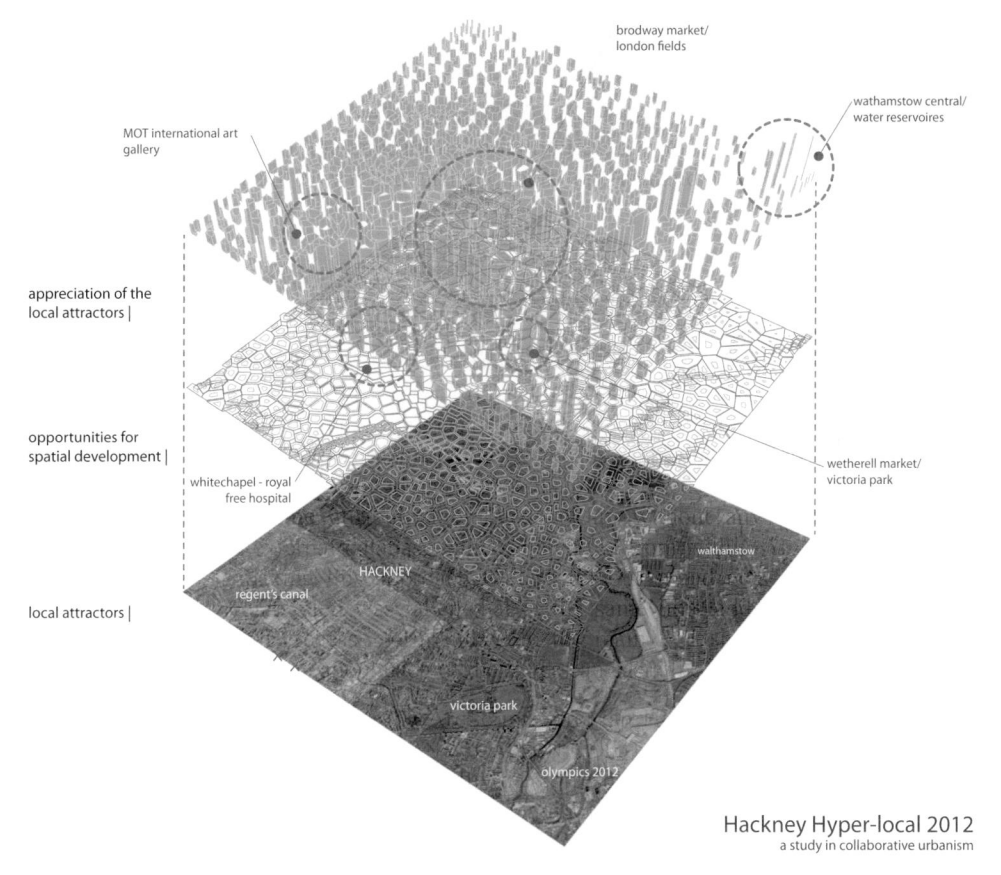

Fig. 4: Generative map of the spatial attractions. The map is divided in three layers, local attractors, opportunities for spatial development and appreciation of the attractors. New drivers of suburban regeneration are detected on the upper layer.

in GeoDesign processes could be even more proactive. The presented model describes the complex relationships that surround a project of urban change like the Olympics site, letting the users acknowledge their position regarding the proposal.

A future step of development of the exercise could be the inclusion of real-time data streams used as rating system of the information provided by the parametric map, to show how the model is received by the users on the hyper-local network before its actual construction. The idea can be implemented using recent Grasshoppers' plug-ins like Ghowl and Pachube, and integrating the model within a social network. A more "dynamic" platform of spatial analysis will then be fostered through the display of the contents in constant update with a virtual forum where new projects can be assessed and rated by rulers and users.

This strategy could provide also a feasible answer to the issue of displacement processes, for which places attractive for certain social groups maybe not attractive anymore if they are transformed. Transformation in space and time can be forecasted and virtually pre-determined in GeoDesign applications. Social media and organized networks recreate in fact a continuous exchange between real life and virtual envisioning. The use of the information detected through such media for GeoDesign purposes should be considered an instrument to acquaint the spatial awareness of the resources available for urban regeneration.

References

AA.VV (2008), *Environment and planning: Planning & design.* Pion Ltd. University of California, 35, p. 649.
Berger, A. (2006), *Dross-Scape. Wasting Land in Urban America.* New York, Princeton Architectural Press.
Bunshoten, R. (2006), *Touching the Second Skin.* In: Game, Set and Match: No.2: the Architecture of Co-laboratory, ed. by Oostheruis, K. & Feireiss, L. Rotterdam, Episode, 598-611.
Guattari, F. (1995), *Chaosmosis: an Ethico-Aesthetic Paradigm.* Trans. Paul Bains and Julian Pefanis. Sydney, Power Publications.
Rossiter, N. (2006), *Organized Networks, Media Theory, Creative Labour, New Institutions.* Rotterdam, NAi.
Scholz, R. W. (2011), *Environmental literacy in science and society: from knowledge to decisions.* Cambridge, Cambridge University Press.
Waldheim, C. (2006), *The Landscape Urbanism Reader.* New York, Princeton Architectural Press.

Visualization in Landscape Architecture

Survival in Dangerous Landscapes – A Game Environment for Increasing Public Preparedness

Ian BISHOP, John HANDMER, Angelinie WINARTO and Eric McCOWAN

1 Introduction

In developing our processes for design, planning and management of the landscape we are increasingly recognizing the temporal dimension (BASTIAN et al. 2006, DENG et al. 2009), although (STEPHENSON 2010) argues that landscape analysis is still dominated by static perceptions. While progressive changes are being recognized in the landscape literature, sudden changes are seen to be in the domain of other disciplines. Landscape changes, on a temporal scale of hours rather than years, are typically destructive in some form and frequently lead to loss of life (human, stock and wild), habitat or property. We think of these rapid changes as rare and abnormal occurrences and consequently they are inadequately considered in landscape planning. Often flood maps, fire risk maps or seismic zones find their way into land suitability analysis and sometimes they have an effect on development patterns, but even in countries with sophisticated technologies and planning regimes people still settle in areas of high risk. Brisbane, the third largest city in Australia was extensively flooded earlier this year because large areas of floodplain are built upon. Less than a year earlier devastating fires killed 173 people mostly within commuting distance of Melbourne. Just as a flood plain can expect floods, so a dry schlerophyll forest can expect fires. They are a part of the landscape. If we are unable to mitigate the risks associated with these landscape then we must seek ways to adapt to them and give the landscape residents the best chance of understanding the processes at work, their potential effects and how to best respond when sudden change occurs. The remainder of this paper is focused on fire as a landscape element, but much the same approach could be taken to community awareness of other hazards.

A fundamental part of Australia's community bushfire policy is the "Prepare, stay and defend or leave early policy" (2009 VICTORIAN BUSHFIRES ROYAL COMMISSION 2010). Implementation requires that those at risk know what to do and know what to expect. So far, material available to householders explains what to do and how to do it, but is weak on what to expect during a fire. Those who have been through the experience often comment that they were taken by surprise by the noise, smoke and heat of the fire. Worse, this sometimes leads people who have decided to stay to change their minds and flee at the last moment – a dangerous action.

This gap in awareness is reflected in much current risk communication literature (BURNS et al. 2010, OLSEN & SHINDLER 2007). Publications continue to make recommendations for public hazard education little different from recommendations made decades ago. Brochures and written material cannot provide an indication of the experience. Part of the difficulty for residents is that they (a) are not sufficiently aware of the necessary preparations, (b) have no direct experience of what it will be like as the fire front approaches then arrives at their home, or (c) do not understand the many different ways in which a house may be made vulnerable and begin to burn. While the preparation may be set out in detail in a brochure, they are not necessarily remembered in critical situations. Ways are needed to provide a vicarious, and hence safe yet memorable, experience of the arrival and passage of a fire front.

Increased awareness can only improve the efficacy of the relationship between resident and landscape. The next section gives a brief background on the vulnerability of houses in fire prone landscapes. We then detail the objectives of the game environment, the development tools and procedures and a sense of the game experience. A brief conclusion looks at options for extension to other landscape learning situations.

2 House Vulnerability in Fire Prone Landscapes

There are three ways in which a bushfire can ignite a building (RAMSAY & RUDOLPH 2003): embers and burning debris carried by the wind; heat radiation from the fire; and direct flame contact. Damage by strong winds, which are typical in fire conditions, can exacerbate the situation by allowing entry for burning debris.

Of these three modes, attack by embers and burning debris appears to be the major means of ignition (BLANCHI et al. 2006). Burning debris is produced as a result of burning of vegetation, buildings and other ignitable materials such as woodpiles and fencing, and is carried by the wind that accompanies bushfires. Showers of burning debris may attack a building some time before the fire front reaches that building, during the passage and for many hours after the front has passed. In contrast, the passage of the fire front with its attendant radiant heat and flame may as little as ten minutes.

Important factors that affect house survivability in a bushfire are landscaping, vegetation and house design (Figure 1). In the garden, certain types of garden and plants (especially deciduous trees) will inhibit the spread of fire across a property better than others. Ideally, buildings should have a fuel-reduced area around them. The idea is to minimise chances of direct flame attack or ignition from the effects of radiant heat. It also serves to provide a relatively clear area around a home to allow for ease of access during any fire fighting operations. Shrubs and trees planted close to the buildings are an obvious, and avoidable, fire risk.

Another important factor is house design. The simpler the shape and design of a house, the less chance there is of it catching alight in a bushfire (SCHAUBLE 2004). In addition, garages and carports can be built as an integral part of the building, thereby avoiding discontinuities and re-entrant corners.

3 Objectives

Our objective was to use a popular computer game engine to create a game-like experience of defending a home during the passage of a bushfire front. Failure to properly prepare, or to become disoriented in the intensity of the fight against embers before and during the passage of the front, or failure to adequately watch for fire after passage the front would lead to destruction of the house. We believed that this experience, with sufficient intensity, would create a greater sense of awareness than reading warning and advice documents. A second objective was to test whether the game did successfully recreate the experience and raise awareness. As yet, we have not been able to do this testing.

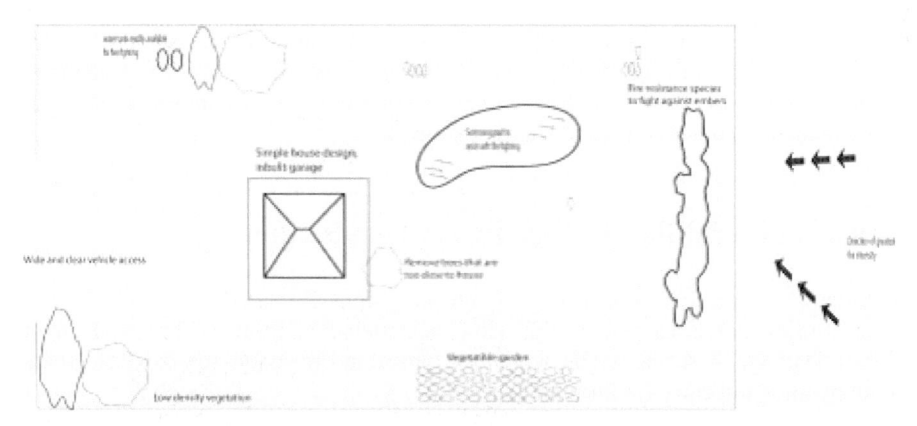

Fig. 1: Suggested layout for small property to increase fire safety (after SCHAUBLE 2004)

4 Development Tools and Procedures

Virtual Reality creates three-dimensional spatial objects in order to create an abstraction of the real world. This virtual representation in turn draws on the user natural perception and memory of space and spatial relationship to create a new reality (BOYD DAVIS 1996). The interactivity and dynamics of Virtual Reality can stimulate the user's engagement with, and understanding of, the real world (GERMANCHIS et al. 2004). By this reasoning, an appropriate modelling environment, to provide a vicarious experience, would come from a popular form of Virtual Reality, a game, or the driver for games, a gaming engine.

The game Crysis is a first person shooter science-fiction 3D computer game developed by video game developer Crytek and published by Electronic Art. Its first version was released in Europe in November 2007. The gaming mode can be single player or multiple player (up to 32 people can play it at the same time online). Crytek's CryEngine SandBox Editor 2 was chosen for this research as this game engine is able to:

- Create and model a true 3D environment and landscape
- Be used on a desktop PC or Laptop
- Allow users to interact with model and create interactive scenarios
- Allow real time movement around the virtual environment
- Provide animation and spatialised sound
- Offer powerful graphics quality without diminishing system performance to an unsatisfactory level (list modified from GERMANCHIS et al. 2004)).

The CryEngine Sandbox Editor comes packaged with the Crysis game. The Sandbox is a real time environment building and editing application. Geometry, surface and sound are authored in external applications and accessible via the Sandbox Editor to create the bushfire virtual environment. In addition, it is also affordable, the most stable, easiest to learn and currently one of the most powerful engines around (GERMANCHIS et al. 2007, TRENHOLME & SMITH 2008).

4.1 Building the Bushfire landscape

The CryENGINE2 Sandbox2 editor includes various tools that allow for a high level of customization. First among these tools is the aforementioned terrain editor, which can import or create and manipulate heightmaps. The terrain then needs to be painted with textures, and the game comes with a full library of textures for user selection. The vegetation placement tool allow the user to select from a series of different trees, which can be custom made, and use those to populate an area with it. Groups of these can be made, which in turn creates more realistic looking areas because trees, small bushes, grass and broken branches can all be grouped and placed randomly in an area, each having characteristics such as size variation, rotation variation and alpha blending. To preserve performance, there is also a brush tool that allow for vegetation to be placed in the same way, only without any physical properties, making it useful for areas in the distance that the player can see but can't access.

A typical three bedroom house plan was chosen from a database and computerised using the AutoCAD (AHP 2010). The house plans were imported into 3DS Max where it was modelled. Other 3D static models were created in 3DS Max and some of the interior and landscape objects were found on-line (Figure 2).

Fig. 2: View of the house in the landscape, and an interior view

4.2 Modelling fire elements

Bushfire attack mechanisms are measured in terms of the heat and intensity of both the flame front and the flux of embers (burning debris and windborne debris), before, during and after the fire front. These parameters can be modelled through the particle effects available in CryEngine SandBox. For example, a sprite is a single picture with frames of an animation tiled across it (e.g. flames licking up from the base of a fire). This sprite can then be used in a new particle effect. From there, different colours may be assigned to it, along with extra 'child' effects and inherent properties. For example, a large fire in this project was formed by modifying a fire animation sprite, adding glow and lighting, sparks, base glow, extra flames (sprites) and noise. These attributes are all saved in an XML file. There were about 7 different types of fires in this project, ranging from small embers (Figure 3) to huge bushfires, all of which were built from different effects. Currently the particles can pass through solids, such as house walls. As well as fire effects, water sprinklers were also designed in such a way that when spraying down, the water only reached the floor and not below it. Extra effects like rising steam and splashes were also added to make the scene more realistic. Particle effects are placed as simple non geometric entities in Sandbox2, and can be activated by a user event or simply remain static. They can also be linked to objects, so that wherever the object goes the effect stays with it.

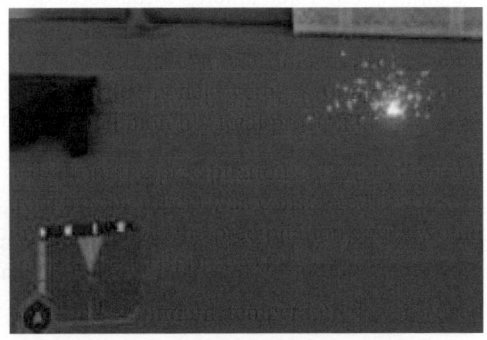

Fig. 3: Ember fire activated in the game (left) and becoming beyond control (right)

4.3 Flowgraphs

The flowgraph editor is a feature of Sandbox2 that allows the user to script events in a similar fashion to programming, only that a visual interface is used instead of code. These flowgraphs can link player movement triggers, keystrokes and actions to other objects, missions/objectives and sequences, allowing the player greater immersion in the game. Logic flow is created by connecting various logic boxes to each other with lines between their input and output gates, and defining their properties and state changes (Figure 4). This allows the user to build complex levels without needing to write C++ code or LUA scripts. The entire mission structure, logic, custom events, triggers and effects of this level were all done within a single XML flowgraph, and used hundreds of logic boxes, each with events flowing back and forth to other boxes based on conditions, qualifiers or player triggers.

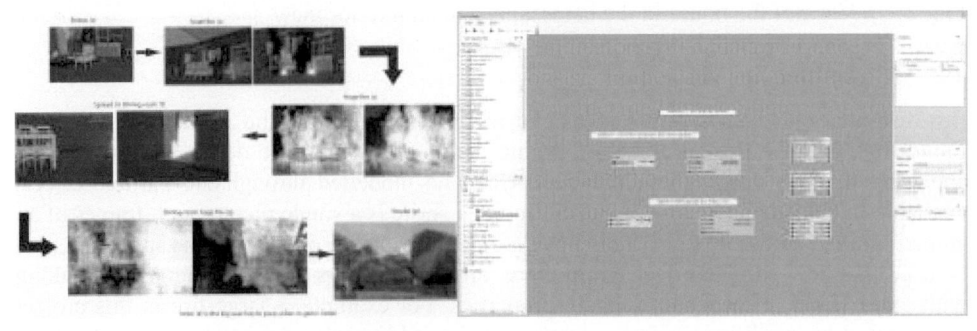

Fig. 4: Flowgraph window and event sequence

A potential scenario is that a burning ember could get into the house through gaps, and this could lead to a fire erupting in the living room. This triggers the creation of a small fire. When the owner of the house is not around to protect the house (i.e. extinguish the small fire in time), the fire erupts into a big fire, engulfing the whole living room. This fire would spread to the dining room, and create a huge wave of fire in the dining room and when it reaches the gas-bottle, an explosion happens. Thus the whole house burns down.

Before a level created in Sandbox2 can be played, it must be readied for engine use. After export, the level may be played in 'pure game mode', as opposed to the quick gameplay option whilst in the editor. However, a few other steps should be taken to ensure the level is playable by most users. Efficiency is important in level design, as this directly affects the frame rate. Too many things going on like lights or explosions will cause a significant increase in the amount of time it takes the engine to calculate each frame, lowering the fluidity of the game.

5 The Game

The developed game was designed to run over two virtual days. On the first day, the resident/player was made aware that the next day would be one of extreme fire danger – dry fuels, high temperatures, strong winds. They would have game time on that day to make preparations in their virtual home. They had the opportunity to:

- Remove a limb from the overhanging tree
- Clear leaves from the gutters
- Remove wooded furniture and debris from around the house
- Fill the bath and water back pack
- Find fire extinguisher and check torch batteries

The second day begins with the resident waking in their bedroom, the news on the TV says that a fire is in their area, soon the sky is turning red and the noise level is increasing. There is still time to fill the bath and retract the curtains but not much else since soon there is smoke around and fire in the hills (Figure 5).

When the fire front arrives – accompanied by intense noise – ember fires will start around, and in, the house. The worse the preparation the more of these fires will start. These can be put out fairly easily using an extinguisher or a back-pack sprayer, but the process will be tiring and personal energy levels will drop. If the fires cannot be put out quickly enough they will of course become larger and much more difficult to quell. The electricity may fail. The water supply may fail. It will become considerably darker during this period.

If the home is still intact after the front has passed it will still be necessary to again check the interior including roof space (dark) and other locations to ensure that all is safe. If all this is done properly the home will be safe.

Fig. 5: Fire approaching, fire consuming

6 Discussion

The game as developed to date covers the situation of a single person who is committed to remaining with their home and dealing with a specific house, garden and fire situation. This initial prototype could clearly be extended in a number of ways.
- Development of the initial single player experience into a multi-player experience to develop a team approach
- Development of a set of homes with different physical characteristics in different landscape conditions
- Different fires scenarios – direction of approach, surrounding landscape, intensity, wind speed etc
- Incorporation of an initial decision phase in which people must choose whether to vacuate or stay and fight for their home
- Addition of a contingency plan if the house catches fire uncontrollably.

At this stage also, the fire simulation is not based at all on physical process models of fire behaviour. These models are complex and still not especially accurate, in addition use of a process model could slow to game to the extent that the game becomes unplayable. The events are however based on observations of behaviour and research into building vulnerabilities (LEONARD et al. 2004). To effectively test the influence of different landscape or building conditions this aspect of the fire behaviour and vegetation and building response would need to be much more rigorous.

The game technology will develop further to include better representation of natural processes (see for example fromdustgame.com) and greater use of multisensory stimuli. At the same time, our knowledge of the processes of natural events in the landscape, including those we think of as 'natural disasters' will improve. Consequently, we can expect greater use of tool such as our game to increase awareness and improve decision-making at all stages of landscape planning, design, management and emergency response.

Acknowledgements

This research was supported by the Cooperative Research Centre for Spatial Information and the Bushfire Cooperative Research Centre. Narelle Irvine assisted with the game development.

References

2009 Victorian Bushfires Royal Commission (2010), *Final Report: Summary*. Parliament of Victoria.
AHP (2010), *Australian House Plans: The biggest and best home plan service in Australia* [Online]. http://www.ozehouseplans.com.au/ [acccessed March 2010].
Bastian, O., Krönert, R. & Lipský, Z. (2006), *Landscape diagnosis on different space and time scales: A challenge for landscape planning*. In: Landscape Ecology.
Blanchi, R., Leonard, J. & Leicester, R. H. (2006), *Bushfire risk at the rural-urban interface*. Bushfire Conference: Life in a Fireprone Environment: Translating Science into Practice. Brisbane, QLD.
Boyd Davis, S. (1996), *The design of virtual environments: with particular reference to VRML*. Support Initiative for Multimedia Applications. London, Advisory Group on Computer Graphics.
Burns, R., Robinson, P. & Smith, P. (2010), *From hypothetical scenario to tragic reality: A salutary lesson in risk communication and the Victorian 2009 bushfires*. Australian and New Zealand Journal of Public Health.
Deng, J. S., Wang, K., Hong, Y. & Qi, J. G. (2009), *Spatio-temporal dynamics and evolution of land use change and landscape pattern in response to rapid urbanization*. In: Landscape and Urban Planning.
Germanchis, T., Cartwright, W. & Pettit, C. (2007), *Virtual Queenscliff: A computer game approach for depicting geography*. In: Cartwright, W. Peterson, M. P. & Gartner, G. (Eds.), Multimedia Cartography. New York, Springer.
Germanchis, T., Pettit, C. & Cartwright, W. (2004), *Building a three-dimensional geospatial virtual environment on computer gaming technology*. In: Journal of Spatial Science.
Leonard, J. E., Blanchi, R. & Bowditch, P. A. (2004), *Bushfire impact from a house's perspective*. Earth, Wind and Fire – Fusing the Elements. Adelaide, Dept of Environment and Heritage, South Australia.
Olsen, C. S. & Shindler, B. A. (2007), *Citizen-agency interactions in planning and decisionmaking after large fires*. USDA Forest Service – General Technical Report PNW-GTR.
Ramsay, C. & Rudolph, L. (2003), *Landscape and Building Design for Bushfire Areas*. Collingwood, Vic, CSIRO Publishing.
Schauble, J. (2004), *The Australian Bushfire Safety Guide: the essential survival guide for every home*. Sydney, Harper Collins.
Stephenson, J. (2010), *The dimensional landscape model: Exploring differences in expressing and locating landscape qualities*. In: Landscape Research.
Trenholme, D. & Smith, S. P. (2008), *Computer game engines for developing first-person virtual environments*. In: Virtual Reality.

Integration of Spatial Outputs from Mathematical Models in Climate Change Visioning Tools for Community-Decision Making on the Landscape Scale

Olaf SCHROTH, Ellen POND and Stephen R. J. SHEPPARD

1 Introduction: Modelling and Integrated Assessment in Landscape Planning

Landscape planning faces increasingly complex and interdisciplinary issues with ecological, social and economic implications on multiple scales such as climate change. The following paper addresses climate change impacts and possible adaptation and mitigation options on a landscape scale and, therefore, as a landscape planning problem. There is a gap between the state of climate research and its current consideration in planning. In this context, GIS and landscape visualization are potential tools for the integration of different models. The Kimberley Climate Adaptation Project (KCAP) case study demonstrates how the spatial outcomes of three mathematical computer models from two different disciplines, i.e. climate science and forestry, were integrated in a participatory scenario building process through a shared geodatabase and 3D landscape visualizations as a spatial platform to integrate quantitative model knowledge and qualitative local knowledge. It is the aim of the paper to demonstrate an approach to the integration of quantitative model knowledge in participatory processes that is based on affordable standard software and suitable for rural communities with limited resources.

In this context, it is worth looking to the developing field of Integrated Assessment (IA). IA has evolved as an approach in environmental sciences to address complex issues across disciplinary boundaries (integrated) and to produce knowledge relevant to policy making (assessment). Key methods of IA are scenario development and analysis, and the use of computer models. In the 1990s, stakeholder participation and the integration of qualitative and normative knowledge received increasing importance, further developing IA to Participatory Integrated Assessment (PIA). SALTER et al. (2010) distinguish a typology of how models can be directly integrated in PIA: through participatory model building, facilitated model use, and interface-driven model design. In participatory model building, workshop participants construct the model together with scientists. Although this is the most participatory approach, it is also the most time and resource intensive (FÖRSTER 2009). Alternatively, scientists could pre-compute expert models, and then communicate the simplified model outcome in the workshops (DAHINDEN et al. 2003). Such facilitated model use provides an opportunity for demonstration of scientific modelling and multiple possible futures (SALTER et al 2010). However, challenges with facilitated models have led to the development of interface-driven models that use graphic interfaces to facilitate communication, e.g. QUEST (CARMICHAEL et al. 2004), and to augment participatory workshops. In this sense, interface-driven models can be used to generate scenarios and inspire more qualitative scenario narratives (SALTER et al. 2010). In this paper, the following research question is addressed: can different facilitated mathematical expert models be integrated in a participatory scenario process using a combination of GIS, geovisualization and 3D landscape visualization? In this context, the term *mathematical model* refers to a model describing processes or phenomena in natural sciences through mathematical language, e.g. climate models or models predicting the spread of forest pests or forest fires. If models combine knowledge from mathematical models and qualitative input, the term *hybrid modelling* is used as well (FLANDERS & SHEPPARD 2010).

2 Methods: Scenarios, Modelling, GIS and Visualization

In 2008, the City of Kimberley was one of two chosen by the Columbia Basin Trust to pilot a year-long climate adaptation project. Kimberley Climate Adaptation Project (KCAP) participants identified a range of potential climate change impacts, assessed local vulnerabilities, and developed adaptation strategies. To help Kimberley address these issues, the Collaborative for Advanced Landscape Planning (CALP) at the University of British-Columbia (UBC) joined the KCAP team to pilot a climate change visioning process using scenarios and 3D visualizations for use in small communities in BC's interior. The CALP-KCAP case study integrated the output of three separate mathematical models by spatializing or rather georeferencing the output, merging it in a geodatabase, and overlaying it in a virtual globe geovisualization and more detailed 3D landscape visualization.

2.1 Participatory scenario process

In general, scenarios are a method used to conceive complex uncertain future developments through plausible pathways or scenario narratives. Scenarios are different from linear predictions as they cover alternative storylines, and ask "what if?" questions. The term scenario is used in ambiguous ways; however, a common distinction is between forecasting and backcasting scenarios. Backcasting starts with a goal for the future and works backwards to determine how that could be achieved (BISHOP et al. 2007). VAN'T KLOOSTER & VAN ASSELT (2006) present an additional list of criteria for distinguishing scenarios through goals, process design and content. The CALP-KCAP process design mixed formal computer-based mathematical modelling tools and more intuitive qualitative methods for stakeholder engagement as recommended in the IPCC reports, including both quantitative and qualitative data. The scenario content was issue and area-based, covering regional down to local spatial scales and long time scales up to 100 years into the future, using forecasting and backcasting techniques.

2.2 Integration of modelling into participatory scenario-building

On the basis of the Mackenzie Basin Impact Study, one of the first integrated participatory regional climate change case studies, COHEN (1996) recommended a stronger integration of mathematical modelling into participatory scenario-building. However, at the time, technological issues and time and resource requirements were too challenging for a fully participatory approach. Förster (2009) researched the integration of modelling into participatory scenario building in great detail and identified multiple barriers to the successful cooperation of experts and lay people in complex model building. Participatory model building is very resource intensive and could exclude lay people who do not have time and interest to familiarize themselves with the scientific background of computer models. If it is pre-computed by external scientists and local stakeholders have no stake in it, a mathematical model is soon perceived as a "black box". In such a case, the model output may railroad the scenario building and hinder successful participation (SALTER et al. 2010).

In the transdisciplinary CALP Kimberley project, scientists' contributions through modelling and stakeholder participation were balanced by first identifying priority community issues in stakeholder workshops as illustrated in examples by CARMICHAEL et al. (2004) and HENRICHS (2003). On that basis, mathematical models were located or prepared for the most pressing community issues (water, forest fire, and mountain pine beetle) and used to inform the following stakeholder workshops. Feedback and qualitative knowledge from the stakeholder workshops was used to review the model output and to improve the models for the next stakeholder workshop. The stakeholder group provided input into scenarios which were presented to the general public in an open house and a local exhibition at the end of the overall KCAP.

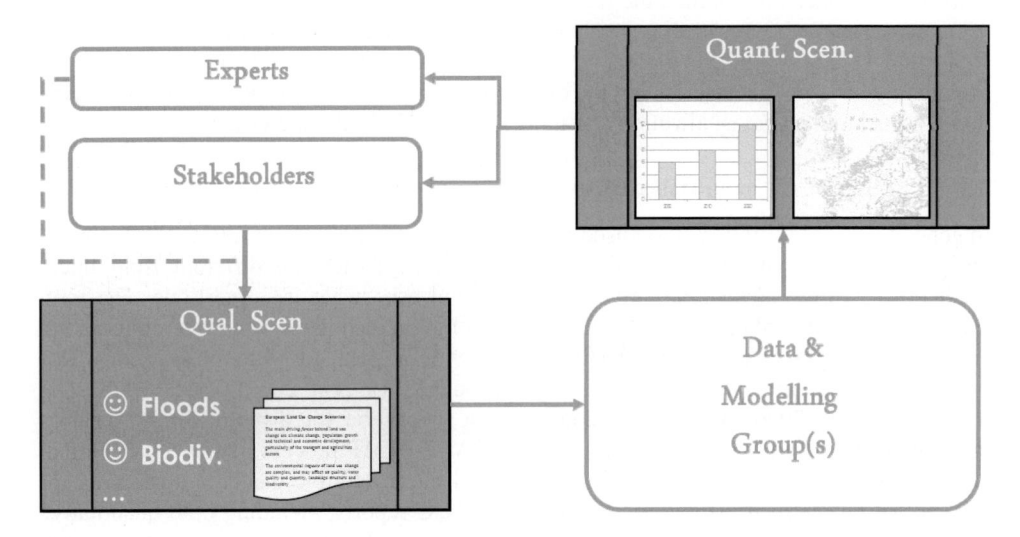

Fig. 1: Combining quantitative and qualitative approaches (HENRICHS 2003)

2.3 Data issues, model-building and geovisualization

The *Hydrological Model* was based on a version of the Canadian Global Climate Model (CGCM3 A1B run 1) run by the Pacific Climate Impacts Consortium (PCIC) and projected future snowpack depth (derived from precipitation) for the region of Kimberley according to the VIC model from the University of Washington. This specific model was chosen because it produced the closest match to historic weather station data, although it should be noted that such data was very scarce in Kimberley. Furthermore, downscaling is still a very complicated process in climate change modelling (NICHOLSON-COLE 2005) and beyond the scope of small rural communities. In this case, only the pro bono support through PCIC made it possible for PCIC to model the changes snowpack from 1960 to 2100. The hydrological model provided snowpack depths as a data spreadsheet with snowpack change for multiple points with x/y/z coordinates. CALP then re-projected and converted the point file into polygons that could be mapped onto the digital terrain model.

Mountain Pine Beetle Susceptibility was modeled in ArcGIS by one of the authors based on the amount of pine in the stand, the age of pine, the density of the stand, and the stand's location according to SHORE & SAFRANYIK (1992). After expert review at the first workshop, the model was updated with the most up-to-date available forestry data by ILMB (2006).

$$S = P \times A \times D \times L$$

with
P is the percentage of susceptible pine basal area,
A is the age factor,
D is the density factor, and
L is the location factor.

The percentage of susceptible pine basal area (P) was calculated as

$$P = \frac{[\text{average basal area/ha of pine} \geq 15 \text{ cm dbh}] \times 100}{[\text{average basal area/ha of all species} \geq 7.5 \text{ cm dbh}]}$$

Forest stand data about species, age, density and location factors came from the Vegetation Resource Inventory (VRI) of the BC Forest Service. Because pine stands older than 80 years show the highest level of susceptibility, infestation is likely to take place as a cycle as presented in an interactive time series. Currently, the mountain pine beetle infestation in the Kimberley area has been projected to see an "exponential growth in mortality" (Forsite 2008) with the peak infestation projected to occur in 2011 (WALTON et al 2008). At the end of the century, when the replanted forest stands reach the critical age of 80 years, the cycle may likely restart – likely further aggravated through climate change.

Forest fires were identified by the working group as one of the biggest current threats for Kimberley. The City's fire chief in cooperation with the City's fire consultant partner provided spatial information about historic forest fires, Forest Fire Condition Classes (FFCC), and fire spread models, including a fire spread model that was calculated for the assessment of evacuation routes. CALP integrated other research on the potential impact of climate change resulting in earlier and longer forest fire seasons, and visualized the outcome of the Farsite *Fire Spread Model* provided by the fire consultant. Figure 2 shows the original digital output of the Farsite model and its re-visualization into an animated series of 3D digital images, each showing an hour of fire spread, for presentation at the final community Open House.

Integration of the spatial model outputs took place in a geodatabase and subsequently, as layers in a Google Earth Geovisualization (Fig. 2 on the right). The different layers of model outputs were thereby put into relation to each other and to current city extent as well as future residential development scenarios. The geovisualizations enabled assessment of vulnerability and resilience of alternative development scenarios.

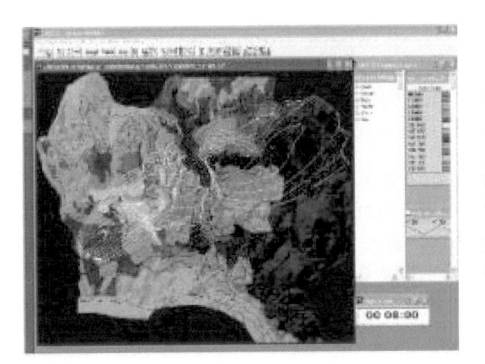

Fig. 2: Farsite fire spread model, visualized as 2D map with the model software by the City's fire consultant (left) and spatialized in Google Earth (right) together with the 3D buildout by CALP.

2.4 Long-term scenario visualization

Alternative long-term future landscape options were visualized based on the current Vegetation Resource Inventory (VRI), and incorporated future projections about climate change and species change (KIMMEL 2009), mountain pine beetle and forest fire risk (see 2.3 Data issues…). As the open source Biosphere3D software specializes in realistic representations of forest stands based on GIS vegetation data, it was used to illustrate plausible climate change impacts on the visual landscape. The level of detail is much higher than it was with Google Earth so that viewers can assess the landscape implications of projected future changes. On the other hand, the high level of detail in combination with the long time horizon may imply so-called "apparent realism" (SHEPPARD 2001).

In order to communicate the uncertainties, three different future options were visualized (see Schroth et al. 2009 for all alternatives). The following scenario shows what projected fire damage, species changes, the shift to grasslands, and the return of pests and diseases might look like by 2100 and beyond if only minor adaptive and mitigative action is taken.

Fig. 3: Future landscape under climate change with minor mitigation or adaptation (SCHROTH)

Due to already severe beetle infestation and high risk of forest fires, both enhanced through climate change, major changes are projected. In addition to the visualizations, technical posters explained the underlying model assumptions for the Community Open House.

2.5 Evaluation methods for visualization effectiveness

Quantitative and qualitative evaluation methods were mixed in a deductive way. First, 38 participants ranked different media they had seen in the final open house forum, i.e., posters, presentations, 2D maps, and digital globes. Second, the same participants were asked to rank spatial, temporal and thematic navigation in Google Earth. Third, 18 participants were videotaped while navigating the planning layers in a Google Earth model. During exploration, the participants applied the "think-aloud" technique, commenting their interactions and impressions. Afterwards, the Google Earth users were interviewed in depth using a qualitative interview script. The full description and analysis of the evaluation is documented in SCHROTH et al. (2009) and will be addressed in more detail in future publications. This paper focuses on the interplay between and communication of the three underlying models only; however, on-going research is evaluating the overall participation process and its longitudinal outcome as well.

3 Policy and Scientific Results

3.1 Policy results

The final KCAP report put forward more than 75 recommendations to the city council. The recommendations include measures for fire-smart housing and forest management, a new water study, renewable energies etc. A series of eight follow-up interviews with city staff and council members showed that some recommendations have been taken further or even implemented in operational decisions or local policies, e.g. a sprinkler bylaw in response to forest fire risk and water supply issues. Geovisualization and scenario-building process recommendations were disseminated for other communities in British Columbia in POND et al. (2010).

3.2 Scientific results

The integration of the spatial output from different mathematical models is the focus of this publication. The case study showed three results. First, the geovisualizations enabled local stakeholders without expert modelling knowledge to assess and verify the model output. In the case of the hydrological model for the snowpack, the spatialization and preliminary 2D visualization of future snowpack levels revealed serious data gaps. As a consequence, the model outputs were refined, but were not used for the 3D model integration; revised downscaling methods were proposed for future research proposals. In 2010, the mountain pine beetle spread throughout the Kimberley watersheds and thus reality is confirming the prior, pessimistic mountain pine beetle susceptibility model output. A quantitative comparison of the model output and actual mountain pine beetle spread in 2010 is suggested for further research.

Second, initial wildfire material including 2D maps had been deemed insufficient with stakeholder review, and improved models and data were visualized for the final scenario building. The 3D visualization added drama to the presentation but also raised more awareness. According to anecdotal evidence, the presentation form was one reason that proactive measures for improved wildfire management gained political and public support locally.

Third, figure 4 shows how the integration of spatial output in a shared geodatabase and visualization platform created new knowledge. The mountain pine beetle susceptibility map, based on the model described in 2.3 was merged with the digitized results from the 1999 Flood Risk Assessment Study for the City of Kimberley (no newer data was available). The layer concept allowed MPB susceptibility and locations of potential flood flows due to debris jams in the creek to be visible at the same time. Viewers could thus understand integrated risks, such as increased downstream flooding from potential MPB deforestation in the watersheds, due to the increased amount of debris from infected forest stands.

Fig. 4: Overlay of mountain pine beetle susceptibility (colored areas) and locations of potential flood flows due to specific potential debris jams in the creek (marked through circles)

4 Discussion and Conclusion

The application of a shared geodatabase to spatialize output from different mathematical models and visualization using different levels of realism in Google Earth and Biosphere3D worked very well as part of the participatory scenario process in the KCAP case study. The research results are consistent with SALTER et al. (2010) who suggest that scenario development within a participatory integrated assessment provides an extended peer review for models if more local scales are addressed. It is still difficult to balance the necessary complexity and simplification of models in the process but the combination of posters and geovisualizations worked well for participant comprehension, and there were few hints that the modelling had been perceived as "black box". Although community-driven, the integration did not happen in an immediately interactive process, but model outputs and visualizations were pre-computed (facilitated model use). With further development steps in the technology, communities with more resources will aim for fully integrated models as described in CARMICHAEL et al. (2004). Until then, the described procedures, integrating multiple different modelling results as layers within a geodatabase and a virtual globe as a shared visualization platform, can provide a suitable solution for small rural communities with limited resources.

References

Bishop, P., Hines, A. & Collins, T. (2007), *The current state of scenario development: an overview of techniques*. In: Foresight, 9 (1), 5-25.

Carmichael, J., Tansey, J. & Robinson, J. (2004), *An integrated assessment modeling tool*. In: Global Environmental Change Part, 14 (2), 171-183.

Cohen, S. J. (1997), *If and So What in Northwest What Canada: to the Future of the Mackenzie Basin? Could Climate Change Make a Difference to the Future of the Mackenzie Basin?* In: Arctic, 50 (4), 293-307.

Dahinden, U., Querol, C., Jäger, J. & Nilsson, M. (2003), *Citizen Interaction with Computer Models*. In: Kasemir, B., Jäger, J., Jaeger, C. C. & Gardner, M. T. (2003), Public Participation in Sustainability Science: A Handbook, 105-125. Cambridge, Cambridge University Press.

Flanders, D. & Sheppard, S. R. J. (Chapter co-authors) (2010), *Future Perspectives in Hybrid Modelling*. In: Kimmins, J. P., Blanco, J. A., Seely, B., Welham C. & Scoullar, K. (Eds.) (2010), Forecasting Forest Futures: A Hybrid Modelling Approach to the Assessment of Sustainability of Forest Ecosystems and their values. London, Earthscan Ltd., Chapter 10, 213-236.

Forsite Consultants Ltd. (2008), *Mountain Pine Beetle Timber supply Impact Assessment for the Cranbrook and Invermere TSAs*. Victoria, BC Ministry of Forests and Range.

Förster, R. D. (2009), *Overcoming implementation barriers of scientific simulation models in regional developments*. Dissertation Nr. 18409, Zurich, ETH Zurich.

Henrichs, T. (2003), *Environmental Scenario Analysis – Overview and Approaches*. European Environment Agency (EEA), Global Environmental Change and Food Systems (GECAFS) Meeting, 18.-19. August 2003: Comprehensive Scenarios "Initial Workshop", Reading, UK.

Integrated Land Management Bureau ILMB (2006), *Mountain Pine Beetle Hazard Rating Documentation version 1.2* ILMB Project #: P07-0199, Southern Interior Forest Region.

Kimmel, E. (2009, *Climate Change Adaptation and Biodiversity.* Vancouver: ACT Adaptation to Climate Change Team, Simon Fraser University.
http://www.sfu.ca/act/documents/ACT_Biodiversity_Background_Report_final.pdf.

Nicholson-Cole, S. A. (2005), *Representing climate change futures: a critique on the use of images for visual communication.* In: Computers, Environment and Urban Systems, 29, 255-273.

Pond, E., Schroth, O., Sheppard, S. R. J., Muir-Owen, S., Liepa, I., Campbell, C., Salter, J., Flanders, D. & Tatebe, K. (2010), *Local Climate Change Visioning and Landscape Visualizations: Guidance Manual.* Vancouver: CALP/UBC.
http://www.calp.forestry.ubc.ca/wp-content/uploads/2010/02/CALP-Visioning-Guidance-Manual-Version-1.1.pdf.

Salter, J., Robinson, J. & Wiek, A. (2010), *Participatory methods of integrated assessment – a review.* Wiley Interdisciplinary Reviews: Climate Change, 1 (5), 697-717.

Sheppard, S. R. J. (2001), *Guidance for crystal ball gazers: developing a code of ethics for landscape visualization.* In: Landscape and Urban Planning, 54 (1), 183-199.

Shore, T. L. & Safranyik, L. (1992), *Susceptibility and risk rating systems for the mountain pine beetle in lodgepole pine stands.* Victoria, Pacific Forestry Centre.

Schroth, O., Pond, E., Muir-Owen, S., Campbell, C. & Sheppard, S. R. J. (2009), *Tools for the understanding of spatio-temporal climate scenarios in local planning: Kimberley (BC) case study.* Bern: Swiss National Science Foundation SNSF.
http://www.calp.forestry.ubc.ca/wp-content/uploads/2010/02/Schroth_2009_Final_SNSF_Report.pdf.

van 't Klooster, S. A. & van Asselt, M. (2006), *Practising the scenario-axes technique.* In: Futures, 38, 15-30.

Walton, A. (2008), *Provincial-Level Projection of the Current Mountain Pine Beetle Outbreak.* Mai 7, 2008. http://www.for.gov.bc.ca/hre/bcmpb/.

Aesthetic Evaluation of Forest Landscape within the Training and Experimental Forest Range (TEFR) Yundola

Emil GALEV

Abstract

The research is focused on understanding the scenic beauty of landscapes in the context of environmental planning and management. Our landscape preferences are thought to be influenced by many factors: age, gender, ethnicity, regionality, recreational activity; some researchers even maintain there is an evolutionary basis behind certain landscape preferences. But of these factors, our dominant culture and history have played major roles in shaping our preferences for landscapes that are natural in character. The aesthetic evaluation of landscapes in the research is made by objective characteristics of the existing topography and vegetation. Data are taken from the map or text materials containing information about the terrain. The dominance elements and variable factors of landscapes appear in varying degrees, depending upon the viewing distance. The research automates aesthetic evaluation of forest landscapes using GIS.

Keywords: Visual impact, scenic beauty, aesthetic, landscape preferences.

1 Theoretical Prerequisites for Aesthetic Evaluation of Landscapes

A number of research exist in which various methods for visual evaluation of the landscape are used. All these studies show that such an evaluation of landscape resources is a very important moment in determining the potential of recreational areas. Through spatial analysis, photographic, visual or psychological evaluation, individual territorial units should be classified to determine their emotional performance, despite the subjective element that can not be avoided.

The method of SEUNG-BIN (1984) is expressed in statistical analysis of evaluations of interviewed people who were shown pictures of 12 urban areas.

Survey methods are often used in evaluating the aesthetic qualities of landscapes. According to ROSENTHAL & DRIVER (1983) most of the respondents mainly appreciate the opportunity to enjoy beautiful scenery and is particularly marked overall demand for peace, solitude and rest in nature. According to ABELLO & BERNALDEZ (1986), all these surveys show that the aesthetic criteria of people depend on the nature, age, gender and their education and grades that they give the landscapes depend on their personal preferences for various forms of recreation. There are even those studies which have been specifically designed to prove weak authoritativeness and objectivity of the results of such inquiries. They apply the visual evaluation method of landscape using two groups of observers. The first group was previously aware of the existence of some clearly visible damage in the landscape and the other does not. The results show that dark observers did not notice the existing visible damage and provide better evaluation of these landscapes. Exactly this was conducted by BUHYOFF (1982) experiment. "Gap" according to him is mainly due to the fact that the sites assessed are too large and it can no longer pay attention to all details and particulars, and the fact that the eye of a non-specialist is not trained to see everything.

According to COOPER & MURRAY (1992) a constructive method for visual evaluation of sites should include a description, analysis and classification of areas to create a structure within which to cover all landscape components. The biggest problem in the development of quantitative methods to evaluate the visual impact by BUHYOFF & RIESENMANN (1979) is to determine the coefficient of importance of individual landscape components in the overall evaluation. UNWIN (1975) describes three stages in the evaluation of landscape: "measurement" of the landscape, formulation of landscape values through the survey of people's preferences, and finally an evaluation of the visual qualities of the landscape. Most sophisticated models in this regard he says are psychophysical which use first psychological impact, and after that objective quantitative and qualitative parameters of the landscape. The creation of such a model requires three sets of data: photos, survey data on people's preferences to landscape and landscape parameters.

The method of SHAFER, HAMILTON & SCHMIDT (1969) for determining psychophysical preference of people to the countryside is to predict how they will appreciate the natural landscape. Most important characteristics for the aesthetic appeal of landscape according to the authors are taken into account. Proportions are calculated between the quantitative values of landscape characteristics in practice. Changing these proportions within a specific landscape creates a feeling of depth and perspective. Based on a mathematical formula involving perimeters and areas of forests, open spaces and water areas the authors define three types of ground cover: plant, non-vegetable and water, and outline the following areas at a distance.

WHERRETT (1997) automates this model using GIS and conduct surveys to identify people's preferences for visual images of landscapes. The results showed that weather conditions and different focal lengths, where photographs were taken on the ground are not significant, but seasonal characteristics of vegetation and architectural elements have a significant influence in shaping those preferences.

CHIUSOLI (1977) offers a valid method to estimate parametric values of landscape and visual appeal of the plant component of the landscape called "integrated analysis of the landscape". It is based on analyzing aerophotos and panoramic images of the study area. By comparing the data obtained the author determined percentage ratios between the different landscape components. These ratios vary widely, thus achieving a just estimate. According to the author it has not yet developed a unified methodology for "parametric" visual evaluation of plant components in the landscape, because in practice the evaluation of its appearance is associated with too many subjective criteria. Therefore he considers the most appropriate first to analyze the landscape using aerophotos and territory be divided into homogeneous zones according to the most common characteristics of plant cover, and then to determine their area ratio. Appearance of landscapes, revealing to be monitored by the ground that what they learn from any point outside or inside them is totally different, so panoramic photographs reflect the real picture is revealed to him. Therefore the author considers most appropriate both method of analysis to unite and after processing the data from aerophotos to create a series of panoramic images for areas with established aesthetic values.

PELT (1980) also recognizes that the perception of the landscape of the casual observer is implemented by the land and therefore pay particular attention to principles of felling and afforestation on different relief forms in order to avoid adverse visual effects resulting from the creation of unsustainable or geometric outlines of woodland. FORESTRY COMMISION (1994) examined much more detail this issue and defined some guiding principles of forest landscape design, designed to preserve the visual value of plantations and open spaces.

In Bulgaria most commonly used criteria for aesthetic evaluation of natural environment is developed by BULEV (1977). Evaluated as the unit area, he used a square side length, depending on the scale of the graphic material. For each of the square sections are determined grade evaluation, depending on the presence or absence in his range of different landscape elements (forests, rivers, rocks, agricultural areas, roads, power lines, etc.). The same criteria used BEZLOVA (1989) and adds them to apply locally for its development. She assesses areas as follows: dynamic of the relief, mosaic structure of plant cover, engineering network, availability of natural phenomena, natural sites and protected areas, and visual-spatial relationships. Then she sum of the ballroom evaluations as a percentage of the maximum value and then groups territories.

In conclusion we can say that experiments, theories and summaries of the visual landscape evaluation has not yet reached the necessary universality of theoretical knowledge in order to establish a common scheme which will only be evaluated.

2 Analysis of the Most Important Natural Components which Determine Aesthetic Properties of Landscapes within the Territorial Scope of the TEFR Yundola

In conclusion of the analysis can be concluded that forest landscapes in the Yundola region can take a significant number of visitors. They should therefore be classified according to the opportunities offered for recreation. Then it is necessary the natural potential to be evaluated but differentiated for individual recreational activities, and these activities can be codified and classified in different levels of aggregation. The most synthesized unit having territorial scope must be the "forest subdivision", but in terms of recreational activities, must be the specific recreational activity.

In analyzing of the individual characteristics of relief and forest vegetation, first was reported their impact on recreational activities and established the practical feasibility of each of them as an evaluation indicator, depending on the impact that have on the main recreational activities. In this respect, are shaped some fundamental conclusions concerning the question of evaluation of recreational forest landscapes in general and of research subject in particular.

General conclusions:

1. When conducting landscape-recreation research is required to analyze taxological data of forest vegetation.
2. Analysis developing and design can be achieved only by additional field studies conducted during different seasons.

Specific findings:

1. In almost all parts of the Forestry range, the taxological data of forest stands evidence of their high productivity as well as of their very good outstanding artistic and aesthetic qualities and recreational function. Therefore:
a/ it can be expected that greater influence in recreational evaluation of the site will have a factor "relief" where the differences are very prominent;
b/ it is most appropriate to take into account only those taxological indicators that most influence the formation of the external appearance of the forest landscape, as well as fo its spatial structure.
2. The majority of forests in the area of the Forestry are accessible in all its parts. The development of mobile communications will make them more accessible and this will create prerequisites for economic development in general and for leisure in particular.
3. The main recreational activities practiced within the research area are: walking and stationary recreational in the nature environment, hiking, sunbathing, picking wild berries and mushrooms, villa holiday, outdoor games and winter sports.
4. In conclusion it should be said that forest landscapes in the vicinity of Yundola must first be classified according to their recreational opportunities and then to be evaluated all available resource potential that can be used for purposes of recreation, but differentiate for individual recreational activities. These activities themselves can be codified and classified in different levels of aggregation. The most synthesized unit in terms of territorial coverage should be "forest subdivision", but in terms of recreational activities should be "specific recreational activity".

3 Evaluation Mechanism

In this paper the aesthetic evaluation of landscapes is defined as a grouping of predefined territorial units in some grade categories according to their positive or negative aesthetic qualities defined by pre-selected indicators and criteria. The indicators and criteria are also systematic and have been elected in accordance with the conditions set by the main objective of the research or development project, for the purposes of that evaluation takes place. Aesthetic evaluation is based on the specifics of the landscape and is determined by visually dominant natural and anthropogenic components.

It serves primarily to determine the visual qualities of open spaces, and in particular their advantages or disadvantages as places to stay static. Significant role in its forming play the relief, the forest vegetation and somewhat aquatic components of landscape, but in many cases could be setting some anthropogenic components.

The factors which most contribute to the aesthetic impact of forest vegetation and broad criteria for aesthetic evaluation of forest stands are classified in Table 1.

indexes	**gradation of categories and the most common criteria for an appreciation**
multilevelness and passability	**the passable stands** in the most general case, form a picturesque framework of open spaces and create greater psychological comfort as for the people watching them on side, and those who pass through them
structure	**the block spatial location** of trees definitely have a strong aesthetic impact on people especially during passing through forest plantations
average height	the forest stands with an average height **over 10 m** caused a strong emotional experience because it goes beyond the human scale
dendrologic richness of forest stands	the forest stands in the composition of which are involved **more than 2 tree species** create more expressive emotional and psychological effects arising from greater volume diversity in the space
presence of much higher trees and single tree species occurring in forest stands	**the much higher trees** as well as **the single tree species occurring** in forest stands presence a greater diversity in in the spatial structure and coloring of forest stands

Table 1: The most common criteria for a high aesthetic evaluation of forest vegetation

point of view		recreational activities		visual passability			structure		average height /m/				dendrological richness				presence of much higher trees and single tree species occurring			
	subclasses	kinds	impassable	hardly passable	passable	uniform	group	below 5	5-10	10-20	20-30	over 30	1 tree species	2 tree species	over 2 tree species with predominance	over 2 tree species without predominance	without any of these	with presence only of much higher trees	with presence only a single species occurring	with presence of both
from the inside	marches	transit													x	x	x	x	x	x	x
		combined with other activities													x	x	x	x	x	x	x
	staying	statical				x	x														
		dynamical																			
from the outside	marches	transit																			
		combined with other activities	x	x	x	x	x														
	staying	statical																			
		dynamical				x	x											x	x	x	x
number of possible kinds activities:			0.5	1.5	6.5	2.5	5	0.5	2.5	5	7.5	8	3.5	4.5	6	7	3	5	4.5	6

appropriate condition   satisfactory condition   unsuitable condition   x irrelevant

Table 2: Componential assessment for aesthetic valuation of the forest stands

The Table 2 specifies the number of appropriate subgroups of most widely practiced recreational activities in certain values of taxological indicators. It is reported the fact that forest stands have a different visual impact when have been seen from side and when have been viewed as an immediate environment for recreation.

The indicator "passability" characterizes the possible of physical and visual intrusion into forest areas and depends on the structure of forest stands expressed by the location and by the different combinations of main component parts of the forest flora. Therefore it presents in both aspects of evaluation. The indicator "construction" determines primarily spatial structure of the forest stands, but has a major role in shaping their external appearance and diversity of the forest landscape. The average height is a very important indicator of psychological comfort of the recreational environment, which is determined by those in human genetic effects to the surrounding area determined by the so-called "human scale". The dendrological richness, and the presence of much higher trees and single tree species occurring in forest stands are a prerequisite for a greater vertical uneventness of forested areas and for greater variety in their coloring.

Factors contributing to the greatest extent for the aesthetic impact of open spaces are systematized in Table 3. Vertical and horizontal indentation of the relief considered separately determine the possibility of visual perception of space. Joint expression of these two factors determines the depth of the visible prospects, as the maximum values of this indicator are obtained by high values of vertical relief indentation and low values of horizontal relief indentation, which creates prerequisites for the detection of more distant panoramic views. The extent of interception of the horizon is determined largely by terrain features, but after reading the above parameters remain only the characteristics of forest vegetation, which can be a framework of perspectives or can be a barrier preventing their detection. The number of visible landscapes depends primarily on diversity of forest vegetation surrounded open spaces and determines in the most a picturesque variety in the foreground of the landscape. The ratio between perimeter and area of landscapes contributes much to the diversity of plastic-volume relationships. For the uniqueness and attractiveness of the mountainous landscape of the utmost importance are also the *degree of indentation of the visible horizon* and the presence of natural phenomena.

To assess the visual impact of wooded areas is used species composition, but from an aspect called dendrological richness. Forest stands were divided into four groups depending on the number of tree species involved, whether they share in the total stock: forest stands consisting of one tree species; forest stands consisting of two tree species; forest standss consisting of three or more species with predominance of one of them; and finally consisting of three or more species without predominance. As a positive quality is reported the presence of much higher trees and single tree species occurring in the species composition. The passability, the construction and especially the average height of the forest stands are also taken into account in determining the visual evaluation.

indexes	gradation of categories and the most common criteria for an appreciation
vertical indentation of the relief	**strongly indented terrains** provide better opportunities to detect distant panoramic views, but **less indented terrains** are favorable to adopt landscape forgroun
horizontal indentation of the relief	**slightly** separated, in a horizontal attitude landscapes have a positive psychological impact because it allows the visual perception of vast spaces
extent of interception of the horizon	territory **low** on the horizon interception create better conditions for visual perception of landscape
number of visible landscapes	**higher values** of this parameter determine the variety of sights and scenery of the landscape
depth of the visible prospects	**maximum values** of this indicator are obtained by high vertical and low horizontal relief indentation and contribute to better visibility of landscapes
degree of indentation of the visible horizon	**high values** contribute to diversity, attractiveness and emotional-psychological impact of landscape
presence of obstacles	**low values** contribute to better visibility of landscapes
presence of natural phenomena	**presence** of interesting rock formations or other sculptural relief forms and phenomena influences strongly on visual perception and creates a unique and exotic landscape
ratio between perimeter and area of landscapes	**higher values** determine a great landscape diversity

Table 3: The most common criteria for a high aesthetic evaluation of the open spaces

classes	subclasses	kinds	recreational activities	vertical indentation			horizontal indentation			extent of interception of the horizon			number of visible landscapes		
				high values	average values	lower values	high values	average values	lower values	low	average	high	2	3-5	over 5
passages	transit	walking													
		by car, motorcycle or other vehicle					x	x	x				x	x	x
		cycling													
		combined													
	combined with other activities	collection of herbs, mushrooms and berries					x	x	x	x	x	x	x	x	x
		hunting and fishing								x	x	x	x	x	x
stay	statical	contemplation, conversation, etc.													
		sun and air baths													
		deployment and camping													
	dynamical	skiing, mountaineering and other sport activities								x	x	x	x	x	x
		outdoor games and other recreational activities													
number of possible kinds activities:				4	7.5	10	3	6	8.5	8	5	2	1	4.5	7

appropriate condition ■ satisfactory condition □ unsuitable condition □ x irrelevant

Table 4: Componental assessment for aesthetic evaluation of the open spaces

		average height /m/	below 5				5-10				10-20				20-30				over 30				
			consisting of one tree species	consisting of two tree species	consisting of three or more species with predominance of one of them	consisting of three or more species without predominance	consisting of one tree species	consisting of two tree species	consisting of three or more species with predominance of one of them	consisting of three or more species without predominance	consisting of one tree species	consisting of two tree species	consisting of three or more species with predominance of one of them	consisting of three or more species without predominance	consisting of one tree species	consisting of two tree species	consisting of three or more species with predominance of one of them	consisting of three or more species without predominance	consisting of one tree species	consisting of two tree species	consisting of three or more species with predominance of one of them	consisting of three or more species without predominance	dendrological richness
		passability																						
without much higher trees and single tree species occurring	presence of much higher trees and single tree species occurring	impassable	10	11	12.5	13.5	12	13	14.5	15.5	14.5	15.5	17	18	17	18	19.5	20.5	17.5	18.5	20	21	uniformly
			12.5	13.5	15	16	14.5	15.5	17	18	17	18	19.5	20.5	19.5	20.5	22	23	20	21	22.5	23.5	in groups
		hardly passable	11	12	13.5	14.5	13	14	15.5	16.5	15.5	16.5	18	19	18	19	20.5	21.5	18.5	19.5	21	22	uniformly
			13.5	14.5	16	17	15.5	16.5	18	19	18	19	20.5	21.5	20.5	21.5	23	24	21	22	23.5	24.5	in groups
		passable	16	17	18.5	19.5	18	19	20.5	21.5	20.5	21.5	23	24	23	24	25.5	26.5	23.5	24.5	26	27	uniformly
			18.5	19.5	21	22	20.5	21.5	23	24	23	24	25.5	26.5	25.5	26.5	28	29	26	27	28.5	29.5	in groups
with presence only of much higher trees		impassable	12	13	14.5	15.5	14	15	16.5	17.5	16.5	17.5	19	20	19	20	21.5	22.5	19.5	20.5	22	23	uniformly
			14.5	15.5	17	18	16.5	17.5	19	20	19	20	21.5	22.5	21.5	22.5	24	25	22	23	24.5	25.5	in groups
		hardly passable	13	14	15.5	16.5	15	16	17.5	18.5	17.5	18.5	20	21	20	21	22.5	23.5	20.5	21.5	23	24	uniformly
			15.5	16.5	18	19	17.5	18.5	20	21	20	21	22.5	23.5	22.5	23.5	25	26	23	24	25.5	26.5	in groups
		passable	18	19	20.5	21.5	20	21	22.5	23.5	22.5	23.5	25	26	25	26	27.5	28.5	25.5	26.5	28	29	uniformly
			20.5	21.5	23	24	22.5	23.5	25	26	25	26	27.5	28.5	27.5	28.5	30	31	28	29	30.5	31.5	in groups
with presence only of single species occurring		impassable	11.5	12.5	14	15	13.5	14.5	16	17	16	17	18.5	19.5	18.5	19.5	21	22	19	20	21.5	22.5	uniformly
			14	15	16.5	17.5	16	17	18.5	19.5	18.5	19.5	21	22	21	22	23.5	24.5	21.5	22.5	24	25	in groups
		hardly passable	12.5	13.5	15	16	14.5	15.5	17	18	17	18	19.5	20.5	19.5	20.5	22	23	20	21	22.5	23.5	uniformly
			15	16	17.5	18.5	17	18	19.5	20.5	19.5	20.5	22	23	22	23	24.5	25.5	22.5	23.5	25	26	in groups
		passable	17.5	18.5	20	21	19.5	20.5	22	23	22	23	24.5	25.5	24.5	25.5	27	28	25	26	27.5	28.5	uniformly
			20	21	22.5	23.5	22	23	24.5	25.5	24.5	25.5	27	28	27	28	29.5	30.5	27.5	28.5	30	31	in groups
with presence of much higher trees and single tree species occurring		impassable	13	14	15.5	16.5	15	16	17.5	18.5	17.5	18.5	20	21	20	21	22.5	23.5	20.5	21.5	23	24	uniformly
			15.5	16.5	18	19	17.5	18.5	20	21	20	21	22.5	23.5	22.5	23.5	25	26	23	24	25.5	26.5	in groups
		hardly passable	14	15	16.5	17.5	16	17	18.5	19.5	18.5	19.5	21	22	21	22	23.5	24.5	21.5	22.5	24	25	uniformly
			16.5	17.5	19	20	18.5	19.5	21	22	21	22	23.5	24.5	23.5	24.5	26	27	24	25	26.5	27.5	in groups
		passable	19	20	21.5	22.5	21	22	23.5	24.5	23.5	24.5	26	27	26	27	28.5	29.5	26.5	27.5	29	30	uniformly
			21.5	22.5	24	25	23.5	24.5	26	27	26	27	28.5	29.5	28.5	29.5	31	32	29	30	31.5	32.5	in groups

- 25 – 32.5 — stands with high diversity and greatly emotional impact
- 17 – 24.5 — stands with adequate diversity and emotional impact
- 10 – 16.5 — stands with uniform character and less emotional impact

Table 5: Evaluation of forest stands to their visual impact

Thus, forested areas are grouped into three groups according to the visual impact of plantations due to their external appearance (table 5):

The results of forest stands assessment, as well as the open spaces assessment are presented on maps (fig. 1) accompanied by photographs. Grouping of landscapes is made mainly based on visual characteristics of the terrain and vegetation component. Based on the results of these study it have been made a number of conclusions necessary for the

development of functional zoning of the area. The aesthetic qualities of the natural conditions are assessed in the following indices:
for the forest stands:

- average height;
- passability;
- construction;
- dendrological richness;
- presence of much higher trees and single tree species occurring;
for the open spaces:
- vertical and horizontal indentation of the relief;
- degree of the horizon shelterness;
- number of visible landscapes;
- passability, construction, dendrological richness, and presence of much higher trees and single tree species occurring in the surrounding tree forest stands.

In conclusion we can say that the determination of aesthetic value of landscape is very complex process including the description, analysis and evaluation, expressed in the grouping of territorial units defined set of criteria associated primarily with sensory experiences.

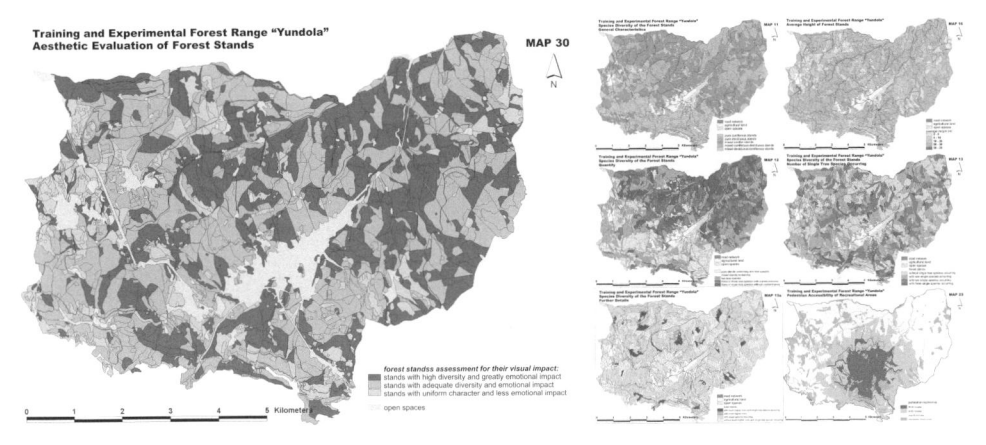

Fig. 1: Aesthetic evaluation of forest stands

References

Appleton, J. (1975), *The experience of landscape*. London, John Wiley. 293 p.
Buhyoff. G. J., Wellman, J. D. & Daniel, T. C. (1982), *Predicting scenic quality for mountain pine beetle and western spruce budworm damaged forest vistas*. In: Forest Science, 827-838.
Cox, T. R. (1985), *Americans and their forests: romanticism, progress, and science in the late nineteenth century*. In: Journal of Forest History, 29, 156-168.

Daniel. T. C. & Boster, R. S. (1976), *Measuring landscape esthetics: The scenic beauty estimation method* (U.S.D.A. Forest Service Research Paper 167). Ft. Collins, CO: Rocky Mountain Forest and Range Experiment Station.

Kaplan, R. & Talbot, J. F. (1988), *Ethnicity and preference for natural settings: a review and recent findings*. In: Landscape and Urban Planning, 15, 107-117.

Lyons, E. (1983), *Demographic correlates of landscape preference*. In: Environment and Behavior, 15 (4), 487-511.

Ribe, R. G. (1991), *The scenic impact of key forest attributes and long-term management alternatives for hardwood forests*. In: McCormick, L. H. & Gottschalk, K. W. (Eds.), Proceedings, 8th Central Hardwoods Forest Conference; 1991 March 4-6 University Park, PA. Gen. Tech. Rep. NE-148. Radnor, PA: U.S. Department of Agriculture, Forest Service, Northeastern Forest Experiment Station, 35-54.

Schroeder, H. W. (1987), *Dimensions of variation in urban park preference: a psychophysical analysis*. In: Journal of Environmental Psychology, 7, 123-141.

Zube, E. H., Pitt, D., G., Evans, G. W. (1983), *A lifespan developmental study of landscape assessment*. In: Journal of Environmental Psychology, 3, 115-128.

Location-aware Mobile Devices and Landscape Reading

Ruben JOYE, Joris VERBEKEN, Steven HEYDE and Harlind LIBBRECHT

Abstract

This paper examines how 'smartphones' – a type of advanced and location-aware mobile phone (e.g. Apple iPhone® or phones running Google Android®) – can be useful for landscape architecture and planning, and more specifically how they can share knowledge about our surroundings using 'Mobile Augmented Reality' (MAR). We examined how a smartphone-based information system can influence reading and understanding the landscape, and to what degree it influences landscape valuation and imaging by its users. A quantitative survey was conducted with students to determine the educational possibilities of such tool. The main part of this paper however, is more technical in nature. Using the 'Layar® Reality Browser' as a framework for our MAR, we wanted to facilitate the management of a 'Layar® 3D' system. We did so by building a graphical front-end to the administrative part of the system by using a combination of ESRI ArcGIS® and Microsoft Access®. This to facilitate adding, editing or removing 'Points Of Interest' (POI), even to people with a limited technical background..

1 Introduction

With smartphones (e.g. Apple iPhone® or phones running Google Android®) becoming more popular every month, in the third quarter of 2010 accounting worldwide for 19,3% of overall mobile phone sales compared to 9,85% in 2009 (COZZA et al. 2010, GARTNER 2010), it's safe to conclude that location-aware mobile devices are getting more and more widespread. These types of devices – equipped with both GPS technology and mobile internet connectivity – are taking GIS information mobile, making it possible to associate digital media to a geographical location (VARNELIS & FRIEDBERG 2008). Handheld location-aware mobile devices are becoming the interface to the 'geospatial Web', that delivers on the spot georeferenced information to its users. This allows people to be present in both the physical and networked (digital) place (ITO 2008). 2008 – the year in which for the first time in history, mobile access to the internet exceeded desktop computer-based access (ITU 2009) – turned out to be the start of an internet revolution, quickly named 'the mobile web'.

An interesting application for these location-aware devices is 'Mobile Augmented Reality' (MAR): the superpositioning of rich media elements (audio, video, images and even 3D-models) on top of a real-time view from the built in camera lens of the portable device. The technical aspect of a MAR-system has already been considered by JOYE et al. (2010) and the opportunities for education in landscape architecture have been discussed by VERBEKEN et al. (2010).

A prebuilt and free client-application called 'Layar® Reality Browser' is available for both the Apple iPhone® and phones running Google Android®. One of the harder parts is setting up your own Layar® POI server that communicates with the Layar® client on the smartphone. Luckily there are a few different development tools available online (CAMERON 2010), with PorPOISe (DE SMIT 2010) – a PHP based POI-server – being one of the most comprehensive. Some adjustment to the programming code is needed, in order to have it connect to the record with the POI's you want to show. This can either be an XML formatted file, or a MySQL® relational database.

PorPOISe (version 1.0a was used) comes prebuilt with a textual web interface called 'PorPOISe server dashboard' (Fig. 1) that allows authorized persons with no programming skills to add new 'points of interest' (POI's). However, when you're planning to have numerous POI's, this can be cumbersome since you need to look up (and enter) the lat-lon-coordinates for each POI manually.

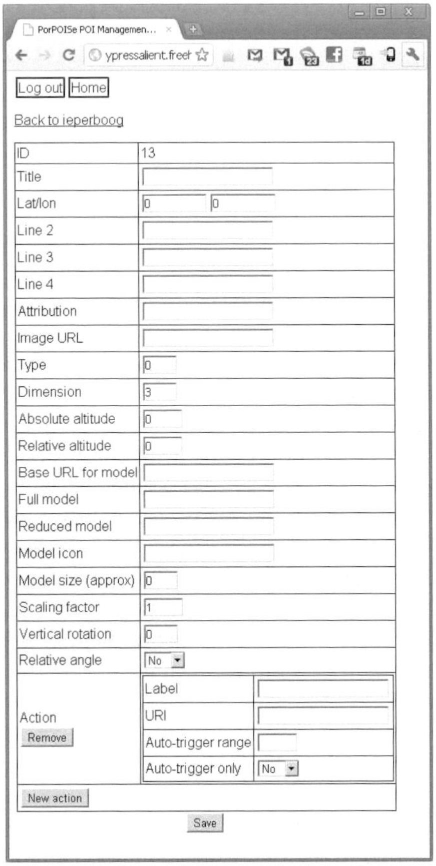

One of the main reasons for fractional use of landscape visualization tools (in general) – according to BISHOP & LANGE (2005) – is the lack of user-friendliness for easy manipulation (BISHOP & LANGE 2005). It was our intention to meet this constraint by improving the usability, by linking the underlying MySQL® database with ESRI Arcmap. A new POI can be added more easily by use of and at the same time it becomes possible to import both the spatial as well as the attribute data of already existing GIS-datasets (e.g. ESRI shapefiles). It appears we are not alone in seeing great value in a tighter integration of GIS-data in the Layar® Reality Browser. A similar idea was carried out by EMGE & PRASAD (2010). Details on their approach are missing though.

Fig. 1: Web interface of the 'PorPOISe server dashboard'

The need for a better way to manage POI's arose from the fact that an arrow-less GPS-based touristic route will be created at the end of our landscape research project. In order to evaluate our new workflow involving ESRI ArcMap, we've put together a small test case. As part of the research project – which studies change in a former World War One region – a pilot study on the subject of landscape evaluation has just been completed in close cooperation with historians from the 'Memorial Museum Passchendaele 1917' (MMP) and the Flemish heritage institute (BOSTYN et al. 2010). This preparatory research work served as a base for the landscape walk.

The second part of the paper will focus on how location based information can influence landscape perception and -valuation. During the process of mental imaging, the observer interprets the perceived landscape. Therefore, individual landscape valuation is not only contextual and time bound, but also strongly personal or subjective (JACOBS 2006) since the plethora of visual stimuli gets filtered, before final imaging takes place. Unconsciously, a selection of stimuli is performed in terms of usefulness for the given situation, partially based on prior knowledge (cognitive aspects) (DIJKSTRA & KLIJN 1992). For instance – in the case of cultural heritage landscape valuation – an observer with little to no knowledge about the landscape and cultural heritage will be able – at best – to distinguish the main landscape structures, but will overlook more detailed information (COETERIER 1995, VAN DEN BERG & CASIMIR 2002). As landscape experience and -valuation is much more than just an aesthetic consideration (DIJKSTRA & KLIJN 1992), this detailed information may have a strong influence on one's final assessment of the whole. By pointing the observer's attention to these details (e.g. cultural remnants), as well as providing background information (e.g. about historic events that took place), the appreciation of landscape experience will be more balanced, more strongly founded, and surpass a purely aesthetic appraisal.

A quantitative survey, was carried out with a group of 20 students enrolled in a one-year advanced study in landscape development. They gave their opinion on both the ease of use of the smartphone system, the educational possibilities, as well as their impressions of the physical environment and how it had changed their appraisal and overall assessment.

2 Material & Methods

2.1 Technical

After installing the ODBC MySQL® Connector (version 5.1.8. was used), linking a shapefile to a MySQL® database is easily feasible from within ArcGIS, using ESRI ArcCatalog to setup a 'database connection' by means of an 'OLE DB Connection'. The additional columns and corresponding values that are stored in the external MySQL® database get added to the attribute table in ArcMap without any problems. But the drawback is that this external database cannot be edited from within ArcMap (as this is a read-only process), making this approach useless for our stated goal. To overcome this problem, an intermediate step was taken that at the same time adds functionality to the workflow. A Microsoft Access® form linked to the main POI shapefile was created, in which all the necessary data about each of the POI's can easily be entered.

This was done by making use of the ArcScript 'ArcMap Hyperlink to Filtered Microsoft Access® Form' provided by CALLAHAN & CARSON (2003) on the ESRI support page. This filters all of the entries in order to show only the corresponding data in a clear Access form (Fig. 1). Although programmed to work in conjunction with ArcGIS 8.2 and Microsoft Access® 2000, this script works just fine using ESRI ArcGIS 9.3.1 and the newer versions of Microsoft Access® (both 2007 and 2010 have been tested).

The use of this script requires working with a personal geodatabase, which can be created in ESRI ArcCatalog. Once we have a personal geodatabase set up, we add a new (point) feature class which we name 'POI'. As coordinate system 'WGS 1984' – the international standard for use in cartography, geodesy, and navigation – is chosen. The resulting feature class has two mandatory fields: 'OBJECTID' and 'SHAPE'. We add an extra field 'accessid' (text) which will mirror 'OBJECTID', but is necessary to link with our Microsoft Access® form since the ArcScript requires the data type of the field to be a text value.

A minimal MySQL® database for use with PorPOISe – that enables basic Layar® functionality – consists of one table named 'POI' with the under mentioned structure. Note that in the overview schema (Fig. 2) we added the three fields necessary for use with ArcGIS (indicated by an asterix), but that these do not negatively influence the functioning of either PorPOISe or Layar®.

POI	Action	Object	Layer
id	id	poiId	layer
attribution	uri	baseURL	refreshInterval
imageURL	label	full	refreshDistance
lat	poiId	reduced	fullRefresh
lon	contentType	icon	showMessage
line2	method	size	id
line3	activityType		
line4	params		
title	closeBiw	**Transform**	
type	showActivity	angle	
doNotIndex	activityMessage	rel	
showSmallBiw	autoTriggerRange	scale	
showBiwOnClick	autoTriggerOnly	poiID	
layerID			
dimension			
alt			
relativeAlt			
SHAPE(*)			
OBJECTID(*)			
accesid(*)			

Fig. 2: Overview of PorPOISe database schema with the five tables making the structure for use with Layar®   (*) = added for use in conjunction with ArcGIS

If you want to make use of more advanced features in Layar® like 2D/3D objects and actions, you need to add additional information to the POI's: what is the URL to the 3D-model, should it be scaled, rotated, … Are there actions connected to the POI like a link to an external webpage, audio, video and should these trigger automatically or not? All this information gets stored in three tables separate from the POI table. A fourth additional table was introduced to support the Layar® v4 API features (DE SMIT 2010), totaling five tables: 'POI', 'Object', 'Transform', 'Action' and 'Layer'. This last table isn't of much interest to us, and can be left blank to have the Layar®-client use the default values.

The functionality of each of these fields has already been discussed in detail by (DE SMIT 2010, WANG 2010, WANG 2011) and therefore is not repeated in this paper. Albeit, the meaning of each field was included in the final version of our Microsoft Access® form as a tooltip displayed when the mouse hovers over a field.

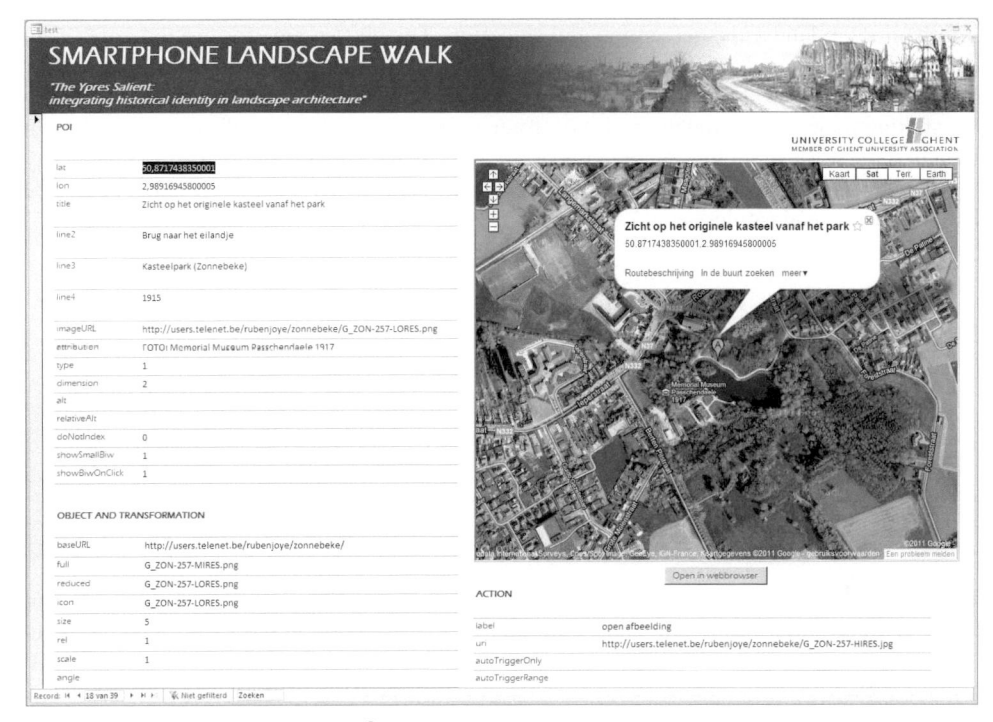

Fig. 3: The Microsoft Access® form linked to the ESRI Arcmap document. This form combines data from five different tables and constitutes the MySQL® database which PorPOISe uses to serve to the Layar® client-application.

To get the proper database schema for all the required tables, a SQL script file 'database.sql' comes with PorPOISe. This file was executed to our external MySQL®-server using the MySQL® Workbench (version 5.2.31a) which easily created all the proper tables and data types settings for the different fields. Having already installed the ODBC MySQL® Connector, we were also able to connect to our MySQL®-server using Microsoft Access®, and have the database schema imported to our personal geodatabase (opened as a Microsoft Access® document). The first thing we need to do for it to work, is to add a new connection by using the Microsoft ODBC Data Source Administrator which can be found in the 'control panel' under 'administrative tools'. Once this connection has been added under the 'file-DSN' tab, it will become available in Microsoft Access®. With our personal geodatabase opened in Microsoft Access, we chose 'external data' and picked 'ODBC-database'. This allows to import the database schema, and have a functional database in accordance with the PorPOISe structure. The process of importing a MySQL® database in Microsoft Access® is described more in depth in a 'White Paper' from SUN MICRO-SYSTEMS (2009). After defining the relations between the different tables, and executing the query, a Microsoft Access® form was created (Fig. 3).

2.2 Content & use

To put this workflow into practice, a small quantitative survey was conducted. Not only to test the ease of use, but at the same time to evaluate the influence of location based information on landscape perception and valuation. The chosen area for our test case, is an old castle park which has undergone complete destruction during the First World War. Now the castle park serves a public function as a museum. In future developments this will be the starting point of a complete landscape route using a smartphone application. We chose a total of 39 historical photos and drawings (Fig. 4-7) that accurately depict the gradual decay of the beautiful center of 'Zonnebeke' (near Ypres) as it underwent successive bombardments and ruthless attacks.

Fig. 4: Soldiers rowing in the pond, with the former castle and church in the background (1915-1916) © MMP 1917

Fig. 5: The original castle and ancient church, in the foreground the pond with bridge (1915-1916) © MMP 1917

Fig. 6: The church and the park show heavy traces of destruction (1915-1916) © MMP 1917

Fig. 7: The devastated region surrounding the castle park (near the end of 1917) © MMP 1917

A group of twenty students (on average aged 22) was taken on a 20 minute walk in the park. As stated in the introduction, these were students enrolled in a one-year advanced study in landscape development, most of them being graduated landscape architects. Of the twelve students who filled out the survey, the male-female ratio was exactly half-half. This quantitative survey looked at not only the ease of use of the smartphone system, but more particularly if and how the use of this system influenced their landscape valuation. While it does give a good insight in their experiences, drawing hard conclusions based on these twelve respondents would be imprudent. On top of that, weather conditions that day where quit harsh: cold winds and rain resulted in smudgy trails. Much to our surprise, none of the students present had a compatible phone. As a consequence, only two smartphones were available to be shared with the entire group.

The survey itself consisted of twenty-two questions divided into three categories:

- their mobile phone habits: which type of phone do they have, is there an intention to buy a smartphone, who pays for their phone, what is their budget,...
- their general experiences on educational innovations: is technologic innovation key to good education, is there immediate educational use for the application,...
- the actual use and experience of the system: was the user interface easy to use, has it changed your way of looking at your surroundings, do you value the area higher on a cultural-historical level, would you take a complete tour with the smartphone once finished,...

3 Results

3.1 Technical

With the personal geodatabase correctly set up as described in 2.1, we started off with a Blanc ArcMap document. The projection was set to our local geographic coordinate system (Belge Lambert 72), and a georeferenced aerial photo of the area was added. Next, the point feature class 'POI' from our personal geodatabase was added to the map. As this data source was set up for use in WGS 84, a system warning about the mismatch is shown. Transformation from WGS 84 to Belge Lambert 72 can be applied from ArcMap. The option to do so is available from within this warning message, where you can choose to have the feature class aligned properly without distortion of the aerial photo.

New POI's can be added by using the 'Sketch tool'from the editor toolbar to create new features. After having added some new features, we need to update the attribute table. The auto increment values in the 'objectid' field need to be copied to both the 'accesid'-field (for the ArcScript to function) and the 'id'-field (for PorPOISe and Layar® to function). The field calculator can be used for this to have the values copied instantly. Contrary to ArcMap, Layar® needs to have the longitude and latitude values of each POI in the attribute table. Similar to the previous step, these values can be calculated, now using the 'Calculate Geometry' tool in the attribute table. Using 'WGS 84' as the coordinate system, the correct decimal degrees can be calculated for both the 'lat' and 'lon' field in the attribute table.

Once these fields have been populated, the hyperlink tool in the tools palette can be used to jump to the filtered form in Microsoft Access®. There, all the necessary information on each POI can be added or updated. Having the latitude and longitude position of each POI as values in the database, also results in being able to display a map with the location of the active POI from within the Microsoft Access® form (Fig. 3).

The final step in physically being able to access this information on a mobile device, is updating the MySQL®-database on our server. A free and easy way to do this, is by using a freeware tool called 'Access To MySQL®' by the company Bullzip. It's as easy as selecting the tables in your personal geodatabase that you wish to transfer, entering your server connectivity credentials and hit 'Run Now'. Commercial solutions for synchronizing a local database automatically are available as well (e.g. DBSync for Access and MySQL®).

Finally, exporting the POI-information in the linked tables as an XML-file is a possibility as well. Attention needs to go to the fact that this needs to comply with the XML-structure set forward by PorPOISe, which requires some tweaking of the exported file.

3.2 Content & use

The Layar® client applications makes viewing the entries on the smartphone possible in three different ways: 3D-live view (Fig. 8), list view (Fig. 9), or on a map (Fig. 10).

Fig. 8: 3D-live view

Fig. 9: POI's viewed in list mode

Fig. 10: Map view

As for the results of our quantitative survey it's important to again emphasize that our sample of 20 students was on the small size, thereby not generating any true generalizable outcomes. The result of this quantitative survey should be interpreted as the opinion of a small group rather than an unambiguous conclusion.

A third of the respondents agreed that the limited size of the smartphone screen is a significant bottleneck to the system. The fact that we had only two devices which had to be passed around, might have contributed to this. Tablet pc's like the Samsung® Galaxy TAB (running on Google Android® as well), can respond to this feeling of shortcoming, as they offer a greater immersive experience due to the significant larger screen size (MAARTEN 2010).

Half of the respondents found that the location specific information they got through the system impacted their landscape perception. A third had no opinion, and two respondents disagreed. There was more consensus on the question whether or not it resulted in a higher valuation of the landscape in terms of cultural heritage, where ten out of twelve (83%) agreed.

Two thirds of the group felt that the augmented reality technology as it is now, is still too premature to have practical use in education. Their critical attitude is also reflected in a more general question about educational innovation. Half of them believes that using multimedia or technology doesn't necessarily result in a better understanding of the learning content or helps them to keep focused.

Whereas three quarters thinks digitalization and technological innovation results in a bigger democratizing of education and knowledge gathering, one might question whether the threshold for obtaining the information does not increase, since it requires having a smartphone. Already 40% has a mobile data plan, but only a third has a smartphone. All of them running Symbian (Nokia's mobile operating system), which at the time of our landscape walk was still unsupported by Layar®. Just recently the Layar® platform has been ported to the operating system Symbian, opening up augmented reality to be used on a selection of Nokia® smartphones (LAYAR 2011).

No more than 15% is prepared to pay upwards to €200 for a new mobile phone, the critical price for an entry level smartphone capable of running the Layar® reality browser. Following price (39%), their main concern is the design (28%) rather than more functional aspects like standby time (11%) or number of downloadable applications (17%). Until prices start to drop even more, it seems unlikely that the technology has chance of being adopted as an everyday tool by these students on short term.

We found no pronounced differences between the responses of male vs. female students when it comes to their landscape experience. A prominent differentiation we did find was that male students are willing to spend more money when buying a new mobile phone, but that they simply do so less often. None of the female students would spend more than €200 on a new phone, while a third of the male students indicate to be willing to pay over €200. Some of them even more than €500. All of the male students use their phones for at least two years, while two thirds of their female counterparts indicate getting a new phone around every year. Additionally the male students showed greater confidence in educational opportunities and believed more strongly that location aware information by means of mobile devices has greater impact when compared to a brochure or billboards.

An encouraging fact is that all of the respondents indicate that – once this system is further developed and contains more information about the larger region of Ypres – they would make use of this location based service in order to explore and learn more about the area.

4   Conclusions and Outlook

The images we placed in our database, were so called "2D billboard images". They had a location, but in some cases students found it hard to tell in what direction the photo was taken. A way to achieve a higher user understanding would be to work with 'experience domes' in Layar®. These are large spheres (3D-models) that have a photo texture applied to the inner surface. Walking into one of these 'virtual caves' fills the entire screen, and depicts only the part of a larger panoramic image that corresponds with your current view direction. This makes it much easier for the user to connect the present with the virtual image. Since the images we had at our disposal were mostly regular photos with limited field of view, the results using the 'virtual cave' principle would have been poor. When it comes to visualizing future intervention however, this approach becomes of great value. As opposed to the limited number of polygons a 3D-model for use with Layar® may contain (JOYE et al. 2010), the level of complexity in the texture of a 'virtual cave' (or dome) is of no importance for the application.

Although the management of POI's to be displayed in Layar® proved very functional, and could be integrated in familiar workflows (i.e. ESRI ArcGIS and Microsoft Access®), the initial setting up in order to have everything streamlined requires advanced computer knowledge that cannot be assumed from neither students nor landscape professionals.

The difference in geographic coordinate system between the aerial photo and the point feature class results in minor deviation. In the smartphone applications this inaccuracy is negligible, especially since the GPS-signal itself manifests higher imprecision.

The numb task of having to have the latitude and longitude values calculated in the attribute table, and copy values from one field to another field using the 'field calculator' before being able to switch to Microsoft Access, could be more automated. Ideally, these values should be updated automatically each time a new feature gets added (or an existing one modified).

Further research and development –conducted by programmers – would be extremely valuable in terms of making the entire process of creating a Layar® and adding POI's even more intuitive and user-friendly. Looking deeper into ways of integrating different georeferenced data standards of different kinds by means of easy to use extensions to for instance ArcMap would be much appreciated. To that respect the recent announcement of what could develop into an open standard for augmented reality – based on a combination of HTML and KML – developed by Georgia Tech looks quit promising (CHRISTOPHER 2011).

Lastly, the effects of using location based services on smartphones in terms of landscape perception and –valuation should be studied more in depth. A larger test group is needed to draw more founded conclusions. In addition to that, comparing other influential parameters (e.g. having a local expert as a tour guide) with a smartphone application would be interesting.

Since learning to 'read the landscape' – a skill that in essence can only be acquired in the field – is an essential part of a landscape architect's education, such system could be used to enrich landscape walks for students. In addition, this also allows students to repeat the walk individually at any time.

References

Bishop, I. & Lange, E. (2005), *Communication, Perception and Visualization.* Visualization in Landscape and Environmental Planning: Technology and Applications, ed. by Bishop, I. & Lange, E. London, Taylor & Francis

Bostyn, F., Heyde, S., Naessens, J., Blieck, K. Libbrecht, H., Verbeken, J. De Mûelenaere, S. & Joye, R. (2010), *Towards a valuation methodology for cultural landscapes on a local scale: Pilot Study for the municipalities Zonnebeke and Passchendaele (internal report).* Memorial Museum Passchendaele 1917, University College Ghent (Department Bioscience and Landscape Architecture), VIOE.

Callahan, K. & Carson, S. (2003), *ArcMap Hyperlink to Filtered Microsoft Access Form.* From http://arcscripts.esri.com/details.asp?dbid=13081.

Cameron, C. (2010), *Layar Creation Tools: Series Wrap-Up.* Retrieved 12/01/2011, from http://site.layar.com/company/blog/category/tools/.

Christopher, M. (2011), *Turning Augmented Reality into an Open Standard.* Retrieved 08/03/2011, from http://www.technologyreview.com/computing/35065/.

Coeterier, J. F. (1995), *De beleving van cultuurhistorische objecten. Een verkennend onderzoek in de Meierij van Den Bosch.* Wageningen, DLO-Staringcentrum.

Cozza, R., Milanesi, C., Gupta, A., De La Vergne, H. J., Zimmermann, A., Lu, C., Sato, A. & Huy Nguyen, T. (2010), *Competitive Landscape: Mobile Devices, Worldwide, 3Q10,* 17.

De Smit, J. (2010), *PorPOISe: Point-of-Interest server for Layar.* From http://code.google.com/p/porpoise/.

De Smit, J. (2010), *Using PorPOISe With Existing Databases.* From http://code.google.com/p/porpoise/wiki/UsingPorPOISeWithExistingDatabases.

Dijkstra, H. & Klijn, J. A. (1992), *Kwaliteit en waardering van landschappen.* Wageningen, DLO Staring Centrum,

Emge, T. & Prasad, S. (2010), *Viewing ArcGIS data in Augmented Reality Application.* Retrieved 12/01/2011, from http://blogs.esri.com/Dev/blogs/apl/archive/2010/04/15/Viewing-ArcGIS-data-in-Augmented-Reality-Application.aspx.

Gartner (2010), *Gartner Says Worldwide Mobile Phone Sales Grew 35 Percent in Third Quarter 2010; Smartphone Sales Increased 96 Percent.* Retrieved 12/01/2011, from http://www.gartner.com/it/page.jsp?id=1466313.

Ito, M. (2008), *Introduction.* In: Networked publics, ed. by Varnelis, K. Cambridge, Massachusetts, MIT Press, 1-14.

ITU (2009), *The world in 2009: ICT facts and figures, a decade of ICT growth driven by mobile technologies.* Geneva, ITU World Telecommunication.

Jacobs, M. (2006), *The production of mindscapes: a comprehensive theory of landscape experience.* Dissertation, Wageningen University, Wageningen

Joye, R., De Mûelenaere, S., Heyde, S., Verbeken, J. & Libbrecht, H. (2010), *On the Applicability of Digital Visualization and Analysis Methods in the Context of Historic Landscape Research.* In: Digital Landscape Architecture 2010, ed. by Buhmann, E. et al. Berlin/Offenbach, Wichmann.

Layar (2011), *Layar Developers Update: Layar for Symbian is here!* Developer Newsletter.

Maarten (2010), *Tablets are great to play Immersive Augmented Reality.* Retrieved 12/01/2011, from http://site.layar.com/company/blog/tablets-are-great-to-play-immersive-augemented-reality/.

Sun Microsystems (2009), *A Visual Guide to Microsoft Access Front-Ends with MySQL.*
Van den Berg, A. E. & Casimir, T. (2002), *Landschapsbeleving en cultuurhistorie: een theoretische en empirische verkenning van de invloed van cultuurhistorie op de beleving van het landschap.* Reeks belevingsonderzoek nr. 2. Wageningen University and Researchcenter, Wageningen.
Varnelis, K. & Friedberg, A. (2008), *Place: The Networking of Public Space.* Networked publics, ed. by. Varnelis, K. Cambridge, Massachusetts, MIT Press.
Verbeken, J., Joye, R., De Mûelenaere, S., Heyde, S. & Libbrecht, H. (2010), *Mobile Augmented Reality in Landscape Architecture: Opportunities for Education.* ECLAS Conference 2010, Istanbul, Turkey.
Wang, X. (2010), *Layar 4 API changes.* Retrieved 12/01/2011, from http://layar.pbworks.com/w/page/27177748/Layar-4-API-changes.
Wang, X. (2011), *GetPOIs-JSON Response.* Retrieved 12/01/2011, from http://layar.pbworks.com/w/page/28473525/GetPOIs-JSON%20Response.

Visualization of the Climate Change with the Shift of the So-called Moesz-line

Anna CZINKÓCZKY and Ákos BEDE-FAZEKAS

1 Introduction

The need to mitigate to climate change has been in focus for over a decade. The issue was addressed in various forms (ranging from static images to virtual reality 3D models equipped with powerful computer graphics) at different levels aiming to influence all sectors from children to adults, from policy makers to high level academia.

Since the climate change is so complex and fundamentally effects our environment, the issue has to be readdressed over and over at different levels and in a variety of forms to avoid boring repetitions and a novelty sensation helps to keep the awareness at a high level. In our present paper we would like to use a less known concept of the so called Moesz-line to model the effects of climate change.

Gusztáv Moesz (1873-1946) was a botanist who was engaged in mycology, museology and obtained an international reputation in these fields. In his main paper which was published more than 100 years ago, he discovered that certain plants share a common northern habitat border and this coincides with the line of vine cultivation. The line represented in Fig. 1 – is the so called Moesz-line – that lies in the territory of Slovakia but at that time belonged to Hungary and is named after him (MOESZ 1911).

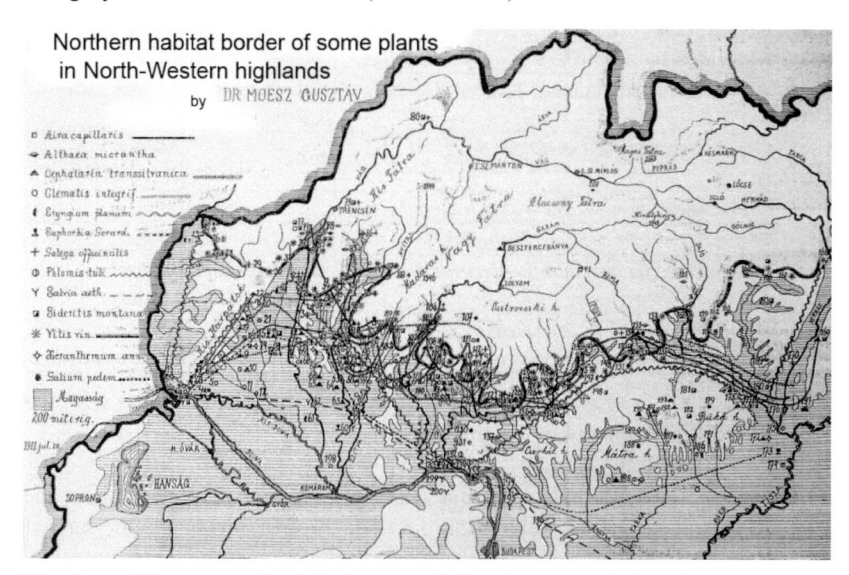

Fig. 1: The scanned original map of the 12 taxa and the vine cultivation area (MOESZ 1911, with translating and some retouching). The hard-to-perceive map has only a historical importance nowadays.

2 Materials and Methods

2.1 Background

The Moesz-line is seldom used in international scientific publications due to its local importance. However, extending the northern cultivation line of grape to the East and to the West, one can obtain the extension of the Moesz-line, and it will still coincide the northern habitat border of some other species such as Grape Hyacinth (Muscari botryoides, SOMLYAY 2003). Hence, the Moesz-line modelling has an importance for the entire Central European region since it includes species outside of the Carpathian Basin. Originally Moesz included 12 plants in the habitat-observation in addition to the grape cultivation (Vitis vinifera). Table 1. shows the original names of the plants that were investigated by Moesz together with their current scientific names and the sources of the habitat maps (MOESZ 1911, MEUSEL 1965, TUTIN 1964, EUFORGEN 2009).

Name used by G. Moesz	Accepted scientific name/ Common name	Synonyms	Source of distribution
Aira capillaria	Aira elegantissima/*Hairgrass*	Aira elegans	Moesz (Meusel)
Althaea micrantha	Althaea officinalis/*Common Marshmallow*	Althaea taurinensis	Moesz (Meusel)
Cephalaria transsilvanica	Cephalaria transsylvanica	Scabiosa transsylvanica	Moesz
Clematis integrifolia	Clematis integrifolia		Moesz(Flora Europaea)
Eryngium planum	Eryngium planum/*Sea Holly*		Moesz
Euphorbia gerardiana	Euphorbia seguieriana		Moesz (Meusel)
Galega officinalis	Galega officinalis/*Goats Rue/French Lilac*		Moesz
Galium pedemontanum	Cruciata pedemontana/*Piedmont Bedstraw*	Valantia pedemontana	Moesz
Phlomis tuberosa	Phlomis tuberosa/*Jerusalem Sage*		Moesz (Meusel)
Salvia aethiopis	Salvia aethiopis/*Mediterranean Sage*		Moesz
Sideritis montana	Sideritis montana/*Hairy Ironwort*		Moesz
Xeranthemum annuum	Xeranthemum annuum/*Everlasting Flower*	Xeranthemum squarrosum	Meusel (Moesz)
Vitis vinifera	Vitis vinifera/*Common Grape Vine*		Moesz (area of cultivation)

Table 1: The list of the 12+1 species originally used for drawing the Moesz-line with the source of the distributional map used in our research

For the landscape design the importance of the Moesz-line lies – not mainly in the investigation of the original 12 plants – but in the fact that Moesz-line shows the habitat area of the vine cultivation which plays a key role in agriculture. Moreover, Moesz-line also demonstrates the northern border of some other important species that have been added later to this concept. Among these are the Chestnut (Castanea sativa, BARTHA 2007), Pubescent Oak (Quercus pubescens, csapody 1932, KÁRPÁTI 1958, KÉZDY 2001, BARTHA

Pubescent Oak (Quercus pubescens, csapody 1932, KÁRPÁTI 1958, KÉZDY 2001, BARTHA 2002), Bladder Senna (Colutea arborescens, CSIKY 2003), Service Tree (Sorbus domestica, VÉGVÁRI 2000), and some pear species (Pyrus magyarica, Pyrus nivalis ssp. orientalis, Pyrus pannonica, Pyrus slavonica, TERPÓ 1992). The species besides the previous ones that still have a significant application in landcape design are Tatar Maple (Acer tataricum), Smoke Tree (Cotinus coggygria), Manna Ash (Fraxinus ornus), Mahalab Cherry (Prunus mahaleb) and Turkey Oak (Quercus cerris). For the landscape practitioners the following plants have less importance: Grape Hyacinth (Muscari botryoides), Monkey Orchid (Orchis simia), Vetch (Vicia sparsiflora) and Wild Grape (Vitis sylvestris). In our research the distribution of Smoke Tree, Manna Ash, Turkey Oak and Wild Grape were modeled based on the maps of Meusel, and the cultivation area of Chestnut was obtained using the EUFORGEN digital database.

According to the global climate-change models, the weather in the Carpathian basin will become warmer and drier, and the extreme precipitation is expected to occur in the hotter period of the year (BARTHOLY 2008). This will bring a very challenging situation for the landscape designers who have to face and be prepared to deal with this issue well in advance. The landscape designers can slightly change the macroclimate, they can chiefly adapt to it and modify the microclimate. So the emphasis has to be put on the proper adaptation. To implement the best adaptation policy, it is essential to know the expected climate of a future period. In case of the landscape designers, horticulturists and dendrologists in means that they have to know the expected natural vegetation and possible adaptations. For some larger plants (trees), the development and growing period is approx. 30 years, so it is high time that we addressed this issue with some easy-to-understand but effective visualization technique to help the landscape designers in optimal plant selection for expected future climate. Such visualizations are essentials for experts who can further influence the urban planners, authorities and general public (SHEPPARD 2005). In our research, we will model the shift of the Moesz-line which can be served as an alternative for geographically analogous regions (HORVÁTH 2008a).

2.2 Modelling approach

The expected shift of the so called Moesz-line can be modelled in several ways. Table 2. contains 3 such possibilities showing advantages and disadvantages of each method.

Name	Description	Advantage	Disadvantage
Line modeling	Modeling the shift of the Moesz-line as the northern border of a fictive distribution	relatively accurate, clear, evident	relatively slow
Distribution modeling	Modeling the shift of the distribution of plants belonging to the Moesz-line and then redrawing the future line	follows the original method of Moesz, has a complex, detailed result	very slow, subjective
Isotherm modeling	Modeling the shift of the minimum temperature isotherm of January (winter months) correlates with the Moesz-line	fastest of the three methods, there is no need for digitalizing distributions	uses only a few climatic parameter(s), inaccurate, questionable

Table 2: Methods of the Moesz-line modeling used in our research

All 3 models are based on the REMO ENSEMBLES RT3 climate model which contains data of a 25-km horizontal grid cell distribution of Europe (170×190 pixels). The reference period is the data between 1961-1990 and the forecasted periods are – according to one of the IPCC SRES scenarios called 'A1B' – the years between 2011-2040 and 2041-2070. The modeling was performed by ArcGIS software.

As a preparation, the original maps of Moesz were digitized and georeferenced, using 20-25 control points (country borders and rivers). For modeling the shift of the habitat-zones, it was necessary to digitize the habitats of each plant, since only the EUFORGEN data contained spatial coordinates. We did not consider the entire habitat regions of plants, only the northern segments of the grids were taken into the account. Since only the northern habitat borders were modeled, this concision did not modify the result. For both methods it was necessary to interpolate the discrete data of the climate model and convert it into continuous functions. Three parameters of the climate model were used are monthly average temperature, monthly minimum temperature and monthly total precipitation.

All the temperature data for 12 months were used. From the precipitation data only the total rainfall in the vegetation period (from April through September) was considered because it was validated by similar models. (Due to the climate change, the precipitation zones would shift to the north with different rates than the temperatures) (BEDE-FAZEKAS 2011).

Since the northern habitat borders was modeled, the minimum temperatures were used (1×24 parameters) together with the upper and lower values of the total rainfall in the vegetation period (2×1 parameters), so altogether 26 different logical conditions had to be fulfilled for a given spatial point to satisfy the climatic conditions. It is summarized in a mathematical formula in the equation below.

$$\left(\prod_{i=1}^{12} I(\langle T_{mean}\rangle_i^f \geq \langle T_{mean}\rangle_{min}^r)\right) * \left(\prod_{i=1}^{12} I(\langle T_{min}\rangle_i^f \geq \langle T_{min}\rangle_{min}^r)\right)$$
$$* I\left(\sum_{i=4}^{9} \langle P\rangle_i^f \leq max\left(\sum_{i=4}^{9} \langle P\rangle_i^r\right)\right) * I\left(\sum_{i=4}^{9} \langle P\rangle_i^f \geq min\left(\sum_{i=4}^{9} \langle P\rangle_i^r\right)\right) = 1 \qquad (1)$$

In the equation (1) the $I(\lambda)$ takes the value of 1 if the condition for λ is true, otherwise it takes the value of 0. The symbol r means the reference period, f stands for future period, i is the running variable.

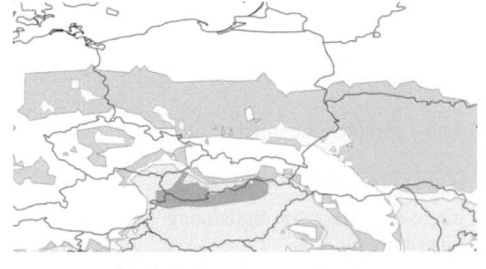

Fig. 2:
An example of the distribution modeling (*Sideritis montana*)

From the individual areas of habitats (Line modeling) we have selected 26 extreme values that belong to all the parameters. (25 minimums, 1 maximum) and we performed the modeling for the two future periods (Fig 2.). This model displays those areas where the plant can find those climatic conditions which are necessary for its existence. Since the original areas were chosen from the real habitat zones, the northern border will give a boundary line for future natural areas of occurrence. We have not dealt with edafic and microclimatic investigations. By modeling the reference period, we wanted to display the difference between the real observations and predictions of the modeling.

In the second type of modeling (Distribution modeling) we have analyzed the possible distribution of the 18 species. The third method, the Isotherm modeling has the closest resemblance to the hardiness zones. In our research however, we have only used the average minimum temperatures of January in each year instead of the absolute minimum temperature of the winter. Most likely that for the submediterranean flora the absolute minimum temperature has a higher importance, but by no means does it describe the proper climatic requirements.

3   Results and discussions

The result of Line modeling is displayed on Fig. 3. The modeled Moesz-line for the reference period follows the original Moesz-curve which shows a good coherence result regarding the spatial resolution of the model. However, the predicted change of the Moesz-line for the period of 2011-2040 – surprisingly – does not show a great shift to the north. Moreover, to the east of the line connecting the Slovakian cities of Rimavská Sobota and Tisovec, the Moesz-line is expected to shift slightly to the south in the period of 2011-2040, which is rather unexpected. From the east to Rožňava the original line cannot shift to the north.

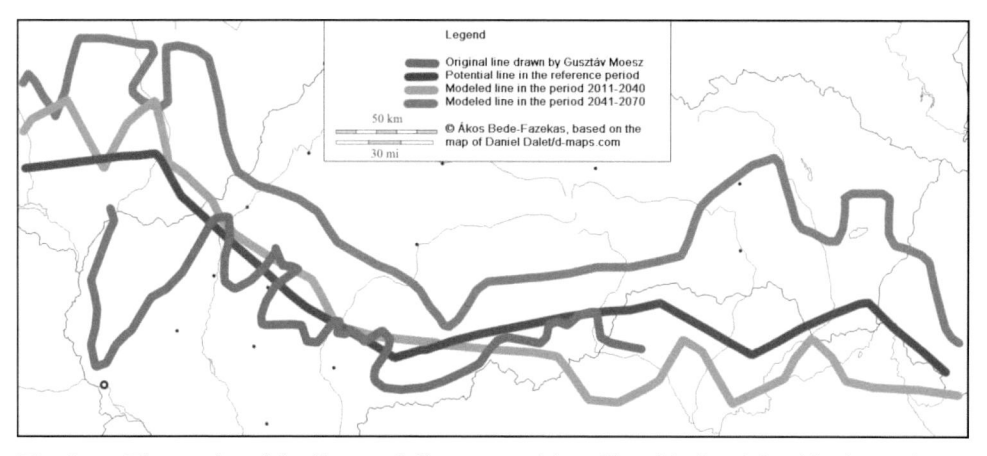

Fig. 3:   The results of the line modeling zoomed into Slovakia (explained in the text)

Nevertheless, in the long run for the period of 2041-2070, the results clearly show the expected shift of the Moesz-line to the north and the line will be transformed to 2 or 3 different sections. First of all, in the Carpathian Mountains it will move to a higher regions

(Fig. 3.) and from the north to the Carpathian Mountains some parts of Poland will belong to this climate zone. As a natural inverse effect, an anti-Moesz-line will be formed that will be bounded from the south by the Carpathians. The results coincide with the modeling of geographically analogue regions (HORVÁTH 2008a).

The southern part of the new Moesz-line (2041-2070) will connect Brno (Czech Republic), Zlín, Trenčín (Slovakia), Zvolen, Lučenec, Kosice, Homenne, Soiva (Ukraine) and Bako (Romania). The northern part of the Moesz-line for the period of 2041-2070 will connect Berlin (Germany), Poznan, Warsaw, Garwolin, Włodawa, Novohrad-Volinszkij (Ukraine) és Bila Cerkva. The anti-Moesz-line will join Dresden (Germany), Gmina Bolesławiec, (Poland), Rybnik, Częstochowa, Krakow and Lviv (Ukraine).

The 2nd method (Distribution modeling) – as it was expected – has resulted in a more detailed shift of the Moesz-line in future. Some species have been separated from the others, the habitats of some species have shifted to the north of the Carpathians as „early" as 2011-2040 (Althaea officinalis, Galega officinalis, Sideritis montana, Vitis vinifera, Fraxinus ornus, Quercus cerris). The other group of plants (Aira elegantissima, Clematis integrifolia, Cruciata pedemontana, Eryngium planum) expected to shift over the Carpathians between 2041-2070. The rest of the plants have remained only on the southern parts of the Carpathians even between 2041-2070.

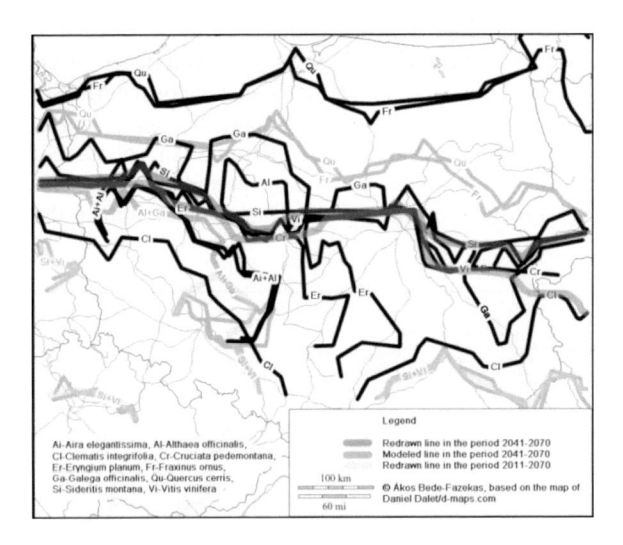

Fig. 4:
The modeled distribution of the species (gray: 2011-2040, black: 2041-2070), and the redrawn Moesz-lines (explained in the text)

The results are shown on Fig. 4. It can be stated that the 12+1 original plants that determined the Moesz-line have produced a more coherent shift of habitats, than those plants which were later added to the concept of Moesz-line. These habitat zones of latter species such as the Turkey oak (Quercus cerris) and the manna ash (Fraxinus ornus) is expected to shift to the north in the greatest extent, and only these two species will find pass the line of Carpathians continously. In addition to this, one can observe, that most of all the common grape vine (Vitis vinifera) and the mountain tea (Sideritis montana) will follow the northern line obtained by the first method between 2041-2070.

Comparing this to the result of the first method, it can be said that between 2011-2040 the Moesz-line is expected to pass over the Carpathians, although the observed plants will only form some isolated or disconnected habitat zones. On the other hand, the Line modeling method did not predict the future occurrence of the Moesz-line over the Carpathians. The Slovakian segments of modeled line are not displayed separately, since they produce almost identical results with the Line modeling method.

The Isotherm modeling (third method) has produced weaker results than it had been expected. The isotherm corresponding to the average minimum temperature in January (-3,86 °C) which mostly coincides with the Moesz-line in the reference period, in some segments it has reached over the Carpathians. The position of the curve is not parallel with the mountains, moreover it is perpendicular with the Carpathians. So we can conclude that Isotherm modeling is not very useful for predicting the future shift in vegetation habitat – no matter if we use the winter minimum or a monthly minimum temperature. Due to the above mentioned problems, we do not consider the Isotherm modeling to be a feasible way to predict habitat shift in future.

4   Conclusion

We have shown three different methods to predict the future shift of the Moesz-line due climate change. Table 3 combines the evaluation and comparison of the methods.

Name	Usefulness	Gives results according to expectations
Line modeling	Yes	No for 2011-2040; Yes for 2041-2070
Distribution modeling	Yes	Yes, for both periods
Isotherm modeling	Questionable	No. It gives a rather different result than we had expected

Table 3: The evaluation of the methods used in our research

It can be said that the shift of the line to north is relatively small between 2011-2040, while for the period of 2041-2070 it is more significant and coincides with other studies. Hence, we think that the results could be used in landscape architecture in long term planning. We can observe the expected change of climate in the next 60 years according to the A1B climate scenario.

As an overall conclusion, we can say that the Line modeling and Distribution modeling produced very similar results for 2041-2071. However, for 2011-2040 period only Distribution modeling predicted possible habitat zones for the investigated plants to the north of the Carpathians. In spite of this, the second approach does not yield to significantly more information. Nevertheless, it is rather time consuming and troublesome to model so many different species separately. Hence, the only main reason to use Distribution modeling is the tradition and scientific respect towards Moesz's works, there seems to be little practical reason. The isotherm modeling produced some doubtful predictions. Hence, we found the first method to be the most effective and reliable. The shortcomings of the methods is that it selects only a few parameters that describe the climate. The further improvement of the modeling could be based on more advanced Statistical Methods (logistic regression, K-means clustering), or Artificial Intelligence (decision trees, neural network algorithms).

Acknowledgements

Special thanks to Mária Höhn and to Levente Horváth for their assistance. The research was supported by Project TÁMOP-4.2.1/B-09/1/KMR-2010-0005. The ENSEMBLES data used in this work was funded by the EU FP6 Integrated Project ENSEMBLES (Contract number 505539) whose support is gratefully acknowledged.

References

Bartha, D. (2002), *A molyhos tölgyek (Quercus pubescens agg.) botanikai jellemzése*. In: Erdészeti Lapok, 137 (1), 7-8.
Bartha, D. (2007), *A szelídgesztenye (Castanea sativa) botanikai jellemzése*. In: Erdészeti Lapok, 142 (1), 14-16.
Bartholy, J. & Pongrácz, R. (2008), *Regionális éghajlatváltozás elemzése a Kárpátmedence térségére*. In: Harnos, Zs. & Csete L. (Eds.), *Klímaváltozás: környezet – kockázat – társadalom*. Budapest, Hungary, Szaktudás Kiadó Ház.
Bede-Fazekas, Á. (2011), *Impression of the global climate change on the ornamental plant usage in Hungary*. In: Acta Univ. Sapientiae Agriculture and Environ. 3 (1), 211-220.
Csapody, V. (1932), *Mediterrán elemek a magyar flórában*. Dissertation, Szeged, Hungary, Szegedi Tudományegyetem.
Csiky, J. (2003), *A Nógrád-Gömöri bazaltvidék flórája és vegetációja*. In: Tilia, 11 (1), 167-301
Euforgen (2009), *Distribution map of Chestnut (Castanea sativa)*. www.euforgen.org.
Horváth, L. (2008a), *A földrajzi analógia alkalmazása klímaszcenáriók vizsgálatában*. In: Harnos, Zs. & Csete L. (Eds.), *Klímaváltozás: környezet – kockázat – társadalom*. Budapest, Hungary, Szaktudás Kiadó Ház.
Horváth, L. (2008b), *Földrajzi analógia alkalmazása klímaszcenáriók elemzésében és értékelésében*. Dissertation, Corvinus University of Budapest, Budapest, Hungary.
Kárpáti, Z. (1958), *A természetes növénytakaró és a kertészeti termesztés közti összefüggés Sopron környékén*. In: Soproni Szemle, 12 (3), 30-54.
Kézdy, P. (2001), *Taxonómiai vizsgálatok a hazai molyhos tölgy alakkörön (Quercus pubescens s. l.)*. Dissertation, Nyugat-Magyarországi Egyetem, Sopron, Hungary.
Meusel, H., Jäger, E. J. & Weinert, E. (1965), *Vergleichende Chorologie der zentraleuropäischen Flora. Band I. (Text und Karten)*. Jena, Fischer-Verlag.
Moesz, G. (1911), *Adatok Bars vármegye flórájához*. In: Bot. Közlem., 10 (5-6), 171-185.
Sheppard, S. R. J. (2005), *Landscape visualisation and climate change: the potential for influencing perceptions and behaviour*. In: Environ. Science & Policy, 8(6), 637-654.
Somlyay, L. (2003), *A Muscari botryoides (L.) Mill. hazai alakkörének rendszertanichorológiai vizsgálata*. Dissertation, Debreceni Egyetem, Debrecen, Hungary.
Terpó A. (1992), *Pyrus taxa in Hungary, and their practical importance*. In: Thaiszia, 2 (1), 41-57.
Tutin, T. G., Burges, N. A., Chater, A. O., Edmondson, J. R., Heywood, V. H., Moore, D. M., Valentine, D. H., Walters, S. M., Webb, D. A., Akeroyd, J. R., Newton, M. E. & Mill, R. R. (1964), *Flora Europaea*. Cambridge, UK, Cambridge University Press.
Végvári, Gy. (2000), *Sorb apple (Sorbus domestica L.) selection in Hungary*. In: Acta Horticulturae, 538, 155-158.

Hollywood Landscapes – An Exploration of Hollywood Styled Visual Effects Techniques for Landscape Visualisations

Zane EGGINTON

1 Introduction

The process of filmmaking has been carefully developed over the last 100 years into a complex art form consisting of many areas of expertise. To understand how to make an effective landscape visualisation we need to look at how a film is put together as a whole and not just the software tools being used. A typical production can be broken down into three major phases; preproduction, production and post production and over the years as films have become more and more technical the visual effects departments have become tightly integrated into each stage. For feature film visual effects techniques to be adopted into a landscape visualisation workflow we need to understand that integration and why it's important. We will look at the major techniques along with some of the digital technologies used to create the spectacular imagery we've become familiar with.

2 Preproduction

During preproduction the entire project is planned out including; scripting, story boarding, shot lists, budgeting, locations, equipment and the organising of cast and crew. This is also when visual effect departments are briefed and it's a time for them to work out how they can achieve the desired shots and feed back to the director/producer and their team as to what can or can't be achieved.

As for a landscape architect this would be the time in which the design process takes place and could include the preparation of digital assets typically exported from CAD programs. Key aspects of the production are outlined (target audience, themes, issues, etc) and the overall intention for the finished production is identified. This may be as far as the landscape architect is involved in process, perhaps deciding to hand the project over to a production company to produce the finished product.

2.1 Script

For a typical production the script is either written specifically for the film, or is adapted from other sources, a novel for example. The script outlines the locations, characters, dialogue, action and sometimes very basic editing notes, e.g. fade to black, text added, and audio notes. For a visualisation project the script should tell a story that reinforces the underlying intent of the project. Translating your design process into a story serves another purpose; memory retention. By telling a story we reshape information into meaning making it easier for the viewer to understand and recall. The final script will act as the framework that will navigate the audience through the project showing the reasons for the design and the process that the design went through to reach a conclusion. It should identify current issues with not only the site directly but also of the surrounding area and then show how these issues have been solved.

2.2 Genre

Genre can be defined as an empirical category that serves to name, differentiate and classify works on the basis of recurring configurations of formal and thematic elements they share (MOINE 2008). Genre is a very powerful tool to describe the underlying intent of a design, for example a visualisation of a skate park could use the action film genre which involves fast cuts, lots of close-ups, rapid movement, hand held camera movement, and fast paced energetic music. Genre sets up a framework to work within, so matching the appropriate Genre to a visualisation is very important.

2.3 Characters

The characters are a vital component when communicating with an audience via an audiovisual presentation. Trying to persuade by stating the features and specifications is meaningless until you add a human to the mix (DUARTE 2010). During social learning we witness the actions of another individual and the reward or punishment they receive. In the brain of the observer a representation of the action and it's consequences is formed as if the observer had experienced the episode them self (PINEDA 2009). It's this learning by observation that we are ultimately trying to achieve in a visualisation.

In a Hollywood blockbuster the characters are typically the most important element, often more significant than the story itself. Many people will see a film purely based on the characters. In contrast many landscape visualisation typically presents characters in the form of static human figures that simply serve to give scale to the scene. But who are the characters in a landscape visualisation and how can we use them?

We could think of the landscape itself as the lead character, it's who we came to see, it's the element we hope to relate to and want to empathise with. The existing site and to a degree it's surroundings can be thought of as the actor playing the part. Perhaps the audience is already familiar with this actor, maybe they have a preconceived idea of who this actor is, similar to that of a famous Hollywood actor playing a part in a film. Are there stereotypes we need to address? How do we introduce our character? What is the journey/transformation that our lead is going through and what filmic techniques can we use to illustrate this?

Other characters we can utilise in a typical project are the users of the site, the client/council, the designer, and perhaps the builders & engineers. These characters can be used to tell a story either directly (through dialogue for example) or as a navigation device, both emotionally and physically. If we see a character walk through a gate and the next shot we see the character enter a garden we know how they got there, if they smile we know they approve and this emotion is transferred to the viewer.

2.4 The Shot

The way in which the camera is used to frame a shot is often misunderstood in visualisations. There are some basic techniques that need to be understood to produce a coherent film that communicates effectively with an audience. The most important shot is the establishing shot, this is often the first shot presented and is often a high angle wide showing where the film is located. Most shots have little or no moment allowing the viewer to study the shot. A dolly shot (where the camera moves slowly through a scene) gives an image depth as objects close to the camera move faster than those further away. Close-ups show detail and by tilting up or down we can give power to an object or make it look insignificant.

The more realistic the camera movement the more believable the project will look. In a typical film/TV programme you'll see a large number of shots being used in quick succession to describe a location, seldom will you see a camera flying around at pace as is often the case in visualisations (a fly through). This rapid movement does not allow the viewer any time to study a shot at there own pace.

The way in which the scene is presented (lighting, colour, tilt, pace, sound, etc) can influence how the viewer perceives these spaces; dark, grimy, and unpleasant or are they happy, relaxing, and tranquil? At this stage it's important to work out what shots will be required and how they are to be achieved (i.e. rendered, real and composited, still images, etc). For many it means transforming ones self from a viewer of imagery to a presenter.

2.5 Story Boards

Story boarding is a way of visualising the film in sequence, much the same way as reading a comic book. It acts as way to communicate the intention of each shot to the cast and crew. A camera operator will be looking at it to work out their framing, makeup will want two know who's in the background and who will be in a close up. The visual effects supervisor will want to know when they need to be on set and so on. If the production isn't making sense or telling the required story at this stage changes can be made without the need to reshoot, or in the case of a visualisation, re-rendered which can be very time consuming and often expensive. It will also dictate where detail will be required in digital assets. By concentrating on producing shots outlined in the storyboard time can be saved by avoiding modelling and rendering items that will not be 'in shot'. For example low polygon models could be used for wide-angle shots and then separate highly detailed models of a specific object created for close-ups.

2.6 Previsualisation

A popular technique in modern productions is the development of a 'pre-viz' animation. This is a rough animated version of the storyboard and helps the director and crew make decisions by presenting an animated version of the shot they are trying to achieve in production and postproduction. Landscape visualization projects could easily adopt this practice as most CAD and 3D modelling applications allow the user to produce rendered animations of varying quality. Low quality renders (e.g.OpenGL renders) can be quickly produced and edited with stills, sketches and other media to give a rough version of the final product.

3   Production

Armed with a script and storyboard we enter the production phase of a project. It's during this phase that all of the material needed for the film is captured. This includes footage on location and/or in the studio, location audio, dialogue, and perhaps still photography. It can be argued on whether or not rendered material is considered part of production or if its postproduction, it really depends on the project. Generally speaking if the rendered content is intended to augment a project/shot then it would be classified as post production (e.g. compositing a large mountain into a background), however if its an animated feature (e.g. Pixar's Toy story) then the rendered components would be thought of as being included in the production phase. Due to the significance of the rendered components in a landscape project it should be considered as part of the production phase.

3.1   Cinematography

During production one of the key roles is that of the Director of Photography (or DoP). The DoP is primarily the head of the camera and lighting departments and as a whole is responsible for Cinematography. It's the DoP's job to get the 'look and feel' of the film as dictated by the Director. Even though many visualisation jobs do not use 'real' footage, or have the budget to artistically light a location it's rules and conventions are perhaps the biggest key to producing a successful project. Lighting, colour, shot composition and camera movement are key for any visualisation (also see section 2.4, 'The Shot'). Therefore the person creating the rendered shots for a visualisation could be considered as a virtual DoP and for those that are interested in producing high quality renders, a study of cinematography is vital.

3.2   Audio

Emotion, story, atmosphere and materiality can all be communicated audibly to an audience although its use in visualisations is often limited to a single music track or even completely non-existent. There are two main categories of audio, diegetic and non-diegetic. Diegetic sound is related to what we see on the screen, footsteps, dialogue, atmosphere (i.e. wind, rain, etc), non-diegetic sound is added audio such as music (unless it's in the shot, e.g. a busker), voice overs and sound effects (e.g. the notorious stabbing audio in Alfred Hitchcock's *Psycho*).

3.3   HDRI Domes

High dynamic range images (HDRI also known as HDR images) are images that contain a larger than normal exposure range. Typically these are created by taking multiple images at different exposures settings then saved in a specialised file format (e.g. .hdr, .exr, ima). This information can be used to create an image that has extra detail in both the shadowed areas as well as the highlights. For example a typical photo of the sky would have the same pixel value for all white objects including the clouds and the sun even though the sun is much brighter, in an HDR image the sun would have a higher pixel value than the clouds. These images can be used to adjust the image in software at a later stage or used to create a new image with reduced contrast using various methods.

An HDRI Dome is a 180/360 degree HDR image created with specialised hardware or generated from multiple HDR images. In the film industry these images are most commonly used to recreate the lighting of a particular location within rendering engines and compositing software. This can be a powerful tool in landscape as we can produce 3D renders of a design using the lighting from that location (fig. 1). This would have huge advantages in areas that have very complex lighting, urban spaces for example often have tall buildings with different surfaces. These surfaces can have very different properties, some scatter the light, others absorb, some surface may be mirrored and reflect light. At night the light sources can be incredibly difficult to reproduce in a digital model with perhaps hundreds of different light sources illuminating a scene. However it's important to keep in mind that all though this technique gives a realistic lighting model it is not necessarily an accurate model and should not be used to calculate shadow lines.

Fig. 1: 3D model rendered into a HDR scene

3.4 Keying

Is the process of algorithmically extracting an object from its background and combining it with a different background (OKUN et al. 2010). This process is easily achieved by shooting a subject in front of a green/blue background then importing the footage into one of many software applications. Most editing programmes can achieve this however more control can be achieved in specialised compositing applications (section 4.5). Attention to detail is very important at this stage of production. Lighting conditions, framing, props, etc must all be carefully dealt with so that it matches with the background and foreground elements added during postproduction. This process is used extensively in Hollywood blockbusters and its use in landscape visualisations brings life and scale into a scene.

4 Post Production

All of the images captured during production are manipulated in some form and edited into a finished product. Most of the visual effects in modern feature films are created at this stage.

4.1 Workflow

The workflow in an effects heavy film can be very complicated, utilising a large number of specialized software packages and sources of material. In Fig. 2 a simplified diagram represents the major data paths from acquisition to final output. Each node represents a set of digital tools and nodes that have been grouped together represent the possibility of a single tool performing the combined task. For example many compositing packages include tracking tools or have 3rd party plugin options available however in most cases motion capture is performed with a specialised application (e.g., *SynthEyes* or *PF Track*).

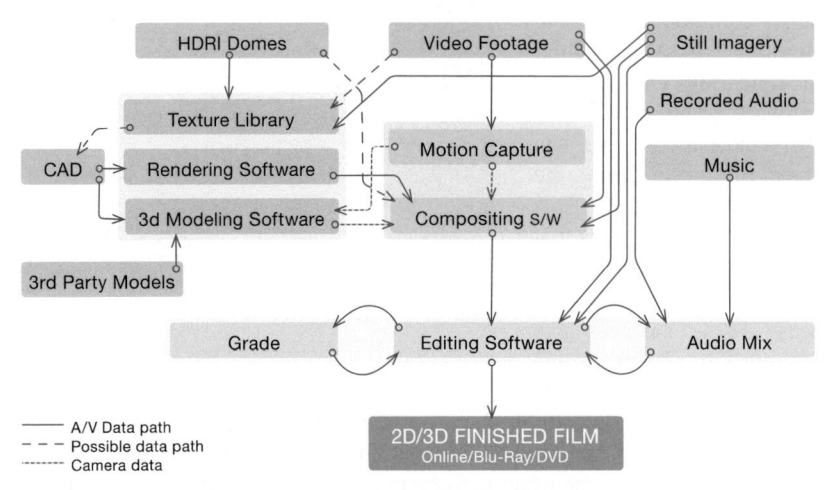

Fig. 2: Simplified digital workflow for a VFX production

4.2 Matte Painting

Matte painting has it's roots in theatrical productions where painted scenes where used as backdrops. In the 1930's Cinematographers modified this technique by using back projected footage of scenes shot on location with the actors performing in a studio environment (THOMPSON & BORDWELL 2003). This technique was most commonly used for shots in moving cars. Later on this technique would be performed in postproduction using optical printers and is now achieved with specialised compositing software (see section 4.5).

Modern matte paintings utilise many different methods to create them. Some of these are still derived from paintings and drawings that are typically modified with digital tools (Photoshop for example). However they can also be created from physical models or more commonly 3D rendering software such as e-on software's Vue (Fig. 3).

4.3 3D Modelling and Rendering

Many of the software packages used in the film industry aren't completely unfamiliar to landscape architects however the way in which they are used can be. In landscape and architectural visualisations there is a tendency to use a single software application to model and render a design, perhaps using Photoshop to modify it in some way. Some may go as far as using a 3rd party plugin renderer or exporting a CAD model into a separate application for further modelling, animation, texturing and rendering (e.g. 3DS Max, Maya, Lightwave, etc). In the VFX industry 3rd party renders, either as a standalone product or as a plugin are very common, in fact many VFX houses modify or even create their own rendering engines (Pixar's Renderman for example). From the renderer separate 'passes' are generated and then composited in a separate application. These 'passes' are images (or sometimes separate layers) that when combined create a finished image. These passes typically include; colour (diffuse), shadows, reflections, refractions, specular highlights, etc and often other useful information such as depth maps, object masks, XYZ normals, etc. The major advantage to this technique is the ability to edit each pass in post without the need to re-render, for example to reduce the sharpness of a shadow. It also aids in the compositing process when including material from other sources. We could use an object mask, a tree for example, to clip a shot of someone walking through a scene to make it appear as if the have gone behind it, something that would be very difficult to do with a 'flattened' render.

For landscape architects one application stands out for rendering highly detailed digital environments; e-on software's Vue (as seen in Fig. 3). Models can be easily imported from CAD software and textures can be swapped out for materials that utilise fractal algorithms that are resolution independent (as opposed to image based texture that are made of pixels). In fact Vue utilises fractal math to generate terrains and plants, which allows for an almost limitless amount of detail.

Fig. 3: Digitally created scene rendered in e-on software's Vue xStream.

4.4 Motion Capture

One of the biggest challenges in creating a VFX shot is dealing with the camera movement. Moving the camera however is a powerful cinematic tool. It can show where objects are relative to each other by panning (see section 2.4 'The Shot') or we can give our shot depth by using a dolly shot which moves elements in the foreground differently on screen than those in the background. This gives the viewer a much better understanding of the scene than a still shot. For example let's say we need a shot similar to that in Fig. 3 but with a handheld styled camera movement. The easiest solution is to use a 'locked off' shot, meaning that the camera does not move, composite our sculpture over the plate (background image) and then shake the entire frame. However the problem with this is the lack of parallax. If we move to the left we expect to see more of the left side of the sculpture. The result would look more like a faulty television than actual camera movement. We could film the background on location with a handheld camera but that would require us to move our rendered sculpture in exactly the same way by moving the virtual camera in the rendering software. This is where 3D motion tracking software enters our workflow (Fig. 2).

Motion tracking software primarily performs two functions, tracking the movement of the object(s) and/or determining the movement of the camera. The art of motion tracking could easily fill an entire book however let's look at our above example. What we need to do is work out how our camera is moving in a 3D space then send this data to our rendering program and use it to create a virtual camera that mimics the real one. Once we have the rendered footage we can composite the rendered footage over our background footage and if all goes well the two should look seamless. Of course it is a little more complicated than that to get a real looking shot however this is the basic workflow. Modern motion tracking software goes one step further though. Not only can they determine the movement in a shot they can recreate the depth in a shot and even build a textured 3D model from the footage.

4.5 Compositing

In 1857 Swedish-born photographer Oscar G. Rejlander combined the imagery from 32 different glass negatives to produce a single, massive print titled *The Two Ways of Life* (BRINKMANN 2008). This image represents one of the earliest forms of compositing. Compositing in the film industry consists of the combining of imagery acquired during the production phase along with other imagery and data (e.g. digital renders, matte paintings, motion capture data, etc). Compositing in landscape architecture is typically referred to as 'photoshopping' an image. However one of the major problems with Photoshop is its linear and inflexible workflow, an image is opened and a step by step work flow leads your to a final product. In the film industry this task is handled in a very different manor due to the complexity of working with moving images, large teams, many sources of image production and their many iterations. The workflow needs to be flexible and reusable and in most productions a 'Pipeline' is designed to handle the flow of data throughout the postproduction phase. Most of the leading compositing software used for visual effects shots utilise a 'Node' based approach, which resembles a flow diagram where each step of the processes (including the inputs) can be modified (Fig. 4). The workflow can be used for multiple frames of an animation or other similar shots therefore providing a very efficient and flexible workflow.

For a landscape visualisation this has some very big benefits. A well setup pipeline can accommodate changes to the original design and the entire postproduction workflow can potentially regenerate high quality finished shots with little effort. For example if an image has been created from a CAD model and the client asks for changes to the design, the alterations can be within the CAD software, the existing 3D model is replaced and fed into the rendering software which generates new rendered passes that update images being utilised in the compositing software which can produce new finished shots.

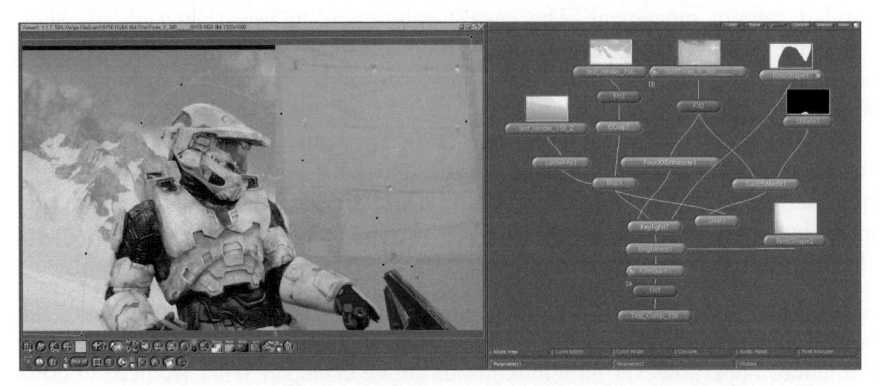

Fig. 4: Example of a node based compositing software, Apple Shake

4.6 Editing

Editing is more than just a process of combining shots into a film. The pace, angles, and montage of shots communicates with the audience in a language that most understand very well however the process can in fact be very complicated. There are four major categories of editing: (1) chronological editing (2) cross-cutting or parallel editing (3) deep focus and (4) montage (HAYWARD 2000). The simplest and most commonly used of these is chronological editing where we focus on the order of events one after another. This is often seen in Hollywood features however we may occasionally see the crosscutting or parallel technique employed as a flashback or shift in time/space.

Deep focus editing tends to have much longer shot lengths with a lower number of cuts and is less common in feature films. It's this style of editing that could be used to represent a more realistic view of a landscape. Perhaps we want to show the contemplative qualities of a particular site, to achieve this we could show very long shots of features within the site. This allows the viewer to observe the scene in a more realistic manor, with little manipulation by the filmmaker. Montage on the other hand is perhaps one of the most manipulative editing styles. It's not so much about the content of each shot but the interactions between them. If we see an environmentalist smiling and then a shot of a courtyard we assume our design is environmentally friendly without actually saying so. Montage works like language by describing a scene with images rather than words.

It's these editing styles that give the editor the ability to warp time and space and put a spin on almost any shot. Most productions will use a combination of these styles when appropriate and how these styles are used make the art of editing a powerful tool in storytelling.

4.7 Colour Grade

Colour grading is where the edited project is given a particular 'look and feel'. This could be giving the shot a warm tone to invoke a sense of romance or perhaps a cold de-saturated contrasting look to make something seem undesirable or menacing. It's also where each shot is manipulated to match closely with the surrounding shots. This stage of the production is highly manipulative on an emotional level. A well graded production not only looks better but it has the ability to influence the emotion of the viewer. If you have a redesign for an existing site you might want to suggest that the current site is dull, boring, perhaps almost toxic looking, and that the redesign is warm, inviting and soft. To achieve this the opening would be heavily de-saturated with lots of grain and contrast, then the redesigned site would have well balanced colour with a soft focus, lightened shadows and a overall subtle orange/red hue.

4.8 3D

One trend that has occurred in recent feature films is the use of 3D cinematography (also known as stereoscopy). 3D films are nothing new, the first 3D movies where created in the early 50's (MENDIBURU 2009) but the technology used to create and deliver these films has advanced significantly in recent times hence it's gain in popularity. The use of 3D technology enhances the user experience by offering a far more immersive product. The extra dimension also invokes a 'Wow' factor, which can put a positive spin on almost any project. The creation of 3D imagery is entirely achievable for landscape visualisations, as the technology simply requires a separate left and right eye image. It can complicate the compositing process, but the tools to achieve this task do exist and is currently an area of rapid development. Most 3D modelling applications have built in tools to help produce these images however it can also be achieved by rendering from two cameras positioned slightly apart (known as the inter-ocular separation). One aspect that needs to be considered when choosing 3D over 2D is the introduced complexity to deliver the finished product. Unfortunately (at time of writing) the software required to encode 3D Blu-ray disks is very expensive however it is possible to deliver 3D content via Youtube or digital files. This technology is very dependent on the viewer's own hardware and experience so presently it's most suited for an organised presentation.

5 Conclusion

Rapid advances in the visual effects seen in modern feature films have lead to the creation of worlds and vistas that have become so realistic it may seem impossible for a typical landscape visualisation to replicate. However the techniques and the technology required to achieve these shots are available to the landscape profession, and have been for some time. What is required to increase the effectiveness of these visualisations is an understanding not only of the technical aspects but also the subtleties of filmmaking. It is these persuasive mechanisms that act as a tool to independently manipulate a viewers perception in ways that no other medium is capable of. In fact its here that we may need to consider the shift from portrayal to propaganda. Is what we are presenting to an audience a fair representation of the project or is it manipulating perception to a point that distracts from reality.

References

Brinkmann, R (2008), *The Art and Science of Digital Compositing: Techniques for Visual Effects, Animation and Motion Graphics.* Second Edition, Elsevier, Inc.
Duarte, N (2010), *Resonate: Present Visual Stories that Transform Audiences.* Wiley.
Hayward, S. (2000), *Cinema Studies, the key concepts.* Routledge.
Miller, F. & Rodriguez, R. (2005), *Sin City: The Making of the Movie.* Troublemaker Publishing.
Mendiburu, B (2009), *3D Movie making.* Elsevier.
Moine, R. (2008), *Cinema Genre.* Blackwell Publishing.
Okun, J. & Zwerman, S. (2010), *Visual Effects Society Handbook*, 865, Focal Press.
Pineda, J. (2009), *Mirror Neuron Systems.* Human Press.
Thomson, K. & Bordwell, D. (2003), *Film History – An Introduction.* McGraw-Hill.

Modeling Nightscapes of Designed Spaces – Case Studies of the University of Arizona and Virginia Tech Campuses

Mintai KIM

Abstract

This paper examines two methods for modeling the interaction between designed spaces and nighttime light pollution. Nightscapes are becoming more important because of their potential effects on energy use and sustainability, public health, ecology, astronomy, safety and security, and placemaking.

Despite its growing importance, however, consideration of the nightscape is often an afterthought in the design process. Except for engineering drawings of lighting systems and occasional atmospheric renderings of nighttime scenes, nightscapes are not depicted. Considerations for nighttime users and for energy savings through sustainable nightscape design practices are lacking. Although several factors may contribute to this omission, including a failure to address nightscape design in landscape architecture education, the lack of appropriate tools for studying the effects of various design alternatives on light pollution is a critical issue.

This study compares two methods of modeling and depicting light pollution. The case study sites are the campuses of the University of Arizona and Virginia Tech both located in the United States. The author has been collecting lighting data for both campuses for the past few years. At the University of Arizona campus in Tucson, Arizona, high-resolution nighttime aerial photographs were taken using a helicopter. By compositing these images, the nightscape of the entire campus was represented. In contrast, the nightscape of the Virginia Tech campus in Blacksburg, Virginia was modeled using a completely different method. Lighting data from approximately 2,000 sampling points on a 10-meter grid were collected on the ground using light meters. Using the ArcGIS spatial analyst tool, a comprehensive picture of the campus nightscape was acquired.

The nightscape modeling method used on the Virginia Tech campus proved to be quicker, simpler, and less expensive compared to the method used for the University of Arizona campus. Undergraduate students taking a land analysis course collected and analyzed the data and were able to model the nightscape inexpensively. Another advantage of the method used at Virginia Tech is the ability to collect data that is difficult to gather with aerial images, such as exact locations of light fixtures, surface materials used on the ground or on the buildings, proximity to buildings, existence of different types of vegetation, and overhanging trees.

1 Introduction

Although a significant portion of human activity takes place at night, little attention has been paid to the design of nighttime spaces by landscape architects, except for lighting design research. Much of this research has focused on (1) the aesthetics of lighting design or (2) the efficacy of bright lighting for crime reduction. More recently, sustainability concerns have brought attention to light pollution issues. This, in turn, has shifted the emphasis in lighting design from advocacy for brighter lights to the reduction of light pollution and energy costs. For example, in the United States, LEED (Leadership in Energy and Environmental Design) sustainable design guidelines identify light pollution reduction as an important factor for sustainability. In addition, light pollution has been shown to affect human and ecosystem health (SHAFLIK 1997, BORG 1996, BELL 1999).

Currently, one of the most common design responses to the problem of light pollution is the use of full cut-off lights in outdoor spaces (fixtures that do not emit light above the horizontal plane). Although these lights can successfully reduce light pollution, they are only one part of a comprehensive strategy of light pollution mitigation that addresses not only lighting fixture design but also a broader set of factors in the designed environment. For example, the effects of surface material reflectivity on light pollution are rarely addressed in the research literature, despite the fact that even with the use of cut-off lights, light that strikes reflective surface materials in the designed environment will still be reflected upward.

One reason for this rather one-dimensional approach to the reduction of light pollution may be the lack of methods for measuring light pollution that also provide sufficient detail about the built environment. This has hampered efforts to provide an understanding of the relationship between design elements and light pollution. More specifically, practical methods for measuring light pollution on the ground lack the detail necessary to see the relationship between the design of a space and the light pollution it produces. Over the last decade, many have studied artificial night lights using satellite imagery (ELVIDGE et al. 1997, CINZANO et al. 2000, CINZANO et al. 2001, CHALKIAS et al. 2006). However, the resolution of the images used in these studies was too coarse to correlate light pollution with space design.

As possible alternatives, this study examines two additional methods for collecting light pollution and space design data: high-resolution nighttime aerial imagery acquired using a low-flying aircraft and a ground survey method. The high-resolution imagery developed with these methods will allow an understanding of the relationship between materials used in designed spaces (and ultimately, the overall design of spaces) and light pollution.

2 Methods

This study compares two methods for modeling and representing light pollution. Both methods were developed by the author (an aerial photography engineer was consulted on Method 1) after finding that high-resolution light pollution imagery did not exist.

The case study sites selected are two U.S. university campuses: the University of Arizona in Tucson, Arizona, and Virginia Tech in Blacksburg, Virginia. The author has been collecting lighting data for both campuses for several years.

The nightscape of the University of Arizona was modeled using aerial photography (Method 1). High-resolution nighttime aerial images were acquired using a helicopter, gyro-mounted camera equipment, and aerial imagery expert services. By compositing the nighttime images, the nightscape of the entire campus was represented.

As an alternative, the nightscape of the Virginia Tech campus was modeled using lighting data collected on the ground (Method 2). Lighting levels were collected using light meters at almost 2,000 sampling points on a 10-meter grid. Using the ArcGIS spatial analyst tool, a comprehensive picture of the nightscape was then acquired.

2.1 Method 1

The study site for Method 1 is the University of Arizona campus. This campus is fairly flat, with an average slope of about 1%. The elevation ranges from approximately 2,415 feet to 2,465 feet above sea level. The majority of light fixtures on the campus are cut-off lights.

To capture high-resolution, high-accuracy nighttime images, three flights were conducted by an Aerial Archives' photographer (June 1, 6, and 7, 2005). The first was a test flight over the Stanford University campus in Palo Alto, California on the evening of June 1. This flight was designed to identify the kind of nighttime photography most useful for capturing high-resolution nighttime images. The second and third flights were flown over the University of Arizona campus on the evenings of June 6 and 7. Visibility was excellent on both evenings, but wind direction and velocity were far more suitable on June 7 for creating the kind of imagery most useful for the study.

Imagery from the second and third flights was created using a Pentax 67II film camera mounted on a Kenlab gyroscopic stabilizer and equipped with a Pentax 105mm f/2.4 lens. The images were captured at 5,000 feet above ground. The sky was clear, and no rain had fallen for several days. Sunset was at 7:28 p.m. MST on June 7, 2005, and Astronomical Twilight (when the sun is 18 degrees under the horizon) was at 9:08 p.m. MST. The moon also set at 8:42 p.m. MST. Thus, dark conditions were maintained between 9:00 p.m. and midnight. In addition, the lighting at all campus athletic facilities was turned off in order to get the optimum result. Finally, although the data for Method 1 were collected during summer, the trees in Arizona are generally short and their leaves are small enough not to affect the readings.

2.2 Method 2

Method 2 used a ground-level survey conducted with a light meter. Lighting levels were measured along a 10-m sampling grid on the Virginia Tech campus. The total number of sampling points was 1,955 (Fig. 1).

Fig. 1: Study Area on the Virginia Tech Campus. Sampling points are at 10-meter intervals

The study area, which is a small part of the campus, is about 20 hectares (195,588 m^2). Elevation ranges from 2,032 ft (619 m) to 2,084 ft (635 m). The site slope is not as flat as the University of Arizona campus but is generally under 10% slope (93% of the site). Unlike the University of Arizona, non-cut-off lights are still used.

Extech EasyView 30 Light Meters were used for the data collection (Fig. 2). Data were collected on moonless nights after astronomical twilight, when the sun was at least 18 degrees below the horizon, in order to remove the influences of moonshine and sunshine.

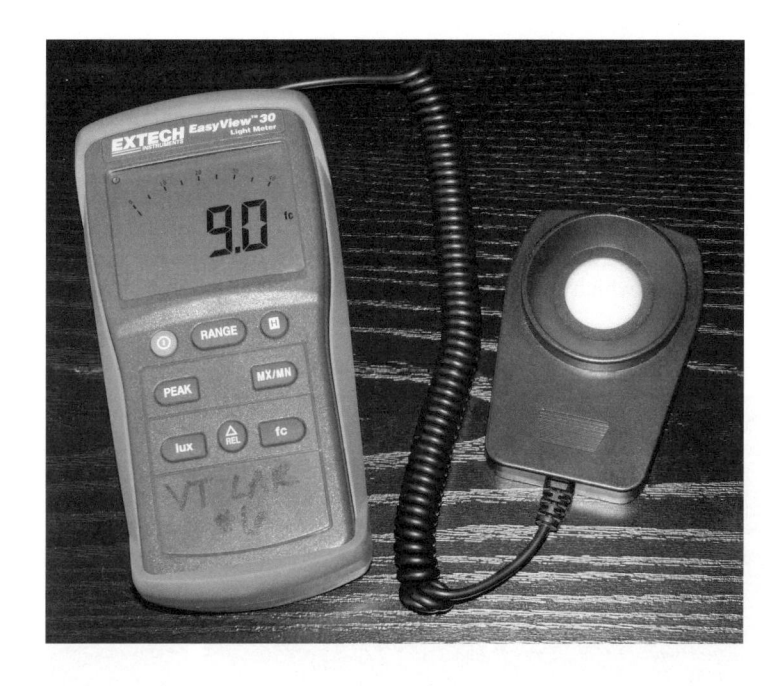

Fig. 2:
Extech EasyView 30 Light Meter

This resulted in a very small window of opportunity each month. In addition, the sampling points were collected only during cloudless nights. Two readings were collected for each sampling point. One was collected with the light meter pointing upward to catch direct light from the light fixtures at 30 cm from the ground. The other reading was taken with the light meter also held at 30 cm from the ground, but this time pointing down to measure reflected light. This second measurement is comparable to the data collected from the aerial method.

The same two readings were also collected for the same points during the daylight in order to calculate surface material reflectivity. Reflectivity was calculated by the ratio of downward light reading over upward light reading at 30 cm above the ground.

Other data collected for each sampling point were surface material and sky cover (%) by trees and structures. To control the influence of the tree canopy, data collection was done before or after tree growth seasons. These data were recorded on printed data sheets by observation during the daylight while reading the reflectivity of the materials. Because of the minimal number of lights located on building rooftops and access issues, data on rooftop lighting were not collected.

3 Results

3.1 Method 1

A 25-cm pixel resolution nighttime image was generated from the Method 1 data. Its resolution was high enough to study the relationship between light pollution and space design, unlike the other imagery currently used to study light pollution. For example, the DMSP (Defense Meteorological Satellite Program) OLS (Operational Linescan System) produces a global nightscape coverage at only 2.7-km resolution. The high-resolution data collected through Method 1 enable further study, such as comparing the image with land cover types, building types, and light fixture types captured from daytime aerial photographs and site visits.

Methods 1 revealed that 8.8% of the total study area is influenced by lights (Fig. 3). The remainder of the study area did not register any signature. Of the total lighted area, parking-related areas represented about 25% and pathways 18%. Materials for both the parking areas and pathways were concrete, which has a relatively high albedo. The brightest patches in Fig. 3 are parking structures.

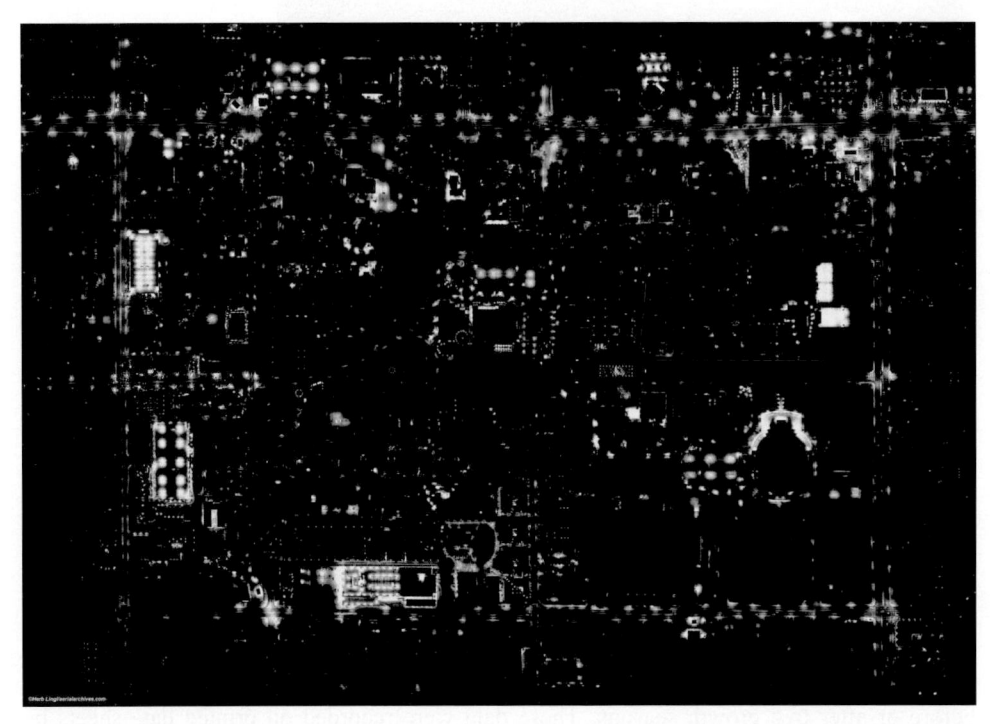

Fig. 3: University of Arizona Campus Captured with Method 1. Brightest areas are parking-related areas with bright concrete surfaces. Most of the campus is served by cut-off lights.

3.2 Method 2

Method 2 resulted in a ground-based light pollution map. Through a spatial analyst tool in ArcGIS, the sampling point data was converted to raster data in order to show the light pollution distribution. Downward- and upward-lighting maps were created (Fig. 4). Although these data, based on a 10-m grid, are spatially coarser than the data collected through Method 1, they provide better spectral information. For example, it was possible to record the light readings upward and downward during the day and at night.

As in the image generated from Method 1, the brightest areas in Fig. 4 are parking-related. The difference is that some of the asphalt-covered parking is brightly lit here whereas the concrete-covered parking is brightly lit on the University of Arizona campus.

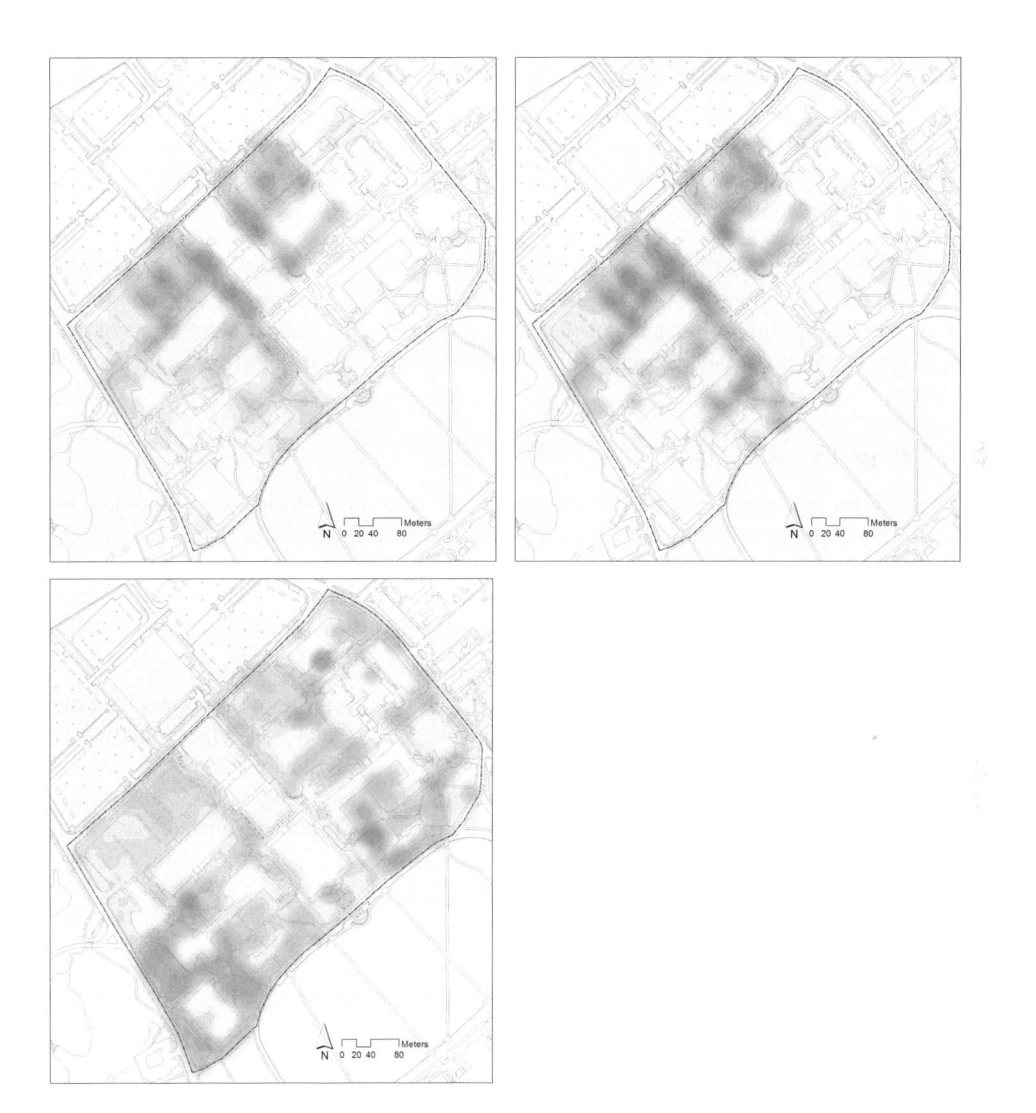

Fig. 4: Virginia Tech Campus Captured with Method 2. Top left image shows downward dings and top right image shows upward readings. Bottom image shows reflectance of ground materials.

4 Discussion

Both methods 1 and 2 provide detailed understandings of the campus nightscapes. Having said this, however, each method has its advantages and disadvantages.

4.1 Ease of Acquiring Data

It is easier and possibly faster to acquire an aerial image of light pollution than to collect ground data. However, post-processing of the aerial image to correct for errors took a long time. In addition, because the work had to be done by professional consultants, the researcher had less control over the data processing.

In contrast, it took many hours to collect light readings for the sample points in Method 2. However, once this process was completed, it was less time consuming to compile and construct the light pollution images. Furthermore, the researcher had complete control over data collection and processing. Finally, the cost for Method 2 was significantly lower than for Method 1.

4.2 Completeness of Data

In terms of the completeness of the data collected and its usefulness in studying the relationship of light pollution to the built environment, each method has its advantages and drawbacks. Method 1 provided a dataset that covered the entire campus, but it could provide only a relative brightness map. In other words, the map reveals only the relative brightness of a pixel compared to the other pixels in the image, rather than absolute brightness values for each pixel.

Method 2 provided an accurate reading (in foot candles) of sampling points, but the coverage was limited to the sample points. Brightness over all other areas had to be interpolated. Thus it was difficult to compare land cover types against light coverage or against lighting fixture types. Although this paper does not report the additional data, the same types of readings (upward and downward lights at night and during the day) were collected at the light fixtures. When combined with the sampling point data, this will provide a more complete sampling of the study site.

In terms of providing the most complete data, a combination of methods 1 and 2 would be ideal, provided that the data could be collected at the same time. It would be unrealistic, however, to expect to collect all 2,000 sample points in one night. If the methods are combined, the number of sample points has to be greatly reduced to be used as ground-truth data.

4.3 Canopy Cover

Unlike Method 2's ground readings, the aerial photography used in Method 1 cannot read light under trees and structures. Because light is reflected off of tree canopies and overhead structures, these factors need to be considered.

5   Conclusion

To date, no efficient models for representing nighttime scenes exist. This paper presents two attempts to model nighttime light pollution and explores the advantages and disadvantages of each. Both methods provide high-resolution images of nighttime lightscapes that were previously unavailable. Previous nighttime images were of very low resolution and, thus, not suitable for study of relationship between light pollution and designed spaces. The methods developed in this study to model nightscapes are not expensive, although Method 1 costs more than Method 2. These methods can be used to model the effectiveness of various light pollution reduction measures in sustainable design.

With the high-resolution images developed in this study, the author was able to make a preliminary analysis on some aspects of the relationship between light pollution and the built environment. For example, the models indicated that light pollution reduction cannot be efficiently achieved through the simple adoption of cut-off lights. Rather, the materials used on the ground and walls affect light pollution because of their reflectivity. American sustainability criteria (LEED) do not cover the ground conditions of lights. The criteria do not provide enough guidance to reduce light pollution. Lights need to be adjusted for different materials used on surfaces.

Further study is needed on the reflectivity of various materials. Although low-reflective materials are preferred for light pollution reduction efforts, the thermal characteristics of materials must be considered. Darker materials that do not reflect as much light tend to heat up and may contribute to the urban heat island effect. The next phase of this study is to examine these thermal and reflective characteristics of materials.

References

Bell, S. (1999), *Tranquility mapping as an aid to forest planning.* UK Forestry Commission, Information Note.
Borg, V. (1996), *Death of night.* In: Geographical Magazine, 68, 56.
Chalkias, C., Petrakis, M., Psiloglou, B. & Lianou, M. (2006), *Modeling of light pollution in suburban areas using remotely sensed imagery and GIS.* In: Journal of Environmental Management, 79, 57-63.
Cinzano, P., Falchi, F., Elvidge, C. D. & Baugh, K. E. (2000), *The artificial night sky brightness mapped from DMSP operational linescan system measurement.* In: Monthly Notices of the Royal Astronomical Society, 318, 641-657.
Cinzano, P., Falchi, F. & Elvidge, C. D. (2001), *The first world atlas of artificial night sky brightness.* In: Monthly Notices of the Royal Astronomical Society, 328, 689-707.
Elvidge, C. D., Baugh, K. E., Kihn, E. A., Kroehl, H. W., Davis, E. R. & Davis, C. (1997), *Relation between satellite observed visible – near infrared emissions, population, and energy consumption.* In: International Journal of Remote Sensing, 18 (6), 1373-1379.
Shaflik, C. (1997), *Environmental effects of roadway lighting.* Technical paper prepared at University of British Columbia, Dept. of Civil Engineering, pp. 9.

Discrimination of Distance-dependent Zones in Landscape Views

Agnieszka OZIMEK, Piotr ŁABĘDŹ and Kornelia MICHOŃ

1 Introduction

A great variety of techniques, including written descriptions, plans, perspective views and panoramas, orthogonal projections or axonometric views and, at least, 3D models are used in order to represent the landscape. Apart from the latter (which have become more popular recently, due to the development of IT), scenic photographs are the most comprehensive for the general public, because they refer to human perception, imitating visual sensations of the observer (KAPLAN & KAPLAN 1989).

GeoDesign may be interpreted as the planning environment that puts the design into the context of its geographic location (FLAXMAN 2010, ERVIN 2011). It equips representatives of different disciplines with adequate tools to merge together their knowledge into a common collection (DANGERMOND 2009). Taking this fact into account, landscape images can be widely used in GeoDesign, as an informative layer, expanding the database related to the specific place.

In landscape analyses often some spatial metrics of image components are used to determine parameters of the analysed scene (SHAFER & MIETZ 1970). These numerical characteristics proved very useful in several tasks, like the studies of landscape preferences, specification of its visual resources, acceptation of landscape transformations, changes monitoring or prediction of future development scenarios (SMARDON et al. 1986, PALMER 1997, 2004). In the research mentioned above serious problems occur due to geometrical distortion, inseparably connected with the perspective projection. Therefore, a comparison between characteristics of objects can give biased results, if their depth in the scene differs. On the other hand, elements of a given view (buildings, trees, rocks) are comparable, regarding their dimensions, shapes or patterns, when they are positioned in the same distance-dependent zone.

The aim of this paper is to present some image processing techniques, appropriate to picture segmentation, for the purpose of automatic or semi-automatic detection of distance-dependent sectors (Fig. 1). In this process a photograph should be split into several sub-images, including adequate areas (e.g. a distant residential area or a forest). The surface of the

Fig. 1: A landscape photograph with the re-coloured distance-dependent zones

selected fragment can be used as a binary mask (MALINA & SMIATACZ 2005), in order to conduct further analyses affecting only this part of the picture (for the exemplary residential zone – to calculate the typical size or shape of buildings).

2 Task analysis

2.1 Distant-dependent zones

Shafer and Mietz distinguished eight zones that can be grouped with regard to distance: immediate (the foreground), intermediate and distant ones, as well as the background and the sky (SHAFER & MIETZ 1970). The taxonomy proposed focused on the analyses of the most beautiful wild landscapes. In everyday practice, particularly, in our European circumstances, landscape architects usually deal with semi-cultural or cultural landscapes (i.e. countryside or urban areas) (SMARDON et al. 1986). Therefore, this systematisation should be supplemented with zones embracing invested terrains, divided according to various types of developments (residential, commercial, industrial), their scale and density. In this case, distance-dependent fragmentation should be maintained, as well (BUHMANN et al. 2011).

From these image parts, two should be excluded from calculations in the initial phase of image analysis: the sky and the foreground. The sky is not objective of this research; moreover, its percentage in the image may depend on the position of the camera and on photographs framing (RIBE et al. 2002). The background includes objects that play no significant role at the landscape scale; however, if this part includes negative elements, the view is usually estimated low (BUHMANN et al. 2011). In addition, the foreground contains a huge number of details and much variation; it is therefore difficult to distinguish among the elements in support of scenic research.

2.2 Distant-dependent features of scenic photographs

We can point out distance-dependent features that are specific for landscape photographs (Fig. 2), the most important of which include:
- the blue haze,
- colour intensity,
- colour diversity,
- contrast between the adjacent pixels.

While the first of these variables increases proportionally with distance, the rest of them – decrease. The characteristics mentioned above will be exploited in a first attempt to conduct automatic or semi-automatic image segmentation.

Fig. 2: Distance-dependence features specific for the landscape photographs (haze, saturation, colour diversity, contrast)

3 Methods

From a whole range of scenic photographs, one was chosen for this illustration, which was taken in a cloudy day. The similarity in colour between distant mountains and the sky makes the distinction of these zones problematic (Fig. 2). In the research all the distance-dependent features have been employed.

Fig. 3: A blue channel of the image

The photograph usually registered in the RGB mode, can be easily split into colour channels (red, green, blue) in almost all graphical environments. In case of scenic views, image segmentation based on the blue channel is often effective, because of the light dispersion (a haze), which results in the increase of the values of blue component with distance (Fig. 3). An image histogram (the diagram showing the number of pixels at every level of intensity) may be analysed regarding local minima (Fig. 4). They usually indicate thresholds for image binarization (conversion into black-and-white mode) (GONZALES & WOODS 2002, JAYARAMAN et al. 2009).

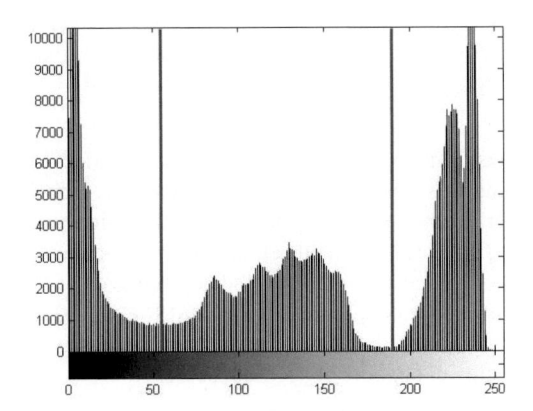

Fig. 4: A histogram of the blue channel with local minima (level 54 and 197)

An image representing colour intensity (Fig. 5) was obtained as a result of image conversion from RGB to HSV (hue, saturation, value) mode and the second channel separation (GONZALES & WOODS 2002, JAYARAMAN et al. 2009).

Colour diversity can be estimated by different indicators, like data variance, standard deviation or z-score method. Another approach consists in the calculation of the difference between pixels' values and the arithmetic mean or the median computed for the whole image. In the research the three variants were examined:
- the absolute difference between every pixel value and the image mean,
- the absolute difference between every pixel value and the image median,
- standard deviation calculated according to the equation:

$$\sigma = \sqrt{\frac{1}{N}\sum_{x=1}^{N}(x_i - \mu)^2}$$

where:

N – the number of pixels,
μ – the image mean.

In contrast calculation the formula was employed:

$$\Delta E^*_{abs} = \left[\left(\Delta L^*\right)^2 + \left(\Delta a^*\right)^2 + \left(\Delta b^*\right)^2\right]^{\frac{1}{2}}$$

where:

ΔE^*_{abs} – stands for the absolute colour difference,
ΔL^*, Δa^* and Δb^* – are differences in colour parameters in CIE L*a*b* colour models.

This colour difference equation was a subject of research, which confirmed that it is applicable not only for industrial purposes, but for analyses of scenic views, as well (BISHOP 1997). In this case a photograph was previously filtered with a high-pass filter, in order to enhance colour difference between the adjacent pixels.

With the aim of taking all four distance-dependent factors into consideration, three grayscale images were generated: the blue channel (Fig. 3), image saturation (Fig. 5) and contrast combined half and half with image colour diversity (Fig. 6). They were saved in grayscale mode (8-bit depth) and used as RGB channels of the new colour space, resembling "false colour compositions". On the basis of the fact that there are three variables, six different allocations are possible. In the eye of beholder their usefulness may be different, since our sensitivity for colours is irregular. From the numerical point of view, they still represent the same content.

In order to check whether this new colour space is efficient and to obtain more objective results, Bayes naive classifier was applied for the specific zone detection (MALINA & SMIATACZ 2010).

Fig. 5: Image saturation (decreasing with distance)

Fig. 6: Image contrast combined with the difference between pixels' values and the image median

After the manual indication of samples representing a given "class" (e.g. distant mountain range) in the landscape view, the parts of this image are classified as belonging to the same class, according to the rule of the maximum likelihood.

4 Results

Fig. 7: Separation of the foreground – a threshold in the blue colour channel (a binary mask).

In the analysed picture the separation of the foreground and the distant elements (mountains and sky) basing on the blue channel and its histogram characteristics brings positive results (Fig. 7). On the other hand, in case of the horizon line, the histogram of the blue channel is not useful, since they have similar characteristics, as far as colours are concerned (Fig. 2 and 3).

The results of three approaches towards estimation of colour diversity are presented in Fig. 8 and 9. The difference between pixels' values and the image median was chosen, as it returns the results with the most profound visual discernment between the fragments located at different depths (Fig. 8).

Fig. 8: A new colour space (R = image saturation, G = contrast + the absolute difference between pixels values and the median, B = blue channel)

In this new colour space the visual distinction between distance-dependent zones is facilitated. This subjective feeling was verified by means of the program that exploits the naive Bayes classifier. The fragments classified as belonging to the indicated class are coloured in saturated yellow, which was absent in the original image (Fig. 10 – 14).

Fig. 9: A new colour space (R = image saturation, G = contrast + standard deviation of pixels values, B = blue channel)

The attempt of distant rocks selection in the original image brought unsatisfactory effects. It was impossible to choose adequate parameters in order to distinguish the mountain range (Fig. 10). As a result of distance-dependent factors adoption, detection of a given fragment became more precise (Fig. 11). Some errors appearing in the top left hand corner (misclassified fragments of the sky) may be eliminated after supplementing the algorithm with the rule of adjacency. Starting from the area indicated as the sample, the program checks the neighbouring pixels and removes detached areas (Fig. 12).

Figures 13 and 14 show the results of this algorithm application for different landscape views. In the first of them an attempt was made to distinguish the spruce groves in front of the lake. In the second image the sky was chosen to be classified as the most differentiated area (clouds and colour gradient).

5 Conclusions and Outlook

Thanks to the idea of using grayscale images, which represent distance-dependent features, as the RGB colour channels, the quality of the photograph can be improved in such a way that detection of the foreground, the intermediate and distant zones, as well as background and the sky, is facilitated. Differences in the values of the blue channel, image saturation, its

Fig. 10: An attempt to detect the distant mountain range in the original image

Fig. 11: Detection of distant mountains in the image displayed in the new colour space

Fig. 12: The result of misclassified fragments elimination

contrast and colour variation, help to expose individual characteristics of areas that look similar in the original photograph. However, it should be kept in mind that this transformation does not introduce new information into the picture, but only depicts it in the more useful way.

Accordingly, in some photographs automatic distinction of sub-images is impossible, due to the identical colours of different objects. In these cases the algorithm should be supplemented with the module that checks the pixels neighbourhood and removes remote areas. However, in numerous views, detached fragments belong to the same object. In Fig. 11 elements from the foreground cut through the forests and meadows located in the intermediate distance. The groups of trees and residential buildings are dispersed in the Fig. 13. As a consequence of the adjacency rule application, the program did not mark the rows of trees at the lakeside. Taking this fact into consideration, the decision on remote fragments removal should be left to the user.

Fig. 13: Distinction of groves located in front of the lake (with checking adjacency)

Fig. 14: Detection of the sky

Fig. 14 depicts unsolvable problem of pixels' values identity. The colours of the distant hill and the sky are so similar that even after image conversion into the new colour space they are numbered among the same class.

The tolerance factor makes the class definition fuzzier, which guarantees more precise results. When the class is more homogeneous, the value of the tolerance factor should be low. On the contrary, while the data are diverse, it should be higher. At present, this calibration must be made manually, basing on the user's judgement. The excelled effects are achieved when the samples of the class are pointed at image fragments that have extreme features, in respect of colours.

The detection of the zones containing dispersed objects with the high colour variety meets cardinal difficulties. The types of areas mentioned above appear not only in the foreground, but also in the more distant zones, like for instance dispersed development, fragmented fields and meadows. The question still remains, how to precise fuzzy boundaries of zones containing these kinds of objects. It is worth to notice, that this decision may influence the results, e.g. calculating the intensity of residential area.

Another approach toward distance zones estimation may consist in the application of 3D cameras. The distance between the objects and the camera can be calculated, based on the geometrical dependencies. Due to its complexity, this problem should be a subject of the individual research.

References

Bishop, I. D. (1997), *Testing perceived landscape colour difference using the Internet*. In: Landscape and Urban Planning, 37, 187-196.

Buhmann, E., Palmer, J., Pietsch, M & Mahadik, S. (2011), *Managing the Visual Resource of the Mediterranean Island of Gozo, Malta for Tourists – A Studio Approach for International Conversion Students, Bridging Different Levels of English*. DLA Proceedings 2011, http://www.kolleg.loel.hs-anhalt.de/landschaftsinformatik/fileadmin/user_upload/_temp_/2011/Proceedings/201_Buhmann_2011_L_E.pdf (accessed 14.01.2012).

Dangermond, J. (2009), *GIS – Designing our future*. In: ArcNews, 31 (2), 6-7 http://www.esri.com/news/arcnews/summer09articles/gis-designing-our-future.html (accessed 14.01, 2012).

Ervin, S. (2011), *A System for GeoDesign*. DLA Proceedings 2011 – Online Version, http://www.kolleg.loel.hs-anhalt.de/landschaftsinformatik/fileadmin/user_upload/_temp_/2011/Proceedings/305_ERVIN_2011May10.pdf (accessed 14.01, 2012).

Flaxman, M. (2010), *Fundamentals of Geodesign*. In: Peer Reviewed Proceedings of Digital Landscape Architecture 2010, Berlin/Offenbach, Wichmann, 28-41.

Gonzales, R. C. & Woods, R. E. (2002), *Digital Image Processing*. Prentice Hall.

Jayaraman, S., Esakkirajan, S. & Veerakumar T. (2009), *Digital Image Processing*. Tata McGraw Hill.

Malina, W. & Smiatacz, M. (2005), *Metody cyfrowego przetwarzania obrazów (Methods of Digital Image Processing)*. Oficyna Wydawnicza EXIT.

Malina, W. & Smiatacz, M. (2010), *Rozpoznawanie obrazów (Image recognition)*. Oficyna Wydawnicza EXIT.

Kaplan, R. & Kaplan, S. (1989), *The Experience of Nature: A Psychological Perspective.*, New York, Cambridge University Press, 16-17.

Palmer, J. F. (1997), *Stability of landscape perceptions in the face of landscape change*. In: Landscape and Urban Planning, 37, 109-113.

Palmer, J. F. (2004), *Using spatial metrics to predict scenic perception in changing landscape*. In: Landscape and Urban Planning, 69, 201-218.

Ribe, R. G., Armstrong, E. T. & Gobster, P. H. (2002), *Scenic vistas and the changing policy landscape: Visualizing and testing the role of visual resources in ecosystem management*. In: Landscape Journal, 21, 42-66.

Shafer, E. L. & Mietz, J. (1970), *It Seems Possible to Quantify Scenic Beauty in Photographs*. U.S.D.A. Forest Service Research Paper Ne-162. http://www.fs.fed.us/ne/newtown_square/publications/research_papers/pdfs/scanned/OCR/ne_rp162.pdf (accessed 14.01, 2012).

Smardon, R., Palmer, J., Felleman, J. et al. (1986), Foundations for Visual Project Analysis. New York, A Wiley-Interscience Publication.

Geovisualization of the Garden Kingdom of Dessau-Wörlitz

Michael POKLADEK, Detlef THÜRKOW, Christian DETTE and Cornelia GLÄSSER

1 Introduction

The Garden Kingdom of Dessau-Wörlitz is part of the UNESCO world cultural heritage since the year 2000. It came into existence in 1758 due to considerable enlightening reformation efforts by Prince Leopold III Friedrich Franz duke of Anhalt-Dessau (1740-1817). Today's appearance of the landscape of the Garden Kingdom corresponds ideally to the definition of a "historical cultural landscape" (see REICHHOFF 1996, WEISS 2005). The creative highlights are landscape parks with structures marked by the influence of classicism and neo-Gothicism. They are all effective as a combined work of architecture, gardening and fine arts. 145 km² of the aesthetic change of territory are still preserved to this day.

It was declared as a historical monument in 1979 and is therefore protected (KÜSTER 2010). The landscape is supposed to have the effect on the visitors of being created by nature and not by an artist. Numerous view axes combine the landscape elements and parks visually (TRAUZETTEL et al. 1998).

The federal state of Saxony-Anhalt is obligated to the UNESCO and therefore must process the physiographical, infrastructural and urban situation, as well as undertake the monument preservation of this unique cultural landscape as a whole. Moreover, the state shall develop prospects for further development. The Kulturstiftung DessauWörlitz (federal cultural foundation) serves as a coordinator for conservation, research and maintenance of the cultural world heritage "Garden Kingdom". One of the most important elements of monument preservation and planning for the specialists of the Kulturstiftung is the "Denkmalrahmenplan" (LDA/KS DESSAUWÖRLITZ 2007). It serves as the fundamental base for this project and was a national, as well as international innovation when it was completed in 2007. Up to this point no such intense acquisition and assessment of a historical cultural landscape, which qualifies for protection, existed in terms of monument conservation.

Today historic gardens are visualized in numerous projects using virtual reality. Examples provide the projects of PAAR & BLAIK (2008) and MORAVCIK & KUBISTA (2008). As early as 1996 a 3D visualization of the Garden "Wörlitzer Anlagen" was generated for the exhibition "Weltbild Wörlitz – Entwurf einer Kulturlandschaft" (see URL1). The constantly growing potential of geovisualization is already being implemented to visualize cultural assets in monument planning (e.g. URL2). To further qualify the results of various planning processes the new media can be used for the Garden Kingdom. The focus of this project is the special support and upgrading of the "Denkmalrahmenplan" of the Garden Kingdom of Dessau-Wörlitz with the integration of virtual reality.

The heterogeneity of the cultural/historical matters and the size of the area require an adaption and further development of the existing range of methods. An entire analysis and presentation of the historical development of the Garden Kingdom based on virtual landscape scenarios shall be made accessible through the interdisciplinary collaboration of landscape planners, monument preservation experts, geo computer scientists and cartographers.

This article regards itself as a report of progress on important goals, methods and implementations, which are a basis for multi-scale and multi-temporal geovisualization.

2 Motivation and Objectives

In addition to the results of the "Denkmalrahmenplan", the changes in landscape of the Garden Kingdom will be shown from the initial situation at the end of the 18th century up to the present in three dimensional, virtual scenarios. Geographical visualizations are done in different *scale levels*, *time levels* and *levels of details*. They are useful for different target groups. Specialists of the Kulturstiftung shall be supported in their work with the "Denkmalrahmenplan". The virtual scenarios provide interactive analyses of the view axes and the natural scenery within the park. They provide a way to identify conflict potentials between different matters of regional planning, monument planning and environmental planning and serve to develop alternatives more efficiently. The visualized alternatives include for example conflicts of interest in specialist planning of flood prevention versus the target planning of the protection of historical monuments. Moreover, selected results constitute an important source of information for interested individuals. Once these results are integrated into widely used map services, such as Google Earth, they allow for globally available "virtual excursions" at World Heritage site and may serve the tourist marketing of the area. To finalize, they generate synergies in environmental education, since those visualizations can be used as visitor information systems via the new media. The acquisition of the landscape takes place on a multi-scale level in several periods. Different scale levels are assigned to different levels of detail (see Table 1 and Figure 1).

Visualization	Scale	Time Levels	Level of Detail[1]
Vis1: Overview of visualization Garden Kingdom	1:25.000	1820, present	LoD1
Vis2: Garden with surrounding cultivated landscape areas	1:10.000	1800, present	LoD2
Vis3: Visualization of the garden, based on results of the "Denkmalrahmenplan"	1:2.000	1800, 1830, present	LoD3

Table 1: Visualization content, map and time scales and level of detail

Furthermore, theoretical approaches can be derived from the developed cartographical methods and the applied principles for the processing of historical and topographical maps (GLÄSSER et al. 2010). These recommendations for action shall contribute to the production of scientific guidelines for the development of multi-temporal and multi-scale geovisualizations of cultural assets.

---

[1] See Level-of-Detail-convention for Building Quality Levels, DÖLLNER & BUCHOLZ 2005.

Fig. 1: LoD- Steps for the acquisition of the individual historic structures. Example used is the Luisium (left: Box model LoD1; middle: non-photorealistic facade model medium level of detail LoD2; right: non-photorealistic facade model high level of detail LoD3).

3 Data and Methods

Comprehensive data exists for the analysis of historical stages (i.e. recent and historical cartography, status plans, pictures, documents). Spatial reconstruction of the cultural landscape is done based on historical and recent cartography, literature and knowledge of the local experts. The creation of the visualization combines the usual processing of data. Examples include the geodetic adaption of base data, the multi-temporal evaluation of maps, of geo-objects and the presentation of the chosen time steps in virtual scenarios, based on the requirements of the "Denkmalrahmenplan".

3.1 Data Basis and Processing

According to the planned scale levels (see Table 1), recent and historical cartography was researched. As a result, the data in the grey areas of Table 2 meet the requirements of the area-wide coverage and the informational content of Vis1, Vis2 and Vis3. To adapt historical maps to the actual geodetic reference system it is necessary to use the officially available geo base data. The key challenge is the spatial processing and homogenization of all data (DETTE et al. 2010).

3.2 Digital Elevation Model (DEM) and Digital Surface Model (DSM)

In order to map ground surface in virtual scenarios, ATKIS®-DEM40 is primary used for the medium scale level, while ATKIS®-DEM1 is used for the high scale level (see Table 2). These models represent the present shape of the relief. A range of differences in former time steps to the present geo-morphological situation can be determined from historical documents. It is necessary to adapt the models to describe the historical geo-morphological situation. Primary examples are the changing development of the stream courses of the rivers Elbe and Mulde, as well as changes in the dike system. Numerous GIS-based functions and algorithms served to determine the approximation of the relief to historical situations. Therefore, Focal Mean and Lee Filter processes, linear and inverse distance weighting interpolation methods were used, which took into account the special circumstances of the flat meadow grounds (LEE 1980, HENGL et. al. 2004, FLORINSKY, 2012). In terms of software, editing took place mainly in Esri ArcGIS/ArcInfo.

	Data	Scale	Source[2]	Content, Further Information
Vis1	Deckersches Kartenwerk[3]	1:25.000	A	Land Use 1816/1821; DECKER, 1816, ZÖGNER, 1981
	Digital Topographical Map 25 (DTK25)	1:25.000	B	Topography (2010)
	ATKIS®-DEM 40	about 1:50.000	B	Relief, Digital Elevation Model with geometrical resolution 40 × 40 meters (2007)
Vis2	Denkmalrahmenplan	1:10.000	C	Map "Historische und verloren gegangene Landschaftselemente des 18. Jahrhunderts"; Map "Historische Elemente der Kulturlandschaft"; Map "Historischer Landschaftsbestand nach Zeitschichten"; LDA/KS DESSAUWÖRLITZ, 2007
	Digital Topographical Map 10 (DTK10)	1:10.000	B	Topography (2010)
	ATKIS®-DLM	1:10.000	B	Land Use from Digital Landscape Model (2008)
	Digital Ortho Photos (DOP20)	1:10.000	B	True color Aerial Photos with 20 cm spatial resolution (2010)
Vis3	Denkmalrahmenplan Bestandsplan Luisium	1:2.000	C	Map "Present Geo-objects Luisium", Map "verloren gegangene historische Landschaftselemente", Map "Elemente der Entstehungszeit", LDA/KS DESSAUWÖRLITZ, 2007
	Map Eyserbeck	not specified	C	Map of the garden "Luisium" about 1790
	„Charte der Feldmarken Naundorf und Jonitz"	not specified	C	Kretzschmar, 1828
	ATKIS®-DEM1, DSM1	about 1:1.000	B	Relief, Digital Elevation Model and Digitital Surface Model with high resolution 1 × 1 meters (2010)

Table 2: Primary spatial information, scales and sources of the visualization products

---

[2] Sources: [A] – Staatsbibliothek zu Berlin, Preußischer Kulturbesitz; [B] – Landesamt für Vermessung und Geoinformation Sachsen-Anhalt; [C] – Landesamt für Denkmalpflege und Archäologie Sachsen-Anhalt, Kulturstiftung DessauWörlitz.
[3] Deckersches Kartenwerk: Begin of the Prussian surveying between 1816 and 1821. Mapping of parts of the provinces Brandenburg, Saxony and Anhalt in the scale 1:25.000. Administration by Majore F. von Rau und Carl von Decker.

3.3 Scan Image and Referencing

According to KRESSNER (2008), if not already digitally available, analog base data has to be scanned in high resolution (600 dpi), followed by an image correction using image editing programs. In order to provide a spatial-related extraction of information, the cartography has to be georeferenced. A simple realization using a few points of reference can be guaranteed by recent plans, which do not have a geo-relation yet (analog maps of the "Denkmalrahmenplan", see Table 1), due to their high accuracy. Object related methods for referencing are mainly used for historical data material. In the park sector, historical buildings served as the primary reference objects. However, in the area of open landscapes, the georeferencing was hindered. The geodetic quality of those sectors is limited and lacked reference objects. Considering the size of the region, the implementation of a simple reference method using the first polynomial function suited the need of "Deckersches Kartenwerk". The area has been adapted to the recent geo base data of the national survey using about 50 points of reference (see Table 2). Despite the numerous existing reference objects, the maps of Eyserbeck and Kretzschmar (1828) feature huge differences according to the preservation of the position of the geo-objects. Sectors featuring high spatial accuracy, as well as high biases, can be identified (see Figure 2).

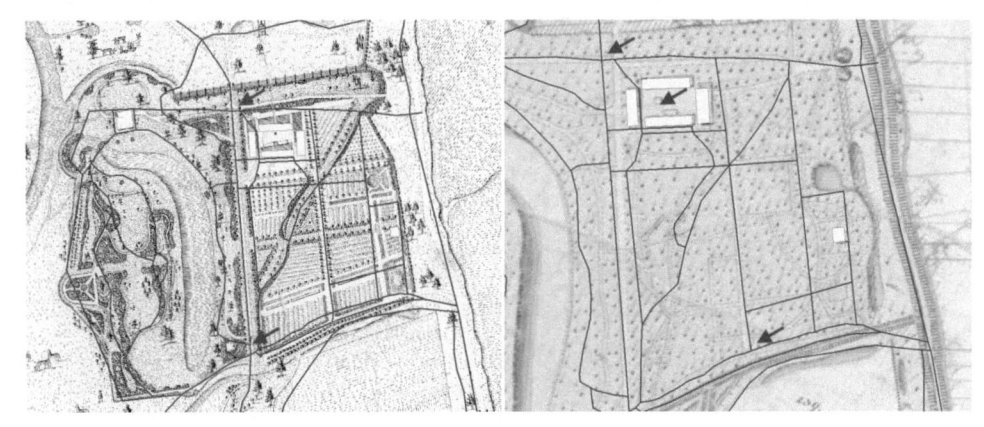

Fig. 2: Georeferenced maps with different spatial precisions of the geoobjects – left: Subset of Map Eyserbeck, right: Subset of "Charte der Feldmarken Naundorf und Jonitz"

3.4 Editing and Digitizing

Once all basic data has been converted into the same reference system (DHDN 4), the map content was digitized (QuantumGIS). Due to the already mentioned differences in quality (recent vs. historical material), the corresponding information for land allocation was not always transferred directly. A knowledge based adaption of the topology of the geo-objects had to be done at this point.

3.5 3D Objects and Texturing

The geometry of construction and vegetation was taken from the vectorized information for land allocation, which served as a base for the modeling of 3D objects. The derivation of the heights, according to the resolution of detail, ranged between generalized and accurate. To fit Vis1, the floor plans of the buildings had been extruded within Esri ArcGIS Desktop with randomized heights as a simple block model, using a random function within a Visual Basic Script (urban section 5 to 15 meter, rural sector 4 to 8 meter). Based on the point and polygon geometry of the vegetation, self-made trees modeled with Autodesk 3ds Max were integrated. Finally, the creation of VR-scenarios was done with Autodesk LandXPlorer. The 3D objects, which are part of the Vis3 and have a scale of 1:2000, have been created with 3ds Max, based on pictures of facades and existing building plans. In order to texturize historical buildings, a technique was used, which conveyed indecisiveness and uncertainty. Therefore, textures had been generated in a non-photorealistic style and implemented in 3ds Max (see DÖLLNER & BUCHHOLZ 2005 and MAASS & DÖLLNER 2006). Additionally, the method of "texture baking" was employed. It allowed in a gentle way to integrate information about lighting in textures. Thus, they do not need to be rendered in real time which saved system resources (see MURDOCK 2011). Vis3 vegetation, particularly trees, was modeled in the 'TreeCreator', a plug-in for the Unity game engine. This plug-in is generating procedural tree models (DEUSSEN 2003). The "Denkmalrahmenplan" itself has three different signatures for trees (broad-leafed tree, conifer, fruit tree). Referring to this configuration these types of trees are used in the Vis3 as well. The models of Vis2 had been derived from the generated LoD3-Objects of the Vis3, by reducing the resolution of detail. An example would be a simple rectangle rotating about 90 degrees to simulate a 3D effect for plants (MUHAR 2001). All 3D objects were then integrated into the previously prepared cartographic ground plane. In case of the VIS1, this was the "Deckersches Kartenwerk", subordinated by DEM40. A 3D-reconstruction of a historical landscape situation, using a historical map as a base, led to a high degree of authenticity and had already been successfully employed in urban sectors (DETTE et. al 2010). In order to visualize details, adequate textures had been chosen and implemented from the libraries of Autodesk 3ds Max to match the extracted information for land allocation.

3.6 Interface and Presentation

To suit the specific standards of the experts, an independent interface for planning processes of the "Denkmalrahmenplan" was necessary to develop a more specialized user interface (see SCHROTH et al. 2011). Therefore Unity game engine was used. The user interface accounts for a systematic integration of navigation functions, lighting situations (see GOLDSTONE 2009, GREENWOOD et al 2009) and particular menus with texts, captions, clues about view axes and further expert information. Using the fbx-format, 3D-objects were imported into Unity in order to create a presentation of data for experts. However, it is reasonable to employ a globally available and familiar interface to reach interested individuals. Reduced objects can be exported from 3ds Max to Google SketchUp using the Collada format. The export of the virtual scenarios in kmz files eventually allows visualization with Google Earth. Selected scenarios will be uploaded to the 3D Warehouse and will thus be integrated into Google Earth by default. Figure 3 outlines the characterized workflow of data and methods.

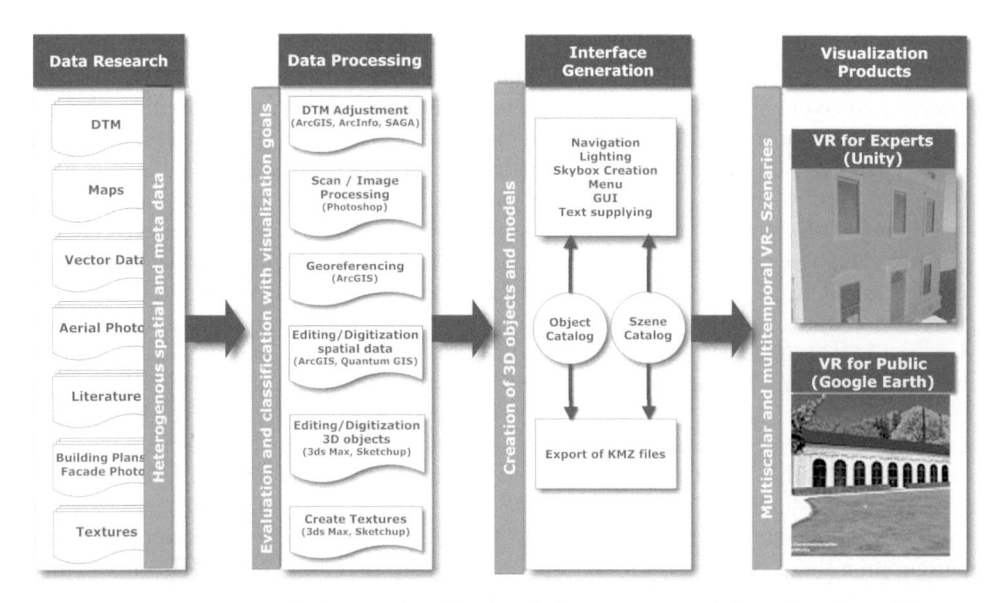

Fig. 3: Components of the geovisualization infrastructure and the related workflow of data processing

4 Results

This chapter presents preliminary results of the work in progress. Selected examples will be exemplary for the described peculiarities within the workflow. The first example represents an extract of the VIS1. The DGM40 overlaid with the "Deckersches Kartenwerk" serves as a base for this. The attached buildings represent the city of Dessau in the year of 1820. About 180.000 cartographic signatures of trees are integrated into this particular extract, which stresses especially the linear elements of the alleys. Thus, the relationship between urban and rural sectors, which existed at that time, can be experienced more clearly.

Figure 5 represents preliminary results of high resolution 3D building models within the LoD3 system. The left side shows a building (i.e. "Schlangenhaus") created with 3ds Max. On the right side is a final import into Unity (i.e. "Luisium") and a corresponding tree model. In order to visualize the models, a lighting system and a skybox system had been integrated. Moreover, an egocentric control system had been generated and evaluated in terms of suitability.

Within the Garden Kingdom, there are, as already mentioned, numerous axes of view among the architectural highlights of the landscape. Figure 6 describes the visual axis through a line of trees to the "Schlangenhaus". The mapped example was taken from the visualization which had been created for the public within Google Earth.

Fig. 4: Visualization of Dessau with surrounding cultivated landscape areas (LoD1, LandXplorer)

Fig. 5: 3D model of the "Schlangenhaus" and the "Luisium" (LoD3, 3ds Max, Unity3D)

Fig. 6: Axe of view to the "Schlangenhaus" (LoD3, Google Earth)

5 Outlook

In order to support the planning of the "Denkmalrahmenplan" multitemporal and multi-scale geographic visualizations are in development as part of an interdisciplinary project. They are designed to support the experts of the Kulturstiftung DessauWörlitz in planning processes for maintenance and further development of the world cultural heritage "Garden Kingdom". The illustration of historical situations in virtual reality will help to qualify the target state, which was defined during the planning process. This approach constitutes a novelty within the "Denkmalrahmenplan". Additionally, the created 3D scenarios can be used for further sectoral planning and environmental education respectively. Due to the integration of selected extracts into Google 3D Warehouse, detailed information on planned reconstruction measures was made retrievable for the general public.

In the first instance, the final implementation of all VR modules is the primary concern. Furthermore, based on the generated workflow, the cartographic development process to support the "Denkmalrahmenplan" will be put into theory, followed by an extensive evaluation and a field trial. Recommendations for action are to be derived from the generated methods, which add to automation of the development process of geo-visualization of historical, cultural landscapes.

6 Acknowledgements

The interdisciplinary research in this project is the result of collaboration of landscape planners, monument preservation experts, geo computer scientists and cartographers. We are grateful for the help and contributions by many experts. We especially thank Dr. Thomas Weiss, Ludwig Trauzettel and Sebastian Doil of the Kulturstiftung DessauWörlitz for their support.

References

Decker, v. C. (1816), *Das militairische Aufnehmen oder vollständiger Unterricht in der Kunst, Gegenden, sowohl regelmäßig als nach dem Augenmaaße, aufzunehmen*. Berlin.

Dette, C., Schulze, M. & Gläßer, C (2010), *Rückblick, Einblick, Ausblick – Die Geschichte der Saline-Insel in Halle (Saale)*. In: Kartographische Nachrichten, 60, 244-251.

Döllner, J. & Buchholz, H. (2005), *Expressive Virtual 3D City Models*. XXII International Cartographic Conference, Spain.

Deussen, O. & Lintermann, B. (2005), *Digital design of nature : computer generated plants and organics*. Berlin, Springer X.media.publishing.

Florinsky, I. V. (2012), *Digital Terrain Analysis in Soil Science and Geology*. Elsevier.

Gläßer, C., Thürkow, D., Dette, C. & Scheuer, S. (2010), *Development of an integrated technical-methodical approach to visualize hydrological processes in an exemplary post-mining area in central Germany*. In: ISPRS Journal of Photogrammetry and Remote Sensing, 65, Special Issue 3 "Visualization and Exploration of Geospatial Data, 275-281.

Greenwood, P. & Sago, J., Richmond, S. & Chau, V. (2009), *Using game engine technology to create real-time interactive environments to assist in planning and visual assessment for infrastructure.* In: Anderssen, R. S. et al. (Eds.), 18th World IMACS Congress and MODSIM09 International Congress on Modelling and Simulation. Modelling and Simulation Society of Australia and New Zealand and International Association for Mathematics and Computers in Simulation, 2229-2235.

Goldstone, W. (2009), *Unity game development essentials.* Birmingham.

Hengl, T., Gruber, S. & Shrestha, D. P. (2004), *Reduction of errors in digital terrain parameters used in soil-landscape modelling Original Research Article.* In: International Journal of Applied Earth Observation and Geoinformation, 5 (2), 97-112.

Kressner, L. (2008), *Georeferenzierte Altkarten aus Mecklenburg als Basis moderner Kulturlandschaftsforschung mittels Geoinformationssystemen.* In: Kartographische Nachrichten, 3, 129-135.

Küster, H. J. & Hoppe, A. (2010), *Das Gartenreich Dessau-Wörlitz. Landschaft und Geschichte.* München.

LDA/KS DessauWörlitz (Eds.) (2007), *Denkmalrahmenplan Gartenreich Dessau-Wörlitz. Historische Kulturlandschaften. Historische Siedlungen. Historische Gartenanlagen.* Landesamt für Denkmalpflege und Archäologie Sachsen-Anhalt und Kulturstiftung DessauWörlitz, Halle (Saale) und Großkühnau.

Lee, J. S. (1980), *Digital image enhancement and noise filltering by use of local statistics.* In: IEEE Transactions on Pattern Analysis and Machine Intelligence, 2, 165-168.

Maaß, S & Döllner, J. (2006), *Ein Konzept zur dynamischen Annotation virtueller 3D-Stadtmodelle.* In: Kartographische Schriften. Aktuelle Entwicklungen in Geoinformation und Visualisierung, 10. 19-26. Potsdam.

Moravcik, L. & Kubista, R. (2008), *CAD-based Design in Reconstruction of Historical Park in Hajna Nova Ves, Slovakia.* In: Buhmann, E. et al. (Eds.), Digital Design in Landscape Architecture 2008. Heidelberg, Wichmann.

Murdock, K. L. (2011), *3ds Max 2012 Bible.* Indianapolis.

Muhar, A (2001), *Three-dimensional modelling and visualisation of vegetation for landscape simulation.* In: Landscape and urban planning, 54 (1-4), 5-17.

Paar, P. & Blaik, R. (2008), *Time-slicing Wrest Park – Interactive 3D Visualisation of a Historic Garden.* In: Buhmann, E. et al. (Eds.), Digital Design in Landscape Architecture 2008. Heidelberg, Wichmann.

Reichhoff, L. (1996), *Historische Kulturlandschaften des Landes Sachsen-Anhalt.* In: Naturschutz im Land Sachsen-Anhalt, 33 (2), 3-14. Halle.

Schroth, O., Pond, E., Campbell, C., Cizek, P., Bohus, S. & Sheppard, S. R. J. (2011), *Tool or Toy? Virtual Globes in Landscape Planning.* In: Future Internet 2011, 3, 204-227.

Trauzettel, L., Quilitzsch, U., Speler, R-T., Bode, U., Savelsberg, W., Froesch, A., Görgner, E. & Alex, R. (1998), *Das Luisium im Dessau-Wörlitzer Gartenreich.* München.

URL1 – Klein, M., (1996), *Weltbild Wörlitz: Entwurf einer Kulturlandschaft.* – Institut für Neue Medien, Frankfurt am Main, http://www.inm.de/projects/woerlitz/ [accessed 10/2/12].

URL2 – Visuelle Lausitz e.V., *Virtuelle Visualisierung. Schloss und Park Branitz.* http://www.visuellelausitz.de/rb/branitz/index.html [accessed 10/2/12].

Weiss, T. (2005), *Unendlich schön. Das Gartenreich DessauWörlitz.* Kulturstiftung DessauWörlitz (Hrsg). Berlin, 12-19.

Zögner, L. (1981), *Preußens amtliche Kartenwerke im 18. und 19. Jahrhundert.* Frankfurt/Main.

Concordance between Photographs and Computer Generated 3D Models in a Michigan Highway Transportation Setting

Shawn PARTIN, Jon Bryan BURLEY, Robert SCHUTZKI and Patricia CRAWFORD

1 Introduction

Investigators are interested in discovering valid constructs concerning human perceptions between the actual physical landscape and representations of landscapes such as black and white photographs. Seminal studies discovered that respondent's perceptions covaried with actual physical landscapes, black and white photographs, color photographs (SMARDON, PALMER & FELLEMAN 1986), and later with video; yet were significantly divergent from perceptions about hand drawings. Respondents evaluated hand drawings based upon the quality of the rendering as opposed to the quality of the physical environment the drawing represented. Thus investigators employed photographic simulations to study respondent preferences for various landscape treatments and avoided the use of drawings. The literature in visual quality research is vast and extensive, beyond the scope of this short research investigation; however, recent studies by LU et al. (2012), BURLEY et al. (2011) and MO et al. (2011), illustrate the current state-of-the-art in visual quality perception, modelling, theories, and applications.

Over the last 20 years, 3D computer landscape depictions have made advancements in ease of use and in expressing the complexities of the environment, as illustrated by the diffusion and adoption of Google SketchUp by many professional planning and design firms. Some investigators have studied validity testing between computer simulations and the actual physical landscape (OH 1994, BERGEN et al. 1995, BISHOP & ROHRMANN 2003). We were interested in determining whether respondents would perceive 3D landscape depictions in a manner similar to drawings or similar to photographs for highway settings in Michigan. We believe this research will add to the body of knowledge that exists about which forms of media are effective in presenting landscape designs. The focus of this study is to discover whether Google SketchUp images are perceived in a manner similar to drawings or to photographs. And if they are perceived similar to photographs, then they can be used to study planning and design treatments with respondents.

2 Material and Methods

An important aspect of our study was to use Q methodology as opposed to Likert scales in assessing respondent preferences. The origins of the Q sort method first begin with the introduction of Q methodology sent in a letter to Nature by William Stephenson (1902-1989) in 1935 (BROWN 1993). Stephenson, both a psychologist and a physicist, is considered the "father" of Q methodology and of the Q sort method. Although introduced in 1935, Stephenson further discusses and describes Q methodology in publications such as

"Correlating Persons Instead of Tests," Foundations of Psychometry: Four Factor Systems," and The Study of Behavior: Q technique and Its Methodology (BROWN 1993). Additional discussions and descriptions of Q methodology and the Q sort method have been presented by Stephen Brown in a variety of texts, but most inclusive being "A Primer on Q Methodology" (BROWN 1993) as well as by PITT & SUBE [ZUBE] (1979) and SWAFFIELD & FAIRWEATHER (1996) in which they describe how Q methodology is used, in detail, in their particular studies. Early uses of the Q sort method, a key component of Q methodology, involved personality assessment. This was done by asking a panel of trained professionals to each give their opinion of a respondent's personality. In order to accomplish this, each panelist was given a deck of cards with personality descriptors on them. The panelists were then asked to organize the deck of descriptors into piles from "least like the respondent to most like the respondent." This process is known as Q sorting. Once all panelists had sorted their decks, the results from each panelist were compared, averaged, and analyzed. From this process, an assessment of the respondent's personality emerged (PITT & SUBE [ZUBE] 1979). Although the process of Q sorting has basically remained the same over the years, its application has been found useful across many disciplines. A quick search of recent literature has revealed examples of Q methodology's use in psychology (Westin et. Al. 1997), planning (SWAFFIELD & FAIRWEATHER 1996), child psychology (BUCKLEY et. al. 2002), landscape planning and visual quality (DEARDEN 1984), education and literature (BOSCOLO & CISOTTO 1999), workforce education (MCKNIGHT 2008), sociology (PREVITE et. al. 2007), nursing (AKHTAR DANESH et. al. 2008), and mobile communications (LIU, 2008).

For our study, we hypothesized that the SU images will receive similar scores as the corresponding photographs. The null hypothesis predicts that SU images and corresponding photographs will receive scores that are extremely different. In our study, we constructed 25 pairs of images comprising of photographs of Michigan highway environments and matching Google SktechUp images of the same landscapes (Figures 1, 2, and 3). In an attempt to further enhance the results of this experiment, 50 seeded images from BURLEY (1997) are included in the study as well. It is hypothesized that the seeded images will receive scores that have strong concordance with scores received in the original study. The null hypothesis predicts that seeded images will receive scores that have no concordance or poor concordance with the scores received in BURLEY'S (1997) original study. The 50 seeded images gave a broader visual context and setting for the respondents to view the SketchUp and corresponding photographs.

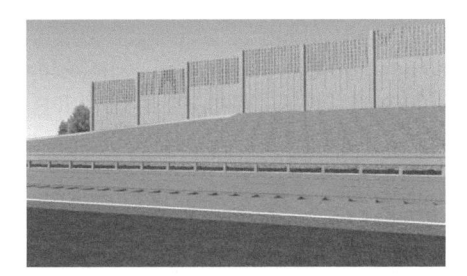

Fig. 1: An example of paired images. On the left is the computer image and on the right is the photographic image. Copyright ©2011 Shawn Partin, all rights reserved, used by permission.

Fig. 2: An example of paired images from a highway construction setting. On the left is the computer image and on the right is the photographic image. Copyright ©2011 Shawn Partin, all rights reserved, used by permission.

Fig. 3: An example of paired images from a new highway rest area building. On the left is the computer image and on the right is the photographic image. Copyright ©2011 Shawn Partin, all rights reserved, used by permission.

The images were seeded into a group of 50 photographic landscape images from a study by BURLEY (1997) containing a broad array of North American landscape images ranging from highly preferred images to highly un-preferred images. This resulted in a total of 100 images (50 seeded photographic images, 25 SketchUp images, and 25 photographs paired to the SktechUp images) presented to Michigan respondents in the fall of 2011, employing Q-sort techniques (BROWN 1993).

Each participant was asked to sort 50 images, randomly selected from the 100-image set. The 100-image set was divided into two, randomly selected, groups of 50 images without replacement. Once both groups had been sorted, two new groups were created again using random selection. Sorting continued until the full 100-image set had been sorted for a total of seven times. From the group of 50 images, five 10-image subgroups were created at random. Participants were given one subgroup at a time and asked to sort the scenic quality of the images from *most aesthetically pleasing* (or high scenic quality) to *least aesthetically pleasing* (or low scenic quality). A score of 1 represented high scenic quality while a score of 10 represented low scenic quality as compared to the other nine images in the subgroup. Each image received a score between 1 and 10; no two images were able to receive the same score within a subgroup. The score was determined by the order in which the participant placed the printed images on a tabletop. Fifteen minutes were allotted to sort each subgroup, if a participant completed the sorting process before the fifteen minutes was up, they were allowed to move to the next subgroup. The participant was finished when all five subgroups had been sorted. The entire 50 image group took roughly fifteen minutes to sort on average. Once a participant finished sorting a 10 image subgroup, the order was recorded by the researcher and the participant was given the next subgroup. When all subgroups had been sorted and recorded, the participant was allowed to take a cupcake. At this time, participants were allowed to ask questions about the study if they desired. If questions were asked, they were answered away from current sorting participants so as not to influence decisions made by current sorters. We evaluated the respondent scores of the paired images with Kendall's Coefficient of Concordance (DANIEL 1978) to test for significant similarity in the rankings of the paired images. The general value for W was found using Equation 1.

$$W = (12*S)/[m^2 n(n^2-1)] \qquad (1)$$

Where: W = Kendall's coefficient of concordance
S = (sum of the ranks – mean)^2
m = number of sets (SU Images and Photographs)
n = number of images in each set

The value for W was then adjusted for ties using Equation 2.

$$W = (12*S)/[m^2 n(n^2-1) - m*\Sigma(t^3-t)] \qquad (2)$$

Where: W = Kendall's coefficient of concordance adjusted for ties
S = (sum of the ranks – mean)^2
m = number of sets (SU Images and Photographs)
n = number of images in each set
$\Sigma(t^3-t)$ = the sum of the number of observations in all sets of rankings tied for a given rank

Finally, the chi-square value was determined using Equation 3.

$$X^2 = m(n-1)W \qquad (3)$$

Where: X^2 = chi-square value
m = number of sets (SU Images and SU Photographs)
n = number of images in each set
W = Kendall's coefficient of concordance adjusted for ties

The X^2 value was then compared to "A Table of Percentage Points of the X^2 Distribution" in DANIEL (1978, p. 452).

3 Results

Table 1 presents the scores and the ranks of the SketchUp images and the paired photographs. This table is used to calculate Kendall's Coefficient of Concordance. We calculated a X2 value of 40.164 with 24 degrees of freedom and this number was compared to a X2 table. Since 40.164 is greater than 39.364, the null hypothesis is rejected and the alternate hypothesis is accepted. The two sets are in concordance to p<0.025. We also evaluated the respondent scores of the seeded images with the predicted scores from Burley (1997). We calculated a X2 value of 86.290 with 49 degrees of freedom and this number was compared to a X2 table. Since 86.290 is greater than 79.490, the secondary null hypothesis is rejected and the secondary alternate hypothesis is accepted. The two sets are in concordance to $p<0.005$.

SU Image	Raw Score	Rank	Photo	Raw Score	Rank
1	40	7.5	26	32	2
2	41	9	27	29	1
3	64	25	28	54	22
4	61	23.5	29	62	25
5	53	19.5	30	55	23.5
6	39	6	31	38	6.5
7	44	11	32	50	17
8	50	16	33	46	12
9	51	17.5	34	50	17
10	56	22	35	50	17
11	51	17.5	36	48	15
12	49	15	37	51	19.5
13	53	19.5	38	51	19.5
14	34	3.5	39	38	6.5
15	40	7.5	40	33	3
16	32	1	41	41	8.5
17	48	13.5	42	45	10
18	34	3.5	43	37	5
19	33	2	44	46	12
20	54	21	45	41	8.5
21	38	5	46	46	12
22	61	23.5	47	55	23.5
23	45	12	48	47	14
24	43	10	49	52	21
25	48	13.5	50	35	4

Table 1: Raw scores and ranks for SU Image and Photo image pairs

4 Discussion and Conclusion

The results show there is significant concordance between SU images and paired photographs to a p-value of <0.025, suggesting a strong link between 2D images of a 3D model and photographs. As noted by BERGEN et al. (1995), "Photographs have been found to be adequate surrogates for actual scenes." Therefore, it can be inferred that 2D images of a 3D model may also be acceptable surrogates for actual scenes. Additionally, the results imply that the average person, when presented with an SU image of a design, whether proposed or existing, will more than likely see an image of a design rather than an image of a model. To clarify, when presented with an image of a model, people will base their opinion of the design on the content of the image rather than on the quality of the SU model.

It is possible that errors may have occurred during this study. Potential errors may include unintentional bias when creating SU images that may incline a participant to favor either the SU image or the photograph, inconsistent direction given to participants if said participant asked questions during the instructional or sorting period, regional bias based on the location of the testing area, and bias to participants who like cupcakes (as participants who completed the survey were given cupcakes). Unintentional bias may have occurred during the construction of SU images based on the researcher's technological capabilities or preferences. Every effort was taken to create SU images in the likeness of the corresponding photo to the fullest reasonable extent possible. SU models were created to what the researcher believed to be a reasonable limit where a model of a proposed design would typically be created within the constraints of the SU photo. To clarify, no additional "artistic flare" was intentionally added and no additional detail was added where it did not already exist in the corresponding photo. Initial directions were given consistently to each participant at the start of each session. However, several participants asked individualized questions or asked for the directions to be repeated. If it was deemed by the researcher that answering a particular question would affect the data, the question was not answered. However, it may be possible that by repeating directions or answering questions that seemed unrelated to the results, this may have skewed that particular data set. Future studies could avoid this by only providing written or prerecorded instructions. There is unavoidable regional bias in this study based on the location of the data recording site. Data was recorded in Okemos, Michigan at the Meridian Farmer's Market. The intention of choosing this location was to get a variety of participants from a variety of backgrounds. While it seems this was achieved, it is possible that citizens who do not attend farmer's markets could be considered biased against. For a comprehensive study, future research should be conducted at a variety of venues at multiple locations around the globe. Finally, since cupcakes were given as an incentive for participants to supply their input for the study, it is possible that a potential participant who does not like cupcakes may not have participated due to the lack of proper incentive. Future studies could include a wider variety of incentives for participants.

Other future studies should include testing using other 3D modeling software such as 3D Studio Max, Revit, or other industry equivalents. Future studies could also include an additional level of rendering detail by applying advanced textures and lighting such as those found in SketchUp Podium, Kerkythea, or Maya. Still other studies may include the use of video animations or interactive 3D models vs. walking through an actual site.

We concluded that Michigan residents perceive highway landscape photographs similar to Google SketchUp images. This means that planners and designers can use 3D computer generated models of the quality produced in the study to examine landscape treatments and respondent preferences. In addition, we also conclude that it is possible to use a set of previously studied images to provide a setting to study a smaller set of images. We encourage other investigators to explore different sets of respondents, 3D software, and environmental settings to test for concordance.

References

Akhtar-Danesh, N., Baumann, A. & Cordingley, L. (2008), *Q-methodology in nursing research: a promising method for the study of subjectivity*. In: Western Journal of Nursing Research, 30 (6), 759-773

Bergen, S., Ulbricht, C., Fridley, J. & Ganter, M. (1995), *The Validity of Computer-Generated Graphic Images of Forest Landscape*. In: Journal of Environmental Psychology, 15, 135-146.

Bishop, I. & Rohrmann, B. (2003), *Subjective responses to simulated and real environments: a comparison*. In: Landscape and Urban Planning, 65, 261-277.

Boscolo, P. & Cisotto, L. (1999), *Instructional strategies for teaching to write: a Q-sort analysis*. In: Learning and Instruction, 9, 209-221.

Brown, S. (1993), *A Primer on Q Methodology*. In: Operant Subjectivity (April/July), 16 (3/4), 91-138.

Buckley, M., Klein, D., Durbin, C., Hayden, E. & Moerk, K. (2002), *Development and validation of a Q-sort procedure to assess temperament and behavior in preschool-age children*. In: Journal of Clinical Child and Adolescent Psychology, 31 (4), 525-539.

Burley, J. B. (1997), *Visual and ecological environmental quality model for transportation planning and design*. In: Transportation Research Record, 1549, 54-60.

Burley, J. B., Deyoung, G., Partin, S. & Rokos, J. (2011), *Reinventing Detroit: grayfields – new metrics in evaluating urban environments*. In: Challenges 2, 45-54.

Daniel, W. (1978), *Applied Nonparametric Statistics*. Boston, Massachusetts, USA, Houghton Mifflin.

Dearden, P. (1984), *Factors influencing landscape preferences: an empirical investigation*. In: Landscape Planning, 11, 293-306.

Liu, C. (2008), *Mobile phone user types by Q methodology: an exploratory research*. In: International Journal of Mobile Communications, 6 (1).

Lu, D., Burley, J., Crawford, P., Schutzki, R. & Loures, L. (in publication, 2012), *Quantitative methods in environmental and visual quality mapping and assessment: a Muskegon, Michigan watershed case study with urban planning implications*. Advances in Spatial Planning, Intech, Rijeka, Croatia.

McKnight, M. (2008), *Applying the Q sort method: a qualitative classification of factors associated with the organization training support inventory (OTSI)*. In: Online Journal of Workforce Education and Development, III (2).

Mo, F., Le Cléach, G., Sales, M., Deyoung, G. & Burley, J. B. (2011), *Visual and environmental quality perception and preference in the People's Republic of China, France, and Portugal*. In: Internat. Journal of Energy and Environment, 4 (5), 549-556.

Oh, K. (1994), *A perceptual evaluation of computer-based landscape simulations*. In: Landscape and Urban Planning, 28, 201-216.

Pitt, D. & Sube [Zube], E. (1979), *The Q-Sort method: sse in landscape assessment research and landscape planning*. National Conference on Applied Techniques for Analysis and Management of the Visual Resource, Scientific Article A2592.

Previte, J., Pini, B. & Haslam-McKenzie, F. (2007), *Q methodology and rural research*. In: European Society for Rural Sociology, 47 (2).

Smardon, R., Palmer, J. & Felleman, J. (1986), *Foundations for Visual Project Analysis.*, New York, USA, John Wiley & Sons.

Swaffield, S. & Fairweather, J. (1996), *Investigation of attitudes towards the effects of land use change using image editing and Q sort method.* In: Landscape and Urban Planning, 35, 213-230.

Westin, D., Muderrisoglu, S., Fowler, C., Shedler, J. & Koren, D. (1997), *Affect regulation and affect experience: individual differences, group differences, and measurement using a Q-sort procedure.* In: Journal of Consulting and Clinical Psychology, 65 (3), 429-439.

Virtual Reality in Landscape Architecture

Integrating Bird Survey Data into Real Time 3D Visual and Aural Simulations

Ed MORGAN, Lewis GILL, Eckart LANGE and Martin DALLIMER

1 Introduction

URSULA, Urban River Corridors and Sustainable Living Agendas, is a major interdisciplinary research project examining the complex problem of sustainable development of urban riverside landscapes. It is generating large amounts of data which is to be used to aid the process of sustainability analysis, both from simulation models and from ecological surveys. Much of this data is "geospatial" but not directly visual and so there is interest in how such data sets can be visualised. HEHL-LANGE (2001) shows how abstract information on the distribution of species of wildlife can be visualized through colour-coding the 3D terrain model, also in combination with draped true color ortho-photography. TRAPP et al. (2009) describe a method of doing this with projective textures. Data-driven colourization will certainly be useful within the URSULA project, but whilst it is effective in integrating the data into the 3D visualisation it also obscures the photorealism present in the visualisation model.

Ecologists are surveying the rivers in Sheffield for existing wildlife and this data could be represented by colouring the visualisation model. However, for birds, which are one of the more charismatic elements of urban biodiversity, other methods may be more appropriate. For example, many birds are vocal and make sounds (e.g., calls and songs). Such vocalisations differ between species and are readily identifiable by experts and can also be distinguished by members of the public. Therefore a different approach to colouring, such as to introduce sounds into the visualisation to represent the data distribution, could be of particular relevance to bird survey data sets. This technique also makes sense as it translates the survey data back into a form which adds to the realism of a 3d real-time visualisation rather than diminishing it. Bird diversity and abundance are also good indicators of biodiversity, not least because for many species in the UK, cities can hold important populations of nationally declining species (FULLER et al. 2009). Although the number of species and individuals can be affected by the type and quality of the urban form (CHASE et al. 2006), the impact of urban developments on birds are rarely incorporated into the decision making process in a way that is readily understandable by members of the general public.

Serious games using interactive 3D landscape visualisation have been produced (MACH 2009), and have been shown to be an effective way of communicating a lot of sophisticated data and information to the end user of the game. In this paper, we describe methods of translating the bird survey data into bird sounds within a real time visualisation so that the sounds are originating from areas in which the birds were spotted. We then describe a serious game which evolved around the bird sound integration which allowed children to walk-through a virtual model of Sheffield riverside and spot different species of animals including some birds.

LANGE (2005) considers the inclusion of abstract and non-visual phenomena in visualisation in order to better communicate possible consequences of alternative future scenarios. As already stated, URSULA is producing much abstract data which could be beneficial to represent in visualisation models. Realism has been shown to have an effect on the perception of the viewer (LANGE 2001), and more accurate modelling of birds with 3D representations in the model, both visual and aural, should add to the level of acceptability for a viewer. To this end one particular data set, bird survey data, was selected and various ways of representing this in the visualisation model are described, firstly through colourisation, and then through the use of audible bird models. This approach is extended by introducing other animal representations within the visualisation model in order to create a serious wildlife observation game.

2 Converting Bird Survey into Virtual Bird Sounds

The following section describes the process undertaken to convert bird survey data into virtual bird sounds that play within existing virtual models.

2.1 Existing visualisation workflow

URSULA has already an established visualisation workflow (MORGAN et al. 2010) through which it has constructed several detailed 3D urban models, using visualisation software called Simmetry3d to allow real time walkthroughs of landscape models. This software is capable of playing sounds within a visualisation, both positional and directional with 6 different attenuation models, and also of animating objects within the virtual model. It also has an "action" based system which allows for animations or sounds to be triggered either when the user approaches an object or when an object is selected manually. It also has a plug-in API that allows extra features to be easily added without needing to change the host program itself.

2.2 Bird survey data

Six visits to the case study site in Sheffield were made during spring and early summer (April to June) 2009. This period corresponds to the breeding season and is therefore the time when birds are most active and readily observable. On each occasion the location, number, gender, activity and species of all birds was mapped using an approach similar to that recommended by BIBBY et al. (2000) for determining the territories of breeding birds. Subsequently bird locations were transferred to a GIS. For each species, the Density function of the Spatial Analyst extension in ARCView was used to construct a density surface. High values on this surface represented areas where bird species were more likely to be encountered. The surface could be output from GIS as a grey-scale image which could be used within the visualisation software; one of which is shown in figure 1.

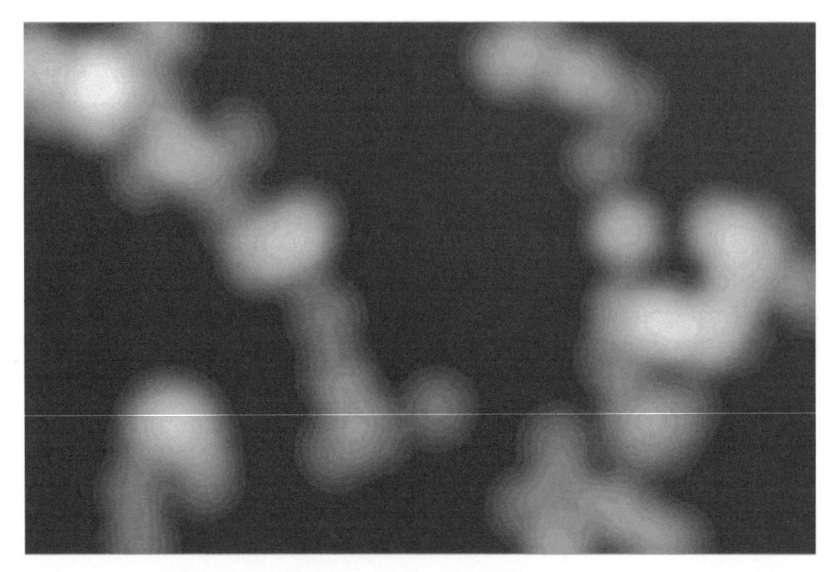

Fig. 1: Example of bird survey data – blackbirds

2.3 Colourisation technique

As a first attempt at conveying the bird survey data in the visualisations, the grey scale image was used to shade the visualisation model according to where the image was lighter/darker.

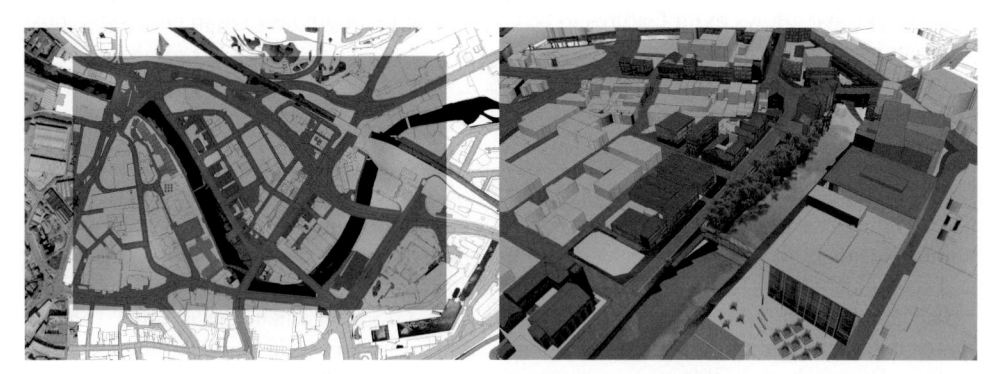

Fig. 2: Colourisation based on bird survey data – green represents increased chance of a sighting of a black bird

This produced an interesting way of visualising the dataset albeit in a rather cryptic fashion. A screen shot of this technique is shown in figure 2.

2.3 Integrating birds into the visualisation model

The data to be used to "visualise" a particular bird species was as follows:
- A density surface image (DSI) for the particular species
- One or more sound files (.wav) for each of the bird's calls
- A 2D (billboard) or 3d model of the bird, dependent upon availability of models

Each bird model was imported into the visualisation model, along with its sounds, and its DSI and then tagged with meta-data which linked these items together. There were two mechanisms experimented with, described as follows:

1. A bird model could be configured to play a sound when a person came close enough to hear it; a bird could be positioned in the visualisation model either roughly where the DSI suggested or by placing it exactly where it had been spotted in one of the surveys. This was essentially a manual approach, but quite easy to achieve, taking only a few seconds to place a bird in the correct spot. This method was extended to animate the bird along a pre-defined path. This was necessary as some birds were only spotted flying during compilation of the survey data.

2. A plugin to Simmetry3d was written to control the bird sounds in the real-time walkthrough viewing mode, which automatically mapped a bird DSI to positions where a bird sound could possibly be heard from, and then randomly positioned a bird, and made it sing based on the generated positions, the number of birds observed, a bird call sound, and the frequency at which they were to sing.

2.4 Wildlife Observation Game

A serious game was created as an extension to the bird visualisation which added other species found in UK urban river environments. This effort was motivated by an opportunity to do some outreach work through an invitation to participate and to engage with children at the "Wildlife of our Waterways" event in Sheffield. The overall event was held in the Weston Park Museum in Sheffield, a major museum focusing on natural and cultural history, as well as in the adjacent Weston park. There were many events to engage children, presented in a fun and welcoming format and focusing on and around wildlife in urban rivers.

A serious game was created as an extension to the bird visualisation which added other species found in UK urban river environments. This effort was motivated by an opportunity to do some outreach work through an invitation to participate and to engage with children at the "Wildlife of our Waterways" event in Sheffield. The overall event was held in the Weston Park Museum in Sheffield, a major museum focusing on natural and cultural history, as well as in the adjacent Weston park. There were many events to engage children, presented in a fun and welcoming format and focusing on and around wildlife in urban rivers.

A game concept was conceived which would allow children to walk around a virtual environment, with the task of identification of the wildlife in this virtual model; children would be given a paper checklist of wildlife they could spot along with points for each one depending on difficulty. The manual positioning method mentioned in the previous section was used to position various species into an existing urban river virtual model. This extended the work to introduce birds into the visualisation, by adding some static and animated 2d(billboard) and 3d representation of various animal species, and by using a pair of virtual binoculars to allow the children to zoom in and better identify the animals they spotted. Movement was controlled through a hand held game controller.

The aims of the game were as follows:
- Raise awareness of biodiversity
- Show children the possibilities of nature in the urban riverside environment
- Help children to identify specific species
- Provide a fun environment in which the previous aims could take place

The game was presented inside the museum using a laptop and data projector. The ages of children attending ranged from four to eleven, and about 30 children attempted to spot some virtual wildlife in about a 4 hour period. Figure 3 shows the game in action on the day, where parents and their children could sit down in front of the projector screen and control their path through the virtual environment.

Fig. 3: Some participants playing the game on the day

3 Results

The workflow developed to incorporate animals and their sounds proved quick and effective, providing the ability to create a serious game using it. The game proved to be a hit with the children; most were familiar with a game controller and accepted the virtual environment without issue. It was played both individually and with groups of up to three children. Every child who participated determinedly stuck at the game to find all the species listed. Not all the children were familiar with all the species listed, but this proved not to be a problem, and by the end of the game they were able to recognize and name these species. Some guidance was necessary to enable some of the participant children to find the gamut of species, and this was provided by parents or URSULA team members. The points system worked well to keep interest in the game and provided a sense of achievement once all species had been found.

Figure 4 shows a screen shot taken when the virtual binoculars were in use; the children enjoyed using them, and they allowed for a much better view of the smaller animals. The binocular effect was achieved by narrowing the viewing angle of the virtual camera by a factor of ten, and then overlaying a binocular-shaped mask on the view. In addition to this, the sensitivity of the controls for looking left, right, up and down also needed to be scaled by a similiar amount to allow for finer scale adjustment of the view direction.

Fig. 4: A view of a kingfisher through the virtual binoculars

4 Conclusions and Outlook

A simple approach for translating bird survey data into both static and animated bird sounds has been described, along with its extension into a "serious game" which allowed children to walk through a virtual environment and learn to recognise different species of wildlife. This work highlights the potential for engagement and education of children using virtual landscape environments with objectives. Including bird sounds in the visualisation does provide an indicator of biodiversity and we intend to investigate some predictive modelling strategies which could alter the density surfaces in response to changes to the physical environment.

Acknowledgements

This paper is based on work undertaken within the URSULA project, funded by the UK Engineering and Physical Sciences Research Council (Grant number: EP/F007388/1). The authors are grateful for this support, but they would like to note that the views presented in the paper are those of the authors and cannot be taken as indicative in any way of the position of the funders or of colleagues and partners. Any remaining errors are similarly those of the authors alone.

References

Bibby, C. J., Burgess, N. D., Hill, D. A. & Mustoe, S. H. (2000), *Bird Census Techniques*. 2nd Edition. London, Academic Press.

Chace, J. F. & Walsh, J. J. (2006), *Urban effects on native avifauna: a review*. In: Landscape and Urban Planning, 74.

Fuller, R. A., Tratalos, J. & Gaston, K. J. (2009), *How many birds are there in a city of half a million people?* I)n: Diversity and Distributions, 15.

Hehl-Lange, S. (2001), *Structural elements of the visual landscape and their ecological functions*. In: Landscape and Urban Planning, 54 (1-4).

Lange, E. (2001), *The limits of realism: perceptions of virtual landscapes*. In: Landscape and Urban Planning, 54.

Lange, E. (2005), *Issues and Questions for Research in Communicating with the Public through Visualizations*. In: Buhmann, E. et al. (Eds.), Trends in real-time landscape visualization and participation. Heidelberg, Wichmann.

Mach, R. (2010), *Serious Gaming? Raising Awareness for Retention Measures and Flood Disasters*. In: Peer Reviewed Proceedings of Digital Landscape Architecture, ed. by Buhmann, E. Et al. (Eds.), Berlin/Offenbach, Wichmann.

Morgan, E., Gill, L., Lange, E. & Romano, D. (2010), *Rapid Prototyping of Urban River Corridors Using 3D Interactive, Real-time Graphics*. In: Peer Reviewed Proceedings Proceedings Digital Landscape Architecture, ed. by Buhmann, E. et al. (Eds.), Berlin/Offenbach, Wichmann.

Trapp, M. & Döllner, J. (2009), *Dynamic Mapping of Raster-Data for 3D Geovirtual Environments*. 13th International Conference on IEEE Information Visualization, IEEE Computer Society Press.

The Communication Value of Graphic Visualizations Prepared with Differing Levels of Detail

David BARBARASH

1 Introduction

This study investigates viewer attitudes regarding the communication effectiveness of computer graphic visualizations (CGVs) with differing levels of visual detail. The emphasis is on levels of detail generated through the application of geometrical refinement, material appearance and texture, and lighting effects in computer generated 3D models. The viewer response group is intended to approximate that associated with the project review process common to an American public Town Planning Board meeting. The purpose is to determine which levels of illustrative content, in this case created from a digital 3D model, are perceived by lay respondents as containing adequate visual detail to foster an informed opinion regarding the merits of specific design elements for a proposed development project. A secondary purpose is to determine the production efficiency of preparing CGVs of differing visual quality relative to their usefulness in communicating design intent.

Other researchers have investigated topics of graphic communication. DANIEL & MEITNER (2001) found that only very high levels of graphic detail are acceptable when examining aspects of scenic beauty in the landscape. However, these conclusions apply mainly to landscapes with panoramic views and significant levels of natural or rural scenery. In contrast, LANGE (2001) determined that any level of detail could be valid in the right situation, though his work focused on large scale views from a land planning perspective. Studies by BISHOP & ROHRMANN (2001) found that computer generated animation is not a valid substitute for real life experiences, and yet they concluded that the animations have some value as a vicarious experience of the real world for certain scene elements only.

In addition to providing for decision making, designers and their visualizations often have to demonstrate design aspects that can be unpopular to the interested public. The impacts of these possibly contentious elements must be shown in sufficient detail to provide fair representation. Neto stated that "the public were much more open to options normally considered unacceptable when they saw what these options would look like in real life" (NETO 2006, 348).

In order to display a project's scope, there must be enough detail in an image to show the project's context in the existing landscape, major amenities and organizational elements, and any areas of conflict. This has been a contentious issue in previous studies, and research has come to conflicting conclusions in measuring abstractness versus realism in levels of detail. Lange found that "even simulations with a lower degree of realism can still contain the most important information needed for a specific purpose" (LANGE 2001, 165) and Hall determined, "The impression of realism does not necessarily require correct imagery in terms of geometric detail as long as the general behavior is reasonable; that high image complexity is primary in creating the perception of realism; that subtle shading and surface detail are key in creating the perception of realism" (HALL 1990, 195). These studies state that any level of detail can be valid in the right situation, but others refute this idea, finding that only the highest possible level of realism is a valid representation of scenic beauty as shown in Daniel and Meitner's 2001 study. During his 2006 Ribeira study, Neto determined that both realistic and abstract representations should be used to communicate designs to both specialists and laypeople.

Other studies single out specific scene elements that must contain a higher degree of realism as in Bishop and Rohrmann's study comparing animation to a real life walkthrough of a site, stating; "most viewers accepted the presentation as reasonably valid" despite rating the "vegetation, [and] colors [as] not fully convincing" (BISHOP & ROHRMANN 2003, 275) Neto supports this idea by claiming that with current computer technology, it is "impossible to simulate our real experience of a space thoroughly" and that we should instead "try to 'catch a likeness' that reveals a key aspect of a prospective design" (NETO 2006, 350) when creating CGVs.

2 Methods

2.1 Computer Generated Visualizations

An imaginary site and project was created to minimize any possible real world bias of survey respondents (NIMBY effects, traffic concerns, etc.). The model was drafted using AutoCAD 2007, then processed in 3D Studio Max 2008 to create the 3D geometry, material textures, and lighting. Three sets of seven images were created from this model, each set representing a unique level of rendered detail. (See Fig. 1)

1. Image Set A – Low Detail (massing) Model – All designed elements are represented by basic geometry with minimal modeled detail. Colors were applied to objects to represent proposed materials without the use of texture mapping. A single direct spotlight provided illumination for the scene while generating shadow mapped shadows. Images were rendered using 3D Studio Max's default scanline renderer.
2. Image Set B – Mid-Detail Model – Design elements are recognizable with basic detail modeled into buildings, curbs, etc. Objects have image based textures applied to represent their proposed materials while vegetation is represented by mid-resolution (150 ppi) image maps placed in a billboard X. A single direct spotlight acted as the sun and cast advanced ray traced shadows, with less intense "fill lights" included to brighten areas of the scene without casting shadows. 3D Studio Max's standard methods of calculating radiosity were used to generate the lighting and shadows.
3. Image Set C – High Detail Model – Design elements are modeled to their actual intent and to the best ability of the author (but were limited by the computer at the time of the study). Objects have image based textures, bump maps, and other custom maps (specular, reflection/refraction, etc.) to best represent real world materiality. 3D Studio Max's sunlight system was used along with the third party V-Ray renderer, with global illumination (GI) enabled to simulate real world lighting and shadows.

Seven vantage points were chosen, representing views commonly used to display a design for presentation to a public review board when seeking project approval. Views were rendered though 3D Studio Max's standard camera, simulating a photograph taken through a 55mm lens, providing an approximate 45° field of view. These viewpoints were shared

Fig. 1: Views at three levels of detail (Image Sets A, B, & C from right to left)

between the three levels of detail when creating the static 2D rendered images to ensure similar experiences between sample groups. The time spent performing each task of the modeling, texturing, lighting, and rendering process was recorded individually for use in a time/value analysis. The rendered CGV's were printed on 8.5"x11" 28lb. bond on a high quality inkjet printer, one image per sheet and were loosely bound in sets with a paper clip.

2.2 The Survey

This study's intent was to survey a sample group thought to represent people likely to be participants in a public town board meeting for project review and approval. Due to scheduling constraints, access to actual town board meetings was not possible, so volunteers were recruited from people using public libraries. Interested passers-by were asked to participate without discrimination in order to collect a random sample. After reading the instructions provided on the survey instrument, participants were randomly assigned a single set of images; either A, B, or C. Respondents were asked to view one set of computer generated visualizations from a single level of detail in order to avoid preference bias between the three levels of detail, as people are likely to choose the highest level of detail; the "prettier picture", when able to compare across detail sets (LANGE 2001, 173).

The survey (see Fig. 2) aimed to measure whether respondents believed that they were able to understand the design intent displayed in the computer generated visualizations and whether they found the images to be an acceptable communication device. It consisted of both qualitative and quantitative questions in order to obtain a measurable scale of acceptance for each level of detail, as well as a set of open ended questions to allow respondents to elaborate on what they believed was most effective or lacking in their set of images. Quantitative responses were measured on a five (5) point Likert scale, with one (1) representing strong disagreement, and five (5), a strong agreement of the question statement. An answer of NA (not applicable) was provided for respondents who felt that they could not respond to a statement based on the materials provided.

Specific elements in the images were called out in pairs of questions on the survey, broken down to the buildings, traffic patterns, vegetation, and wayfinding elements. One question dealt with arrangement of scene elements, and the other with their appearance. Separating the two allowed this study to analyze whether representational detail contributes to, or interferes with image comprehension and decision making.

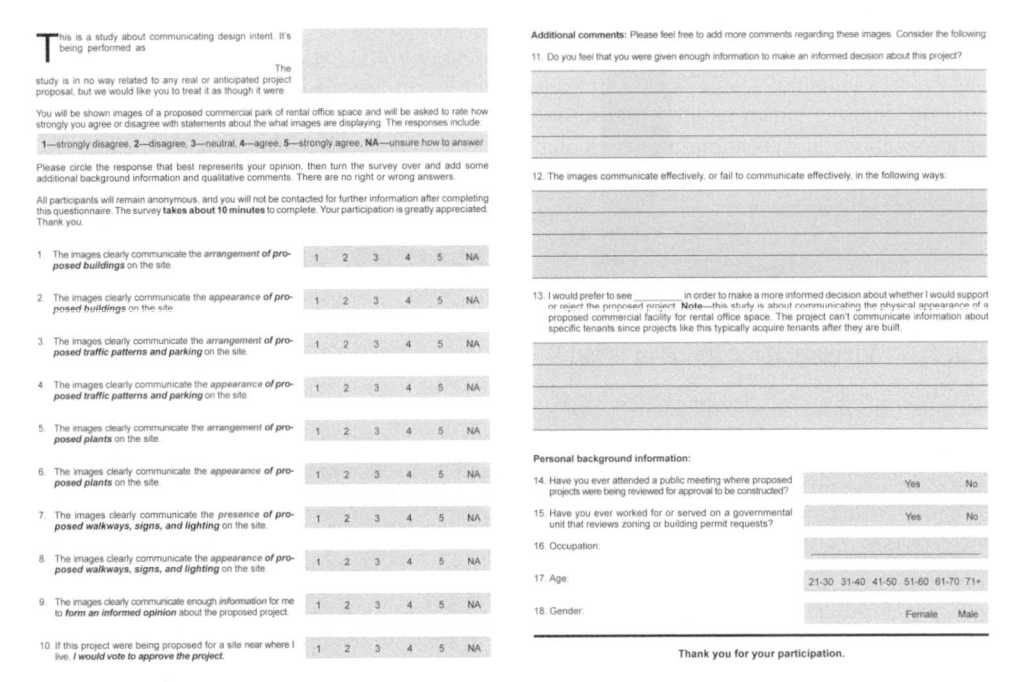

Fig. 2: The studies two page survey form

3 Results and Findings

3.1 Quantitative Results

A three-way analysis of variance (ANOVA) was done over the first eight Likert questions for all three sample groups and a single factor ANOVA was run for questions 9 & 10 (see Fig 2). The analysis indicates there is a significant difference between image sets A-B and A-C, but not between B-C (P-value <.01); with Groups B and C showing nearly identical means (4.09 and 4.07 respectively) and Group A being quite a bit lower at 3.74. This indicates that overall the people in Group A thought the graphics they were viewing were less effective at communicating various aspects of the project being depicted than did those in Groups B and C.

The analysis indicates that there were no significant differences between the sample groups for most questions. However, there were significant differences (P-values < .01) for questions two and nine (see Fig. 2). This indicates that the people in Group A thought the graphics they were viewing were less effective at communicating project content than did those in Groups B and C. One would expect that understanding and communication of design elements would increase as greater levels of detail were shown, but instead the data shows that Image Set B (the mid-detail image set) was preferred by survey respondents.

Responses by gender and surveyed age group did not show a statistically significant difference within or between image sets, though male respondents tended to be slightly less approving of the images overall.

There was no statistically significant difference in viewer responses for the arrangement of scene elements, while the appearance of the buildings in Image Set A received a significantly lower rating from the other levels of detail. While the appearance of the vegetation in the low detail set did not show a statistical difference from the other sets, respondents wanted to see more plant material (based on responses to open ended questions) despite vegetation amounts and locations being identical across all levels of detail. Image Set A also revealed a significant difference in whether people felt that the images contained enough information for them to make an informed decision on the project.

The results show some unexplainable discrepancies. For example, the geometry used for roads and parking, vegetation, walkways, lights, and signs in Image Sets B and C were identical. There was enough modeled detail in the mid-detail set that it was not worth the modeler's time to improve upon them for the high detail set outside of material adjustments. Despite this, Image Set C received noticeably higher marks for traffic patterns, parking, walkways, signs, and lighting, while Image Set B saw a higher mean in reference to the vegetation. Each of these discrepancies occurs in the questions regarding the arrangement of site elements; responses dealing with appearance were rated much closer between Image Sets B and C.

With the exception of vegetation in Image Set B, interpretation of the arrangement of scene elements was rated higher than that of their appearance. This demonstrates that project design information could be gathered independently of the graphic quality of representative elements. Representative detail did have a noticeable effect on whether people were able to make an informed decision on the project or in voting for its approval, as shown in Image Set A's low ratings for questions 9 and 10.

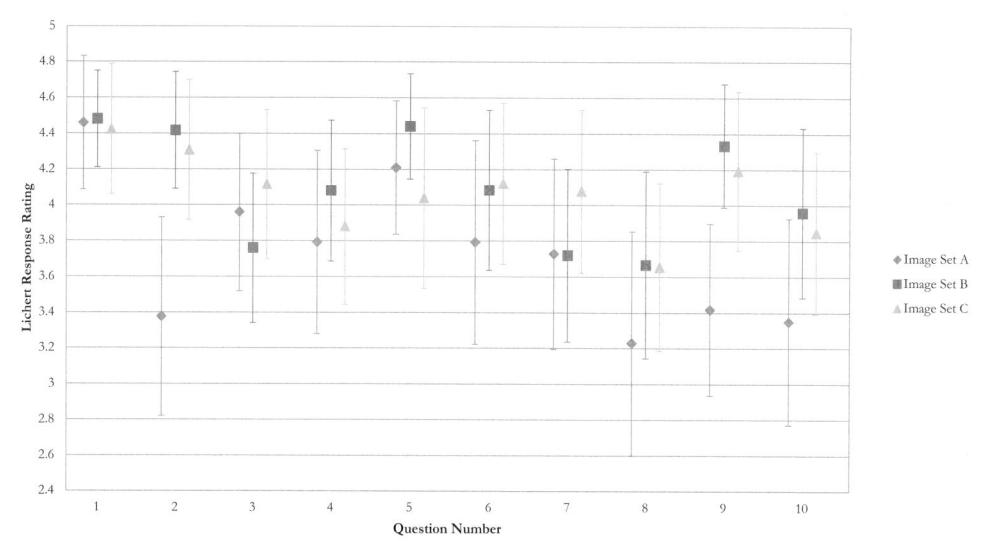

Fig. 3: Mean response per question with 95% confidence intervals

3.2 Qualitative Comments

In creating the 3D models, it was decided to keep off-site elements to a minimum to keep survey respondents focused on the proposed improvements. Adjacent buildings were modeled simply and quickly to provide context, but the ground plane was purposely ignored. However, the most common comment from all groups was that more off-site context would help them in making decisions. If a designer is required to provide more off-site context in the 3D model beyond the conceptual massing level, not only would model creation time increase, but rendering time as well. This could force a reduction in model quality with the increased computer memory load of off-site details, textures, and lighting solutions.

Another often requested element was for supporting analysis diagrams, surveyed traffic, and pedestrian figures. This indicates that respondents were being conscientious about the project evaluation, but the data requested was outside the scope of what this study sought to discover. Despite the instructions included on the survey form stating the imaginary nature of the project and the purposeful lack of proposed tenant information, many respondents wrote that they could not vote for project approval without knowing the types of businesses that would reside in the buildings.

Other comments for the low detail CGVs (Image Set A) asked for more detail, specifically in the buildings, and that there did not seem to be enough plants shown. For the mid-detail images (Image Set B), only one person requested more detail, with most other comments asking for building tenant information and more lighting, both on buildings and in the landscape.

Respondents to Image Set C were able to look beyond what was shown in the model and notice things that were missing, such as ADA ramps, handicap parking spaces, HVAC equipment, vehicular directional arrows, and building lighting. Comments for the other detail sets focused more on the site surroundings or a lack of detail in the modeled elements.

3.3 Time and Value Assessment

Digital Image Creation Time Invested (Hours, minutes, seconds)				
	CAD Setup	Image Set A	Image Set B	Image Set C
CAD Drafting	4:00:00	4:00:00	4:00:00	4:00:00
Line Clean-up	1:15:00	1:15:00	2:00:00	2:30:00
Terrain Modeling		2:30:00	3:30:00	3:30:00
Building Modeling		0:15:00	2:45:00	5:15:00
Landscape Modeling		0:30:00	1:15:00	3:15:00
Entourage Modeling		0:30:00	1:30:00	4:00:00
Materials		0:30:00	1:30:00	1:30:00
Lighting		0:30:00	0:45:00	1:15:00
Post-Processing		0:15:00	0:30:00	1:00:00
Total Time	5:15:00	10:00:00	17:45:00	26:15:00
Rendering Time		0:01:36	0:13:13	76:36:15
Total Per Image Set		10:01:36	17:58:13	102:51:15

Fig. 4: Time invested per element and per level of detail

Initial set-up for the 3D models used Autodesk's AutoCAD 2007 to create the basic 2D line work for the project. This necessitated some clean-up of the lines after importing into 3D Studio Max as some of the smooth curves created in CAD were misinterpreted, changing their radii. This optimization took 5 hours 15 minutes.

The low detail model elements took 4 hours 45 minutes to model and 1 minute 36 seconds to render all seven final images, making a total of 10 hours 1 minute 36 seconds for completion of image set A after adding in the CAD setup time. The largest amount of time went into the terrain, 2 hours 30 minutes, which includes the roads, curbs, sidewalks, and landscaped areas.

Image set B, displaying the mid-detail model added an additional 7 hours 45 minutes to the project time with a total of 17 hours 45 minutes for modeling alone. Instead starting from scratch, the lower detail elements were reused and refined to create the greater level of detail. Here the most time was spent modeling the buildings at 2 hours 45 minutes, though an additional hour was needed to refine the terrain, meaning that if this were the intended level of detail, then the terrain modeling would still require the greatest amount of time.

In creating the rendered CGVs, the mid-detail model used a different method of lighting than the low detail model. Radiosity approximates light reflection and refraction by shooting parallel "light beams" from a light source and calculates the angle of deflection, color pickup, and intensity loss as they hit a modeled element depending on the material applied to its surface. These calculations need only be performed once, and then any number of rendered images can be produced from the model. The radiosity calculations took 27 minutes 50 seconds for image set B, and when combined with the time spent rendering each of the seven views, the total rendering time was 43 minutes 43 seconds.

The high detail model required similar refinement as the mid-detail model did, though again, elements were re-used from the earlier models instead of starting over. Total modeling time for image set C was 26 hours 15 minutes, with the largest investment in time given to the buildings. The terrain modeling was deemed sufficiently detailed enough to re-use the mid-detail elements without need for refinement in geometry, though the materials were changed. The third party plug-in V-Ray was able to use global illumination (GI) to calculate lighting for the scene. GI simulates real world conditions better than radiosity alone, providing smooth lighting gradients and shadows. The increased realism came at a heavy price, with the total rendering time for the seven high detail images at 76 hours 36 minutes 15 seconds. If the computer used for the modeling of this project was a professional machine configured for 3D modeling, then 3D trees and people could have been used, increasing the level of detail at the cost of greatly longer render times. As it was, the computer was unable to handle the processing demands.

This study attempted various methods of creating 3D geometry, along with different lighting setups and material settings. Reviews of models in progress, and problems encountered in model processing, led to changes in direction or further refinement of geometry that added unforeseen time to the modeling process. The time associated with this learning curve was not included in the time invested for the time-value analysis. Because the design and the site were imaginary, the model and the rendered views were able to change as the focus of the project progressed.

An office creating 3D content is likely to have computers built specifically for the task with workflow and graphic style standards that would streamline this type of work. It is also likely that they would have a library of purchased and re-purposed 3D content. If this study had access to an existing source of content, the models for this project could have been created much faster, allowing a designer to focus on increasing project specific detail instead of having to start from scratch for each element. Lighting rigs and materials are likely to have been standardized (and approved), and the ability to re-purpose not only saves time up front, but reduces the need for a large amount of time wasting test renders and changes.

3.4 Results

Though the comments seem to show otherwise, Image Set B received nearly identical marks (a .028 difference between the means) to those of Image Set C when analyzing the Likert responses despite the enormous differential in time invested. It should be noted that Image Set B did seem to communicate better than Image Set C, especially in questions 9 & 10 (see Figures 2 & 3) despite the difference in levels of detail. The mid-detail set (B) of images took less than 18 percent of the time needed to produce compared to the high detail set (C). The comments show that the low detail set was lacking in a number of areas and would likely only be used inside of an office as a massing study model.

The time investment for creating the high detail set of images seems to be cost inefficient given the survey results compared to the mid-detail set, as it required 8 hours 30 minutes more of a designers time, and nearly 73 hours of additional computing time to create the rendered computer generated visualizations. This time could be reduced with faster computers, but that requires a larger investment in hardware and time differences between the two levels of detail would likely remain constant.

The results of this study show that the public would be willing to accept a lower level of detail for some content (traffic patterns and entourage elements in Image Set A), but desire greater levels of detail for others (buildings and lighting in Image Sets B & C); that there is no single consistent level of acceptable detail. This is stated with the understanding that this study's results are based on a single presented design and further study is necessary to support these findings, especially as computers are able to create ever more convincing representations of reality.

If a client or a review board were to request greater detail in a set of images, the comments and data show that a designer's time would be best spent refining architecture, being sure to include more character defining elements like doorknobs, entry signs, building lights, and railings. Following this would be site entourage elements like walkways, lighting, signage, and public utilities.

Key References

Al-Kodmany, K. (1999), *Using visualization techniques for enhancing public participation in planning and design: process, implementation, and evaluation.* In: Landscape and Urban Planning, 45, 37-45.

Appleton, K. & Lovett, A. (2003), *GIS-based visualization of rural landscapes: defining 'suffient' realism for environmental decision-making.* In: Landscape and Urban Planning, 65, 117-131.

Appleton, K., Lovett, A., Sunnenberg, G. & Dockerty, T. (2002), *Rural landscape visualization from GIS databases: a comparison of approaches, options and problems.* In: Computers, Environment and Urban Systems, 26, 141-162.

Bishop, I. D. & Rohrmann, B. (2003), *Subjective responses to simulated and real environments: a comparison.* In: Landscape and Urban Planning, 65, 261-277.

Daniel, T. C. & Meitner, M. M. (2001), *Representational validity of landscape visualizations: the effects of graphical realism on perceived scenic beauty of forest vistas.* In: Journal of Environmental Psychology, 21, 61-72.

Hall, R. (1990), *Algorithims for realistic image synthesis.* In: Computer Graphics Techniques: Theory and Practice, ed. by Rogers, D. F. & Earnsha, R. A. New York, Springer.

Karjalainen, E. & Tyrvainen, L. (2002), *Visualization in forest landscape preference research: a Finnish perspective.* In: Landscape and Urban Planning, 59, 13-28.

Lange, E. (2001), *The limits of realism: perceptions of virtual landscapes.* In: Landscape and Urban Planning, 54, 163-182.

Muhar, A. (1999), *Three-dimensional Modelling and Visualization of Vegetation for Landscape Simulation.* In: Landscape and Urban Planning (Our Visual Landscape).

Neto, P. L. (2006), *Public Perception in Contemporary Portugal: The Digital Representation of Space.* In: Journal of Urban Design, 11 (3), 347-366.

Oh, K. (1994), *A perceptual evaluation of computer-based landscape simulations.* In: Landscape and Urban Planning, 28, 2001-216.

Orland, B., Budthimedhee, K. & Uusitalo, J. (2001), *Considering virtual worlds as representations of landscape realities and as tools for landscape planning.* In: Landscape and Urban Planning, 54, 139-148.

Prince, S. (1996), *True Lies: Perceptual Realism, Digital Images, and Film Theory.* In: Film Quarterly, 49 (3), 27-37.

Tress, B. & Tress, G. (2003), *Scenario visualization for participatory landscape planning – a study from Denmark.* In: Landscape and Urban Planning, 64, 161-178.

Who's Afraid of Virtual Darkness – Affective Appraisal of Night-time Virtual Environments

Joske M. HOUTKAMP and Alexander TOET

1 Introduction

We investigated to what extent simulated darkness determines the affective appraisal of desktop virtual environments (VEs).

Computer simulations have become indispensable tools to communicate design and planning impacts and to investigate human perception of built environments (e.g. TAHRANI & MOREAU 2008). Desktop VEs are also increasingly deployed to study the effects of environmental qualities and interventions on human behaviour and feelings of safety in built environments (COZENS et al. 2003, PARK et al. 2008, PARK et al. 2010). The effectiveness of desktop VEs for these applications depends critically on their ability to correctly address the user's emotional, cognitive and perceptual experience. In the real world ambient darkness elicits fear of victimization (BOX et al. 1988) by concealing potential dangers (BLÖBAUM & HUNECKE 2005, GRAY 1987, NASAR & JONES 1997, WARR 1990). Darkness may turn places that are pleasant during daylight into frightening places after dark (HANYU 1997, NASAR & JONES 1997). The innate fear for darkness which most people have also extrapolates to immersive virtual environments (MÜHLBERGER et al. 2007). Although commercial desktop games sometimes deploy low-key lighting to evoke suspense and dread (NIEDENTHAL 2005), it is not yet known if darkness in desktop VEs can also effectively induce fear related emotional responses, and determine the affective appraisal of the VE.

Affective appraisals are judgments concerning the capacity of the appraised environment to alter an individuals' mood (RUSSELL & SNODGRASS 1987). An affective or emotional reaction refers to an internal state (such as fear) that a person feels in relation to the environment (NASAR 2008). Affective appraisals are not necessarily accompanied by an affective or emotional change (RUSSELL & SNODGRASS 1987) Whether a reaction occurs depends on the strength of the stimulus, the relevance the event or environment has for the viewer, and the viewer's affective state and characteristics (e.g. personality). The affective response can be mapped on the two dimensions pleasure and arousal of Russell's circumplex model of affect (RUSSELL 2003). A conscious experience of one's affective state can be described as an integral blend of these two dimensions (and thus as a single point in this map). For instance, fear is an emotion high in arousal and low in pleasure, directed at a specific object or event. Anxiety is generally considered as a mood, also high in arousal and low in pleasure, but objectless.

Here we report two studies that were performed to investigate how simulated darkness determines the affective appraisal of, and emotional reaction to, desktop virtual environments.

In the first study participants inspected a virtual model of a small Italian village, either in daytime or in night-time conditions, and gave their affective appraisals of the environment afterwards. The results of this study showed only a minor effect of simulated darkness on the affective appraisal of the desktop VE. Probable reasons for this result may have been the presence of reassuring effects like auditory cues suggesting social presence, and the fact that the inspection task lacked any personal relevance. As a result, the simulated darkness probably did not evoke any fear related associations.

We therefore performed a second study in which participants were requested to explore either a daytime or a night-time version of a desktop VE representing a typical Dutch deserted rural area. The lack of social presence and the deserted nature of the VE served to evoke victimization related associations. In some conditions an attempt was made to enhance the personal relevance of the simulation, and thereby its emotion inducing capability, by leading the participants to believe that the virtual exploration tour would prepare them for a similar tour through a corresponding real environment. The emotional state of the participants, their mood and feelings of presence in the VE, and their affective appraisal of the VE were measured through self-report.

In this paper we will discuss the results of both studies and their implications for the validity of desktop VEs as an appropriate medium for both etiological (e.g. the effects of signs of darkness on walking behaviour and fear of crime) and intervention (e.g. effects of street lighting) research.

2 Experiment 1: Reconnaissance of a Small Italian Village

Darkness is known to induce unpleasant feelings like fear and anxiety, and this effect is enhanced by prior exposure to a social stressor (GRILLON et al. 2007). We therefore hypothesized that after a stressful experience, participants would score a night-time version of a desktop VE on Russell's (RUSSELL 2003) pleasure-arousal scale as significantly less pleasant and more arousing than its equivalent daytime version.

2.1 Methods

We first exposed 52 young male participants (aged between 18 and 32 years) either to a non-stressful control (reading) task, or to a task that elicits stress: the validated psychosocial Trier Social Stress Test (TSST: KIRSCHBAUM et al. 1993, WILLIAMS & HAGERTY 2004). Then they explored either a daytime or a night-time version of a virtual environment (the Italy level of Counter-Strike®), representing a small village with typical Italian architecture, with narrow streets, steep stairs, and with a market place in its centre surrounded by houses. While exploring, the participants heard a simulation of the sound of their footsteps, as well as all sorts of background noise (music, a singing voice, a passing airplane, wind, rumour) through their headphones. Participants were instructed to imagine that they were military scouts situated in an unknown village, whose task it was to perform a reconnaissance of this village. This task served to ensure that they would perform a thorough visual inspection of the VE. The participants were unaware that their affective appraisal of the environment would be tested at a later time. Room lighting remained on for the participant who explored the daytime VE, and was turned off for those who explored the night-time VE. To objectively assess their anxiety level we measured free salivary cortisol and heart rate. A validated Dutch translation of the state self-report scale (VAN DER PLOEG 2000) from the Spielberger State-Trait Anxiety Inventory (STAI: SPIELBERGER 1985) was administered to assess how anxious participants felt, both before and after the experiment. The affective appraisal of the VE was measured with a semantic questionnaire (RUSSELL & PRATT 1980). Full details of this experiment are given elsewhere (TOET et al. 2009).

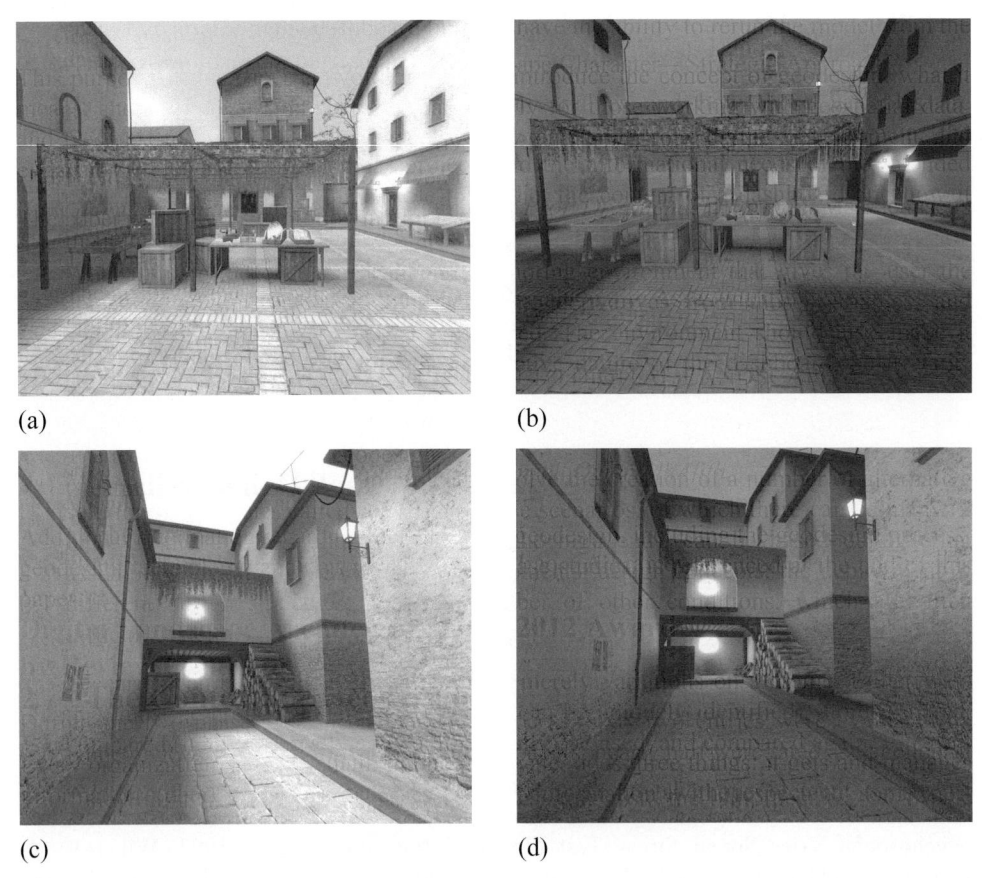

(a)

(b)

(c)

(d)

Fig. 1: Screenshots of a village square (a,b) and an alley (c,d) in the daytime (a,c) and nighttime (b,d) virtual environment

2.2 Results and Discussion

The TSST effectively induced distress: free salivary cortisol levels, heart beat rates, and scores on the TSAI were all higher after the TSST than before. We found that on the pleasure-arousal scale used for assessing the appraisal (RUSSELL & PRATT 1980) the desktop VE was considered less pleasant and more unpleasant by the simulated nighttime lighting conditions, but not more arousing. Also acute prior stress did not evoke the expected appraisals, nor did it elicit higher anxiety in the participants during the task in the

VE. Since a prior affective experience can bias the appraisal of the corresponding affective quality of an environment in a direction opposite to the value of the prior experience (RUSSELL & LANIUS 1984), the unpleasant TSST experience may have shifted the appraisal of the VE in the direction of pleasantness, thus counteracting a possible darkness effect. Another reason for the lack of an overall effect of darkness may be the rather friendly atmosphere of the VE. Although the experimental area contained several locations of entrapment (blocked escape) and concealment (blocked prospect), which are factors known to induce fear (FISHER & NASAR 1992a, FISHER & NASAR 1992b, NASAR & FISHER 1993, NASAR & JONES 1997), the presence of simulated artificial lighting focussed on restorative details may have had a reassuring effect (NIKUNEN & KORPELA 2011, NIKUNEN & KOPPELA 2009), just like some auditory cues such as music and singing voices which could be heard at some moments during the simulation and which may have suggested social presence. In addition, the unfamiliar task probably lacked any personal relevance, and therefore failed to induce any emotions. As a result, the simulated darkness probably did not evoke any fear related associations.

3 Experiment 2: Exploring a Deserted Dutch polder

In this second experiment participants were requested to explore either a daytime or a night-time VE representing a typical Dutch deserted rural area. The lack of social presence and the deserted nature of the VE were meant to evoke victimization related associations. In some conditions additional information served to enhance the personal relevance of the simulation. We hypothesised that the night-time VE would be appraised as less cosy and more tense than its daytime equivalent, and that increased personal relevance of the VE would enhance its emotion inducing capability and thereby indirectly amplify the effects of simulated darkness on the affective appraisal of the VE.

3.1 Methods

A sample of 72 female volunteers (aged between 17 and 32 years) were requested to explore either a daytime or a night-time version of a VE representing a deserted Dutch polder landscape (Levee Patroller: HARTEVELD et al. 2007) and to draw a map of the area afterwards. The VE contained a small village, with some houses, roads and grasslands next to a large canal. A levee protected the village from the canal. The original version of the simulation was used, which contained no living creatures and only a few dynamic elements (e.g. rain, clouds, undulating water surfaces and a moving gate: HOUTKAMP et al. 2008). In the daytime condition the environment was lit by the sun. In the night-time condition streetlights along the road and stars in the partly clouded sky provided the only illumination. The lack of social presence and the deserted nature of the VE served to evoke victimization related associations. Young females were selected as participants since this group is particularly susceptible to fear of darkness (WARR 1984, WARR 1990). The map drawing assignment merely served to ensure that the participants would cover most of the area, and that they would not simply stay in one part. In some conditions participants were led to believe that the virtual walking tour would prepare them for an unaccompanied tour through a corresponding real environment, in lighting conditions that would either be the same or the opposite of those shown in the simulation (darkness/daylight). This fictitious follow-up assignment was meant to enhance the personal relevance of the simulation, and thereby its emotion inducing capability. The emotional state of the participants, their mood and feelings of presence in the VE, and their affective appraisal of the VE were measured through self-report. A validated pictorial rating scale (the Self-Assessment Manikin: BRADLEY & LANG 1994) was used to measure emotional state of the participants. A validated translation of the Positive and Negative Affect Scale (PANAS: ENGELEN et al. 2006, PEETERS et al. 1996) was used to measure the mood of the participants. Presence was measured using the Dutch translation of the Igroup Presence Questionnaire (IPQ: SCHUBERT et al. 2001). The affective appraisal of the VE was measured using a differential rating scale that was originally designed to measure ambience in indoor environments with different lighting conditions (VOGELS 2008). It includes four factors (cosiness, liveliness, tenseness, detachment) that are similar to the dimensions arousal and pleasure in Russell's circumplex model of affect (RUSSELL 2003). Full details of this experiment will be presented elsewhere (VREUGDENHIL et al. 2012).

Fig. 2: Screenshots of a road on a dike (a,b) and a path near a canal (c,d) in the daytime (a,c) and nighttime (b,d) virtual environment

3.2 Results and Discussion

The night-time version of the VE was appraised as less cosy and more tense than its daytime equivalent. In both conditions the VE experience was significantly displeasing by itself, while the simulation of darkness had an arousing effect (measured with the SAM). This result agrees with the earlier finding of Rohrman and Bishop (ROHRMANN & BISHOP 2002) that people appraise a night-time VE as more arousing than a daytime VE.

Although the fictitious follow-up assignment reduced the overall positive mood (measured with the PANAS) of the participants, it did not affect their emotional state (measured with the SAM), and also did not influence their affective appraisal of the VE. Participants who experienced the darkness condition without the suggestion of a similar real-world follow-up task showed significantly higher positive affect (PANAS) than participants who had been informed that they would be requested to perform such a task. In combination with the finding that darkness in the VE had an arousing effect, this result suggests that participants found the night-time VE more arousing than its daytime equivalent when the experience had a further personal relevance. Overall, the degrees of presence and involvement experienced by the participants in this study were rather low.

4 Conclusions

A night-time VE was appraised as less pleasant or cosy and more tense than its daytime equivalent. However, simulated darkness had only minor effects on the emotional state of the viewers (e.g. on their arousal level). Darkness also did not elicit anxiety, even after prior stress. In this respect desktop VE representations differ from immersive night-time VE representations, which are capable to elicit fear related responses. There are indications that the personal relevance of a desktop VE may affect the emotional state of the user (the suggestion of a follow-up task reduced positive affect), and thus influence her affective appraisal of the VE. This finding suggests that it is essential to carefully consider the context in which desktop VE representations are deployed for serious gaming and training as well as for instance the assessment of personal safety in urban areas.

Acknowledgement

This research has been supported by the GATE project, funded by the Netherlands Organization for Scientific Research (NWO).

References

Blöbaum, A. & Hunecke, M. (2005), *Perceived danger in urban public space. The impact of physical features and personal factors*. In: Environment and Behavior, 37 (4), 465-486.

Box, S., Hale, V. & Andrews, G. (1988), *Explaining fear of crime*. In: The British Journal of Criminology, 28 (3), 340-356.

Bradley, M. M. & Lang, P. J. (1994), *Measuring emotion: the self-assessment manikin and the semantic differential*. In: Journal of Behavior Therapy and Experimental Psychiatry, 25 (1), 49-59.

Cozens, P., Neal, R., Whitaker, J. & Hillier, D. (2003), *Investigating personal safety at railway stations using "virtual reality" technology*. In: Facilties, 21 (7/8), 188-194.

Engelen, U., De Peuter, S., Victoir, A., Van Diest. I. & Van den Bergh, O. (2006), *Verdere validering van de Positive and Negative Affect Schedule (PANAS) en vergelijking van twee Nederlandstalige versies [Further validation of the Positive and Negative Affect Schedule (PANAS) and comparison of two Dutch versions.]*. In: Gedrag & Gezondheid, 34 (2), 89-102.

Fisher, B. S. & Nasar, J. L. (1992a), *Fear of crime in relation to three exterior site features: prospect, refuge, and escape*. In: Environment and Behavior, 24 (1), 35-65.

Fisher, B. S. & Nasar, J. L. (1992b), *Fear of crime in relation to three exterior site features: prospect, refuge, and escape*. In: Environment and Behavior, 24 (1), 35-65.

Gray, J. A. (1987), *The psychology of fear and stress*. Cambridge, UK, Cambridge University Press.

Grillon, C., Duncko, R., Covington, M. F., Kopperman, L. & Kling, M. A. (2007), *Acute stress potentiates anxiety in humans*. In: Biological Psychiatry, 62 (10), 1183-1186.

Hanyu, K. (1997), *Visual properties and affective appraisals in residential areas after dark*. In: Journal of Environmental Psychology, 17 (4), 301-315.

Harteveld, C., Guimarães, R., Mayer, I. & Bidarra, R. (2007), *Balancing pedagogy, game and reality components within a unique serious game for training levee inspection*. In: Hui, K. et al. (Eds.), Proceedings of the Second International Conference on Technologies for E-Learning and Digital Entertainment (Edutainment 2007). Berlin/Heidelberg, Germany, Springer.

Houtkamp, J. M., Schuurink, E. L. & Toet, A. (2008), *Thunderstorms in my computer: the effect of visual dynamics and sound in a 3D environment*. In: Bannatyne, M. & Counsell, J. (Eds.), Proceedings of the International Conference on Visualisation in Built and Rural Environments BuiltViz'08, IEEE Computer Society, Los Alamitos, USA

Kirschbaum, C., Pirke, K. M. & Hellhammer, D. H. (1993), *The 'Trier Social Stress Test – a tool for investigating psychobiological stress responses in a laboratory setting*. In: Neuropsychobiology, 28 (1-2), 76-81.

Mühlberger, A., Wieser, M. J. & Pauli, P. (2007), *Darkness-enhanced startle responses in ecologically valid environments: a virtual tunnel driving experiment*. In: Biological Psychology, 77 (1), 47-52.

Nasar, J. L. (2008), *Assessing Perceptions of Environments for Active Living*. In: American Journal of Preventive Medicine, 34 (4), 357-363.

Nasar, J. L. & Fisher, B. (1993), *'Hot spots' of fear and crime: A multi-method investigation*. In: Journal of Environmental Psychology, 13 (3), 187-206.

Nasar, J. L. & Jones, K. M. (1997), *Landscapes of fear and stress*. In: Environment and Behavior, 29 (3), 291-323.

Niedenthal, S. (2005), *Shadowplay: simulated illumination in game worlds*. Proceedings of DiGRA 2005 Conference: Changing Views – Worlds in Play, University of Vancouver, Vancouver, Canada.

Nikunen, H. & Korpela, K. M. (2011), *The effects of scene contents and focus of light on perceived restorativeness, fear and preference in nightscapes*. In: Journal of Environmental Planning and Management [Online], 1-16.

Nikunen, H. J. & Koppela, K. M. (2009), *Restorative lighting environments – Does the focus of light have an effect on restorative experiences?* In: Journal of Light & Visual Environment, 33 (1), 37-45.

Park, A. J., Calvert, T., Brantingham, P. L. & Brantingham, P. J. (2008), *The use of virtual and mixed reality environments for urban behavioural studies.* In: PsychNology Journal, 6 (2), 119-130.

Park, A. J., Spicer, V., Guterres, M., Brantingham, P. L. & Jenion, G. (2010), *Testing perception of crime in a virtual environment.* Proceedings of the 2010 IEEE International Conference on Intelligence and Security Informatics (ISI), IEEE Press, Washington, USA

Peeters, F. P. M. L., Ponds, R. W. H. M. & Vermeeren, M. T. G. (1996), *Affectiviteit en zeltbeoordeling van depressie en angst.* In: Tijdschrift voor Psychiatrie, 38 (3), 240-250.

Rohrmann, B. & Bishop, I. D. (2002), *Subjective responses to computer simulations of urban environments.* In: Journal of Environmental Psychology, 22 (4), 319-331.

Russell, J. A. (2003), *Core affect and the psychological construction of emotion.* In: Psychological Review, 110 (1), 145-172.

Russell, J. A. & Lanius, U. F. (1984), *Adaptation level and the affective appraisal of environments.* In: Journal of Environmental Psychology, 4 (2), 119-135.

Russell, J. A. & Pratt, G. (1980), *A description of the affective quality attributed to environments.* In: Journal of Personality and Social Psychology, 38 (2), 311-322.

Russell, J. A. & Snodgrass, J. (1987), *Emotion and the environment.* In: Stokols & Altman (Eds.), Handbook of environmental psychology, New York, USA, John Wiley & Sons.

Schubert, T., Friedmann, F. & Regenbrecht, H. (2001), *The experience of presence: factor analytic insights.* In: Presence: Tele-operators and Virtual environments, 10 (3), 266-281.

Spielberger, C. D. (1985), *Assessment of state and trait anxiety: conceptual and methodological issues.* In: Southern Psychologist, 2 (4), 6-16.

Tahrani, S. & Moreau, G. (2008), *Integration of immersive walking to analyse urban daylighting ambiences.* In: Journal of Urban Design, 13 (1), 99-123.

Toet, A., van Welie. M. & Houtkamp, J. M. (2009), *Is a dark virtual environment scary?* In: CyberPsychology & Behavior, 12 (4), 363-371.

van der Ploeg, H. M. (2000), *Handleiding bij de Zelfbeoordelingsvragenlijst: een Nederlandse bewerking van de Spielberger State Trait Anxiety Inventory [Manual to the Dutch version of the Spielberger State Trait Anxiety Inventory].* Lisse, The Netherlands, Swets Test Publishers.

Vogels, I. (2008), *Atmosphere metrics. Development of a tool to quantify experienced atmosphere.* In: Westerink, Ouwerkerk, Overbeek, Pasveer & de Ruyter (Eds.), Probing Experience. From Assessment of User Emotions and Behaviour to Development of Products. Dordrecht, The Netherlands, Springer Netherlands.

Vreugdenhil, P., Toet, A. & Houtkamp, J. M. (2012), *Effects of simulated darkness on the affective appraisal of a virtual environment.* In: Computers, Environment and Urban Systems (submitted).

Warr, M. (1984), *Fear of victimization: why are women and the elderly more afraid?* In: Social Science Quarterly, 65 (3), 681-702.

Warr, M. (1990), *Dangerous situations: social context and fear of victimization.* In: Social Forces, 68 (3), 891-907.

Williams, A. & Hagerty, B. M. (2004), *Trier Social Stress test: a method for use in nursing research.* In: Nursing research, 53: 277-280.

Digitalized Re-Rendering of a City's Landscape

Dhruv CHANDWANIA and Aarti VERMA

1 Introduction

Pune has gradually evolved into a dynamic city of academic, cultural and economic importance from what it was as only a cultural hub with hum drum of any urban character. It is now the 9th largest metropolis in India catering to almost 3M (en.wikipedia.org) people and still increasing as people are migrating from surrounding areas to find a niche for themselves. One of the reasons Pune is getting popular is because of its comforting weather. Pune is today acknowledged as the knowledge and cultural capital of Maharashtra. The city's character and growth patterns reflect the conglomeration of socio-economic structure, open space structure and its cultural heritage. The recent growth of Information Technology sector has been the impetus towards creating a multi-cultural atmosphere in the city.

2 Response to the Design Ideologies ... Concept Generation

The challenge while formulating responses towards designing urban spaces is diverse. Should architecture take clues from the old patterns generating a compatible form or should it be futuristic? Should it respond to the natural systems, environment, history and context of the place and an evaluation of the existing urban pattern or should it form an identity on its own? With varied social and economic strata to behold, whom should the design be for? What methods would best suit the changing paradigms of design? Should the methodology rework existing methods within architecture – by using already available Computer Aided Design software in the experimental ways, or even by rewriting the structure of the software adapting and catering to a designers need or could it borrow techniques from other industries or fields to generate new architectural strategies?

This is the stage where computational parametric designing plays an important role. Digital Designing can describe how architects can more effectively engage recent technological developments to produce forms of architecture that further generate feedback from their users and with the culture at large. Designers are at a constant pressure of keeping up with developments in cultures and technologies together adhering to flexibility in generating new techniques. Lately, with digital practices, these techniques having been inspired from other emerging fields like economic modeling systems, aerospace and automotive industries, etc to name a few.

'Technologies such as 3 dimensional modeling; software may have inspired innovative designs at their foundation. Technological practices employ feedback from the environment into their design process rather than beginning with a preconceived design and applying it to a context. Thus, the traditional design process of concept, analysis and construction give way to incorporating perceptual feedback between analysis, intervention and exchange with environment. Computers' potential for generating real time feedback and a more dynamic interactivity between design and users is tapped' (RAHIM 2006).

Fig. 1: The iconic Wayag archipelago located between Gag Island and Manuran Island in the Raja Ampat islands

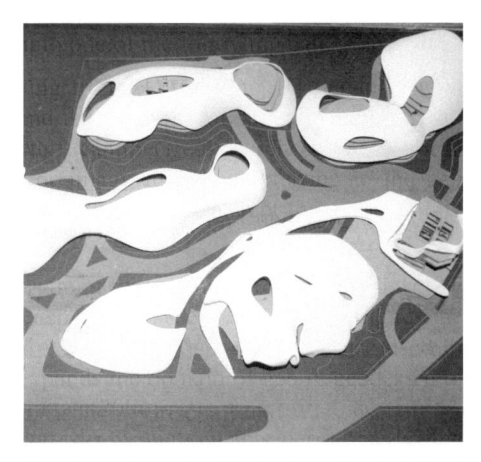

Fig. 2: Housing Project by Snehal Pisal (student from 4th Year B. Arch, BNCA)

Architecture is generally conceived – designed and realized built in response to an existing set of conditions. These conditions may be purely functional in nature, or they may also reflect in varying degrees, the social, political and economical status of a city. The initial phase of any design is the recognition of a problematic condition and a decision to find a solution to it. The above example of a design by student from 4th year B. Arch, got inspired by a rivulet which passed by the site and wanted to use archipelago based massing. The design was developed by using parametric tools in the computers to have relative geometries.

Pune has distinctive varied political, social and economical strata and is growing rapidly towards a cosmopolitan culture. Above factors along with architectural language affect open and public spaces, where such parametric interventions can be suggested. One of such public spaces located at the center of the city is the Sambhaji Park on the banks of Mutha River.

3 Sambhaji Park and Its Extents as a Case Study

Sambhaji Park and its environs were selected as a case study owing to its significant location and the complexities of parameters. Defining parameters included the busy commercial street (J.M.Road) which has high end retail attracting young generation, proximity to one of the biggest bus transport terminus of the city and hence an inflow of commuters from various economic strata, the river edge conditions forming the most important connection with nature, and the cultural medley.

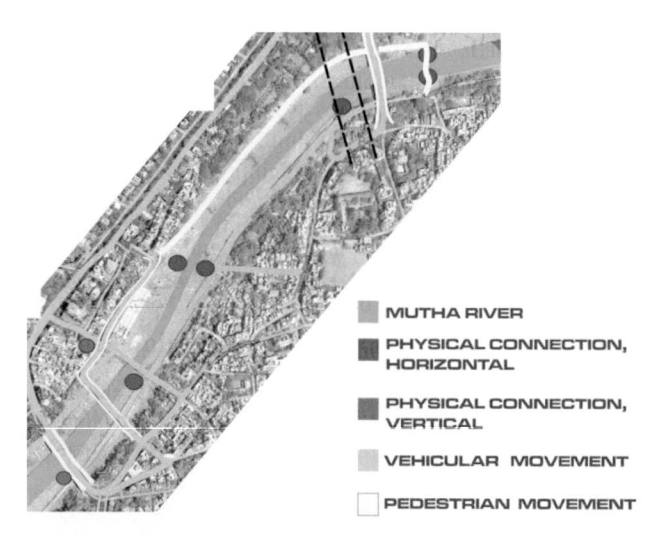

Fig. 3: J. M. Road and vicinity mapping study. HRHC Workshop, BNCA.

Apart from vibrant and dynamic surroundings, it caters to a variety of cultural gatherings in one of the oldest theatre "The Bal Gandharva Rang Mandir" which witnesses activities almost throughout the year. The space also witnesses intermittent activities like exhibitions, circus and cultural gatherings (Melas) once a year.

4 The Design Process

In this case, the study area was divided into four study zones depending on the complexities and uniqueness of the parameters they offered. Each parameter was then assigned a definitive value, depending on its significance. The existing conditions were then documented, defined with respect to its context and analyzed.

4.1 Zone 1: River edge as an open space

The defining parameters included:
1. Value as a significant open space
2. Confluence of two major developments (old and the new)
3. Pedestrian and vehicular connections with a heavy traffic load
4. Ecological zone and a riverine habitat
5. Visual Architectural character of the city skyline
6. Informal activities like food joints, Mela, Ghats and parking spaces

A value of these parameters combined with design concepts evolved dynamic sketches that generated prototypes of recluse within these zones. Following is the detailing of one of the model with its process which shows how the entities have strong inter-relationships through which different diagrams are prepared. These diagrams further follow a system of transformation from 1D to 3D suggesting opportunities of habituation.

Visual Connector_Pavilion: Workshop at BNCA on Digital Architecture.

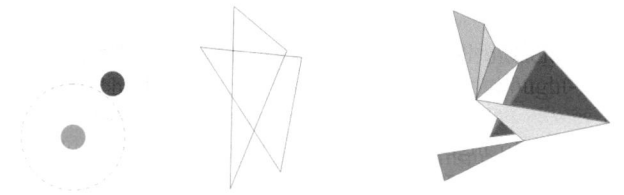

Fig. 4: Stage 1: The base parameters like aesthetics and functions are combined generating series of graphics and suggesting deformation of fabric

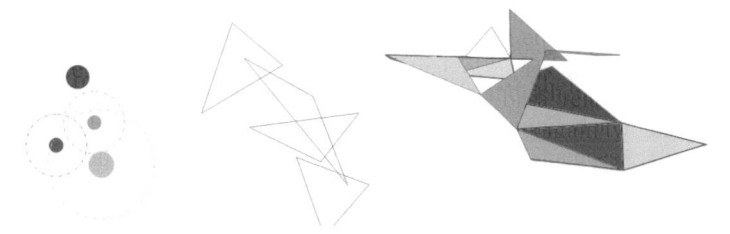

Fig. 5: Stage 2: Addition of new parameters create reaction amongst the entities formulating more dynamic fabrics suggesting habitual functions

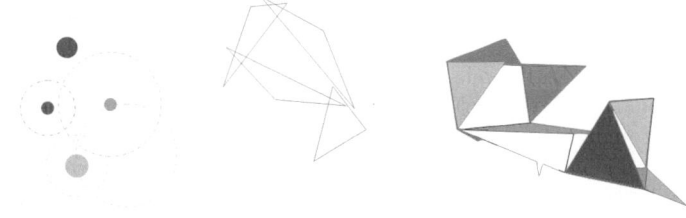

Fig. 6: Stage 3: The frozen iteration from the sequence showing a harmonious frame between two extreme parameters like the old and the new city

Fig. 7: Model by Isha Rowtu (Student from 3rd year B. Arch, BNCA)

4.2 Zone 2: Traffic Corridors

This kind of space usually displays a typical urban character in the form of its dynamic skyline and a continuous movement pattern. Surrounding land use is predominantly commercial, at times a mix of commercial and residential. The skyline depicting a Doppler effect suggested the most defining parameter while attempting to design a recluse for this zone.

Traffic Corridors_Urban Connector: Workshop @ BNCA on Digital Architecture.

Fig. 8: Above: J. M. Road Skyline sketch, HRHC Workshop, BNCA

Fig. 9: Below: Doppler Effect, Graphic

Relationship between open and built spaces has perceptive pattern congruent to that of Doppler Effect. With that as a pretext, the parameters for the following model were purely patterned geometry.

Fig. 10: Model by Meenakshi Dravid (Student from 3rd year B. Arch, BNCA)

4.3 Zone 3: Urban Plazas

The space necessarily needs to be an interactive space for the young generation with a contemporary, dynamic and lively expression. It could cater to many informal activities like just lounging around mostly used during nighttime. The activity specifics could include shopping, recreation and a passive recluse. The urban plazas are places where different activities can overlap/merge. These crossroads can be mapped with public movement and interventions generated from cultural/social contexts. Hence these diagrams suggest responsive areas (both positive and negative) which are considered as parameters while generating design solutions. Such small interventions in culmination bind the spaces at a larger scale. The common thread lies in the cultural importance, social awareness and modernity at par. These spaces have been designed from the data that is gathered with mapping exercises and through a post designing methodology where a strong perception of public domain is considered.

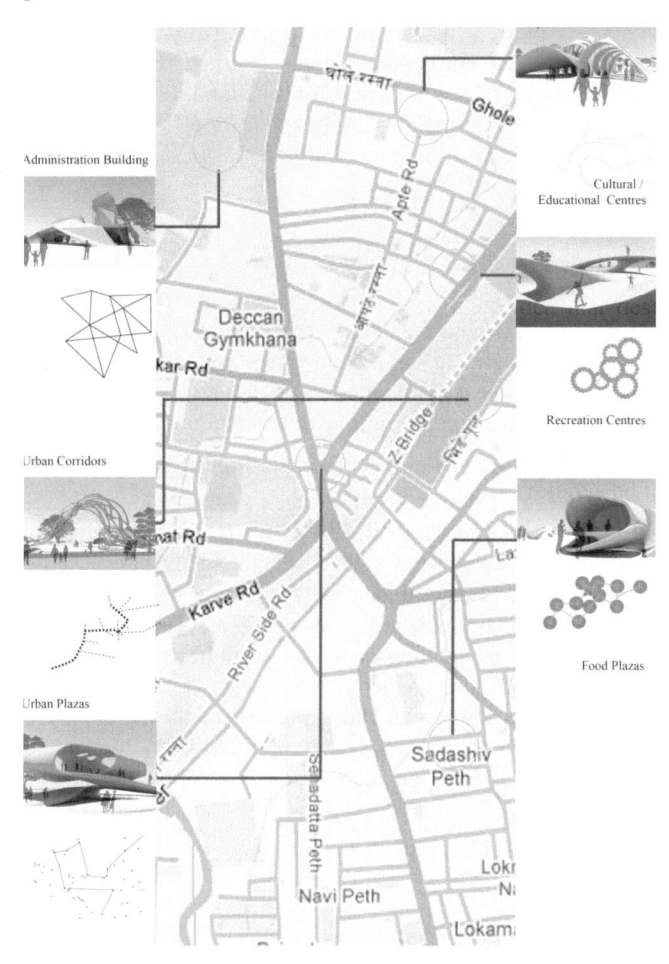

Fig. 11: Possible Interventions of Prototypes

Projects: Bhat Amulya, Wasunkar Priyanka, Dutia Bijal, Kudale Priyam, Puppal Akshada, Changedia Dharti, Raichur Mayuri, Naik Netra (Students from 3rd Year B. Arch, BNCA).

5 Application of these Methodologies

The attempt of redesigning these kinds of spaces was s small endeavor in designing spaces that are more definitive. At a city level, the possibilities are countless. The identification of such spaces commence with the personal open spaces within a plot area, transport corridors of varied scale, formal and informal gathering spaces in the form of *Maidens* (town square, http://en.wikipedia.org/wiki/Maidan) and public parks, interactions of natural elements with the built forms, in short, a complete comprehensive study of open and built spaces also portraying a typical character of the city.

Adding to these dynamics is the regulating governance, people's participation and perceptions, time management and economic policies, a realm that ventures into urban and town planning. A Landscape Architect plays a major role in working on an integrated approach where public sector, the development world, stakeholders, architects and urbanists work together to develop integrated concepts that answer questions beyond just forming an artistic expression of architecture. They can help making landscape truly efficient in order to solve environmental problems that are social, ecological and political in order to create healthy livable environments within the city.

6 Conclusion

This presented work has tried to detail out a new designing methodology, geo designing, where project is concieved from conceptualizing to developement of design through simulations in the computer. These simulations are generated through a series of systamtic programming and thus helps a designer acheive more iterations in less amount of time.

Taking a culturally strong site for intervention that can reinterprete the urban utility with parametric designing has been the main aim of the projects presented above. With reference to Pune city (and that may be true for any other Indian city as well), these parameters change dramatically for every sector or block or at times even within the blocks. Generating prototypes is also governed by the parametric calculations within a time span. With the technological advancements and the cultural globalization, designing a space resilient to change is challenging. At any given time, the design process is always in transition with the changing paradigms and the bigger the scale, the greater these complexities. This paper hence tried to focus on a scaled projects like townships and expansion of the cities where through design approach we can graft urban characters. These modified grafts generated would represent a strong amalgamation of expression of a designer's approach and the existing conditions proposing unified variance.

References

Ching,. F (1996), Form, Space and Order.
http://en.wikipedia.org/wiki/Pune.
McHarg, I. *Design With Nature.*
Rahim, A. (2006), *Catalytic Formations: Architecture and Digital Design*
The International Review of Landscape Architecture and Urban Design (2010), Topos, 73, *City Regeneration.*
The International Review of Landscape Architecture and Urban Design (2010), Topos; 71, *Landscape Urbanism.*
Waterman. T. (2010), *Urban Design.* Basic Landscape Architecture 01.

Standardization and BIM

GIS + BIM = Integrated Project Delivery @ Penn State

David E. GOLDBERG, Robert J. HOLLAND and Scott W. WING

1 Introduction

In 2007, at the American Society of Landscape Architects Conference in San Francisco, Jim Sipes of AECOM presented, "Applying Building Information Modeling to Landscape Architecture." There he shared a recent survey by the American Institutes of Architects showing that three-quarters of architecture firms in the country are using 3D and building information modeling (BIM) in their practices. Moreover, clients are increasingly requiring that BIM be a part of the qualifications when selecting design teams. Unfortunately for landscape architects, the current BIM software and standards were developed for the modeling of buildings—not sites. Sipes advocated that the profession needs to leverage software vendors and standards committees to further develop BIM technologies to incorporate site elements—and suggested that this new technology be called site information modeling. He later added that landscape architects need to be proficient in this 'new' technology or be left behind (SIPES 2007).

The challenge then is if BIM is not designed for site modeling, how do we educate landscape architecture students in BIM? The answer might be in the delivery process and not in the technology.

2 Methods

2.1 BIM and Integrated Project Delivery (IPD)

Owners, designers and contractors are exploring BIM as a way to change the design and construction process to produce more coordinated buildings at lower life-cycle cost with less risk, shorter project schedules and, potentially, facilitate more sustainable designs. Many companies are actively seeking graduates who can effectively work on these types of projects. To meet these evolving demands of the design and construction processes, universities have implemented a variety of courses to expose students to the new BIM software platforms. While BIM is a powerful digital tool, its effectiveness can be severely limited if it is not applied in an efficient and collaborative process. To this end, students should be exposed not only to the new software, but should also have an opportunity to utilize this new software in an integrated collaborative environment to design a project to meet certain specific project performance goals.

In 2007 the American Institute of Architects (AIA) National published the Integrated Project Delivery (IPD) Guide. The Guide defines IPD as a project delivery approach that "integrates people, systems, business structures and practices into a process that collaboratively harnesses the talents and insights of all participants to optimize project results, increase value to the owner, reduce waste, and maximize efficiency through all phases of design, fabrication, and construction" (AIA 2007).

2.2 The Studio Collaborative

The architectural design and construction process is highly interdisciplinary. The accrediting boards of the allied disciplines require collaboration as a learning component. While they do not specify how this collaboration occurs they only ask that is achieved (POERSCHKE 2011). The IPD/BIM Collaborative Studio involves students and faculty from architecture, landscape architecture and four distinct architectural engineering disciplines in a design studio project, which explores BIM technology as a collaborative design tool. The studio is organized around the IPD process. Thus the IPD/BIM Collaborative Studio is providing an opportunity for students to not only become proficient in new digital tools, but perhaps more importantly, exposing them to a more "real world" collaborative design process.

The studio began three years ago as a small size interdisciplinary studio format (eighteen students from the architecture, landscape architecture and architectural engineering departments). Today, the studio has expanded to thirty students (five teams of six). Tasks undertaken include architectural, landscape and engineering design, energy analysis, cost estimating, scheduling, constructability, coordination and clash detection. Course content also includes an overview of BIM and its application to the design and construction process (including organizational and application challenges and potential legal issues), current BIM software as well as BIM trends in the design professions and construction industry.

Landscape architecture students apply their knowledge of site analysis, site engineering, and site design to that of building systems and performances. For example, they learn to appreciate that their goal to provide shading to the site directly impacts the program of mechanical architectural engineer who aims to increase the natural daylight to an interior space.

2.3 Studio Space

2.3.1 Integrated Studio

The physical studio space for a successful IPD/BIM collaboration is a challenge. It is not a matter of appropriate hardware and software, but the physical space. The studio is held in a student computer lab where the hardware and most software are available, but the workstation configuration is not conducive for team collaborations. Workstations in University labs tend to be arranged for individual research or classroom learning – not for team collaboration. Preferred arrangements are planned, but not yet implemented. Another constraint is software. Architectural engineering students working in an architectural and landscape architectural lab do not have access to specific engineering software. The workstations are equipped with the latest versions of Revit Architecture, Revit MEP, Revit Structure, ArcGIS, AutoCAD Architecture, Civil 3D, Ecotect, Navisworks, Project Vasari, and SketchUp. They are not equipped with specialized architectural engineering software such as ETABS, DAYSIM, RISA-2D, STAAD Pro, Trance Trace, eQUEST, and AGi32. Therefore, students wishing to use those specialized products must use alternate labs or load on mobile workstations.

2.3.2 Virtual Studio

To accommodate the needs of the growing studio and to allow greater access to specialized software, students are encouraged to work independently in their preferred spaces, but collaborate with the studio via enabling technologies such as Autodesk's Project Bluestreak, TeamViewer, and Adobe Connect. Each landscape architectural student is equipped with an Asus Slate notebook to facilitate team collaboration. Using the Slates, students can utilize their preferred space and remotely interact with their studio colleagues via displays and multi-touch surfaces.

2.3.3 Presentations

Teams present their projects in three phases, BIM Execution Plan, Schematic Design, and Design Development, to their peers, faculty, clients, and outside professionals (Fig. 1). Professionals unable to travel to the University are accommodated to jury presentations via online collaboration tools such as Adobe Connect. The outside jurors can watch and listen to live presentations and type questions for the teams to respond to. This allows the studio to invite highly qualified jurors to the presentation regardless of their location.

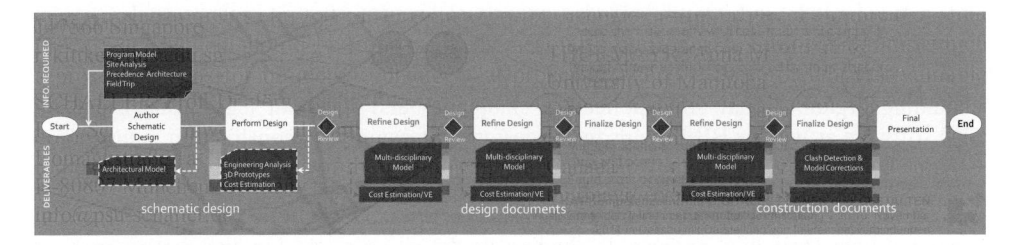

Fig. 1: Student team graphic illustration of their BIM Execution Plan

2.4 Project Selection

The project is carefully selected for the IPD/BIM Studio, as the program must be complex enough for a semester-long endeavor for all disciplines. This is often a challenge for landscape architecture. The original request for proposals for building projects seldom includes a detailed site program for the team to work with. Therefore, site programs are enhanced for the studio.

The availability of a real client, project team and detailed design information for the actual project greatly enriches the experience.

The project should also be local so that students can make regular site visits (although this need may be mitigated with better online maps). Through the success of the 2010 studio offering, the ideal building size is around 20,000 S.F., though buildings have ranged upward to 100,000 S.F. in size. The optimization of building size is primarily to accommodate the time constraints of a semester studio.

In order for the students to work toward an integrated model, their designs must be completed midway through the semester. Otherwise, there is not enough time for the BIM integration and performance analysis to occur.

To date the projects types have all been academic institutions. The first was a prototyped elementary school, the second an on-campus daycare facility, the third an elementary school added to an existing school site, and this year an intramural sports complex addition.

According to Sipes, campuses are one project type where landscape architects could be using BIM. These types of projects involve the integration of buildings with a large site and from a master planning process also typically involves developing an understanding of the condition and capacity of academic and residential facilities and infrastructure; open spaces and landscape features; pedestrian, vehicular, and parking networks; academic needs; and the interface between a university and the larger community (SIPES 2008).

2.5 Team Selection

According to the AIA IPD Guide, "The project team is the lifeblood of IPD. In IPD, project participants come together as an integrated team, with the common overriding goal of designing and constructing a successful project." This is true for all projects, but when complications arise in traditional teams individual discipline groups tend to "batten down the hatches" to protect their financial interest. In contrast, IPD demands that the team works together to resolve the issue. Because of this, the composition of the integrated team and the ability of team members to adapt to a new way of performing their skills within the team are critical (AIA 2007).

For the IPD/BIM Collaborative Studio this semester we have formed five student teams, each with a full complement of disciplines (architecture, landscape architecture and the four AE options: construction, structural, mechanical and lighting/electrical engineering). Our goal was to not only find thirty students, but to find thirty highly-motivated students with at least minimal background in REVIT or other BIM platform programs. Each student submitted their academic credentials along with a statement as to why they wanted to take the BIM Collaborative Studio. To establish the team assignments each students was to complete a survey of three questions: (1) whom do you not want to work with; (2) what is your level of BIM experience; (3) whom do you want to work with? Generally, this has worked well in establishing collegial and BIM competent teams.

2.6 Faculty and External Collaboration

The IPD/BIM Collaborative Studio is essentially three discrete courses collaborating on a single project. There is a course in architecture, landscape architecture, and architectural engineering each assigned a faculty member in the respective department. Additionally, a teaching assistant from architectural engineering and a research assistant from architecture are assigned to this course. Additional faculty members from architecture, landscape architecture and architectural engineering attend studio work sessions and formal project reviews. While input by faculty members at presentations and critiques is valuable, there is a need for on-going professional discipline support for the students in order to gain maximum benefit from the IPD/BIM studio.

Outside practitioners who are working on the same project as the students are invited to participate in both formal presentations and studio critiques. One studio activity that really benefits the teams is an activity we call "speed consulting." In these sessions, professionals from several disciplines (e.g. civil engineering, construction engineering, architecture, landscape architecture, and architectural engineering) move from team to team and provide feedback at a timed interval. This both enlightens the students with professional advice and energizes the practitioners with a satisfaction with the understanding that what they struggle with in their daily practice is also a challenge in academia. Practitioners also provide instructional models on subjects such as code and energy modeling.

A BIM Wiki previously developed at Penn State is also made available as a resource for students in both the BIM Collaborative Studio and other departmental initiatives (BIM Wiki).

2.7 The BIM Integration

The BIM and IPD process is a natural fit. One can say that the IPD process is only possible with the implementation of BIM and the success of BIM is only possible with the IPD process. This is true for the architects and architectural engineers in building projects, but BIM is a challenge for the students in landscape architecture.

Landscape architecture students are highly proficient with CAD, GIS, and 3D modeling, but few join the studio with BIM experience. They learn BIM from their architectural engineering teammates who have learned BIM in their earlier coursework. The landscape architectural limitation is not because of the lack of BIM education, but rather the lack of site-specific BIM software. The tools are not designed for their needs. "While Autodesk provides almost a limitless library of building, structural, and mechanical components, its landscape components are limited" (FLOHR 2010). Autodesk representatives at conferences have proposed Revit Site or Revit Civil, but none have made their way to the retail Revit suite. Therefore, the landscape architecture students have utilized other software applications to contribute to the BIM model including Civil 3D.

Landscape architecture students are using Autodesk Civil 3D to generate topographic site models. These models feature existing and proposed contours for their grading plans and include stormwater management elements. These final TIN models are not easily exported and imported to Revit, therefore they are simplified as 3D polylines and the TIN file is recreated in Revit (Fig. 2). This process works well for creating more detailed context to the architectural building, but does little for the building analysis.

Fig. 2: Civil 3D terrain imported into Revit

2.8 The GIS Integration

Landscape architecture students performed suitability analysis for last year's proposed elementary school by inventorying and analyzing slope, views, linkages, vegetation, and existing infrastructure. The resulting map was converted to vector shapes to drive the location and orientation of the building in the Revit model.

This year's project of an intramural building addition will likely not warrant detailed suitability analysis for siting a building, but context mapping of circulation systems and infrastructure, along with site views to and from building, will be evaluated.

In a campus setting, this project has potential of being a GIS-centered BIM project, where the Revit models are located on the GIS, rather than traditional BIM-centered workflow where the map elements are exported to the Revit models.

2.9 Other Technologies

In addition to GIS and BIM software, students utilize additional software tools for conceptual design and presentation in the IPD/BIM Collaborative Studio. Students take full advantage of the simplicity of SketchUp for conceptual design. SketchUp allows students to quickly generate concepts that are then imported into Revit. The building's structural and mechanical components are designed in Revit and then exported back to SketchUp for further conceptual modeling before being returned to Revit for the final model. In the past, several teams utilized this 'round tripping'– especially those with limited Revit experience (Fig. 3).

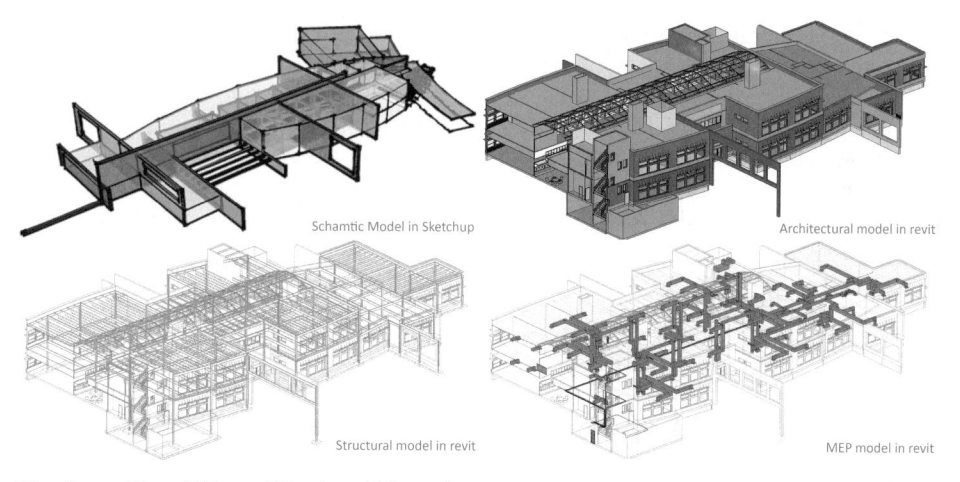

Fig. 3: SketchUp and Revit collaboration

The teams utilize Autodesk 3ds Max to render interior materials and lighting scenarios for their final presentations. Teams also rendered site plans in the past studios with 3ds Max, but never with full integration of the BIM model. Instead, students touch-up their views or simulate landscape plans with the tools of Adobe Photoshop (Fig. 4). This years' studio aims to explore more integrated approaches to site rendering without the need to perform photo simulations. Navisworks is also used for systems integration (clash detection) and 4D modeling.

Fig. 4: Student Civil 3D terrain imported to Revit, Rendered in Revit and enhanced in Photoshop

3 Conclusions and Future Studios

Although the IPD/BIM Collaborative Studio has received a NCARB Award (2010), an ACSA Award (2010), two National AIA Technology in Practice Awards (2010, 2012), and an Autodesk Experience Award (2010), it is not without its shortcomings. The three areas for improvement of this studio are software, interoperability, and space. The software is not so much a shortcoming of the studio, but of the industry. The studio utilizes Revit products for the architectural modeling. Unfortunately, out of the box, Revit products offer very little for the practice of landscape architecture. A lot of work is required to create basic landscape object libraries in order for the software to benefit the work for the landscape architecture students. Perhaps Revit libraries can be purchased or students can utilize other BIM products for their tasks (e.g. Vectorworks Landmark or Land F/X). The interoperability of BIM and GIS data is a challenge. Translation of this data is not perfect. Information is lost and workarounds constantly need to be invented. In an academic enterprise experimentation is welcomed. In practice, experimentation is most likely unbillable. Lastly, the studio requires a more collaborative workspace. Creating virtual collaboration is an option for some days, but having a studio configuration with shared projected displays would foster greater team collaboration.

References

AIA, AIA California Council. *Integrated Project Delivery Guide.*
BIM Wiki, http://bim.wikispaces.com/ARCH+497A+-+BIM+Studio, accessed: 12/2011.
Flohr, T. (2011), *A Landscape Architect's Review of Building Information Modeling Technology.* Landscape Journal: design, planning & management of the land, 30 (1), 169-170.
Poerschke, U., Holland, R. J., Messner, J. I. & Pihlak, M. (2011), *BIM collaboration across six disciplines.* Proceedings of the International Conference on Computing in Civil and Building Engineering, Nottingham University Press.
Sipes, J. L. (2007), *Applying Building Information Modeling to Landscape Architecture.* American Society of Landscape Architects Annual Meeting, San Francisco, CA 2007.
Sipes, J. L. (2008), *Integrating BIM Technology Into Landscape Architecture.* Landscape Architectural Technical Information Series, 1.

The Need for Landscape Information Modelling (LIM) in Landscape Architecture

Ahmad Mohammad AHMAD[a] and Abdullahi Adamu ALIYU[a]

Abstract

Building Information Modelling (BIM) is growing in the Architecture, Engineering and Construction (AEC) Industry, with the new UK Government strategy to mandate BIM by 2016, there is a need to identify the importance of BIM in architectural landscape design to encourage BIM adoption. The use of BIM best practices can lead to efficient and effective BIM collaborative technology and partnering. BIM has the potential to challenge some of the limitations of designing, constructing and managing the built environment. The drive for this paper is a literature review on BIM and landscape architecture, describing current BIM use in landscape architecture using academic search engines and findings from UK Landscape Institute. Landscape architects should be BIM oriented. It is difficult to specifically identify a BIM software for landscape architects, this creates a need for landscape architects to come together and demand software, creating market for software vendors, to realise the manufacturing of landscape BIM software, with more specific landscape software, landscape architects could be able to provide a more innovative design with effective competence and collaborate with other BIM users efficiently. BIM process requires collaboration.

This research might be of importance to practitioners and academics, as it outlines different definitions of BIM, the relevance, need, benefits and challenges of BIM in landscape architecture.

Keywords: BIM; collaboration; construction; design; efficiency; landscape architecture.

1 Introduction

BIM means different things to different people (ARANDA et al 2007, 2008; Succar 2009). This research defines BIM by different BIM champions in the (AEC) Industry. The methodology used in the definition of BIM is based on a robust literature review. It is important to better understand BIM definitions, as there is a UK Government push to mandate the use of BIM in major building projects by 2016 (BIS 2012, JAMIESON 2011). This paper addresses two key questions on BIM; what are the benefits of BIM in a broader case? What is the need for landscape architects to use BIM in their projects? JAMIESON (2011) stated that there is: (1) Global increase in population growth; (2) Increase in infrastructure construction growth rate, with an estimate of "infrastructure construction growth" of 128% and 18% "developed markets" by 2010-2020; (3) By 2020 the global

---

[a] School of Civil and Building Engineering, Loughborough University, United Kingdom.

construction market will be 55% "emerging markets" and 45% "developed markets". (4) An estimate of 70% of populations will be living in urban areas by 2050. These statistics shows that there would be an increase in population, city centre developments, creation of green spaces, infrastructure construction and so on. SHEPLEY (2010) stated that in 2000, there was no website related with green healthcare architecture, while awareness increased to make healthcare facilities greener, with these developments, green areas needs to be designed, created and organised to improve sustainable environmental development, perhaps this creates a need for landscape and urban designers to engage with modern tools to participate in improving the AEC Industry. By 2016 AEC Industries in the UK most use BIM to a certain level in their projects, currently there is no specific BIM platform for landscape architects, and this can hinder collaboration with other BIM compliant professionals. FLOHR (2011) stated that Landscape architects should not be left out of the BIM process.

2 Need for BIM in Landscape Design

Using BIM tools and applications, landscape analysis can be explored, developed and documented for design planning and organisation, as BIM stores object information such as: irrigation pipe lines; areas allocated to specific plants; and list of different plants involved in each landscape project. Annotations used in landscape drawings can be 2D or 3D with information attached to it. BIM can help in planning for both hard and soft landscape elements, with many details involved it can also be used to store data for landscape architects to easily develop and organise detailed information. With the ability for simulation and visualisation, landscape architects should be able to produce detailed plans, walk-through animations and renderings for presentation and exploring the scope and nature of work to make informed decision at early design stages. Plant data description and specifications can be used to make sure the right plants are used in the right places, for example a drought tolerant plant can be used were there are major issues with water availability, sun loving plants used in open areas, and natural growing plants to be used in chemical controlled areas. With lots of landscape element's information (soft elements) such as different plant types, water usage and (hard elements) such as lighting, surface covering, pools, benches, and walk pathways to combine and analyse in a single landscape model, these details will allow the designer to design, report plans and proposals.

According to GOLDMAN (2011), "There exists a belief that BIM software is immature or simply not applicable to the field of landscape architecture. This is the leading barrier to adoption of BIM for landscape". This was also mentioned further in an interview with a member of the UK Landscape institute, he stated that "there is major concern in the landscape industry regarding BIM use, that lack of expertise in the use of Landscape information modelling could effectively remove landscape architects from the supply chain", therefore they would have to take important initiatives to equip their registered member with the necessary skills and information required to fully participate in the BIM workflow chain, he also stated that "Working groups will gain an appreciation of the level of BIM knowledge amongst our registered practices" and that the best software for landscape information modelling were Autodesk Revit and a times Bentley. There are software that can be used for landscaping, but for detailed landscaping that will facilitated functionality, workflow and site planning; there is a need for software vendors to focus on landscape architects to enable collaboration with other BIM users in the AEC Industry. HAYEK et al. (2011) quoted (APPLETON et al. 2002; PAAR 2006) that there is no "universal landscape visualisation solution" perhaps software like ArchiCAD and Autodesk's Revit can be used to partly carry out landscape design tasks.

References	Relevance of BIM to landscape architecture
PIETSCH (2009)	State that "Landscape models can be used to make simulation, explanation, experimentation and communication". A BIM model can also be used to store digital information with landscape planning details for "landscape analysis, landscape assessment, impact assessment, landscape planning"
BOHMS (2008)	BIM allows for partnering/collaboration, sustainability, efficiency and consumer orientation
KALKHAN (2011)	Describes that landscape parameters can be understood through the use of complex model with thematic mapping approaches.
JOSEPH (2011)	Describe that with exception of the building, plants could also contain integral intelligent properties that will foster better construction and maintenance.
WEYGANT, (2011)	State that BIM has created an opportunity for visualisation of exterior components by "creation of accurate site-based components] objects, landscaping, planting and topography".
JACKSON (2010)	BIM is taught in institutions. It is encouraging that BIM technology is designed to integrate information, by facilitating collaboration among different student's disciplines in the construction industry such as planning, facility management, architecture, landscape architecture and engineering.
HAYEK et al. (2011)	State that there is no "universal landscape visualisation solution".
BIS (2012)	"Building Information Modelling and Management BIM(M) is a managed approach to the collection and exploitation of information across a project. At its heart is a computer-generated model containing all graphical and tabular information about the design, construction and operation of the asset"
AUTHORS	To enable landscape architects to use BIM.
AUTHORS	To collaborate with fellow BIM users in the AEC Industry when BIM is mandated by the UK Government in 2016.
AUTHORS	BIM can help in organising landscape data.
AUTHORS	BIM automation can reduce time frame to conduct landscape design tasks.

Table 1: Why BIM use in landscape architecture

BIM and landscape architectural articles are available in a few numbers; Figure 1 shows primary findings from academic search engine that articles relating to BIM and landscape architecture are growing slowly. Figure 2 shows a slow rise in publications from 2002-2011. This research has sampled academic search engines relating to engineering and construction on the Athens page in January, 2012, which are not in any particular order or hierarchy. The aim of sampling data was to find the current development of articles in the field of BIM and landscape architecture. Keywords such as "BIM and landscape architecture" were used and results showed a low rate of growing publications. Findings below raise research questions such as: Why are there few BIM and landscape architectural related articles?; Is there a need for the application of BIM in landscape architecture?; How can landscape architects plan for the UK Government strategy to mandate BIM by 2016?; Can landscape architects work efficiently and effectively within the BIM work flow without specific BIM (landscape) software?

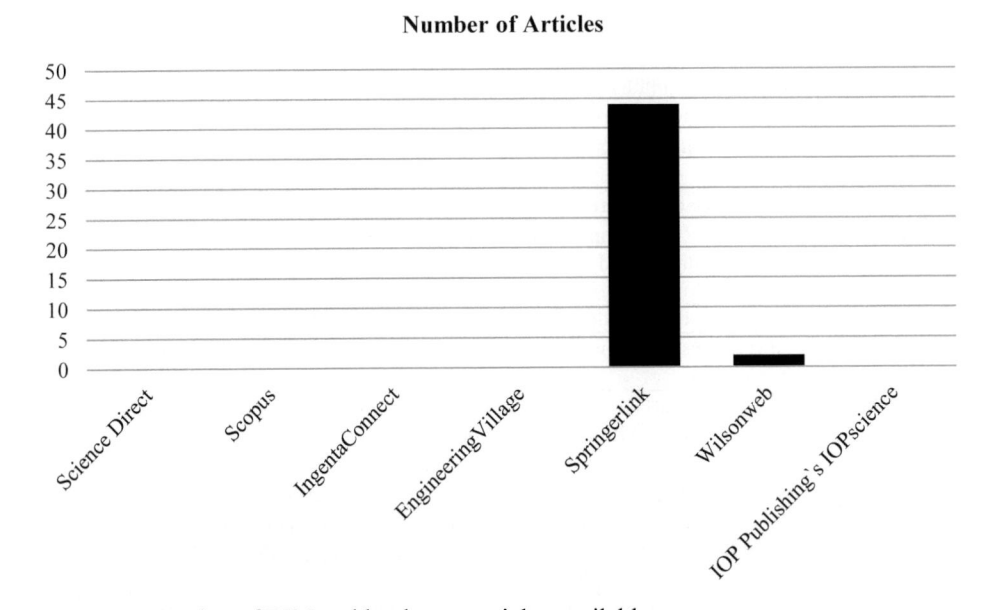

Fig. 1: Number of BIM and landscape articles available

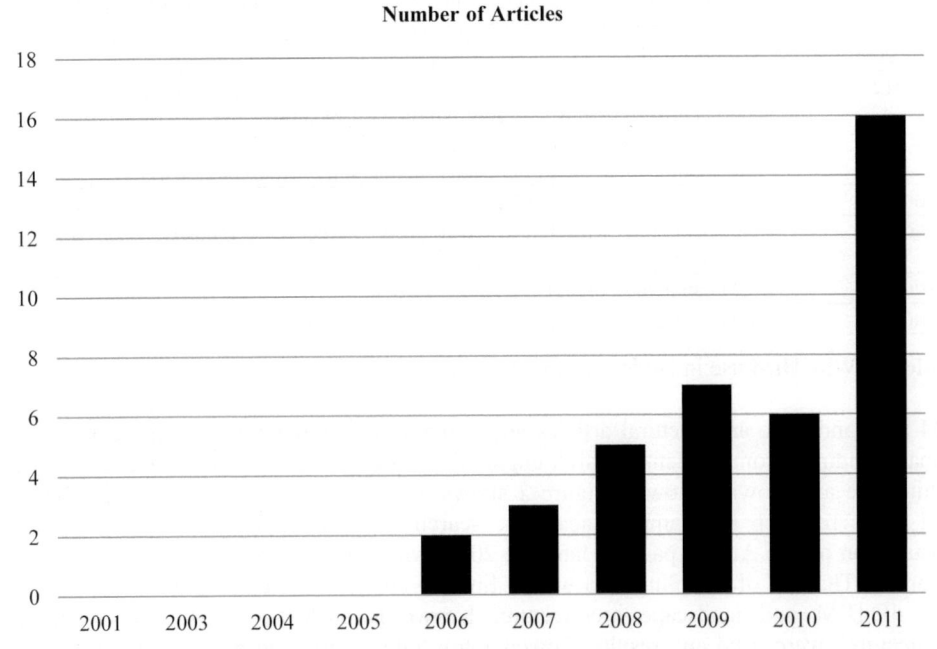

Fig. 2: Number of BIM and landscape architecture related articles from 2002 – 2012

3 Literature review

3.1 Defining "BIM"

HARDIN (2009) agrees with EASTMAN et al (2011) that BIM is defined by various experts and organisations differently. BIM requires technology to be implemented. AUTODESK (2003) stated these technologies as: CAD (traditional drafting); object CAD (3D geometry of buildings) and parametric Building Modelling (real time self-coordination of information). Key words used in these definitions below are: technology, information, building, management, process and geometry.

References	Definitions
1. WEYGANT (2011)	BIM is a technology that has improved the way structures are designed and built"…"A technology that allows relevant graphical and topical information related to the built environment to be stored in a relational database for access and management".
2. HARDIN (2009)	Describe BIM as a process, and also a means of adopting and establishing a new notion of thinking.
3. BENTLEY (2011)	define BIM as a new way of approaching design and document of building projects, the entire life cycle information is considered (design, build and operations) defining and simulating building delivery and operations using integrated tools.
4. STATE OF OHIO (2010)	State that "The term BIM may be used as a noun to describe a single model or multiple models used in the aggregate, the term BIM may also be used as a verb in the context of Building Information Modelling or Management, the process of creating, maintaining, and querying the model".
5. AUTODESK (2002)	Define BIM as an information technology in the building industry that facilitate the creation and operation, management and collaboration for digital database to be captured, preserved and used for building construction.

Table 2: What is BIM?

3.2 Benefits of BIM

(BIM Benefits in Pre-design and design phase): OLOFSSON et al. (2008) highlighted that BIM at the conceptual design stage provide the following benefits; quick visualisation; good decision shore up in project development process; precise automatic updating; diminution of man hours for space programs; increased project team communication; increased confidence of scope of work. BIM is used to produce schematics design details and improve the presentation to clients for easier and better decision making.

(BIM Benefits in Pre-Construction phase): The use of BIM in the construction phase enables scheduling and work flow coordination whilst at the preliminary pre-construction level, cost estimates and constructing virtual logistic of cranes and materials on site (AUTODESK 2003). The 4 fourth dimension is of particular importance as BIM is used in the construction phase to establish and evaluate various construction options. A 4D model can be obtained by adding schedule data to a 3D building design, introducing time as the 4 fourth dimension (AUTODESK 2003). 4D planning is a technique that integrates 3D CAD models with construction activities (schedule) which allows clear visualisation of a construction programme sequence. 4D is noted by AUTODESK (2003) to be developed by comprehending schedule dates from project plan to model, the evaluation of construction sequence, detection of clashes, identifying construction milestones before construction begins to take place.

BIM Benefits in Construction Phase: The main benefits of BIM at this phase as stated by AUTODESK (2003) includes: Scheduling of construction schemes, scheduling what is constructible; Clash detection and reporting; And quality of projects is being analysed and improved by rescheduling.

Benefits of BIM in a broader view: ARANDA-MENA et al. (2008) suggested that execution of BIM at industry level has three outcomes, which are: Technical capabilities: this involves the ability to share information with other consultants and the production of drawings and documents from BIM model; Operational capabilities: this supports design collaboration, reduces errors and allows the ability to design in 3D environment; Business capabilities: completing projects with efficiency, and also reduces information errors. VTT (2006) described the benefits of BIM use, to have various opportunities to professionals in the AEC Industry, these are listed as: Strategic thinking + BIMs = Transformational opportunities; Tactical thinking + BIMs = Informational opportunities; Operational thinking + BIMs = Automational opportunities.

3.3 Limitation of BIM use

Technical and managerial issues are stated as the two main problems of BIM execution. BERNSTEIN & PITTMAN (2005) described Technical issues as: Production of a well-defined construction process model; Adopting formats that enable design data to be computable; Enable accurate data exchange and integration within a BIM model. They also described the Managerial issues as a major factor hindering the progress of BIM implementation; it is related to the issues of BIM implementation cost and use. Due to lack of standardised established BIM implementation manuals, the management of BIM at industry level becomes very difficult to implement. WEYGANT (2011) noted that the main challenges of BIM are: poor data exchange approaches; interoperability; cost of training, hardware and software.

4 Research Methodology

Literature review was used in constructing the basis of this research, defining BIM and also describing its benefits and challenges at various stages. The literature searches involved the use of both online and offline data to gather information on BIM and landscape architecture in the AEC Industry. Data sources used includes:

1. Springerlink;
2. Automation in construction;
3. ProQuest;
4. Google books; and
5. Scholar.

Other academic sources used for the purpose of secondary data collection includes: Books; journals; and conference papers. Keywords such as BIM, BIM and landscape architecture, Landscape information modelling, landscape architecture and landscape modelling were used to find relevant data for the purpose of this research paper. A short interview with a member of the UK Landscape Institute was conducted. This paper addresses questions such as: Is there a need for BIM in the AEC Industry? Is there a need for BIM in landscape architecture? What are the benefits of BIM in a broader case? And what are the uses of BIM in landscape architecture?

Landscape encourage public realm improvements through pedestrians networks, creating sense of place that is embrace by the public, these landscape areas are integrated into the existing urban fabric. Landscape enhances the quality of our build environment, in healthcare facilities, green environments are encouraged as they can improve patient care. Despite the advantages of landscape in the urban fabric, there are many problems affecting the practice in the UK such as economic recession, and the need of BIM application in practising landscape firms. Membership with the UK Landscape Institute is still growing regardless of the economic recession that affected the building industries; this shows a growing interest in landscape architecture. There is a growing rate of UK landscape institute membership by landscape practitioners, with 5214 members in 2007, 5427 in 2008 and 5693 in 2009 (Landscape Institute, 2008-2010).

4.1 Landscape with BIM

Landscape architecture is not totally compatible with all architectural BIM software, software should be developed with landscape application in mind, to enable landscape designers to design effectively, lack of specific software for landscape design can create drawbacks, software at the moment can only be recommended for landscaping, choosing a software can therefore be difficult, it is claimed that there is no one software package that is landscape information modelling ready. Regardless of these difficulties, BIM is still encouraged as technology that will allow the ability for GIS and design landscape in multi-dimensions, to achieve creativity and innovative state of the art designs. Software such as Land FX can be used for planting and irrigation modelling purposes, ArchiCad for presentation, rendering and quantification, it can be possible to calculate the amount of water needed for the entire landscape design when figures are collected, the amount of water for each plant can vary, but an estimate of water required for each plant can be obtained, with these increased modelling information, BIM can facilitate information storing, sharing and modelling capability providing distinctive design opportunities for landscape architects, it also creates an opportunity for the visualisation of exterior building components such as benches, pools, planters and so on, these objects in 3D will contain information that will foster the design process. To achieve a collaborative BIM studio, all professionals including landscape architects have to come together. Collaboration between architects, landscape architects and engineers has being an on-going process for over a decade (FRUCHTER 2003). BIM is widely gaining recognition in the AEC Industry, and its potential is not yet fully exploited and explored. Therefore landscape can be enhanced with continuous BIM explorations. Landscape architects are currently using BIM tools for their various designs, but this software is not 100% compatible with all landscape tasks, as current software is modified to enable landscape architects to use for their designs.

4.2 Common BIM use in Landscaping

Some uses of BIM in landscape include: quantity counts (number of both soft and hard elements); error reduction with organisation of data; smart symbol use (2D and 3D); landscape presentation (plants) before they eventually grow; storing data; Site information modelling; cutting and filling sites (site analysis details); assigning plants types at areas that suit their nature (site analysis details); and exploring and presenting ideas to clients. In a short interview with a member of the UK Landscape Institute, states that BIM software that can be used for landscape architecture includes: Vectorsworks Landmark, Land F/X, LandCADD, SiteWorks, ArchiTerra, AutoCAD Civil 3D, Autodesk's Revit, Grahisoft ArchiCAD.

5 Conclusion

LANDSCAPE INSTITUTE (2010) stated that "When landscape is placed at the heart of the development process, developers profit while businesses and communities reap the economic benefits" perhaps landscape can be a cutting edge service that enhances buildings and the entire built environment. BIM use is important in landscape architecture for the purpose of: automation; collaboration; adding quality to design; saving time and cost; reducing errors; and designing landscape with objects which have additional information attached, allowing object data to be easily collected and managed, this foster the design by enabling designers to make inform decisions at early design stages due to the availability of instant and comprehensive amount of information. There are various benefits to the application of BIM in landscape, but BIM faces a staggering challenge of economical; social; legal; educational; and technical constraints. Interoperability seems the ultimate challenge of BIM, PIETSCH (2011) noted that the problem of information exchange in landscaping modeling planning can be improved by extending existing standards and producing new standards. "BIM and landscape architecture" articles are limited; this shows a need for awareness, to inform landscape architects the importance of the use of BIM in their profession and within the BIM collaboration process. The analytic aim of this paper is to identify BIM: awareness; partnering; need and application. A prescriptive solution is for stakeholders to come together and produce landscape information modeling articles and make available to practitioners. This can increase awareness of "BIM and landscape architecture" in industry.

References

Aranda, M. G., Chevez, A., Crawford, J. R., Wakefield, T., Froese, R., Frazer, J. H., Nielsen, D. & Gard, S. (2008), *Business Drivers for Building Information Modelling*. Brisbane, Australia, Cooperative Research Centre – Construction Innovation.

Aranda, M. G., Crawford, J. R., Chevez, A. & Froese, T. (2009), *Building Information Modelling Demystified: Does it Make Business Sense to Adopt BIM?* In: International Journal of Managing Projects in Business, 2 (3), 419-434.

Autodesk (2003), *Building Information Modelling in Practice: Autodesk Building Solutions* (White Paper). http://images.autodesk.com/emea_s_main/files/bim_in_practice.pdf (accessed on: 12th July, 2010).

Bentley (2011), *About BIM* http://www.bentley.com/en-GB/Solutions/Buildings/About+BIM.htm (accessed on: 13th July, 2011).

Bernstein, P. G. & Pittman, J. H. (2005), *Barriers to the Adoption of Building Information Modelling in the Building Industry*. Autodesk Building Solutions Whitepaper, CA, Autodesk Inc.

BIS (2012), *Report to the Government Construction Clients Board on Building Information Modelling and Management*. Department for Business, Innovation and Skills, UK. http://www.bis.gov.uk/policies/business-sectors/construction/research-and-innovation/working-group-on-bimm (accessed on: 4th April, 2011].

Bohms, M. (2008), *The Building Information Modelling (BIM) Landscape: Facts and Opinions- Action and Results*. http://www.tno.nl/downloads/The%20BIM%20Landscape.pdf (acessed on: 12th July, 2010).

Eastman, C. M., Eastman, C., Teicholz, P., Sacks, R. & Liston, K. (2011), *BIM Handbook: A Guide to Building Information Modelling for Owners, Managers, Designers, Engineers and Contractors*. Canada. John Wiley and Sons.

Flohr, T. (2011), *A Landscape Architect's Review of Building Information Modelling Technology*. In: Landscape Journal, 30 (1), 169-170.

Fruchter, R. (2003), *Innovation in Engaging, Learning and Global Teamwork Experience*. Proceedings of the 4th Joint International Symposium on Information Technology in Civil Engineering, Nashville, TN.

Goldman, M. (2011), *Landscape Information Modeling*. http://www.di.net/articles/archive/landscape_information_modeling/ (accessed on: 4th April, 2012).

Hardin, B. (2009), *BIM and Construction Management Proven Tools, Methods, and Workflows*. Indiana, Wiley Publishing.

Hayek, U. W., Melsom, J., Neuenschwander, N., Girot, C. & Grêt-Regamey, A. (2011), *Interdisciplinary studio for teaching 3D landscape visualization*. Lessons from the LVML.

Jackson, J. B. (2010), *Construction Management, Jump Start: The Best Step Forward a Career in Construction Management*. 2nd Ed. Indiana, Wiley Publishing.

Jamieson, C. (2011), *The Future for Architects*. http://www.buildingfutures.org.uk/assets/downloads/The_Future_for_Architects_Full_Report_2.pdf (accessed on: 6th January, 2012).

Joseph, C. (2011), *BIM Integration of Landscape Objects: Recording Water Consumption for Management and Digital Modeling*. http://bim.wikispaces.com/file/view/WaterRecordingRevit.pdf (accessed on: 6th January, 2012).

Kalkhan, M. A. (2011), S*patial Statistics: GeoSpatial Information Modeling and Thematic Mapping*. United States of America, CRC Press, Taylor and Francis Group.

Landscape Institute (2010), *Why Invest in Landscape*. http://www.landscapeinstitute.org/PDF/Contribute/WhyInvestFinalA4pages.pdf (accessed on: 6th January, 2012).

Landscape Institute Review (2008-2010), http://www.landscapeinstitute.org/PDF/Contribute/LandscapeInstituteAnnualReview200810_000.pdf (accessed on: 6th January, 2012).

Olofsson, T., Lee, G. & Eastman, C. (2008), *Editorial – Case Studies of BIM in Use*. In: Itcon, 13, 244-245.

Pietsch, M., Heins, M., Buhmann, E. & Schultze, C. (2009), *Object-Based, Process-Oriented, Conceptual Landscape Models-A Chance for Standardizing Landscape Planning Procedures in the Context of Road Planning Projects.*

Royal Institute of British Architects (RIBA) (2011), *Building Information Modelling Report.* RIBA Enterprises Ltd. http://www.thenbs.com/pdfs/bimResearchReport_2011-03.pdf (accessed on: 6th January, 2012).

Shepley, M. M. (2010), *Developing evidence for sustainable healthcare.* In: Building Research and Information, 38 (3), 359-361.

State of Ohio (2010), *Building Information Modelling (BIM) Protocol.* State Architects Office, General Services Division.

Succar, B. (2009), *Building Information Modelling Framework: A Research and Delivery Foundation for Industry Stakeholders.* In: Automation in Construction, 18 (3), 357-375.

VVT (2006), *Building Information Models: Inter-Organisational Use in Finland.* VBE2 (1) Research and Development Project, Work Package 1.4, Public Report.

Weygant, R. S. (2011), *BIM Content Development: Standards, Strategies, and Best Practices.* New Jersey, John Wiley and Sons.

Technical Papers

Introducing Geodesign –The Concept

William R. MILLER, Esri
Director of GeoDesign Services

1 Introduction

This purpose of this paper is twofold: First, to introduce the concept of geodesign, what it means, and some of its implications, particularly for those working with geospatial data; and Second, to encourage the reader to play an active role in the development and expansion of this nascent field.

The paper will address the following topics:

- The context for geodesign
- The history of geodesign
- Defining geodesign
- The importance of geodesign
- The nature of design
- Managing complexity
- The technology of digital geodesign
- Creating the future

Additional information regarding the subject of geodesign, including the geodesign process, geodesign technology, and various geodesign case studies, is referenced at the end of this paper.

2 The Context for Geodesign

Every organization, large or small, public or private, does three things: it gets and manages information (data), analyzes or assesses that information with respect to some purpose(analysis), and (based on that information and those assessments) creates or re-creates goods and/or services (design). It is, in fact, the creation or re-creation of goods and/or services that gives most organizations their reason for being.

By and large, geographic information system (GIS) technology, as it's known today, serves organizations quite well with respect to the need to acquire and manage geospatial information. GIS also offers organizations a wide range of geoprocessing functions for analyzing geospatial information. While this is beginning to change, present-day GIS still offers little functionality with respect to an organization's need to create and/or re-create goods and services, that is, its need to do design … to do geodesign.

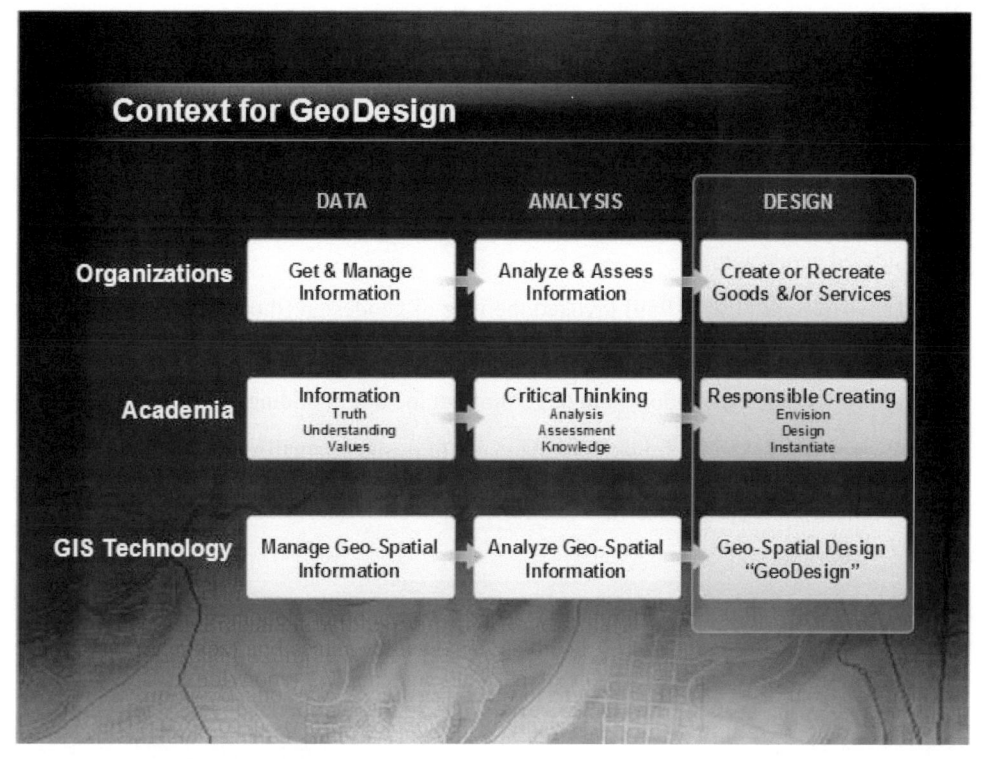

Fig. 1: Context for Geodesign (Image courtesy Esri)

Assuming the average organization spends a third of its operational resources on each of these three segments (data, analysis, and design), a GIS technology company (like Esri) could increase the organization's revenue by 50 percent, without adding any new customers, by simply expanding its products and services to support customers' activities related to design.

3 The History of Geodesign

The history of geodesign can be described as the emergence of geodesign as an *activity* or as the emergence of the *term* geodesign.

3.1 Emergence of the activity

The main idea underlying the concept of geodesign, namely that the context of our geographic space conditions what and how we design (that is, how we adjust and adapt to our surroundings), has been with us since the beginning of time.

Deciding where to locate a tribal settlement, choosing materials to use to construct shelters, developing a strategy for hunting wild animals, deciding where to plant crops, or laying out the plans for defending a settlement from intruders are all geodesign-related activities. That is, the successful design of each of these depends on having adequate knowledge of the relevant geographic conditions and the ability to work with those conditions, as well as respecting the constraints and taking advantage of the opportunities suggested by those conditions.

The corollary is also true: what and how we design has the power to condition or change the context of our surroundings, that is, to change our geographic space. In fact, any design-related activity that depends on or in some way changes the context of our surroundings can be considered geodesign.

Frank Lloyd Wright (1867-1959) invoked the idea of geodesign (though he did not use the term) when he formalized the idea of organic architecture, that is, making the structures and nature one by, for instance, bringing the outdoors in (e.g., through the use of corner windows) and moving the indoors out (e.g., through the use of sliding glass doors).

When Wright was asked by Edgar Kaufmann Sr. to design a small vacation home on Bear Run in rural southwestern Pennsylvania (the home later known as Fallingwater), he had been without a commission for months. He postponed working on the design to a point

Fig. 2: Image of the famous building of Frank Lloyd Wright the "Fallingwater" in Pennsylvania

where many of his disciples began to wonder if he was beyond his prime and perhaps not up to the challenge. That was just about the time Kaufmann called Wright to ask how he was coming along with the design and tell Wright that he was on his way to Taliesin, Wright's studio near Spring Green, Wisconsin. Wright responded by saying he was expecting Kaufmann's visit and encouraged Kaufmann to come at his earliest convenience, which turned out to be about three hours.

Wright then hung up the phone and went to work on the design, his students and staff sharpening pencils as Wright feverishly worked at his drafting table, laying out the design of the house, including floor plans, elevations, sections, and a quick perspective. The basic concept was fully completed by the time Kaufmann arrived later that afternoon (TOKER 2005).

Was Wright doing geodesign? The answer is, most definitely. Wright had the site's geography fully in mind while he was doing the design, giving consideration to topography, the location of the stream and waterfall, the placement of boulders that provided the foundation for the house, views to and from the house, and site-related environmental conditions such as the use of solar access for heating the house in the winter and cold air flow along the stream for cooling the house in the summer. Wright was most definitely doing geodesign.

Richard Neutra (1892-1970), an Austrian architect who had worked with Wright in the mid 1920s, later wrote Survival through Design, one of the pivotal books on the importance of designing with nature. In it, he advocated a holistic approach to design, giving full attention to the needs of his client while at the same time emphasizing the importance of the site, its natural conditions, and its surroundings. Neutra's book predated the environmental movement by 20 years and in many ways contributed to the formation of the Environmental Protection Act of 1970, the year Neutra died (NEUTRA 1954).

Ian McHarg (1920-2001), Scotsman, landscape architect, and educator, is without a doubt, though he never used the term, one of the principal founders of geodesign. His 1969 book Design With Nature not only expresses the value of designing with nature (primarily as related to the fields of landscape architecture and regional planning) but also sets forth a geo-based technique (which was most probably based on Manning's work), viewing and overlaying thematic layers of geographic information to assess the best (or worst) location for a particular land use (MCHARG 1969).

McHarg was also one of the first to advocate a multidisciplinary approach to environmental planning, which until that time had been dominated by narrow views and singular values. Supported by a series of grants while leading the program at the University of Pennsylvania, he was able to assemble a team of scientists and experts from a wide variety of disciplines in the physical, biological, and social sciences (MCHARG 1996).

While McHarg's technique was completely graphical (non-digital), his book gave birth to a whole new way of thinking about regional planning and design. It not only laid out a clear procedure for assessing the geographic context of a site or region but also presented that procedure with a clarity that quickly led to the digital representation of geographic information (as thematic layers) and assessment strategies (e.g., using weighted overlay techniques), which, in time, contributed to the conceptual development of GIS.

It is interesting to note that while McHarg was at the University of Pennsylvania promoting his graphical overlay technique and receiving considerable attention for his book, a substantial body of knowledge related to environmental planning (geodesign) was being quietly developed and accumulated by Carl Steinitz and his colleagues at the Harvard Graduate School of Design.

Carl Steinitz (1938-), working with his colleagues and students over a period of approximately 30 years, developed a complete framework (conceptual framework, design strategies, and procedural techniques) for doing geodesign as applied to regional landscape studies. The Steinitz Framework for Geodesign (STEINITZ 2012), previously called a Framework for Landscape Planning (STEINITZ 1995), advocates the use of six models to describe the overall planning (geodesign) process:

Representation Models	How should the context be described?
Process Models	How does the context operate?
Evaluation Models	Is the current context working well?
Change Models	How might the context be altered?
Impact Models	What differences might the alterations cause?
Decision Models	Should the context be changed?

The first three models comprise the assessment process, looking at existing conditions within a geographic context. The second three models comprise the intervention process,

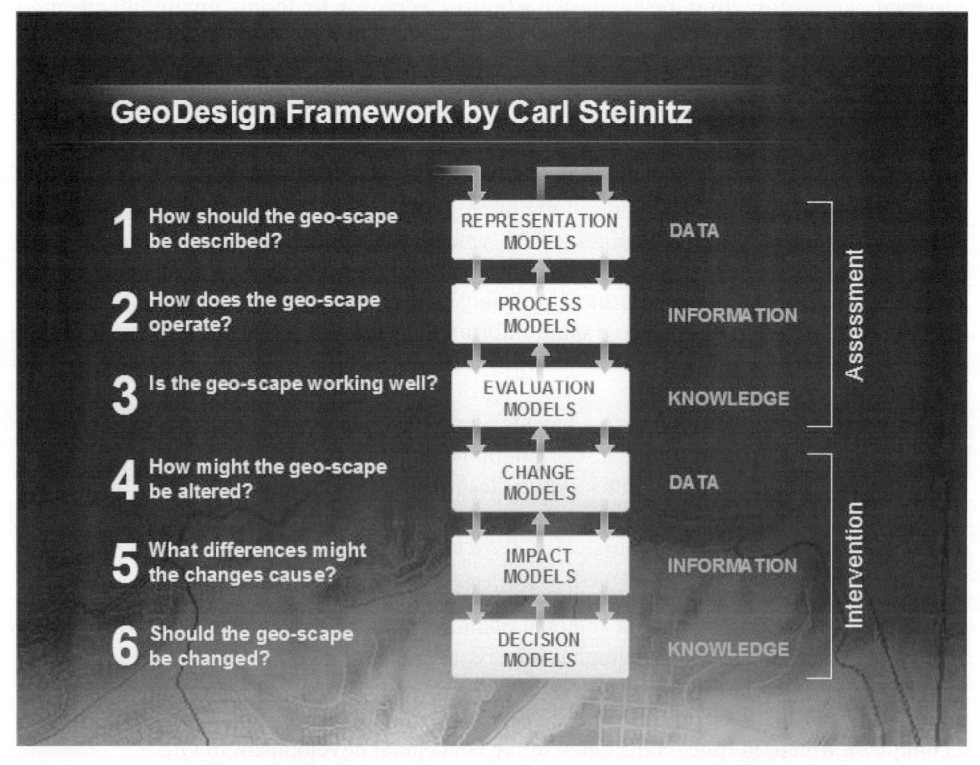

Fig. 3: The geodesign framework – by Carl Steinitz (Image courtesy Esri)

looking at how that context might be changed, the potential consequences of those changes, and whether the context should be changed.

The fourth model, the Change Model, provides the specific framework for developing and creating proposed changes (design scenarios) that are predicated on the science- and value-based information contained in the Representation Models and assessed against that same information in the Impact Models, which is the essence of the underlying concepts of geodesign.

Steinitz's new book, A Framework for Geodesign, soon to be published by Esri Press, delineates the conceptual framework for doing geodesign and will surely become one of the bibles for both practitioners and academics for years to come.

3.2 Emergence of the term

The term geodesign, unlike the activity of geodesign, is relatively new. Klaus Kunzmann provides an early reference to geodesign in his paper, "Geodesign: Chance oder Gefahr?" (1993). He used the term to refer to spatial scenarios. Since then, a small number of geo-related businesses have used geodesign as part of their name.

In approximately 2005, Dangermond and a few others were observing a demo at Esri showing how users could sketch land-use plans in GIS using an extension we had developed for ArcGIS® called ArcSketch™. One of the members of our team was sketching in points, lines and polygons, all defined and rendered to represent various types

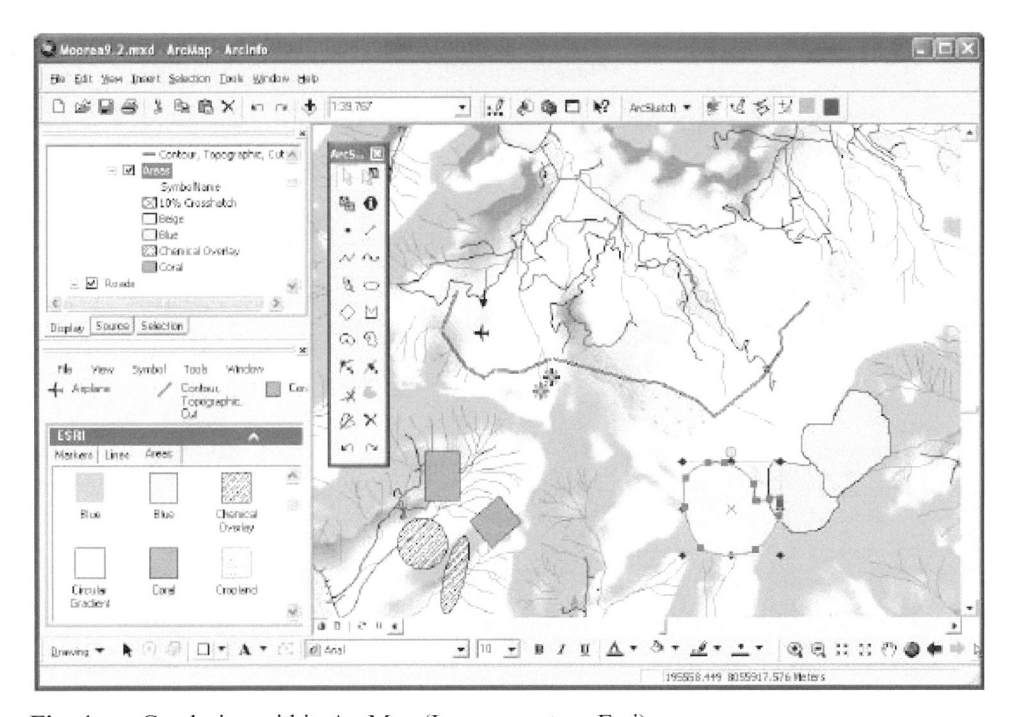

Fig. 4: Geodesign within ArcMap (Image courtesy Esri)

of land use, when "I turned to Jack and said, 'See, Jack, now you can design in geographic space.' Without hesitation, Jack said, 'Geodesign!'" (MILLER 2011). The term stuck and soon became the moniker for Esri's agenda for supporting the needs of designers working in a geospatial environment. More broadly, it has also become the moniker for a whole new wave of thinking regarding the use of GIS as geographic frame work for design.

4   Defining Geodesign

Before proceeding, however, a little deeper look at what is meant when using the term *geodesign* will be beneficial, particularly as it relates to Esri's agenda for supporting designers. The definition of geodesign is derived from two terms, *geo* and *design*. Both of these component terms are subject to a wide variety of interpretations. As such, they need to be clearly defined before attempting to define geodesign.

Defining *Geo*
The term *geo* can be simply defined as *geographic space* – space that is referenced to the surface of the earth (geo-referenced). In general, thinking of *geographic space* brings to mind a 2D geographic space (a flat map) or, for those who are a bit more advanced in their thinking, a 2.5D geographic space – that is, an undulating surface (a relief map). This

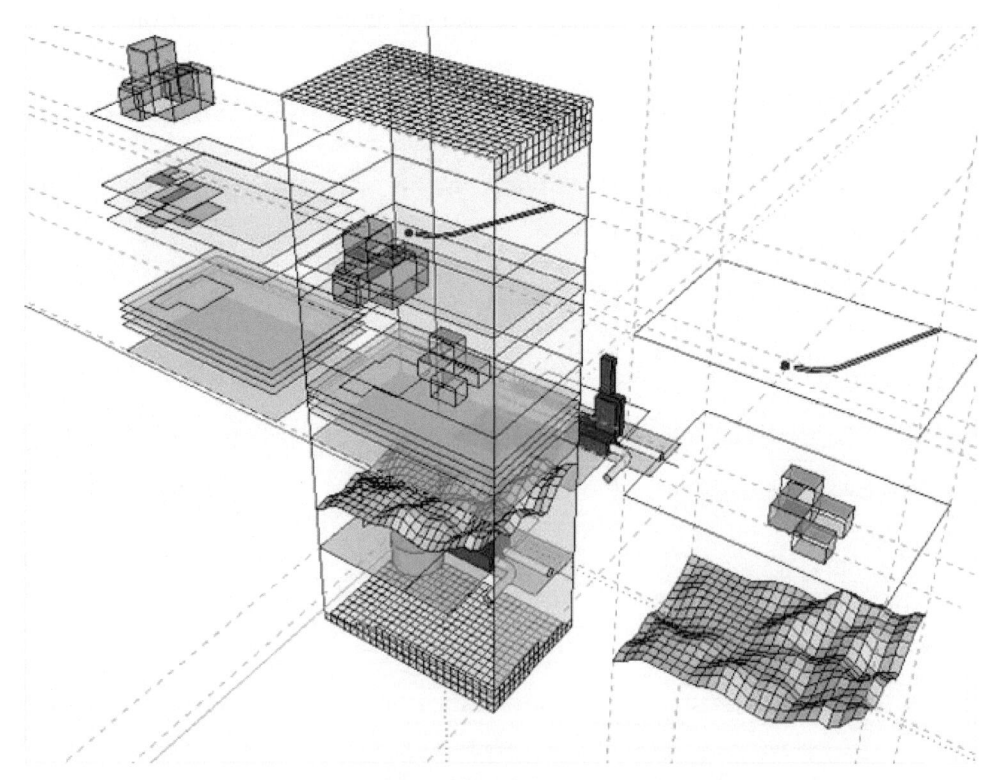

Fig. 5:   Introducing the 3D Geodesign concept (Image courtesy Esri)

thinking could also be extended to include 3D geographic space, providing the ability to geo-reference what lies below, on and above the surface of the earth, including what exists inside and outside buildings, as well as 4D geographic space, giving the added ability to geo-reference time-dependent information such as population growth or the migration of a toxic plume through a building.

These extended views of geographic space (moving from 2D to 3D to 4D), coupled with the idea that most data, at some level, is spatial and that all types of spatial data (physical, biological, social, cultural, economic, urban, etc.) can be geo-referenced, lead to an expanded view of what is typically envisioned, or imagined, when referring to the *geo* portion of *geodesign*. This expanded view is embodied in a new concept that is beginning to emerge within the geospatial community ... that of *geo-scape*.

Geo-scape is the planet's life zone, including everything that lies below, on, and above the surface of the earth that supports life. Geo-scape expands the view of what constitutes the content of geography as well as the dimensional extent of the geographic space used to reference that content. As a consequence, it also expands the domain of geo in geodesign to include everything that supports or inhibits life (MILLER 2004).

Geo in geodesign thus refers to the full spectrum of the earth's life support system and extends thinking to move from

          Land   to   Land, water, air
         Surface   to   Below, on, above the surface

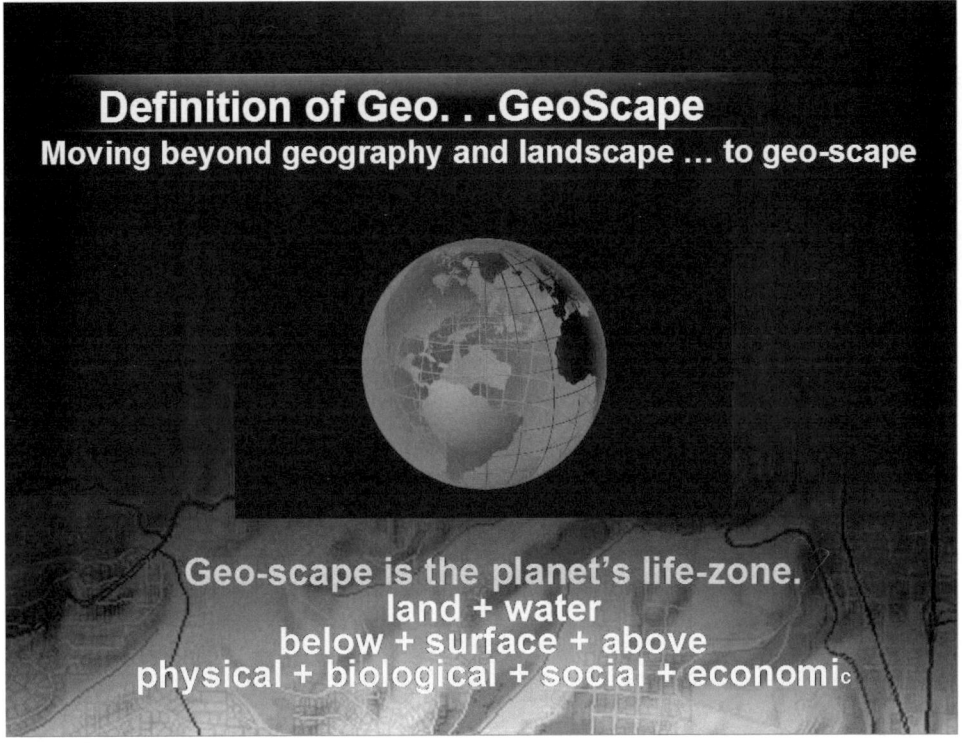

Fig. 6: Definition of Geo ... Geo-Scape (Image courtesy Esri)

Land	to	Land, water, air
Surface	to	Below, on, above the surface
2D/2.5D	to	3D/4D
Rural	to	Rural and urban
Outside buildings	to	Outside and inside buildings
Objects	to	Objects, events, concepts, and relationships

Each of these moves represents a significant transformation in the way people think about geography, geodesign and the use of GIS.

4.1 Defining Design

The word *design*, the second component of geodesign, can be defined as either a noun or a verb. As a noun, *design* generally refers to some object or other entity. As a verb, it usually refers to a process or series of activities.

"Design is the thought process comprising the creation of an entity" (MILLER 2005).
It is first thought, or the type of thought called insight. It is the mental synapse that instantly sees the potential connection between problem and possibility, the capacity for order in the midst of chaos or for improvement amid inefficiency.

Design is also intuition, that form of subconscious thought that leads to a deeper sense of knowing, often in the apparent absence of rational confirmation. Intuition is akin to an

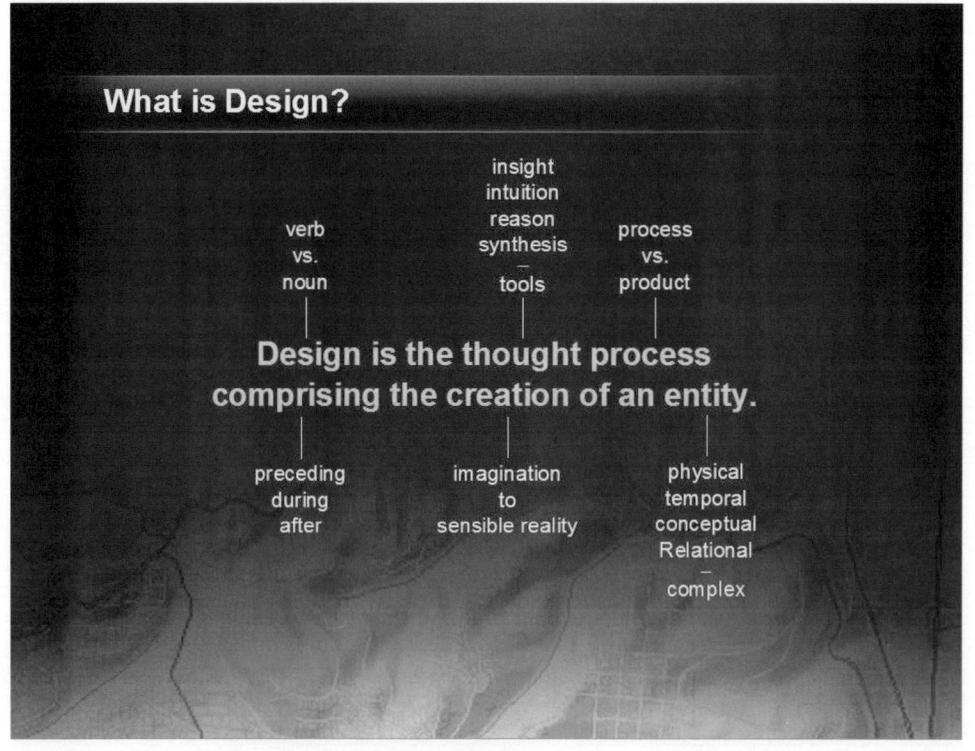

Fig. 7: What is Design? (Image courtesy Esri)

elongated insight that tells us we are on to something. It is the hunch that often underlies efforts to perform rational analysis.

The nature of this process, which is often modeled as a linear sequence of events, is in reality a highly complex, multifaceted set of thought-filled activities. While design is linear in the sense that it is sequenced in time as one moves from initial concept to a completed product, it is also nonlinear. Design thought often jumps in discontinuous steps from one aspect of a problem to another as it searches for a solution. It is also multileveled in the sense that overall systems, subsystems, and even minute details often need to be considered simultaneously.

This comprehensive thought-filled process is directed toward and culminates in creation. That is, it leads to the tangible realization of an entity (the thing being designed) in time and space. An entity can be an object that occupies space, an event that occurs in time, a concept (such as the theory of relativity), or a relationship (such as a treaty between nations). Most entities are complex in that they contain two or more of these entity types.

Any entity can be designed or created with intent and purpose. The total thought process encompassing the creation of that entity – the process that gives it its form, be it physical, temporal, conceptual, or relational – is design.

4.2 Defining the Purpose of Design

It is important to note that the preceding definition of design does not define, or in any way describe, what constitutes *good* design.

The ethic of design, that is, how a design (noun) is determined to be good or bad comes not from the definition but rather from the purpose of design, which at a fundamental level is always the same.

"The purpose of design is to facilitate life" (MILLER 2006).

Simply put, if an entity (the thing being designed) facilitates life, then it is good; if it inhibits life, it is bad; and if it does neither, it is neutral. While this is a very simple ethic, or appears as such at first glance, one must constantly remember two things: what it means to facilitate and what is meant by life.

The word facilitate means to empower, enable, or assist, but not dictate, as was sometimes assumed by the utopian designers of the early 20th century. Utopian design, based on the notion that the designer knows what is best, is really dictatorial design and is often a form of imprisonment in that it shackles its users to a particular behavior pattern or singular point of view. The purpose of good design is not to imprison but rather to enrich (that is, to facilitate) the lives of those using the design (noun).

Fritjof Capra, author of The Web of Life, describes four aspects of life (CAPRA 1996), First, all living systems are open systems. Second, all living systems are interdependent systems. Third, all living systems are self-organizing. And fourth, all living systems make use of some form of feedback (loops, networks, webs) to manage themselves.

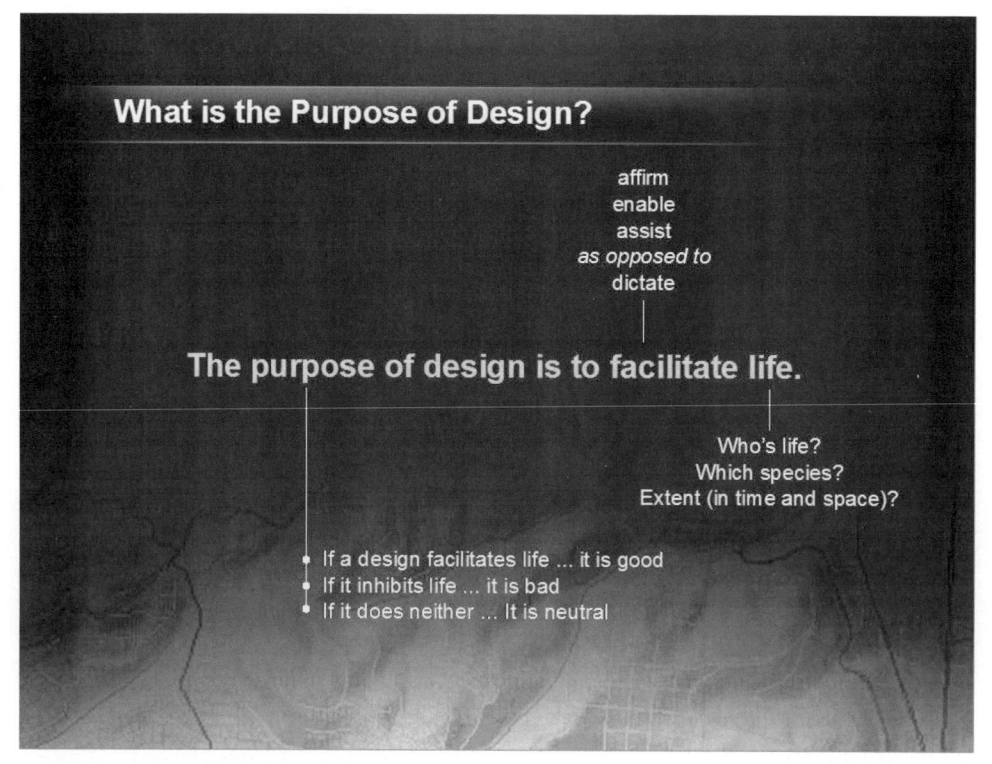

Fig. 8: What is Purpose of Design? (Image courtesy Esri)

Open systems require the input of an energy source, for example, food, oxygen, and sunlight to sustain themselves. They also produce output that if it can be used by another living system, is called product; if not, it is called waste. It is important to acknowledge that all living systems are open and require a continuous input of resources and that they constantly produce some type of output.

As such, living systems are neither independent nor dependent systems but rather interdependent systems that rely on neighboring systems for their survival, supplying their input and processing their output. Carrying these links forward, it is not difficult to see that all living systems are interdependent in one way or another with all other living systems.

Consequently, the question that really must be asked is, "Whose life?"

Are we talking about the life of the designer or the design team? The lives of those commissioning the creation (design) of an entity? The lives of those destined to use the entity? Or the lives of those affected by the use of the entity? Human life? The life of a particular species, or life in general? The question of whose life to facilitate, over what period of time, and to what extent, is very important and often leads to unexpected complexity.

The answers to these questions are not simple, surely not singular, and often not static. In many cases, both the questions and answers, as related to a particular entity, change over time. While this complexity, as it intensifies, has the potential to give one pause, or even overwhelm, it is always important to remember the simplicity of the original statement of purpose – that the purpose of design is to facilitate life. This serves as the foundation for the fundamental question to ask regarding the purpose of the entity being designed.

The answers to these questions for a given project form the design ethic for that project and, in so doing, provide the ability to assess the goodness of the design (noun).

4.3 Defining Geodesign

Given the new definitions of geo and design, they can now be combined to form a definition of geodesign:
Geodesign is the thought process comprising the creation of an entity in the planet's life zone (geo-scape).

Or, more simply, geodesign is design in geographic space (geo-scape). Correspondingly, the purpose of geodesign is to facilitate life in geographic space (geo-scape).

The essential aspect of this definition is the idea that design – the process of designing (creating or modifying) some portion or aspect of the environment, be it natural or man-made – occurs within the context of geographic space (where the location of the entity being created is referenced to a geographic coordinate system) as opposed to conceptual space (creating something in the imagination with no locational reference), paper space (creating something with pencil and paper, again with no locational reference), or even CAD space (where the entities in that space are referenced to a virtual coordinate system as opposed to a geographic coordinate system).

At first glance, this seems to be a trivial point. However, the fact that the entity being created or modified is referenced to the geographic space in which it resides means that it is also, either directly or indirectly, referenced to all other information referenced to that space. This means that the designer can take advantage of, or be informed by, that information and how it relates to or conditions the quality or efficiency of the entity being designed, either as it is being designed or after the design has matured to some point where the designer wishes to perform a more comprehensive assessment.

5 The Importance of Geodesign

This referential link between the entity being designed and its geographic context provides the tangible basis for doing both science-based and value-based design. Additionally, it has the ability to provide operational linkages to a wide variety of domain-specific information and, in so doing, provides the multidisciplinary platform for doing integral design (holistic design).

5.1 Science-Based Design

Science-based design is the creation or modification of an entity within the context of scientific information (including scientific processes and relationships) such that the design of the entity is conditioned or informed by that science as it is being designed. Geodesign, through the use of a common geographic reference system, provides the ability to link geographic entities (those entities that are being designed) to scientific information, relevant to the creation, instantiation, or utilization of those entities.

5.2 Value-Based Design

Value-based design is the creation or modification of an entity within the context of social values (global, community, cultural, religious, etc.) such that the design of the entity is conditioned or informed by those values as it is being designed. As is the case with science-based design, geodesign provides the ability to link geographic entities (those entities that are being designed) to social values relevant to the creation, instantiation, or utilization of those entities, assuming those values are referenced to the same geographic reference system.

5.3 Integral Design

Geodesign not only provides the ability to link the entity being designed to relevant science- and value-based information, but also provides the framework for exploring issues from an interdisciplinary point of view and resolving conflicts between alternative value sets. In this sense, it can be seen as an integral framework for intelligent, holistic geospatial design.

The important point to note, however, is that the act or process of design occurs in geographic space where the entity being designed is geo-referenced to a common geographic coordinate system and, thus, directly or indirectly to other information that is also referenced to that system. This referential link between the entity being designed and information (be it science-based or value-based) gives the designer the ability to design within the context of that information and, in so doing, improve the quality and efficiency of the design process as well as that of the entity (the product of that process).

6 The Nature of Design

Design (the process of designing something) is, in general, not well understood. While most people, particularly those working with GIS, can understand the value or importance of geodesign as described in the previous section, relatively few have been trained in design and lack, at least to some degree, an appreciation of the nature of design and the way designers think and work.

While the responsibility for fully describing the nature of design and all its idiosyncrasies lies beyond the scope of this paper, it will be helpful to understand three characteristics that are fundamental to most design activities: abductive thinking, rapid iteration, and collaboration.

6.1 Abductive Thinking

Abductive thinking is an extension of classical Aristotelian logic, moving beyond what can be logically induced (bottom-up thinking) and/or what can be logically deduced (top-down thinking) to what might be hypothesized, guessed, or imagined beyond what is logical.

Abductive thinking goes beyond logic to that reasoned edge where designers are challenged to, at one end of the spectrum, make their best guess or, at the other end of the spectrum, wildly imagine a possibility beyond reasoned assumption. They are challenged to take that abductive leap and, in so doing, learn from the perceived consequences of that leap.

The nature of design is all about this type of reasoning (or nonreasoning). It is about leaping beyond reason or beyond what might seem reasonable to unforeseen possibilities. As such, while many design decisions and design-related actions are unpredictable, they can often lead to highly vital solutions.

6.2 Rapid Iteration

Design thinking is an iterative process that occurs rapidly, with little patience for context management. It occurs spontaneously. It does not tolerate interruption or diversion and is best supported by tools that require no attention during their use. Designers want to go from the figment of their imagination to some rendition of that imagination with zero impedance.

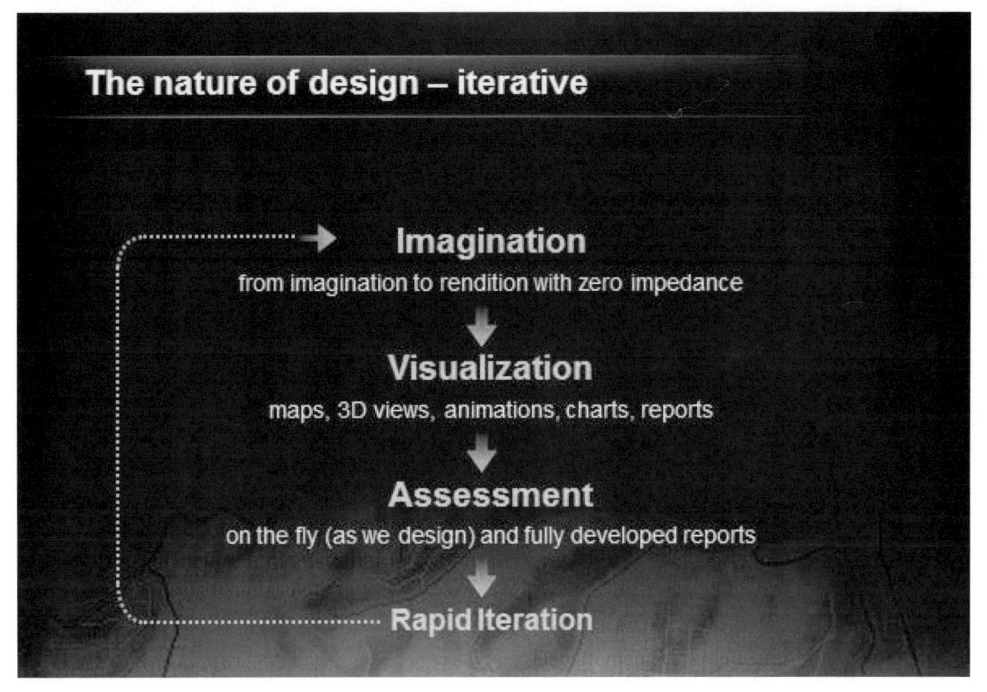

Fig. 9: The nature of design-iterative (Image courtesy Esri)

Design thinking is also exploratory – it is not afraid to try something lacking reasonable support or a pre-established schema. In this sense, it is also highly unpredictable and resists being constrained or inhibited by a particular workflow.

This does not mean, however, that designers are illogical or irrational. Designers are often guided by logical thought processes. However, those processes are typically more abductive by nature than they are strictly logical. They are based more on logical inference (abductive reasoning or making a reasoned guess) than they are on inductive or deductive determinism.

This means that designers want to be free to explore and express their ideas, whatever their basis, with as little resistance as possible. Additionally, they then want to be able to quickly revisit, or make another exploration, each time learning from the results of their exploration.

6.3 Collaboration

The third aspect concerning the nature of design is collaboration. While this may seem by most people to be an obvious functional component of design, indigenous to the design process and the nature of design, to those trained as professional designers – particularly to those trained to see design as an art – it can be an oxymoron.

In truth, however, even to those imbued with the idea that design is a singular activity, most projects – particularly those involving input from many disciplines and design-related professionals – require and could not be accomplished without a high degree of collaboration.

Collaboration of this type involves sharing predesign considerations, ideas, strategies, proposed solutions, assessments, and implementation strategies in a distributed time-space environment. The idea that all involved in the development of a valued design solution for any given project can meet at the same time and in the same space, repeatedly over the life of a project, is rarely valid.

The ability to effectively collaborate, and the tools supporting that collaboration – particularly for larger, more complex projects – become the tools that can make or break the success of a project.

7 Managing Complexity

While the ability to relate an entity to its geographic context can be performed in mental space, as Frank Lloyd Wright did when he designed Fallingwater, the quality and quantity of those relationships are limited to what the human mind can reasonably hold (remember) and manipulate.

Many years ago, Princeton psychologist George A. Miller wrote a paper titled "The Magic Number Seven Plus or Minus Two: Some Limits on Our Capacity for Processing Information" (MILLER 1956). What Miller basically said was that an average person could keep track of seven things in their mind at once. One who was really smart could handle nine. One not so bright could probably handle five.

The reason pencil and paper are so popular is because they extend the ability to explore, assess, manage, and record the information in the mind. As a consequence of their use, people are able to extend their thinking and even pass it on to others. The pencil-and-paper approach, however, as is the case with the mental approach, reasonably limits the number of factors that can be considered simultaneously. It is also a passive environment in that it performs no analysis (other than what occurs in the mind of the designer). The advantage of paper and pencil, however, is that the tools are both historically and intuitively familiar and, as a consequence, extremely easy to use. The disadvantage is that their utility diminishes as the degree of problem complexity increases.

The advantage of the digital approach to geodesign, particularly when one is using GIS, is that it can handle a wide spectrum of spatial complexity. Its disadvantage is that the digital tools, given today's technology, are non-intuitive and relatively difficult to use. The challenge with respect to the development of useful geodesign technology is not only the identification of what tools need to be developed but also the development of those tools so they are easy to use – as easy, it can be said, as using pencil and paper.

8 The Technology of Digital Geodesign

The advantage of a digital environment for doing geodesign can only be realized if that environment is readily accessible and easy to operate and affords the designer the ability to leverage it as an integral component of the geodesign workflow.
While the essential aspect of geodesign lies in the fact that it is predicated on the ability to design (create entities) in geographic space, there are a number of other aspects, or characteristics, that make up the entourage of concepts and capabilities now associate with geodesign. These include the following:

Operational Framework

From an application perspective, an operational framework includes everything the user sees and touches, including hardware (display screens, keyboards, mice, touch screens, styluses, audio devices, interactive tables, tablets, and cell phones) and software (operating systems, application environments, user interfaces, and web-related services).

An operational framework designed to facilitate geodesign needs to provide the user with use patterns that are generally consistent across all supported devices, given the functionality of the device, and provide a software environment that is intuitive and transparent, as well as one that easily supports the functional aspects and workflows associated with each of the characteristics described below.

Data Models

Data models are used to describe entity geometry, attributes, and relationships with respect to how they are defined from a user perspective, which is often domain specific, and how they are structured within the context of a relational database. Most entities, such as a stream or a lake, are represented in the database using standard feature types (e.g., points, lines, polygons, rasters).

Most of the feature types referenced in a GIS are predicated on two-dimensional geospatial geometry, and while they offer the user a powerful and efficient way to represent domain-specific information (using domain-specific data models) in 2D space, they are limited when it comes to representing and analyzing 3D entities, particularly those associated with urban environments such as buildings and other forms of civil infrastructure.

Fig. 10: Thematic Layers (Image courtesy Esri)

Creation and Modification Tools

There are three types of feature creation and modification tools: geometry tools that allow the user to create, replicate, and modify feature geometry; attribute tools that allow the user to assign meaning to the feature; and symbology tools that allow the user to render that feature with cartographic representations that are visually meaningful.

While these tools exist in GIS software, they have been designed to support careful feature editing with respect to the integrity of a well-structured geodatabase as opposed to the rapid creation of features generated by a designer. In general, these tools need to be designed to support greater ease of use, giving them the ability to successfully compete with pencil and paper.

Fig. 11:
Interactive Design using touch screentablests (Image courtesy Wacom)

Inference Engines

Inference engines are used to make assumptions based on the implied intent of the user. For example, if the user is drawing a line that is nearly parallel to the x-axis, the inference engine might assume that it is the user's intent to make that line parallel to the x-axis. This being the case, the inference engine would condition the specification of that line so it was indeed parallel to the x-axis.

There are many types of inference engines: geometry engines, such as the one alluded to in the previous example (SketchUp uses an inference engine of this type to aid in the creation of rectilinear geometry); topology engines used to maintain topological integrity; referential engines used to position features with respect to other features (snapping); and domain-specific engines (used to force compliance with domain-specific standards).

While it is possible to program inferred behavior responsive to some of the generic functions related to data creation, it is less so with respect to the specification of inferred behavior associated with domain-specific functions. For example, it would be relatively easy to create a tool to aid the user when drawing a line intended to be parallel to a previously drawn line. It would be more difficult, however, to create a tool that would interpret the cross-sectional characteristics of a line representing a street centerline, not because the programming is difficult but because it is difficult to know the specific characteristics of the street cross section and how those characteristics change when that line meets another line.

What one really needs is an authoring environment that allows users to create their own domain-specific inference engines, perhaps similar to the behavior of rule-based authoring environments for creating expert systems.

Fig. 12: Applying inference Engines for road allingment (Image courtesy Esri)

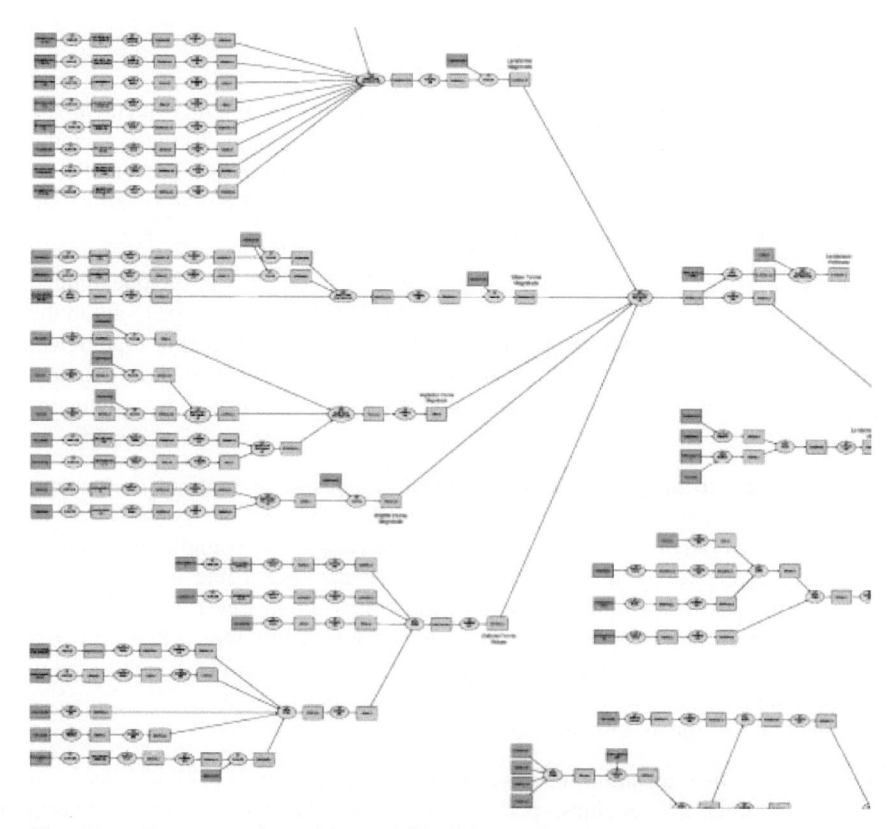

Fig. 13: Geoprocessing with Model builder (Image courtesy Jones and Jones)

Geoprocessing Tools

Geoprocessing tools (models and scripts) are most typically used to generate derived data from one or more geospatial datasets. One of the most powerful features of a geoprocessing model is that the output from one function can be the input to a subsequent function.

With respect to doing geospatial analysis, geoprocessing models are often used to assess the geospatial context of a study area with respect to the area's suitability for, or vulnerability to, a particular set of land uses or land-use management strategies. They can also be used to create impact models designed to assess the probable impacts of the proposed changes.

While the geoprocessing environment in GIS software is very powerful, it has been designed primarily to accommodate geospatial analysis in 2D space as opposed to 3D space.

Additionally, while there are some workarounds in this area, the geoprocessing environment has not been designed to support geospatial simulations (discrete event-based simulation, continuous simulation, or agent-based modeling).

Feedback Displays and Dashboards

Geoprocessing models produce two types of output: geographic displays (usually viewed as maps) and scalar values (such as the area of a polygon or the summed area of a set of polygons), which can be used to derive various types of performance indicators. Feedback displays, often referred to as *dashboards*, are often used to calculate and display those performance indicators.

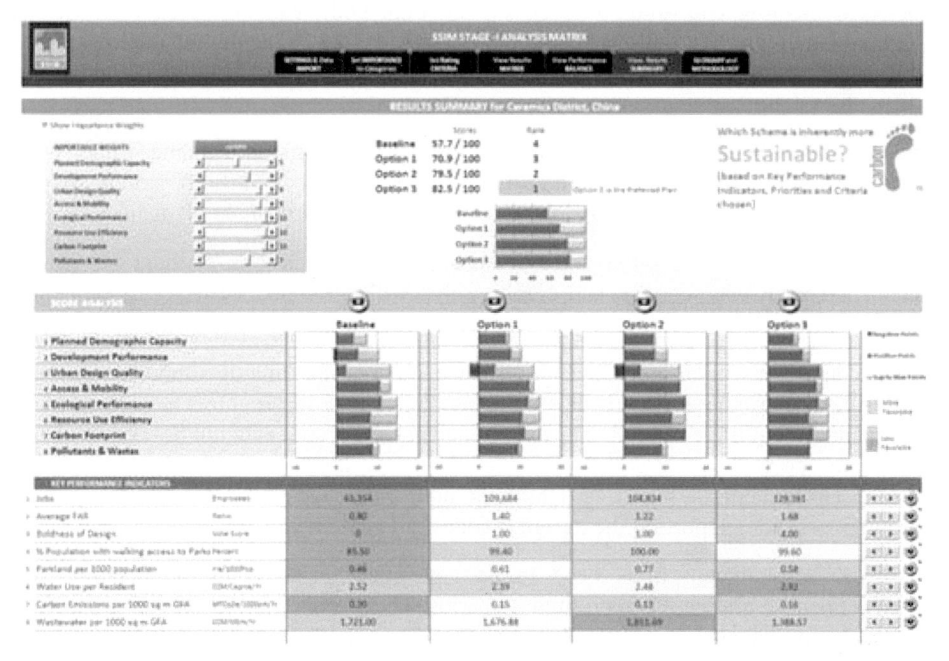

Fig. 14: Comparing alternate plans – via a dashboard (Image courtesy Aecom)

Metaphorically, dashboards can be as simple as a single bar chart or as complicated as the control panel of an airplane. While most dashboards are display-only dashboards, they can also be created as interactive dashboards, thereby giving the observer the ability to change one of the displayed variables and see how it affects the other variables.

From a geoprocessing point of view, this interaction can be associated with parameter tables (such as those normally found in a spreadsheet) or with the geoprocessing model itself. Being directly associated with the model implies that the user can not only change a variable in the dashboard but also have the ability to rerun the model from the dashboard.

Dashboards are created (configured with variables and how those variables are rendered) for a wide variety of purposes depending on the project domain, the characteristics of the particular project, and the informational needs of the intended user. It is thus virtually impossible to predict the content of a dashboard and how it should be displayed.

What is really needed is a dashboard authoring environment that gives the user the ability to select source variables, calculate derivatives from those variables, and condition how those variables are displayed. The environment should also make it possible for the user to specify the interactive nature of the dashboard, be it static, dynamic, or fully interactive.

Scenario Management Tools

Most land-use planning/design projects involve the creation of a number of alternative solutions, sometimes called scenarios. These scenarios, of which there can be many (2 or 3 on the low end and 50 to 60 on the high end), can be based on differing assumptions regarding performance requirements; design concepts; the deployment of different design strategies; and any number of other conditions, which are often difficult to define.

Scenarios can be distinctively different or merely variations on a theme. Either way, they must be properly referenced (so they can be uniquely identified), stored, shared, compared (both graphically and parametrically), revised, and compared again.

Additionally, designers often take, or would like to take, one element (or set of elements) from one scenario and combine that with an element (or set of elements) from another scenario to provide the seed for creating a third scenario.

While there are workarounds, scenario management is not directly supported by most GIS software programs. This is due to the fact that most GIS programs have focused on the management of geospatial data and on the analysis of that data but not on the use of that same geoprocessing environment for designing land-use (or land management) plans. The advent of geodesign thinking now challenges GIS with a new set of requirements for supporting the design (geodesign) process.

Scenario management tools need to be developed to facilitate the creation and management of alternative geodesign scenarios. These tools need to support all aspects of scenario management, including the creation of scenarios and how they are referenced, stored, retrieved, compared, revised, including how portions of one scenario can be combined with portions of another scenario to create a third scenario. The system also needs to manage version control, keeping track of not only the various versions and their variations but also when and why they were created, and who created them.

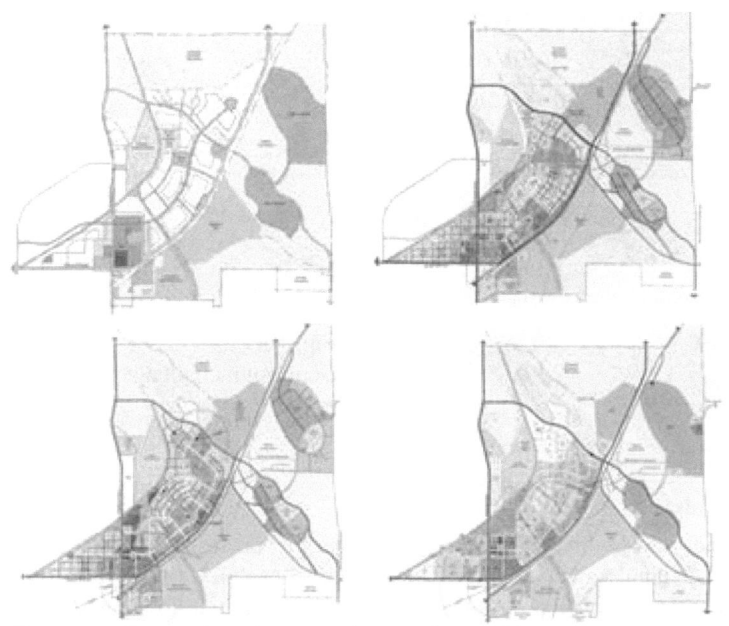

Fig. 15: Comparing alternate plans – via graphics
(Image courtesy Aecom)

Collaboration Tools

Collaboration forms the conceptual basis for working with a team, most particularly during the accumulation of intellectual capital, the application of that capital to the assessment of conditions, and how those conditions affect the creation of something. It also provides the operational context for co-creating that something (e.g., a land-use plan).

One can imagine a group of designers in a room together, with maps and drawings spread over a large table, surrounded by whiteboards filled with concept diagrams, the designers making notes and referencing information on personal digital devices or making notes on paper as they wait to take turns at the whiteboard or presenting their ideas using digital media. Events like this occur all the time in the lives of designers.

The difficulty resides in one's ability to replicate this bricks-and-mortar environment in digital space where space is distributed and time is not always synchronous. Most planning/design projects – particularly those involving multiple disciplines – require collaboration beyond the simple share-review-comment workflow typically supported by document management systems. Creative collaboration typically involves exploration, creation, assessment, modification, presentation, and documentation of alternate design scenarios in both shared and distributed space and in both synchronous and asynchronous time.

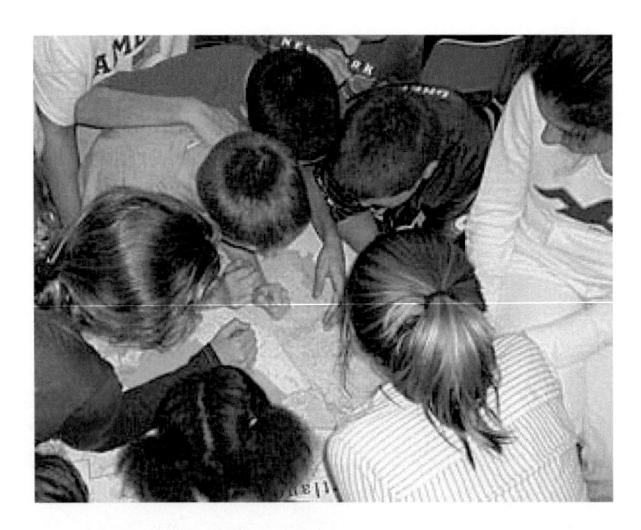

Fig. 16:
Geodesign challenges for participation

Interoperability Tools

Geodesign is a broad field involving many different types of professionals (scientists, planners, architects, landscape architects, engineers, agency representatives, constructors, sponsors, stakeholders, etc.) working in many different domains (the list being too long to enumerate). Given this wide spectrum of professional activity, there are a correspondingly large number of software programs supporting that activity, each domain having its own cluster of software tools supporting various aspects of the design process.

The ability of these various tools (software programs) to conveniently talk to each other can be problematic and time-consuming, often requiring the advanced skills of a highly qualified data gymnast to make it all work. The impedance associated with interoperability issues severely inhibits the work of designers and to this day stands as one of the more significant barriers to the overall design process.

One of the current approaches to interoperability is the use of third-party interoperability tools. While these tools often provide a solution, they require the use of yet another program, which in and of itself can create enough impedance to inhibit the design process.

Designers need transparent interoperability. They need the ability to simply transfer data created in one program into another program without having to do anything. This would be akin to doing projections on the fly. A few years ago GIS, analysts would have to explicitly convert their data from one projection to another. Today, translating from one projection to another is typically done on-the-fly as needed. The same type of behind-the-scenes transparency is needed as designers move data from one program to another.

The challenge to both the developers and appliers of geodesign technology is to understand the nature and importance of each of these characteristics, viewed individually, and how each relates to the other as integral components of a comprehensive geodesign support system.

9 Creating the Future

The future of geodesign depends on the collective understanding of the importance of design, an overall understanding of geodesign and what it means to design in the context of geographic space, a clear understanding of the nature of design and how designers work, and a concerted commitment to develop design-centric (designer-friendly) technologies and workflows supporting all aspects of the design/geodesign process.

This leads to four challenges:

Challenge 1 – Develop a comprehensive understanding of geodesign.

While this paper attempts to lay the groundwork for the development of a shared understanding of geodesign, it is neither comprehensive not does it represent a shared vision. At best, it serves as a catalyst for further discussion and understanding. In this sense, The author expects the responses to this paper to serve the greater geodesign community more than the paper itself. The challenge is to carry this conversation forward and work together to translate respective understandings of what is meant by *geodesign* into a share vision.

Challenge 2 – Develop a design-centric GIS technology.

Perhaps the greatest challenge resides in the capacity of the software development community to absorb and assimilate the unique characteristics (needs) of geodesign and the somewhat idiosyncratic nature of the designer. The programmer's challenge is to create digital frameworks and functionality that truly facilitate the design/geodesign process. This is no small challenge, especially when one considers the designer's desire for zero impedance. The idea of writing design-centric software that is so easy to use that the use of that software is unnoticeable lies beyond the imagination of most programmers, notwithstanding the possible exception of those responsible for the development of Apple's iPhone or iPad.

Challenge 3 – Apply that technology to a wide variety of geospatial design problems.

The success of this work to instantiate geodesign as a credible way of thinking, as an advantageous way to do geospatial design, or as a way to design in geographic space will come from the repeated application of what is now known about geodesign using the tools that are now available (however limited they may be for the moment) to real-world problems. Applying knowledge will help designers learn what works, what doesn't, and what needs to be done to improve the capability to design in the context of geographic space and, in so doing, leverage the science and values co-referenced to that space. The dissemination of this learning through these varied applications will serve to enhance the capacity to improve the quality of work and the vitality of those served by the work.

Challenge 4 – Establish a discipline of geodesign, both in practice and in academia.

Finally, there is a challenge to move beyond the geodesign catchphrase and associated rhetoric to establish a discipline of substance, including values, semantic clarity, and clearly defined processes that can be taught within the context of the various curricula offered by academic institutions and instantiated in professions. While geodesign may or may not become a singular profession, such as architecture or landscape architecture (many argue that it should not), it will surely (or perhaps, hopefully) find its way into the way people design the various entities that affect lives and, in some cases, the very life of the planet.

Regarding the future of geodesign, it is as Abraham Lincoln, Buckminster Fuller, Alan Kay, and Peter Drucker all said, "The best way to predict the future is to create it."

Bibliography

Capra, F. (1996), *The Web of Life*. Anchor Books.
Ervin, S. (2011), *A System for GeoDesign*. In: Proceedings of Digital Landscape Architecture, Anhalt University of Applied Science, 2011.
Flaxman, M. (2010), *Fundamentals of Geodesign*. In: Proceedings of Digital Landscape Architecture, Anhalt University of Applied Science, 2010.
Goodchild, M., Maguire, D. & Rhind, D. (1991), *Geographic Information Systems: Principles and Applications* (2 Vol.). Longman.
Kunzmann, K. (1993), *Geodesign: Chance oder Gefahr?* In: Planungskartographie und Geodesign. Hrsg.: Bundesforschungsanstalt für Landeskunde und Raumordnung. Informationen zur Raumentwicklung, 7/1993.
Miller, G. (1956), *The Magic Number Seven, Plus or Minus Two*. In: The Psychological Review, 1956.
Miller, W. (2004), *Landscape Architecture: Education & Virtual Learning Environments*. In: Proceedings of Trends in Online Landscape Architecture, Anhalt University, 2004.
Miller, W. (2005), *Definition of Design*. Trimtab, Buckminster Fuller Institute.
Miller, W. (2006), *Purpose of Design*. Trimtab, Buckminster Fuller Institute.
Miller, W. (2011), *Personal recollection*.
Steinitz, C. (1995), *A Framework for Landscape Planning Practice and Education*. In: Progressive Architecture, 127.
Toker, F. (2005), *Fallingwater Rising*. Alfred A. Knopf.

Quotes

Drucker, P., *The best way to predict the future is to create it*, www.brainyquote.com
Fuller, B., *The best way to predict the future us to design it*, www.thegreenspotlight.com
Kay, A., *The best way to predict the future is to invent it*. www.smalltalk.org
Lincoln, A., *The best way to predict your future is to create it*. www.goodreads.com

Additional Resources

Artz, Matt, *Geodesign: A Bibliography.* Science and Design (blog), August 13, 2009
   http://gisandscience.com/2009/08/13/geodesign-a-bibliography/
Harvard bio, Professor Howard Taylor Fisher,
   http://www.gis.dce.harvard.edu/fisher/HTFisher.htm
McElvaney, Shannon, *Geodesign for Regional and Urban Planning*, Esri Press, 2012.
Miller, William R., *Definition of Design*, February 11, 2004,
   http://www.wrmdesign.com/Philosophy/Documents/DefinitionDesign.htm
Miller, William R., *Purpose of Design*, July 28, 2004,
   http://www.wrmdesign.com/Philosophy/Documents/PurposeDesign.htm
NCGIA *home page*, http://www.ncgia.ucsb.edu/
NCGIA, *Carl Steinitz*, http://www.ncgia.ucsb.edu/projects/scdg/docs/cv/Steinitz-cv.pdf
Steinitz, C., *A Framework for Geodesign*, Esri Press, 2012
Wikipedia, s. v., *Frank Lloyd Wright*, accessed November 22, 2011
   http://en.wikipedia.org/wiki/Frank_Lloyd_Wright
Wikipedia, s. v., *Geodesign*, accessed November 22, 2011
   http://en.wikipedia.org/wiki/Geodesign

Introducing the KAT Competence Center "Digital Planning and Design" at Anhalt University

Alexander KADER, Stephan PINKAU, Einar KRETZLER,
Claus DIESSENBACHER and Marcel HEINS

1 What is the KAT Kompetenzzentrum DIGITALES PLANEN und GESTALTEN (Competence Center for Digital Planning and Design)?

The KAT-Kompetenzzentrum DIGITALES PLANEN und GESTALTEN is a trans disciplinary and scientific center for applied information technology as well as for information and knowledge management in planning and design.

Fig. 1:
Logo of the KAT Competence Center of universities in Saxony Anhalt

It takes the potentials and activities of the areas of architecture; design; landscape architecture; landscape and environmental planning; digital media; etc., and brings them together with the competences of technologically-oriented fields (computer applications in surveying, computer science, etc.) as well as with business (innovation management, technology transfer, etc.).

The Kompetenzzentrum is located at Anhalt University and is part of „Kompetenznetzwerk für Angewandte und Transferorientierte Forschung" (the initiative "Competence Network for Applied and Transfer-Oriented Research") of the universities in Saxony Anhalt.

Through the cooperation of different areas of expertise and through networks with external partners, industry has an efficient system with which it can access and make use of :

- Research and development
- Planning and design
- Knowledge and technology transfer
- Personnel transfer and exchange
- Further education

2 Goals

The KAT Competence Center pursues the following objectives:
- Making use of the scientific system available to regional and other businesses
- Increasing the utilization rate of the results of scientific works
- Increasing innovation in mid-sized and large firms
- Creating a network to improve cooperation between universities and industry
- Establishing long-term partnerships between science and industry
- Increasing the placement rate of college graduates in companies and other institutions

3 Target Groups

The KAT Competence Center endeavors to first make its resources available to small and mid-sized business in the state of Saxony Anhalt. Additionally, networks and firms working on the national and international levels should be approached in order to successfully market products and services.

4 Services

The KAT Competence Center offers the following services:
- Research services
- Product and process development
- Further education and training
- Planning/concept development
- Coaching/consulting
- Use of resources, laboratories and equipment
- Personnel transfer – placement of college graduates, as well as thesis students and
- intern

5 Expertise

The services provided by the KAT Competence Center reflect its research focuses, and areas of competence (also see "Kompetenz-Cloud"):
- Information and knowledge management (data processing and structuring)
- Interface Man and Machine
- Learning and teaching design
- Lifestyles in the virtual world

Interested scientists can access references and years' long experience. State of the art instruments and equipment are available for their use.

Fig. 2: Network of CAD, GIS and Information technology experts within KAT

Fig. 3: Areas of CAD, GIS and Information technology expertise within KAT – the "Cloud of Competences"

Fig. 4: Presentation of KAT competences at Anhalt University

6 Benefits

Possibilities to improve the performance and competitiveness of enterprises through:
- Current information on advisory services in the area of applied information technologies as well as information and knowledge management in planning and design
- Demand-oriented F&E services
- Cost-effective use of state-of-the-art equipment
- Access to innovation through research and development networks
- Development of advanced marketing strategies for local and global markets
- Securing qualified staff through long-term personnel development in working with
- Anhalt University's Career Center
- Building effective and sustainable cooperative relations

Contact

Contact persons

Prof. Einar Kretzler	Tel. +49 (0) 3496 67-1159	e.kretzler@loel.hs-anhalt.de
Alexander Kader	Tel. +49 (0) 340 5197-1559	a.kader@afg.hs-anhalt.de

Directors

Prof. Einar Kretzler, Prof. Dr. Claus Dießenbacher, Prof. Stephan Pinkau

Staff

Marcel Heins, Alexander Kader, Michael Walter

Vorstellung des KAT-Kompetenzzentrums „DIGITALES PLANEN und GESTALTEN" an der Hochschule Anhalt

Alexander KADER, Stephan PINKAU, Einach KRETZLER,
Claus DIESSENBACHER und Marcel HEINS

1 Was ist das KAT-Kompetenzzentrum DIGITALES PLANEN und GESTALTEN?

Das KAT-Kompetenzzentrum DIGITALES PLANEN und GESTALTEN ist ein transdisziplinäres und fachwissenschaftlich fokussiertes Zentrum für Angewandte Informationstechnologien sowie Informations- und Wissensmanagement in PLANUNG und GESTALTUNG.

Abb. 1:
Logo der KAT-Kompetenzzentren der Hochschulen des Landes Sachen- Anhalt

Es bündelt für diesen Bereich die Potenziale und Aktivitäten der Fachgebiete Architektur, Design, Landschaftsarchitektur, Landschafts- und Umweltplanung, Digitale Medien etc. und verknüpft diese mit den Kompetenzen technologieorientierter Fachgebiete (Geoinformatik, Informatik etc.) sowie des Fachgebiets Wirtschaft (Innovationsmanagement, Technologietransfer etc.).

Das Kompetenzzentrum DIGITALES PLANEN und GESTALTEN hat seinen Sitz an der Hochschule Anhalt und ist Bestandteil der Gemeinschaftsinitiative „Kompetenznetzwerk für Angewandte und Transferorientierte Forschung" der Hochschulen des Landes Sachsen-Anhalts.

Durch die Kooperation verschiedenster Fachgebiete und Vernetzung mit externen Partnern wird Unternehmen ein effizientes System für die Inspruchnahme von Leistungen in den Bereichen:

- Forschung und Entwicklung,
- Planung und Gestaltung,
- Wissens- und Technologietransfer,

- Personaltransfer und -austausch sowie
- Weiterbildung

zur Verfügung gestellt.

2 Ziele

Das KAT-Kompetenzzentrum DIGITALES PLANEN und GESTALTEN verfolgt diese Ziele:

- Nutzbarmachung des Wissenschaftssystems für die regionale und überregionale Wirtschaft,
- Steigerung der Verwertungsquote der Ergebnisse wissenschaftlicher Arbeiten,
- Erhöhung der Innovationsrate in mittelständischem Gewerbe und in der Industrie,
- Schaffung von Netzwerken zur Verbesserung der Zusammenarbeit zwischen Hochschulen und Wirtschaft,
- Etablierung von nachhaltigen Partnerschaften zwischen Wirtschaft und Wissenschaft,
- Erhöhung der Vermittlungsrate von Hochschulabsolventen an Unternehmen und weiteren Institutionen.

3 Zielgruppen

Das KAT-Kompetenzzentrum DIGITALES PLANEN und GESTALTEN richtet sein Leistungsangebot in erster Linie an die kleinen und mittleren Unternehmen des Landes Sachsen-Anhalt. Darüber hinaus sollen auch überregional agierende Netzwerke und Unternehmen angesprochen werden, um marktfähige FuE-Leistungen erfolgreich in der Wirtschaft umzusetzen.

4 Leistungen

Durch das KAT-Kompetenzzentrum DIGITALES PLANEN und GESTALTEN werden angeboten bzw. vermittelt:

- Forschungsleistungen,
- Produkt-/Verfahrensentwicklung,
- Schulung/Fortbildung,
- Planung/Konzeptentwicklung,
- Coaching/Consulting,
- Bereitstellung von Ressourcen, Laboren, Equipment,
- Personaltransfer-Vermittlung von Hochschulabsolventen sowie
- Studierenden für Abschlussarbeiten und Praktika.

5 Kompetenzen

Das Leistungsspektrum des KAT-Kompetenzzentrum DIGITALES PLANEN und GE-STALTEN reflektiert die Forschungsschwerpunkte und Kompetenzfelder (siehe auch: „Kompetenz-Cloud"):

- Informations- und Wissensmanagement
  (Strukturierung und Verarbeitung von Daten),
- Schnittstelle Mensch – Maschine,
- Gestalten lernen und lehren,
- Lebensweisen im Digitalen Raum.

Engagierte Wissenschaftler können innerhalb ihrer Fachgebiete auf langjährige Erfahrungen und Referenzen zurückgreifen. Ihnen stehen modernste Geräte und Ausrüstungen zur Verfügung.

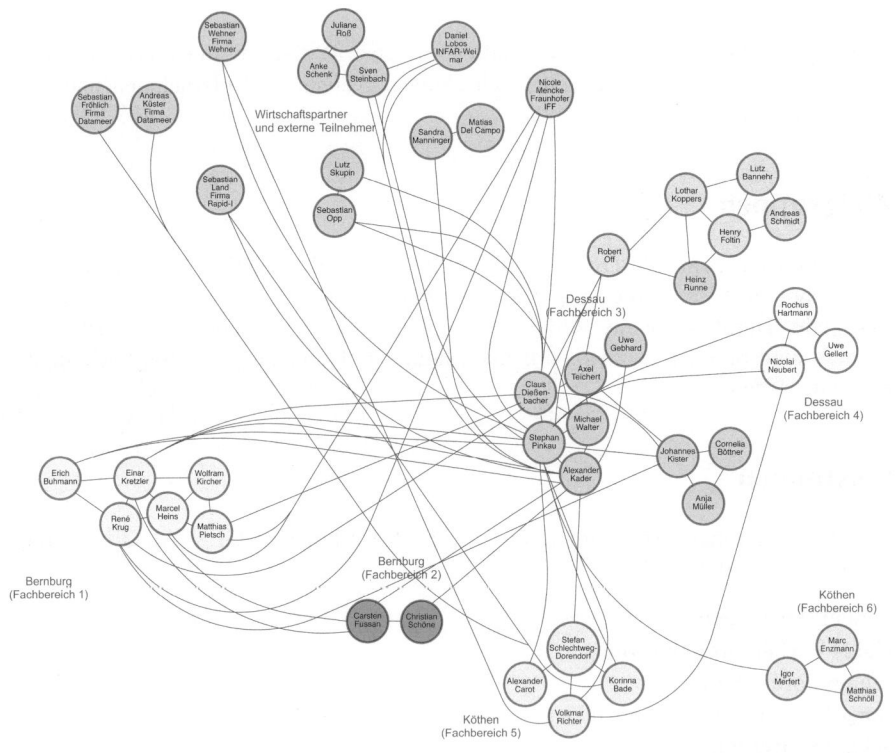

Abb. 2: Vernetzung der CAD-, GIS- und Informatikexperten an der Hochschule Anhalt

Vorstellung des KAT-Kompetenzzentrums „DIGITALES PLANEN und GESTALTEN"   575

Vermessung	Datenerhebung		Simulationen	Virtuelle Welten
Digitales Data Mining	Statistische Datengenerierung		Interface Design	Architektur der Medialen Räume
Klima- / Gebäudeklimaforschung	Wissenschaft universeller Strukturierung		Cave Automatic Virtual Environments	Künstlerische Rekonstruktion
Strukturierung und Verarbeitung von Daten			**Schnittstelle Mensch – Maschine**	
3D-Datenerfassung, Laserscanning	Intelligente Strukturierung		Digitale Werkzeuge in Entwurfsprozessen	Labor für Usability Testing
Wärmebilder	Digitale Vermessung	Arrangieren von Daten	Modellbau	Powerwall
	Demographieforschung		Moderne Utopien	
Entwicklung von Strukturierungsvermögen		Sustainability	Designing Ecologies	Raumproduktion
Medienpädagogik	Materialwissen	Entwicklung von Lernumgebungen	Wie wird man in Zukunft wohnen?	
Gestalten lehren und lernen			**Lebensweisen im Digitalen Raum**	
Kommunikationsdesign	Fortbildung, lebenslanges Lernen	Menschenbild	Das intelligente Haus	Menschenbild und Gebäudetechnik
Gestaltete Objekte als pädagogische Agenten		Interventionen, Strategien		Beratung von Kommunen
	Kommunikationsplattformen, Informationsnetze		Landschaftsentwicklung, Denkmalpflege	

Abb. 3: Arbeitsfelder der CAD-, GIS- und Informatikexperten an der Hochschule Anhalt – die „Kompetenz-Cloud"

Abb. 4: Präsentation des KAT Kompetenzzentrums in Dessau

6   Ihr Nutzen

Möglichkeiten zur Verbesserung der Leistungs- und Wettbewerbsfähigkeit Ihres Unternehmens durch:
- aktuelle Informationen zum Leistungsspektrum im Bereich der Angewandten Informationstechnologien sowie im Informations- und Wissensmanagement in PLANUNG und GESTALTUNG,

- bedarfsorientierte F&E Leistungen,
- kostengünstige Nutzung des hochmodernen Equipment,
- Zugang zu Innovationen über Forschungs- und Entwicklungsnetzwerke,
- Entwicklung fortschrittlicher Marketingstrategien für lokale und globale Märkte,
- Sicherung des Fachkräftebedarfs einer langfristigen Personalentwicklung durch Zusammenarbeit mit den Career Center der Hochschule,
- Aufbau effektiver und nachhaltiger Kooperationsbeziehungen.

Kontakt

Ansprechpartner

- Prof. Einar Kretzler   Tel. +49 (0) 3496 67-1159   e.kretzler@loel.hs-anhalt.de
- Alexander Kader   Tel. +49 (0) 340 5197-1559   a.kader@afg.hs-anhalt.de

Postanschrift

Hochschule Anhalt
KAT-Kompetenzzentrum DIGITALES PLANEN und GESTALTEN
Seminarplatz 2a
06846 Dessau

Leitung

Prof. Einar Kretzler, Prof. Dr. Claus Dießenbacher, Prof. Stephan Pinkau

Mitarbeiter

Marcel Heins, Alexander Kader, Michael Walter

NAEXUS – Virtual Space Scope – A Space of Illusion

Claus DIESSENBACHER and Michael WALTER

Introduction

Reason and motivation for the development of the NÆXUS was the virtual reconstruction of the ancient monastery Heisterbach with the aim to make the architecture of the no longer visible buildings close to reality and immersively recognizable. For more than six centuries, the valley in the north of the Siebengebirge Mountains in the west of Germany near Bonn, was characterized by monastic life. In 1803, secularization ended the history of the monastery and most buildings were destroyed. Today, only a fragment of the ancient choir of the Abbey Church tells us something about the architecture of the past (fig. 1).

Fig. 1:
Fragment of the Abbey Church

In this contect, a prototype of a mobile space of illusion was developed within the scope of a master thesis at the Dessau Institute of Building Design of Anhalt University of Applied Sciences. The aim of the development was a realistic perception of computer-generated architectural spaces and objects. For the design and the construction of a space like this, the following parameters should be used.

- Interactive virtual reality for the study and the presentation of a computer-generated building or space. It should be possible to move freely in the virtual reality. In this case, the use of a navigation device, which is also used for computer games is possible.

- Graphic quality: The graphic quality of the presentation of an interactive virtual reality should be close to reality to receive a high level of immersion.
- Spatial effect to obtain a very realistic spatial effect of a virtual reality, the computer model should be presented in the scale of 1:1 and should cover the human field of vision.
- Low costs: To give many people the opportunity to use the device, the costs should not exceed € 20,000 .
- Mobility: For the presentation of computer-generated models of architecture at different locations, the construction of the space of illusion should be structured modularly with the aim to make a fast assembly possible.
- Simple handling: The handling of the digital equipment should be simple to learn and orientated by commercially available computer engineering.
- Multimedia: The visualization of digital objects should not be restricted to interactive virtual realities. It should also be possible to present more media such as films, panorama pictures, etc.
- Multiple users: The perception of the above described presentations should be possible for many people at the same time.
- Sound: To enhance the feeling of immersion, it should be possible to perceive the media not only with one's eyes but with one's ears as well. Therefore a suitable sound system should be installed.

The developed NÆXUS – Virtual Space Scope is a mobile ellipsoid-shaped large projection system. The ellipsoid space has a cross-section dimension of 6 meters and a height of 3.50 meters. Inside of the figure there is a cylindrical screen of about 240 degrees. By the application of four data projectors, virtual realities can be projected seamlessly on the curved screen. Interactive interface systems, which are also used for computer games, allow the users to go on an immersive journey in real time through the synthetic world. To maximize the presentation of the computer models close to reality, a software is used which has been developed by the game industry (cryengine 3) because the quality of the realistic presentation of objects like buildings, vegetation etc. is very high. Beside the presentation of interactive virtual realities, the system is also able to present animations, movies and panoramas of 360 degrees. Sound equipment supports the perception by acoustic elements such as noise and music. Another potential of the NÆXUS is its flexible application, independent of location. The construction is structured modularly. The oval shape of the NÆXUS is like a skeleton, composed of vertical and horizontal wood elements which are connected by steel corners. Due to the cross-section dimension of 6 meters, twenty people are able to be inside.

Based on the architectural drawings of the architecture historian Sulpiz Boisserée and in cooperation with the Office for Ground Care of Monuments of the city of Cologne and the Heisterbach Foundation, a three dimensional computer model of the monastery and the landscape was developed (fig. 4a-c).

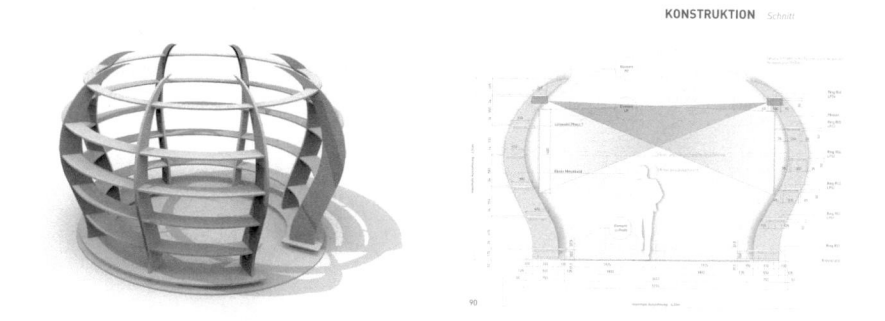

Fig. 2: Construction principle of the NÆXUS

Fig. 3: NÆXUS in action

With the help of the game engine "Cryengine3" from the firm Crytek, the computer model has been visualized and can be interactively perceived in the NÆXUS. The dimension of the NÆXUS and the associated size and curvature of the screen provides a perception of the reconstruction nearly on the scale of 1:1 (fig. 5).

Fig. 4a: Computer model of the abbey church

Fig. 4b: Interior of the abbey church

Fig. 4c: Monastery in the landscape

Fig. 5: Visualization of the digital monastery inside of the NÆXUS

Meanwhile, other applications show the efficiency of the NÆXUS in relation to the visualization of architecture close to reality. For instance, students of architecture of the AFG Department at the Anhalt University of Applied Sciences used the space of illusion to present and to study their architectural ideas. Particularly the spatial, realistic control of work first developed on paper in two dimensions appears, in this way, as a helpful support in the architect's education (fig. 6).

Fig. 6: Students are using the NÆXUS to study and to communicate their architectural ideas

Another future use of the NÆXUS principle is found in the new Romans Museum in the archaeological park in Xanten. There, the ancient, only still partially visible Roman city "Colonia Ulpia Traiana" should be made recognizable again for visitors in a digital space of illusion. The aim of this is, in addition to the immersive perception of the urban space, to be able to go on an exploration tour through the virtual Roman city.

References

Walter, M. & Dießenbacher. C., NAEXUS –Virtual Space Scope Flyer 2010.

info@naexus.com; http://www.naexus.com

NAEXUS – Virtual Space Scope –
Ein Illusionsraum

Claus DIESSENBACHER und Michael WALTER

Anlass und Motivation zur Entwicklung der NÆXUS war die virtuelle Rekonstruktion der ehemaligen Klosteranlage Heisterbach im Siebengebirge mit dem Ziel, die Architektur der nicht mehr sichtbaren Klostergebäude realitätsnah und immersiv erlebbar zu machen.

Über sechs Jahrhunderte prägte klösterliches Leben das Tal nördlich des Petersberges im Siebengebirge bei Bonn. 1803 setzte die Säkularisierung dem ein Ende. Das Kloster Heisterbach wurde aufgehoben und weitestgehend abgerissen. Heute noch sichtbarer Zeuge ist der ruinöse Chor der ehemaligen Abteikirche (Abb. 1).

Abb. 1:
Chorruine der ehemaligen Abteikirche

In diesem Zusammenhang wurde an der Hochschule Anhalt im Dessauer Institut für Baugestaltung im Rahmen einer Masterarbeit im Jahre 2009 ein erster Prototyp eines mobilen Illusionsraumes entwickelt. Ziel der Entwicklung war die immersive also wirklichkeitsnahe Wahrnehmung von computergenerierten, architektonischen Räumen und Objekten. Für den Entwurf und die Konstruktion eines solchen Raumes sollten folgende Parameter Bedingung sein:

- Interaktive virtuelle Welt:
  Für die Studie oder Präsentation einer computergenerierten Gebäude- oder Raumsituation sollte es möglich sein, sich frei in der virtuellen Welt bewegen zu können. Denkbar hierfür ist ein Steuerung- bzw. Navigationsgerät, welches auch für Computerspiele verwendet wird.

- Grafische Qualität:
  Die grafische Qualität der Darstellung einer interaktiv zu begehenden virtuellen Welt sollte realitätsnah sein, um einen möglichst hohen Immersionsgrad zu erreichen.
- Raumwirkung:
  Um eine möglichst realitätsnahe Raumwirkung einer virtuellen Welt zu erreichen, sollte das Computermodell im Maßstab 1:1 und blickfeldübergreifend dargestellt werden.
- Geringe Kosten:
  Um die Nutzbarkeit für viele Interessenten möglich zu machen sollten die Kosten für die Herstellung 20.000,-- Euro nicht überschreiten.
- Mobilität:
  Für die Präsentation von computergenerierten Architekturmodellen an verschiedenen Standorten, sollte die Konstruktion des Illusionsraumes modular aufgebaut sein, mit dem Ziel, einen schnellen Auf- und Abbau an verschiedenen Standorten möglich zu machen.
- Einfache Bedienung:
  Die Handhabung der digitalen Technik sollte einfach zu erlernen sein und sich an dem allgemeinen Umgang mit kommerziell verfügbarer Computertechnik orientieren.
- Multimedial:
  Die Visualisierung von digitalen Objekten sollte sich nicht nur auf interaktive virtuelle Welten beschränken. Filme, Panoramadarstellungen, Standbilder etc. sollten ebenso präsentiert werden können.
- Mehrere Nutzer:
  Die Wahrnehmung der zuvor genannten Darstellungen sollte für mehrere Personen gleichzeitig möglich sein.
- Sound:
  Um einen möglichst hohen Immersionsgrad zu erreichen, sollten neben den Augen auch die Ohren als Wahrnehmungsorgane angesprochen werden. Eine entsprechende Soundanlage wäre somit vorzuhalten.

Das entwickelte „NÆXUS – Virtual Space Scope" ist ein mobiles ellipsoides Großprojektionssystem. Der ellipsoide Raum hat einen Durchmesser von 6,00 m und ist 3,50 m hoch. Im Inneren der Figur befindet sich eine zylindrische Leinwand von 240 Grad. Durch den Einsatz von vier Beamern können virtuelle Welten nahtlos auf die gekrümmte Leinwand projiziert werden. Interaktive Interfacesysteme, welche auch für Computerspiele verwendet werden ermöglichen dem Nutzer eine immersive Reise in Echtzeit durch die synthetische Welt. Zur Generierung einer möglichst realitätsnahen Darstellung der Computermodelle wurde eine Software aus der Spieleindustrie verwendet (Cryengine 3), da die Qualität der realistischen Visualisierung von Objekten (Gebäude, Vegetation etc.) hier sehr hoch ist. Das System ist in der Lage, neben der Präsentation von interaktiven virtuellen Welten auch Animationen, Filme, 360 Grad Panoramen und Videokonferenzen entzerrt darzustellen. Eine Soundanlage unterstützt die Wahrnehmung durch akustische Elemente wie Geräusche und Musik. Ein weiteres Potential des NÆXUS ist der flexible, standortunabhängige Einsatz. Die Konstruktion ist modular aufgebaut. Die ovale Gestalt des NÆXUS zeigt sich in Form eines Skeletts, bestehend aus vertikalen und horizontalen Holzelementen, welche mit Stahlwinkeln verbunden bzw. verschraubt sind. Bedingt durch den Durchmesser von 6,00 m können bis zu 20 Personen in der Form Platz finden.

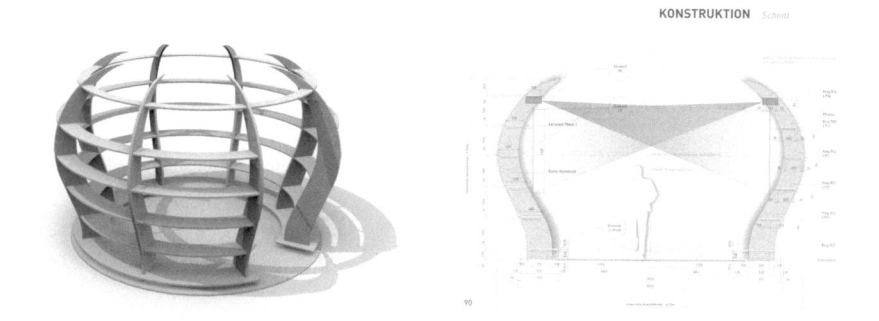

Abb. 2: Konstruktionsprinzip der NÆXUS

Abb. 3: NÆXUS in Aktion

Auf der Grundlage von Architekturzeichnungen des Architekturhistorikers Sulpiz Boisserée und in Zusammenarbeit mit dem Amt für Bodendenkmalpflege der Stadt Köln und der Stiftung Heisterbach wurde ein dreidimensionales Computermodell der historischen Klostergebäude samt Umland erstellt (Abb. 4a-c).

Mithilfe der Spieleengine „Cryengine3" der Firma Crytek wurde das Modell visualisiert und kann in der NÆXUS interaktiv wahrgenommen werden. Die Dimension der NÆXUS und die hiermit verbundene Größe und Krümmung der Leinwand vermittelt eine Wahrnehmung der Rekonstruktion nahezu im Maßstab 1:1 (Abb. 5).

Abb. 4a: Computermodell der Abteikirche

Abb. 4b: Innenraum der Abteikirche

Abb. 4c: Kloster und umgebende Landschaft

Abb. 5: Visualisierung der digitalen Klosteranlage in der NÆXUS

Mittlerweile zeigen weitere Anwendungsfälle die Leistungsfähigkeit der NÆXUS in der realitätsnahen Visualisierung von Architektur. So haben z. B. Architekturstudenten des

Fachbereichs AFG an der Hochschule Anhalt den Illusionsraum genutzt, um ihre Architekturentwürfe zu präsentieren und zu studieren. Besonders die räumliche, wirklichkeitsnahe Kontrolle der, zuvor auf dem Papier in zwei Dimensionen entwickelten Ideen, zeigt sich hierbei als eine hilfreiche Unterstützung in der Architektenausbildung (Abb. 6).

Abb. 6: Studenten nutzen den NÆXUS, um ihre Ideen zu studieren und zu kommunizieren

Eine weitere, zukünftige Verwendung findet das NÆXUS-Prinzip in dem neuen Römermuseum im Archäologischen Park in Xanten. Dort soll die ehemalige, nur in Teilen noch sichtbare römische Stadt „Colonia Ulpia Traiana" in einem Illusionsraum für die Besucher wieder erlebbar gemacht werden. Das Ziel hierbei ist, neben der immersiven Wahrnehmung des städtischen Raumes, selbst auf Erkundungstour durch die virtuelle Römerstadt zu gehen.

Weitere Informationen zu Projekt unter: www.naexus.com.

DLA Awards 2011 and 2012

We are pleased to announce those who have received the awards from the 2011 and 2012 conferences in Bernburg and Dessau, Germany.

Digital Landscape Architecture DLA 2011 Award by Review Committee

The following awards are for the full papers, resulting from the blind peer review by the Digital Landscape Architecture Review Committee.

- **SCIENTIFIC EXCELLENCE** was given to
  Dr. Hans-Georg Schwarz-v.Raumer and *Prof. Antje Stokmann*, University of Stuttgart, Institute of Landscape Planning and Ecology, Germany for the contribution: "GeoDesign – Approximations of a Catchphrase" (Highest possible score of 6.0).

The second highest score of 5.5 was given to the following authors:

Philip Paar and *Prof. Dr. Jörg Rekittke*, National University of Singapore, Singapore: "The Trojan Horse – Teaching Geoinformation Methods to Students of Landscape Architecture".

Prof. Dr. Jürgen Döllner, *Tassilo Glande* and *Matthias Trapp*, Hasso-Plattner-Institut, University of Potsdam, Germany:
"Concepts for Automatic Generalization of Virtual 3D Landscape Models".

Dr. Ulrike Wissen Hayek, *Noemi Neuenschwander*, *James Melsom* and
Prof. Christophe Girot of ETH Zürich, Switzerland for their contribution:
"Interdisciplinary Studio for Teaching 3D Landscape Visualization – Lessons from the LVML".

Howard Hahn, Kansas State University, United States:
"Visualizing Wetland and Meadow Landscape".

Ellen Fetzer, Nürtingen-Geislingen University, Germany:
„Computer-Supported Collaborative Learning with Wikis and Virtual Classrooms across Institutional Boundaries – Potentials for Landscape Architecture Education".

Ed Morgan, Lewis Gill, Martin Dallimer and *Prof. Dr. Eckart Lange*, Sheffield University United Kingdom:
"Integrating bird survey data into real time 3D visual and aural simulations".

Prof. Dr. Roman Lenz, Werner Rolf and *Christian Tilk*, Nürtingen-Geislingen University, Germany:
"Teaching Digital Methods in Landscape Planning – Design, Content and Experiences with a Course for Postgraduate Professional Education".

Frank Roser, University of Stuttgart, Institute of Landscape Planning and Ecology, Germany:
"Is the Beauty of Landscape Computable? A Methodical Approach for GIS-based Visual Landscape Assessment for Large Areas".

We are pleased to announce the following

Digital Landscape Architecture DLA 2011 Poster Awards

The **DLA 2011 Poster Award** for
- BEST APPLICATION OF NEW MEDIA IN LANDSCAPE ARCHITECTURE:
  by vote of the participants, was given to
  Vesna Koscak, Miocic-Stosic, Biserka Bilusic Dumbovic, Vladimir Kusan,
  University of Zagreb, Faculty of Agriculture, Croatia
  for the poster:
  "Digital Mapping and Evaluation of Landscape Character – Strategic Approach".

The **DLA 2011 Poster Award for**
- BEST GIS APPLICATION IN LANDSCAPE ARCHITECTURE:
  by vote of the participants, was given to:
  Guoping Huang, University of Virginia and *Prof. Dr. Michael Flaxman*, Massachusetts Institute of Technology, United States
  for the poster
  "Direction and Content-based Landscape Visual Quality Assessment".

The **DLA 2011 Poster Award for**
- BEST PROFESSIONAL CONTRIBUTION TO DIGITAL LANDSCAPE ARCHITECTURE:
  by vote of the participants, was given to
  Ahmed Alomary, Ali Haider Saad Ali Al-Jameel, Omar AlhAfith,
  Mosul University, Iraq
  for the poster:
  "Future Landscape Teaching Challenges in Iraq – A Framework".

Digital Landscape Architecture DLA 2012 Award by Review Committee

During the two phase 2012 peer-reviewing process, sixty-two entries were evaluated by forty-nine reviewers. Of the forty-two accepted papers, nearly half were rated as excellent.

The following awards are for the full papers, resulting from the blind peer review by the Digital Landscape Architecture Review Committee:

- **SCIENTIFIC EXCELLENCE** was given to
  Amii Harwood, Andrew Lovett and *Jenni Turner*, University of East Anglia,
  United Kingdom for the contribution:
  "Extending Virtual Globes to Help Enhance Public Landscape Awareness"
  (Highest overall score in the DLA 2012 review: 5.5).

The second highest overall score of 5.0 was given by the reviewers to the following fifteen teams of authors:

Ulrike Wissen Hayek, Noemi Neuenschwander, Jan Halatsch, Antje Kunze, Timo von Wirth, Adrienne Grêt-Regamey and *Gerhard Schmitt*, ETH Zurich, Switzerland:
„Concept of a Transdisciplinary Urban Collaboration Platform Based on GeoDesign Ensuring Urban Quality in Agglomerations".

David Barbarash, Purdue University, United States:
"The Communication Value of Graphic Visualizations Prepared with Differing Levels of Detail".

Jörg Rekittke, Philipp Paar and *Yazid Ninsalam*, University of Singapore:
"Foot Soldiers of GeoDesign".

Sigrid Hehl-Lange, Lewis Gill, John Henneberry, Berna Keskin, Eckart Lange, Ian Caleb Mell and *Ed Morgan*, University of Sheffield, Department of Town and Regional Planning, United Kingdom:
"Using 3D Virtual GeoDesigns for Exploring the Economic Value of Alternative Green Infrastructure Options".

Boris Stemmer, Universität Kassel, Germany:
„Collaborative Landscape Assessment and GeoDesign".

Shawn Partin, Jon Burley, Robert Schutzki and *Patricia Crawford*, Michigan State University, United States:
"Concordance between Photographs and Computer Generated 3D Models in a Michigan Highway Transportation Setting".

Songlin Wei, Cyril Fleurant and *Jon Burley*, Zhongkai University, China; Agrocampus Ouest, France; Michigan State University, United States:
"Replicating Fractal Structures with the Reverse Box Counting Method – An Urban South-east Asian Example".

Anna Czinkoczky and *Ákos Bede-Fazekas*, Corvinus University of Budapest, Hungary:
"Visualization of the Climate Change with the Shift of the So-called Moesz-line".

Nikolay Popov, Unitec Institute of Technology, New Zealand:
"Shaping Suburbia – An Agent-Based System for Sustainable Suburban Development".

David Goldberg, Robert Holland and *Scott Wing*, Penn State University, United States:
"GIS + BIM = Integrated Project Delivery @ Penn State".

Sándor Jombach, László Kollányi, József László Molnár, Áron Szabó and *Tádé Dániel Tóth*, Corvinus University of Budapest, Department of Landscape Planning and Regional Development, Hungary:
"Geodesign Approach in Vital Landscapes Project".

Agnieszka Ozimek, Piotr Labedz, Kornelia Michoń, Cracow University of Technology, Poland:
"Discrimination of Distance-dependent Zones in Landscape Views".

Mintai Kim, Virginia Tech, United States:
"Modeling Nightscapes of Designed Spaces – Case Study of the University of Virginia and Virginia Tech Campuses".

Nadezda Stojanovic, Nebojsa Anastasijevic, Vesna Anastasijevic, Mirjana Mesicek and *Sasa Matic*, University of Belgrade, Serbia:
"The Classification of Maintenance Units for a GIS-based Cadastre on Urban Green Spaces".

Diana Santa Cruz Schaack, Fernando Bujaidar and *Gabriel Seah*, HfWU Nürtingen, Germany:
"Beep-Scape".

We thank all the reviewers who helped with their recommendations and we compliment the authors for their **SCIENTIFIC EXCELLENCE!**

The 2012 DLA awards by vote of the participants will be available online at the beginning of June after the 2012 DLA conference at: www.digital-la.de.

Committees Contributing to the DLA Conference

Scientific Program Committee of DLA

Anhalt University of Applied Sciences has been organizing and hosting the annual Conference on Information Technology in Landscape Architecture for thirteen years. The growing international interest in this series of conferences, which focuses on the specific needs of applying and developing information technologies for the field of landscape architecture, has lead to the formal establishment of a Scientific Program Committee. Members of this committee have all been giving academic advice on the development of this conference for a number of years now and/or have served as very important reviewers in the blind review process.

The conference committee advises the Conference Director on essential long-term development issues such as
- the main theme of the annual conference,
- selection of the Local DLA Conference Chair and the location of the next DLA Conference,
- both the content and production of the call for papers and promotional posters,
- selection of relevant keynote speakers, and
- selection of editors of the peer-reviewed proceedings.

The conference committee also advises the Local DLA Conference Chair on
- the conference program based on the blind evaluations of the abstracts, and
- the resolution of conflicts among the evaluations of different reviewers.

The Local DLA Conference Chair rotates among members of the committee. Among the responsibilities of the Local Conference Chair are
- management of the review process,
- local logistic arrangements, and
- national fundraising for the conference.

The DLA 2011 conference was held in Dessau, Germany together with the LE:NOTRE / ECLAS Summer School on "Teaching in Landscape Architecture." The DLA 2012 is being held in Bernburg and Dessau, Germany on "Landscape Visualization, 3D-Landscape Modeling and GeoDesign". The DLA 2013 is scheduled in Bernburg and Dessau again and the DLA 2014 is scheduled to be held at the Science City Campus, Zurich-Hoenggerberg in Switzerland.

The Scientific Program Committee of the annual Digital Landscape Architecture DLA Conference currently includes the following members.

BUHMANN, Prof. Erich, Landscape Architect BDLA
   Chair and Scientific Director of the International Conference on Information Technology in Landscape Architecture: DIGITAL LANDSCAPE ARCHICTECTURE DLA; Chair of the ECLAS Committee on Digital Technology;
   Anhalt University of Applied Sciences, Hochschule Anhalt, Germany,
   http://www.digital-la.de | private: atelier.bernburg@t-online.de

BISHOP, Prof. Dr. Ian
   University of Melbourne, Australia,
   http://www.geom.unimelb.edu.au/cgism/ian.html | i.bishop@unimelb.edu.au

BILL, Prof. Dr. Ralf
   Rostock University,
   ralf.bill@uni-rostock.de | http://www.auf-gg.uni-rostock.de/

BLASCHKE, Prof. Dr. Thomas
   Center for GeoInformatics, University of Salzburg, Austria,
   http://www.zgis.at | thomas.blaschke@sbg.ac.at

DÖLLNER, Prof. Dr. Jürgen
   Computer Graphics Systems, Hasso-Plattner-Institut at the University of Potsdam, Germany, http://www.hpi3d.de | doellner@hpi.uni-potsdam.de

ERVIN, Dr. Stephen M., Landscape Architect FASL
   Assistant Dean for Information Technology and Director of the Computer Resources Group at the Graduate School of Design, Harvard University, USA
   http://www.gsd.harvard.edu | servin@gsd.harvard.edu

FLAXMAN, Prof. Dr. Michael
   Dep. of Urban Studies and Planning, Massachusetts Institute of Technology, USA,
   http://dusp.mit.edu/ | http://futures.mit.edu | mflaxman@mit.edu

HASBROUCK, Prof. Hope
   The University of Texas at Austin, USA,
   http://soa.utexas.edu/people/profile/hasbrouck/hope | hhasbrouck@austin.utexas.edu

HEHL-LANGE, Dr. Sigrid
   The University of Sheffield, United Kingdom,
   s.hehl-lange@sheffield.ac.uk

JØRGENSEN, Prof. Dr. Ian, Landscape Architect
   Forest & Landscape, University of Copenhagen, Denmark,
   http://www.land-3d.com | Iajo@life.ku.dk

KIAS, Prof. Dr. Ulrich
   Weihenstephan University of Applied Sciences, Germany,
   kias@fh-weihenstephan.de

KIEFERLE, Prof. Joachim, Architect
   Hochschule RheinMain, Wiesbaden, Germany,
   http://caad.fab.hs-rm.de | joachim.kieferle@hs-rm.de

KLEINSCHMIT, Prof. Dr. Birgit
Technische Universität Berlin, Germany,
http://www.geoinformation.tu-berlin.de/ | birgit.kleinschmit@tu-berlin.de

KRAMER, Henk
Wageningen UR, Alterra,
henk.kramer@wur.nl

LANGE, Prof. Dr. Eckart
The University of Sheffield, United Kingdom,
http://www.shef.ac.uk/landscape/staff/profiles/eckartlange | e.lange@sheffield.ac.uk

LOVETT, Andrew
University of East Anglia, School of Environmental Sciences, Norwich, United Kingdom, a.lovett@uea.ac.uk

LENZ, Roman
Hochschule für Wirtschaft und Umwelt Nürtingen-Geislingen, HfWU,
Nürtingen, Germany, roman.lenz@hfwu.de

MARLOW, Christopher
Ball State University, Department of Landscape Architecture,
College of Architecture and Planning, Muncie, Indiana, 47306, United States,
marlow@bsu.edu

PALMER, Prof. Dr. James, Landscape Architect FASL
SUNY Syracuse, New York, USA,
palmer.jf@gmail.com

PETSCHEK, Prof. Peter, Landscape Architect
Dep. of Landscape Architecture, HSR Hochschule für Technik Rapperswil, Switzerland,
peter@petschek@hsr.ch

PIETSCH, Matthias – M.SC. GIS. Landscape Architect
Anhalt University of Applied Sciences, Hochschule Anhalt, Germany,
http://www.kolleg.loel.hs-anhalt.de/gis-seminar/index.htm | pietsch@loel.hs-anhalt.de

REKITTKE, Prof. Dr. Jörg, Landscape Architect
National University of Singapore (NUS), Singapore,
http://www.arch.nus.edu.sg | rekittke@email.de

SCHWARZ-V.RAUMER, Dr. Hans-Georg, University of Stuttgart, Germany,
svr@ilpoe.uni-stuttgart.de

STEINITZ, Prof. Dr. Carl
Alexander and Victoria Wiley Research Professor, Graduate School of Design,
Harvard University, USA,
csteinitz@gsd.harvard.edu

STEMMER, Boris, Kassel University, Germany,
stemmer@uni-kassel.de

STREMKE, Dr. Sven, Wageningen University and Research, Wageningen, Netherlands,
sven.stremke@wur.nl

TAEGER, Prof. Dr. Stefan
    Fachhochschule Osnabrück, Fakulät Agrarwissenschaften und
    Landschaftsarchitektur, Germany,
    s.taeger@fh-osnabrueck.de

TOMLIN, Prof. Dr. Dana,
    School of Design, University of Pennsylvania, USA,
    http://www.cml.upenn.edu | tomlin.dana@verizon.net

Review Committee and Thank You from the Editors

Every paper accepted for this publication has been rigorously reviewed by at least two peers from an international panel of scholars. These proceedings combine the peer reviewed papers of the DLA conference 2011 and the DLA Conference 2012, both held in Bernburg and Dessau, Germany. The high standards of the reviewers have assured that the papers in this year's proceedings will advance the theory and application of digital methods in landscape architecture.

With fifty excellent reviewers from the different areas of IT in landscape architecture and an efficient online conference management system hosted by Prof. Joachim Kieferle, Rhein-Main Hochschule in Wiesbaden Germany, we have been able to continuously improve and professionalize the peer review process for the proceedings of the conference series DIGITAL LANDSCAPE ARCHITECTURE DLA. Reviewers were able to retrieve papers from a conference review system and submit commentaries and evaluations directly to a content management system.

For the first step of the blind review process, five blind reviewers were each asked to give an initial evaluation for each of the anonymous abstracts assigned to her/him and place it into one of the following six levels of evaluation:

(1) Reject: Content inappropriate for the conference or has little merit
(2) Probable Reject: Basic flaws in content or presentation or very poorly written
(3) Marginal Tendency to Reject: Not as badly flawed; major effort necessary to make abstract acceptable but content already well-covered in literature
(4) Marginal Tendency to Accept: Content has merit, but accuracy, clarity, completeness, and/or writing should and could be improved in time
(5) Clear Acceptance: Content, presentation, and writing meet professional norms; improvements may be advisable but abstract is acceptable as is
(6) Must Accept: Candidate for outstanding submission. Suggested improvements still appropriate

The reviewer was also asked to recommend the particular conference session most appropriate for that paper. Finally, reviewers were asked for written comments addressed to each author and/or the Conference Chair.

Abstracts with a less than average score of 4 were rejected, while those with an average score of 4 were accepted for short presentations.

For the second step, the blind review of the full papers, only two reviewers with in-depth knowledge of the subject matter were assigned per paper. These second reviewers were usually selected from the original group that had evaluated each initial abstract, and they were asked to consider any recommendations already made.

As a result of this review, each paper was placed into one of the following categories.

(1) Reject for publication w/o complete reworking
(2) Probably reject for publication w/o reworking
(3) Marginal rejection for publication if not reworked
(4) Marginal acceptance for publication with minor corrections
(5) Accept for publication with minor corrections
(6) Fully accept for publication in this form

In the written suggestions, the reviewers were also asked to check for

- missing references to the literature, and
- irrelevant or mistaken assertions.

English language proof-reading was done continuously throughout this period by Jeanne Colgan and, in 2012, Judith Mahnert assisted in the final English editing of this publication.

Each author received the comments of the reviewers and additional comments by the editors for revising the full paper. When necessary, papers were returned to authors for revision until accepted by both reviewers. In some cases, additional anonymous reviewers were also asked to comment.

During the 2011 peer reviewed paper review forty-four entries were reviewed by forty-five reviewers. Twenty-six papers were accepted for peer reviewed publication in 2011.

During the 2012 peer reviewed paper review, sixty-two entries were reviewed by forty-nine reviewers. Thirty-eight papers were accepted for peer reviewed publication in 2012.

Sixty-four groups of authors from the 2011 and 2012 DLA conferences received a final confirmation letter that their contribution would be published in the peer-reviewed proceedings; Buhmann/Pietsch/Ervin (Eds.) (2012): Peer Reviewed Proceedings of Digital Landscape Architecture 2012, Anhalt University of Applied Sciences. Wichmann Verlag Berlin/Offenbach, Germany, May 2012. We are pleased that nearly all positively reviewed authors managed the rigid time schedule to submit the full paper for publication.

We are also happy that eight authors of contributions to the 2011 DLA Conference were invited to publish their papers in gis.SCIENCE – Zeitschrift für Geoinformation, 24. Jahrgang 4/2011, and 25. Jahrgang 1/2012 by Wichmann VDE Verlag Berlin/Offenbach. Three of these papers therefore appear only as abstracts in these proceedings, while the other authors submitted updates of their papers for this publication.

The more than 600 peer-reviewed pages of this publication (only the introductory chapter and the Invited Contributions are, because of their nature, not peer reviewed) will be distributed to all libraries of landscape architecture schools. All professional journals in landscape architecture, urban design and environmental planning will receive an evaluation copy.

We would like to give special thanks to all our colleagues who helped serve on the DLA Review Board to further develop the level of presentations of this conference and the scientific standards of peer-reviewed proceedings of this conference. We thank all of them for the extensive time spent in reviewing the abstracts and the papers for these proceedings.

The quality of the peer-reviewed papers benefits greatly from the extensive advice given in comments by the blind reviewers to the anonymous authors. We are very flattered by the academic support given by our DLA reviewers.

We hope that the reviewers listed below will also find time for the upcoming reviews.

AGIUS, Carol – MAGI, Malta, http://www.maltagi.org/

ALBERT, Christian – Leibniz University Hannover, Germany, albert@umwelt.uni-hannover.de

BECK KOH, Anemone – Landscape Architect Wageningen, Netherlands

BERBEROGLU, Suha – Cukurova University, Landscape Architecture Department, Turkey, suha@cu.edu.tr, http://landgis.cu.edu.tr/index_eng.html

BISHOP, Prof. Dr. Ian – University of Melbourne, Australia, http://www.geom.unimelb.edu.au/cgism/ian.html

BLASCHKE, Prof. Dr. Thomas – Center for GeoInformatics, University of Salzburg, Austria, http://www.zgis.at

BRAMLET, Alison – University of Georgia, United States, alisonsb@uga.edu

BUHMANN, Prof. Erich – Anhalt University of Applied Sciences, Campus Bernburg, Germany, www.landschaftsinformatik.de

BURLEY, Prof. Dr. Jon Bryan – Michigan State University MSU, United States

CANFIELD, Tess – Landscape Architect, United Kingdom, tesscanfield@yahoo.com

CONRAD, Max – Louisiana State University, United States, mconrad@lsu.edu

DIAZ, Joaquin – FH Giessen, http://www.fh-giessen.de

DEVIREN, Dr. A. Senem – Istanbul Technical University, Turkey, sd07landscape@yahoo.com

De VRIES, Jeroen – Van Hall Larenstein, Netherlands

DIESSENBACHER, Prof. Dr. Claus – Department of Architecture, Anhalt University of Applied Sciences, Campus Dessau, Germany

DÖLLNER, Prof. Dr. Jürgen – Hasso Platner Institut, Universität Potsdam, http://www.hpi3d.de

ERVIN, Prof. Dr. Stephen M. – Director Computer Resources, Assistant Dean for Information Technology Harvard Design School, Boston, United States, http://www.gsd.harvard.edu

ESBAH TUNCAY, Hayriye – Istanbul Technical University, Turkey, hayriyeesbah@yahoo.com

FERGUSON, Bruce – University of Georgia, United States, bfergus@uga.edu http://web.me.com/bruckferguso

FIRNIGI, Anett – Corvinus University of Budapest, Hungary
anett.firnigl@gmail.com

FLAXMAN, Prof. Dr. Michael – Massachusetts Institute of Technology, United States,
http://futures.mit.edu

GRUNAU, Jens-Peter – University Stuttgart, Germany, mail@grunau.de

von HAAREN, Prof. Dr. Christina – Institute of Landscape Planning and Nature Conservation, University of Hannover, Germany

HAASE, Prof. Dr.-Ing. Andrea – Master of Architecture Program, Anhalt University of Applied Sciences, Campus Dessau, Germany

HASBROUCK, Prof. Dr. Hope – The University of Texas at Austin, United States
http://soa.utexas.edu/people/profile/hasbrouck/hope

HEHL-LANGE, Dr. Sigrid – The University of Sheffield, United Kingdom,
s.hehl-lange@sheffield.ac.uk

HEINS, Marcel – Anhalt University of Applied Sciences / Hochschule Anhalt, Germany
m.heins@loel.hs-anhalt.de

HERNIK, József – University of Agriculture in Kraków, Poland, rmhernik@cyf-kr.edu.pl

HÖFFKEN, Stefan – TU Kaiserslautern, Germany, s.hoeffken@rhrk.uni-kl.de,
http://cpe.arubi.uni-kl.de/

JØRGENSEN, Prof. Ian – University of Copenhagen, Forest & Landscape, Denmark,
http://www.land-3d.com

KIAS, Prof. Dr. Ulrich – Weihenstephan University of Applied Sciences, Germany

KIEFERLE, Prof. Joachim – University of Applied Sciences Wiesbaden, Germany
http://caad.fab.fh-wiesbaden.de

KIRCHER, Prof. Dr. Wolfram – Master of Landscape Architecture Program, Anhalt University of Applied Sciences, Campus Bernburg, Germany

KLEINSCHMIT, Prof. Dr. Birgit – Technische Universität Berlin, Germany,
http://www.geoinformation.tu-berlin.de/

KOEPPEL, Hans Werner – Bonn, Germany, http://www.koeppel-home.de

KOPPERS, Prof. Dr. Lothar – Anhalt University of Applied Sciences, Dep. Architecture, Facility Management and Geoinformation, Campus Dessau, Germany

LANGE, Prof. Dr. Eckart – The University of Sheffield, United Kingdom,
http://www.shef.ac.uk/landscape/staff/profiles/eckartlange

LAMM, Bettina – Forest & Landscape, University of Copenhagen, Denmark,
http://www.sl.life.ku.dk/

LAMMEREN, Ron van – Centrum Landschap, Alterra, Wageningen Netherlands,
Ron.vanLammeren@wur.nl.

LINDHULT, Prof. Mark – Master of Landscape Architecture Program, University of Massachusetts, United States, http://www-unix.oit.umass.edu/~lindhult/

MACH, Rüdiger – Mach:Idee, Ing.-Buero R. Mach, Karlsruhe, Germany,
http://www.machidee.de

MATTOS, Cristina – ESRI Germany, c.mattos@esri.de

MERTENS, Elke – Hochschule Neubrandenburg, Germany, mertens@hs-nb.de

MOORE, Kathryn – Birmingham City University, United Kingdom, kathryn.moore@bcu.ac.uk

NIEMAN, Thomas – University of Kentucky, United States, tnieman@uky.edu

ORLAND, Brian – Penn State University, United States, boo1@psu.edu, http://larch.psu.edu/faculty/brian-orland

ORTACESME, Veli – Akdeniz University, Turkey, http://peyzaj.ziraat.akdeniz.edu.tr

OZDIL, Taner R. – The University of Texas at Arlington, United States, tozdil@uta.edu, http://www.uta.edu/ra/real/editprofile.php?pid=3021&onlyview=1

OZIMEK, Agnieszka – Cracow University of Technology, Poland, http://www.pk.edu.pl/~aozimek

OZIMEK, Dr. Pawel – Cracow University of Technology, Poland, http://www.pk.edu.pl/~aozimek

PAAR, Philip – TU Berlin, Berlin, Germany, paar@laubwerk.com, http://www.laubwerk.com

PALMER, Prof. Dr. James – SUNY Syracuse NY, United States, palmer.jf@gmail.com

PIETSCH, Matthias – Anhalt University of Applied Sciences, Campus Bernburg, Germany, http://www.kolleg.loel.hs-anhalt.de/gis-seminar/index.htm

PINKAU, Prof. Stephan – Master of Architecture Program, Anhalt University of Applied Sciences, Campus Dessau, Germany, s.pinkau@afg.hs-anhalt.de

REKITTKE, Prof. Dr. Jörg – National University of Singapore (NUS), Singapore, http://www.arch.nus.edu.sg

ROSS, Lutz – Technische Universität Berlin, Germany, http://www.geoinformation.tu-berlin.de

ROOS-KLEIN LANKHORST, Janneke – Centrum Landschap, Alterra, Wageningen Netherlands, Janneke.roos@wur.nl

ROLLEY, Stephanie – Kansas State University, United States, srolley@k-state.edu, http://capd.ksu.edu/larcp/

SANDQVIST, Sofia – Swedish University of Agricultural Sciences, Sweden, sofia.sandqvist@slu.se, http://www.slu.se/en/faculties/nl/about-the-faculty/departments/urd/

SCHROTH, Olaf – CALP / University of British Columbia UCA, Canada, http://www.calp.forestry.ubc.ca/

SECKIN, Cagatay – Istanbul Technical University, Turkey

SHARKY, Bruce – Louisiana State University, United States, bshark2@lsu.edu

SIMON, Alf – University of New Mexico, United States, http://saap.unm.edu

SPEED, Prof. Dr. Chris – Edinburgh College of Art, United Kingdom, http://fields.eca.ac.uk

STREICH, Prof. Dr. Bernd – Department of CAAD and Planning Methods, University of Kaiserslautern, streich@rhrk.uni-kl.de, http://cpe.arubi.uni-kl.de/

STROBL, Dr. Josef – Department of Geography and Applied Geoinformatics University of Salzburg, Austria, http://www.zgis.at

TAEGER, Prof. Dr. Stefan – Fachhochschule Osnabrück, Fakulät Agrarwissenschaften und Landschaftsarchitektur, Germany

TOBIAS, Prof. Dr. Kai – Ecological Planning and Environmental Impact Assessment, University of Kaiserslautern, Germany, tobias@rhrk.uin-kl.de

TOMLIN, Dana – University of Pennsylvania, United States, tomlin.dana@verizon.net

TUTUNDZIC, Andreja – University of Belgrade / EFLA, Serbia, andreja.tutundzic@sfb.bg.ac.rs, http://www.sfb.bg.ac.rs, http://europe.iflaonline.org/

VUGULE, Kristine – Latvia University of Agriculture, Latvia, kristine.vugule@llu.lv, www.llu.lv

WISSEN, Dr. Ulrike Hayek – ETH Zürich, Switzerland, http://www.irl.ethz.ch/plus/index_DE

ECLAS and LE:NOTRE Committee on Digital Technology

Since its founding in 2002, the European Council of Landscape Architecture Schools, ECLAS, has benefited enormously from the activities of the LE:NOTRE Thematic Network and its spiritual father Prof. Richard Stiles from the Technical University of Vienna. With the help of LE:NOTRE, a strong network, a common internet platform, and numerous meetings have all helped to develop a link between the fast growing numbers of landscape architecture schools in Europe and worldwide. Within LE:NOTRE, working groups have been established. In order to assure continuity after the LE:NOTRE project funding comes to an end, ECLAS has established a subcommittee to follow up these working groups, the ECLAS Committee on Digital Information. All fifty members of the former LE:NOTRE IT working group are being asked to continue in this new committee. At the moment, LE:NOTRE is still offering the "Channel Information Technologies" of the LE:NOTRE platform for communication of this committee.

In 2009, Prof. Erich Buhmann was voted as Chair for the newly established Committee on Digital Technology of the European Council of Landscape Architecture Schools ECLAS. The main objective for ECLAS Committee on Digital Technology is the Improvement and further development of Teaching & Learning with Digital Methods & Tools in Landscape Architecture. This was also the main conference theme of the DLA 2011 in Dessau, Germany.

The annual conference DIGITAL LANDSCAPE ARCHICTECTURE DLA will be an important asset for the work of this committee. Further objectives are being discussed. Besides meetings with the Vitero conference system as needed, a second regular personal meeting is provided by the annual ECLAS conferences.

The contents of past and upcoming DLA conferences are stored on the homepage of the Anhalt University homepage www.digital-la.de or
http://www.kolleg.loel.hs-anhalt.de/landschaftsinformatik/dla-conference.html.

The project group "Digital Landscape Architecture Conferences"
of the ECLAS / LE:NOTRE Committee on Digital Technology has been established as a communication platform at www.lenotre.com:
http://www.le-notre.org/pg/pg_details.php?pg_id=112.

Addresses of Authors

AHMAD, A. M.
Loughborough University, UK
Epinal Way Loughborough
LE11 3TU, Leicestershire, United Kingdom
A.M.Ahmad@lboro.ac.uk

ALBERT, Christian
University of Hannover – Gottfried Wilhelm
Leibniz Universität Hannover, Welfengarten 1
D-30167 Hannover, Germany
albert@umwelt.uni-hannover.de

BARBARASH, David
Purdue University
625 Agricultural Mall Drive
47907 West Lafayette, United States
dbarbara@purdue.edu

BILL, Prof. Dr.-Ing. Ralf
Rostock University
Justus-von-Liebig-Weg 6
D-18059 Rostock, Germany
ralf.bill@uni-rostock.de

BISHOP, Prof. Dr. Ian
University of Melbourne
Grattan Street
3010 Parkville, Australia
i.bishop@unimelb.edu.au

BUHMANN, Prof. Erich
Anhalt University of Applied Sciences –
Hochschule Anhalt
Landschaftsinformatik, Solbadstraße 2
D-06406 Bernburg, Germany
la@loel.hs-anhalt.de
http://www.landschaftsinformatik.de
privat: atelier.bernburg@t-online.de

BURLEY, Prof. Jon Bryan
Michigan State University
4598 Skyline Dr.
48871 Perrinton, United States
burleyj@msu.edu

CHANDWANIA, Dhruv
Rahul Towers
Flat no. 102, bldg no. 04
Bhusari Colony, Kothrud
Pune 411038, India

CZINKÓCZKY, Anna
Corvinus University of Budapest
Villanyi ut. 29-43
1118 Budapest, Hungary
annaczinkoczky@gmail.com

DIESSENBACHER, Prof. Dr. Claus
Anhalt University of Applied Sciences
Hochschule Anhalt, Strenzfelder Allee 28
D-06406 Bernburg, Germany
claus@diessenbacher.net

EGGINTON, Zane
Unitec Institute of Technology
139 Carrington Road, Mt Albert
1025 Auckland, New Zealand
zegginton@unitec.ac.nz

ERVIN, Dr. Stephen M.
Harvard University
Massachusetts Hall
Cambridge, MA 02138, United States
servin@gsd.harvard.edu
http://www.gsd.harvard.edu

FETZER, Ellen
HfWU Nürtingen-Geislingen
Schelmenwasen 4-8
D-72622 Nürtingen, Germany
ellen.fetzer@hfwu.de

FORMOSA, Dr. Saviour
University of Malta
Criminology, Rm 113 New Humanities Blk A,
University of Malta
MSD 2080 Msida, Malta
saviour.formosa@um.edu.mt

FRICKER, Pia Christina
Departement of Architecture, ILA,
Prof. Girot
ETH Zurich, HIL H 57.2
CH-8093 Zurich, Switzerland
fricker@arch.ethz.ch

GALEV, Emil
University of Forestry, Sofia
Kliment Ohridsky BLVD
1756 Sofia, Bulgaria
Emil.galev@abv.bg

GARCIA PADILLA, Marcela
Anhalt University
Knobelsdorffstrasse 32
D-14059 Berlin, Germany
margarpa@gmail.com

GLANDER, Tassilo
HPI (Universität Potsdam)
Prof.-Dr.-Helmert-Str. 2-3
D-14482 Potsdam, Germany
tassilo.glander@hpi.uni-potsdam.de

GOLDBERG, D. E.
Penn State University
University Park
121 Stuckeman Family Building
16802 United States
dgoldberg@psu.edu

HAAREN, Christina v.
University of Hannover – Gottfried Wilhelm
Leibniz Universität Hannover
Welfengarten 1
D-30167 Hannover, Germany
haaren@umwelt.uni-hannover.de

HAHN, Howard
Kansas State University
201B Seaton Court
66506 Manhattan, United States
hhahn@ksu.edu

HARWOOD, Amii
University of East Anglia
School of Environmental Sciences, University of East Anglia
Norwich
NR4 7TJ, United Kingdom
a.darnell@uea.ac.uk

HEHL-LANGE, Dr. Sigrid
University of Sheffield, Department of Town and Regional Planning
Western Bank
S10 2TN Sheffield, United Kingdom
s.hehl-lange@sheffield.ac.uk

HEINS, Marcel
Anhalt University of Applied Sciences
Hochschule Anhalt
Landschaftsinformatik, Strenzfelder Allee 28
D-06406 Bernburg, Germany
heins@loel.hs-anhalt.de
http://www.landschaftsinformatik.de

HOUTKAMP, Joske M.
Alterra, Wageningen University Research Center, PO Box 47
NL-6700 AA, Wageningen, Netherlands
Joske.Houtkamp@wur.nl

JOMBACH, Sandor
Corvinus University of Budapest
Department of Landscape Planning and Regional Development
Villanyi ut 35-43
1118 Budapest, Hungary
sandor.jombach@uni-corvinus.hu

JOYE, Ruben
University College of Ghent
Brusselsesteenweg 161
9090 Melle, Belgium
ruben.joye@hogent.be

KADER, Dr. (Polytechnikum Mailand) Alexander
Anhalt University of Applied Sciences
Hochschule Anhalt
Fachbereich 3 – Architektur, Facility Management und Geoinformation
Gropiusallee 38 (Bauhaus, Raum 112)
D-06846 Dessau-Roßlau, Germany
a.kader@afg.hs-anhalt.de

KIM, Mintai
Virginia Tech
14410 Stroubles Creek
24060 Blacksburg, United States
mintkim@gmail.com

KIRCHER, Prof. Dr. Wolfram
Anhalt University of Applied Sciences
Hochschule Anhalt, Strenzfelder Allee 28
D-06406 Bernburg, Germany
w.kircher@loel.hs-anhalt.de

KRAMER, Henk
Wageningen UR, Alterra
Droevendaalsesteeg 3
NL-6708 PB Wageningen
Netherlands
henk.kramer@wur.nl

KRETZLER, Prof. Einer
Anhalt University of Applied Sciences
Hochschule Anhalt
Strenzfelder Allee 28
D-06406 Bernburg, Germany
e.kretzler@loel.hs-anhalt.de

ŁABĘDŹ, Piotr
Cracow University of Technology
Warszawska 24
31-155 Krakow, Poland
plabedz@pk.edu.pl

LEINER, Class
Kassel University
Gottschalkstraße 26
D-34127 Kassel, Germany
Claas.leiner@uni-kassel.de

LENZ, Prof. Dr. Roman
HfWU
Schelmenwasen 4-8
D-72622 Nürtingen, Germany
roman.lenz@hfwu.de

LOVETT, Prof. Andrew A.
University of East Anglia
School of Environmental Sciences
University of East Anglia
Norwich
NR4 7TJ, United Kingdom
a.lovett@uea.ac.uk

MANYOKY, Madeleine
ETH Zurich, PhD Candidate
IRL, HIL H53.2, Wolfgang-Pauli-Str. 15
CH-8093 Zürich-Hönggerberg, Switzerland
manyoky@nsl.ethz.ch

MCLEAN, Ross
Edinburgh College of Art
Lauriston Place
EH3 9DF Edinburgh, United Kingdom
gordonrossmclean@hotmail.com

MERTENS, Prof. Dr. Dipl.-Ing. Elke
Hochschule Neubrandenburg
Reusenort 2
D-17033 Neubrandenburg, Germany
mertens@hs-nb.de

MORGAN, Ed
Department of Landscape
University of Sheffield
219 Portobello
Sheffield
S1 4DP, United Kingdom
Ed.Morgan@sheffield.ac.uk

MÜLDER, Jochen
Lenne3D GmbH
Güntherstr. 98b
D-22087 Hamburg, Germany
muelder@lenne3d.com

OZIMEK, Agnieszka (dr eng arch)
Cracow University of Technology
Warszawska 24
31-155, Krakow
Poland
aozimek@pk.edu.pl

OZIMEK, Pawel (dr eng arch)
Cracow University of Technology
Warszawska 24
31-155, Krakow
Poland
ozimek@pk.edu.pl

PAAR, Philip
Laubwerk GmbH
Geschäftsführer/Managing Director
c/o Beuth Hochschule, Business Innovation
Centre, Kurfürstenstr. 141
D-10785 Berlin, Germany
paar@laubwerk.com
http://laubwerk.com

PÉ, Raffaele
Politecnico di Milano
via Bonardi, 3, Milan
20133 Milan, Italy
raffaelepe@yahoo.it

PETSCHEK, Peter
Hochschule für Technik Rapperswil
Oberseestrasse 10
Postfach 1475
CH-8640 Rapperswil, Switzerland
peter.petschek@hsr.ch

PIETSCH, Matthias
Anhalt University of Applied Sciences
Hochschule Anhalt
Strenzfelder Allee 28
D-06406 Bernburg, Germany
http://www.kolleg.loel.hs-anhalt.de/gis-seminar/index.htm
m.pietsch@loel.hs-anhalt.de

PINKAU, Prof. Stephan
Anhalt University of Applied Sciences
Hochschule Anhalt
Fachbereich 3 – Architektur, Facility
Management und Geoinformation
Gropiusallee 38
D-06846 Dessau-Roßlau, Germany
s.pinkau@afg.hs-anhalt.de

POKLADEK, Michael
Martin-Luther-Universität
Institut für Geowissenschaften
Von-Seckendorff-Platz 4
D-06120 Halle (Saale), Germany
m.pokladek@hotmail.de

REKITTKE, Prof. Dr.-Ing. Jörg
National University of Singapore
School of Design and Environment
Department of Architecture
4 Architecture Drive
117566 Singapore
rekittke@nus.edu.sg

SCHALLER, Prof. Dr. Jörg
Prof. Schaller UmweltConsult GmbH
Domagkstraße 1a
D-80807 München, Germany
info@psu-schaller.de

SCHMIDT, Prof. Rainer
Beuth Hochschule für Technik Berlin
Grünplanung
Raum 118, Haus Beuth,
Luxemburger Straße 10
D-13353 Berlin, Germany
Privat: info@rainerschmidt.com
www.rainerschmidt.com

SCHROTH, Olaf
University of British Columbia (UBC)
2045-2424 Main Mall
V6T 1Z4 Vancouver, Canada
schrotho@interchange.ubc.ca

SCHWARZ-V.RAUMER, Dr. Hans-Georg
University of Stuttgart
Keplerstr. 11
D-70174 Stuttgart, Germany
svr@ilpoe.uni-stuttgart.de

STEINITZ, Prof. Dr. Carl
Harvard University
Massachusetts Hall
Cambridge, MA 02138
United States
http://www.gsd.harvard.edu
steinitz@gsd.harvard.edu

STEMMER, Boris
Kassel University
Gottschalkstraße 26
D-34127 Kassel
Germany
stemmer@uni-kassel.de

STYLIADIS, Athanasios
Kavala Institute of Technology
Xrysovergi & Kassandrou
66100 Drama
Greece
styliadis@ath.forthnet.gr

THURMAYR, Anna M.
University of Manitoba
Department of Landscape Architecture
R3T 0C8 Winnipeg
Canada
thurmayr@cc.umanitoba.ca

TOMLIN, Prof. Dr. Dana
University of Pennsylvania
3451 Walnut Street
Philadelphia, PA 19104
United States
tomlin.dana@verizon.net

WARREN-KRETZSCHMAR, Dr. Barty
University of Hannover – Gottfried Wilhelm
Leibniz Universität Hannover Welfengarten 1
D-30167 Hannover
Germany
warren@umwelt.uni-hannover.de

WISSEN-HAYEK, Dr. Ulrike
ETH Zürich
Wolfgang-Pauli-Str. 15, HIL H 52.2
CH-8093 Zürich
Switzerland
wissen@nsl.ethz.ch

30 May – 1 June
Bernburg and Dessau, Germany

DIGITAL LANDSCAPE ARCHITECTURE
DLA 2013

Hochschule Anhalt
Anhalt University of Applied Sciences

CALL FOR PAPERS and POSTERS
14th International Conference on Information Technologies in Landscape Architecture

The 14th annual International Conference on Information Technology in Landscape Architecture: Digital Landscape Architecture DLA 2013 will be held in Bernburg and Dessau Germany, 30 May – 1 June, 2013. We would like to invite you to contribute to this conference on teaching of new technologies in landscape architecture. Please submit "Extended Abstracts" (minimum one page, maximum two pages text including key references and optional additional selected graphics) as pdf file online via a web-based submission system at

→ www.digital-la.de → DLA Conference 2013 by October 1, 2012

for review and possible inclusion in the program as presentation or as poster.

The conference proceedings are published as fully reviewed papers by Wichmann Verlag, Berlin/Offenbach as following: "Buhmann/Ervin/Pietsch (Eds.): Peer Reviewed Proceedings Digital Landscape Architecture 2013, Anhalt University of Applied Sciences. Wichmann Verlag, Berlin/Offenbach, May 2013".

Main Theme and Suggested Topics:
The DLA Program Committee cordially invites you to submit proposals for original, unpublished presentations focusing on the conference's main theme
"GeoDesign – Teaching and Case Studies of Integrated Geospatial Design and Planning"
in one of the other following areas:

1) **GeoDesign Concepts**
2) **Teaching GeoDesign in Landscape Planning**
3) **Teaching GeoDesign in Landscape Design**
4) **GeoDesign Case Studies in Regional Planning**
5) **GeoDesign Case Studies in Urban Design**
6) **Mobile Devices for GeoDesign**
7) **Landscape Information Model (LIM) and Standardization**
8) **Interactive Virtual Landscapes and Visual Landscape Assessment**

Important dates for the review process are:

Abstracts due:	October 1, 2012
Notification of acceptance:	November 1, 2012
Full manuscript draft due:	December 1, 2012
Reviewed manuscript due:	February 1, 2013
Conference:	May 30 – June 1, 2013

Invited Keynote Speakers:
Prof. Dr. Carl Steinitz, Harvard University, USA – Keynote
Philipp Oswalt, Director Bauhaus Foundation Dessau – Keynote (req.)
Invited Speakers:
Prof. Dr. Jörg Rekittke, National University of Singapore
Prof. Dr. Josef Strobl, University Salzburg, Austria
Prof. Dr. Eckart Lange, University of Sheffield, England
Rainer Schmidt, Landscape Architect, RSL, Munich, Berlin, Peking, Bernburg
Prof. Dr.-Ing. Walter Schönwandt, University of Stuttgart – (req.)
Prof. Dr. Bernd Streich, University of Kaiserslautern, Germany – (req.)
Prof. Dr. Alexander Zipf, University of Heidelberg – (req.)

The conference language is English. The conference will be held at the Bernburg Campus and the nearby Dessau Bauhaus Campus in Germany.
For further information please contact
Anhalt University of Applied Sciences / Hochschule Anhalt (FH),
Dep. 1, Forschungsbereich Landschaftsinformatik, Prof. Erich Buhmann,
Strenzfelder Allee 28, D-06406 Bernburg, Germany
Office: INDIGO Innovationspark, Solbadstraße 2, Anbau Room 005
Conference proceedings: atelier.bernburg@t-online.de //
Conference organization: la@loel.hs-anhalt.de

Local Chair for 2013 and DLA Conference Scientific Director: Prof. Erich Buhmann

The concept of the DLA event is developed by the Scientific Program Committee of the Digital Landscape Architecture DLA conference. The DLA is organized in cooperation with long term partners such as AGIT Symposium und Fachmesse Angewandte Geoinformatik of University of Salzburg, the LE:NOTRE Thematic Network of the Technical University of Vienna and the German Professional Association of Landscape Architects, BDLA. The conference serves as platform for the ECLAS Committee Digital Information.

More information on our series of conferences on Digital Landscape Architecture at
http://www.digital-la.de

Committee
Digital Information of

ECLAS
EUROPEAN COUNCIL OF
LANDSCAPE ARCHITECTURE
SCHOOLS

CALL FOR PAPERS and POSTERS
14th International Conference on Information Technologies in Landscape Architecture